SYMBOLIC COMPUTATION

Managing Editors: J. Encarnaçao P. Hayes

Artificial Intelligence
Editors: L. Bolç A. Bundy J. Siekmann A. Sloman

Automation of Reasoning

1 Classical Papers on Computational Logic 1957–1966

Edited by Jörg Siekmann and Graham Wrightson

Springer-Verlag
Berlin Heidelberg New York 1983

Jörg H. Siekmann
Universität Karlsruhe
Institut für Informatik I
Postfach 6380
D-7500 Karlsruhe, West Germany

Graham Wrightson
Victoria University
Department of Information Science
Wellington
New Zealand

ISBN-13 : 978-3-642-81954-4 e-ISBN-13 :978-3-642-81952-0
DOI : 10.1007 / 978-3-642-81952-0

Softcover reprint of the hardcover 1st edition 1983

2121/3140-543210

Lasst uns rechnen!
G.W. Leibniz, 1646–1716

Contents

Preface

"Kind of crude, but it
works, boy, it works!"

Alan Newell to Herb Simon,
Christmas 1955

In 1954 a computer program produced what appears to be the first computer generated mathematical proof: Written by M. Davis at the Institute of Advanced Studies, USA, it proved a number theoretic theorem in Presburger Arithmetic. Christmas 1955 heralded a computer program which generated the first proofs of some propositions of Principia Mathematica, developed by A. Newell, J. Shaw, and H. Simon at RAND Corporation, USA. In Sweden, H. Prawitz, D. Prawitz, and N. Voghera produced the first general program for the full first order predicate calculus to prove mathematical theorems; their computer proofs were obtained around 1957 and 1958, about the same time that H. Gelernter finished a computer program to prove simple high school geometry theorems.

Since the field of computational logic (or automated theorem proving) is emerging from the ivory tower of academic research into real world applications, asserting also a definite place in many university curricula, we feel the time has come to examine and evaluate its history. The article by Martin Davis in the first of this series of volumes traces the most influential ideas back to the 'prehistory' of early logical thought showing how these ideas influenced the underlying concepts of most early automatic theorem proving programs. The article by Larry Wos and Larry Henschen in the second volume covers the period of 1965 to 1970, when most of the early theorem proving systems emerged; the article by S. Maslov provides an overview of Russian and Eastern European work during this period.

This series of volumes, the first covering 1957 to 1966 and the second 1967 to 1970, contains those papers, which have shaped and influenced the field of computational logic and makes available the classical work, which in many cases is difficult to obtain or had not previously appeared in English. However, the main purpose of this series is to evaluate the ideas of the time and to select those papers, which can now be regarded as classics after more than a decade of intensive research.

The editors' selectivity attempts - as far as it can be done today - a fair evaluation and historical representation of the field. To even more closely emphasize objectivity, an international advisory committee composed of those researchers who themselves have helped to shape the history of automated theorem proving was formed. It consisted of the following people:

P. ANDREWS	(USA)	S. MASLOV	(USSR)
W. BLEDSOE	(USA)	B. MELTZER	(BRITAIN)
M. DAVIS	(USA)	D. PRAWITZ	(SWEDEN)
R. KOWALSKI	(BRITAIN)	J.A. ROBINSON	(USA)
D. LOVELAND	(USA)	L. WOS	(USA)

The selection criteria were formulated as:

Selection Criteria: The selected papers should have:

1) Decisively influenced the discipline.
2) A high standard with regard to form and presentation.
3) A direct relevance to the mechanization by computers.

These criteria were not quite uniformly applied, since some of the very early papers have been included for historical reasons even though they may violate one of these three conditions. Apart from these considerations, criteria 1) to 3) have been applied, however. Criterion 2) for example excluded most technical reports, even though they may have been rather influential at the time, and 3) excluded many purely logical papers (e.g. on decidability results or 'foundation papers'), important as they may have been.

The selection procedure was carried out in the following manner: the editors prepared a listing of all papers that had been published up to and including 1970. Although some papers were included that appeared later in a journal, in each case it had appeared at least as a preprint or departmental report beforehand, and the author was asked for written evidence in case of doubt. This initial listing of papers was marked into four categories:

1) Definite candidates for inclusion
2) Likely candidates, but which do not appear to clearly fulfil all the selection criteria; e.g. the paper was influential but poorly formulated
3) Possible candidates, but which were dubious for one reason or another, e.g. we were not well enough

acquainted with the history or the content of the paper, or the paper may be in a bordering field (e.g. induction) etc.

4) Unacceptable candidates.

The marked listing was then sent to each member of the advisory committee and their suggestions and revisions were collected to provide a new listing of all the papers that emerged as clear candidates for inclusion. This new listing was again sent to each member of the advisory committee. The deduction conference at Les Arcs, France, in 1980 provided an opportunity for the whole committee to meet and to discuss the remaining controversial cases. A new and final listing resulted which for space considerations had been shortened by excluding work on theorem proving in higher order calculi, non-classical logic, and papers on applications of theorem proving (e.g. applications in program synthesis, question answering, problem solving, program verification, robot technology, and programming languages). These unfortunate omissions were necessary to keep the 2 volumes within manageable size, however it is planned to have additional volumes in the series AUTOMATION OF REASONING covering these fields. It is also planned to cover the work in Computational Logic which appeared after 1970.

In order to avoid too much bias - since most members of the committee had been active in the period under consideration - and to ensure a broad acceptance of the selection within the academic community, this final listing was then sent to about a hundred scientists still actively working in the field, requesting their advice and suggestions. The responses were very helpful, particularly concerning some embarassing omissions. Further correspondence, consultation, and additional special refereeing of papers by respective experts, produced the final listing, which again was sent to each member of the advisory committee.

Lastly, the classics of the field, i.e. those papers probably familiar to anyone actively working in the subject, are asterisked (*) as an aid to the reader.

On completion of this final listing, the difficult task of tracing the authors ensued. Each available author was informed on the inclusion of his paper and asked for a page of commentary on errors, historical remarks, and current evaluation. These comments are printed immediately following each paper. The papers are listed according to year of appearance and alphabetically by the first author's name within each year.

Finally each volume contains a bibliography of about 450 publications on computational logic, which - to the best knowledge of the editors - is complete up to and including 1970. Included again is work with direct relevance to computational logic, in particular automated theorem proving and its applications. Excluded are purely logical papers and work on program verification and semantics which has been extensively referenced elsewhere.

Sincere thanks to those who spent so many hours of their private time on evaluation, refereeing, and selection, particularly many thanks to the advisory committee. Without their generous collaboration this task, which took almost three years of evaluation and selection time, would have been impossible.

Jörg H. Siekmann
Graham Wrightson

January, 1983

The Prehistory and Early History of Automated Deduction[1]

M. Davis[2]

In its brief existence, the field of automated deduction has spawned a large literature including at least two treatises. [3] The explosion of interest which has produced the field as we know it today can be traced to [Robinson 1965a] in which the elegance and simplicity of the resolution principle as a basis for mechanized deduction first appeared. But, broadly conceived, the history of automated deduction is really the history of the *idea* of mechanizing human thought. It is the purpose of this article to trace this history from the seventeenth century to 1967. In the process we shall see not only the struggles of early workers in the field, and the origin of some of the specific notions that have played a key role in work on automated deduction, but also the surprising obstacles which certain philosophical beliefs placed in the path of progress. Our emphasis will be on underlying ideas rather than technical details.

We begin with Déscartes' discovery that algebraic methods could be used to develop classical Greek geometry. This discovery was the crucial fountainhead of many developments in mathematics (including the calculus and the entire structure of mathematical analysis). For us, what is significant is that what had seemed in Euclid to be the result of cunning and mathematical ingenuity was now revealed as being accessible to relatively mechanical treatment: introduce a coordinate system, write out the equations, and carry out the required algebraic manipulation. Déscartes himself was quite aware of this aspect of his work. As he said (in [Déscartes 1637]),

> ... it is possible to construct all the problems of ordinary geometry by doing no more than the little covered in the four figures[4] that I have explained. This is one thing which I believe the ancients did not notice, for otherwise they would not have put so much labor into writing so many books in which the very sequence of the propositions showed that they did not have a sure method of finding all ...

The dream of mechanizing thought, of doing for all deductive reasoning what Déscartes had done for geometry, is first to be found in the writings of Leibniz. Although Leibniz' technical work in logic consists of mere fragments, he proposed two enormously ambitious projects:

a *calculus of reason* (calculus ratiocanator) and

a *universal language* (lingua characteristica).

His assessment of the importance of this work is clear from his statement [5]

> For if praise is given to the men who have determined the number of regular solids — which is of no use, except insofar as it is pleasant to contemplate — and if it is thought to be an exercise worthy of a mathematical genius to have brought to light the more elegant properties of a conchoid or cissoid, or some other figure which rarely has any use, how much better will it be to bring under mathematical laws human reasoning, which is the most excellent and useful thing we have.

That Leibniz was thinking explicitly of *mechanizing* reason, is surely suggested by his assertion that using his proposed calculus of reason, the mind "will be freed from having to think directly of things themselves, and yet everything will turn out correctly." [6] As compared with this spectacular prospect, Leibniz' actual technical contributions are most disappointing. In the first place his notions were restricted to propositions whose logical character was to be obtained, following in the footsteps of Aristotle, from their analysis in terms of subject and predicate. But even within the limits imposed by this restriction, it must be admitted that Leibniz' work did not go very far. From our point of view, what he did was to develop a few fragments of Boolean algebra. But his calculi involved only one operation (for which Leibniz used various symbols but ultimately settled on $\oplus$), which, his explanations make plain, he sometimes thinks of as union (or disjunction) and sometimes as intersection (or conjunction). Thus, from the definition, "A is in B if for some N, $A \oplus N = B$," together with the axiom $A \oplus A = A$, Leibniz deduces: A is in A. [7] Reading Leibniz' work on logic leads one to a number of reflections. On the one hand, we are led to realize, from the struggles of a genius like Leibniz, how difficult it was to attain understanding of what for us are the most elementary matters, and how remarkable the achievements of his successors have been. On the other hand, we come to appreciate the enormity of the step that Leibniz was able to take in conceiving of logic itself as a deductive science whose principles can be formally derived from axioms, and which is abstract in the sense of possessing various different interpretations.

If Leibniz' calculus ratiocanator remained a matter of some fragments, his lingua characteristica remained merely a hope. Leibniz envisioned a language, suitable for communication, encompassing science and mathematics, and containing the calculus ratiocanator as a part. The symbols would be ideographic rather than phonetic, and Leibniz specifically mentioned Chinese as an example of this. The scope of Leibniz' project can be gathered from his proposal to begin with a universal encyclopedia. The kind of application he expected to be possible can be judged from his touching picture of men of good will proposing to decide some vexing question in human affairs by transcribing it into the lingua characteristica and then saying "Let us calculate". [8]

A "rational calculus" in Leibniz' sense was constructed by George Boole two centuries later. [9] His work has become so much a part of our education that little need be said about it. He developed a calculus of classes and noted that it could be regarded as an abstract system obtained from suitable axioms. He noted as well that his calculus had a propositional interpretation as well, in which the letters could be thought of as standing for truth values.

The fact that Boole's work had, to an extent, *mechanized* logic was explicitly recognized by the economist and logician Stanley Jevons, who constructed a working machine for verifying Boolean identities in 1869. It apparently resembled in appearance a cash register in which symbols representing specific Boolean combinations are caused to appear by pressing keys. [10]

The present article was developed from a lecture which was intended to honor the centenary of the appearance of Gottlob Frege's "Begriffsschrift" — literally "concept writing" — in 1879, and it was in Frege's work that many of the key forms and concepts of modern mathematical logic first appeared. Frege saw himself as carrying out Leibniz' most ambitious project, as putting flesh on Leibniz' vision of an universal language. Thus, in replying to criticism (which objected to Frege's new and unfamiliar notation), Frege said: [11]

> My intention was not to represent an abstract logic in formulas, but to express a content through written signs in a more precise and clear way than it is possible to do through words. In fact, what I wanted to create was not a mere *calculus ratiocinator* but a *lingua characterica* [sic] in Leibniz's sense.

The *Begriffsschrift* contains a complete development of quantification theory (otherwise known as predicate calculus). It is important to realize why this was such a stupendous achievement. Until Frege, there was enormous confusion over the relationship between the propositional and the class interpretations of Boolean algebra. It was well understood that the calculus of classes sufficed only for a small part of the deductions that occur in mathematics, that, for example, a theory of relations was also needed. It was in the *Begriffsschrift* that first appeared the now familiar conception that it was the propositional interpretation of Boolean algebra which should be regarded as fundamental and that for the purpose of logical analysis complex sentences should be expressed by combining atoms expressing predicates of any number of arguments using Boolean connectives and quantifiers.

Frege's work is directly relevant to our subject not only because of his development of the predicate calculus, but also because of the degree of rigor which Frege attained. It was in the *Begriffsschrift* that for the very first time, the syntax of an artificial language was laid out in a precise and almost merciless way. Thus, the *Begriffsschrift* is not only the direct ancestor of contemporary systems of mathematical logic, but also the ancestor of all formal languages, including computer programming languages.

Unfortunately, as important as Frege's work was, it remained largely unknown, partly no doubt because, having made such a large advance on what had been done previously, readers tended to find it rather obscure and impenetrable.

A decade later, Peano was still working on a much lower level than Frege had attained. But it is Peano's notation that has survived, especially as incorporated in the work of Russell and Whitehead. Peano, independently of Frege, had arrived at a rather similar grammatical analysis of the sentences that occur in mathematical discourse. But Peano had nothing of Frege's syntactic clarity, and, he used quantifiers only hesitantly, referring to them as "abstruse".[12] Worst of all Peano contented himself with a language devoid of rules of inference. Thus, although Peano's *statements* are expressed in his special symbolism, with no words from natural languages intruding, Peano's *deductions* must all be expressed in ordinary language. This was a real step backwards from Frege, who had explicitly put the rule of *modus ponens* [13] forward as being fundamental. On the other hand, Peano not only contributed much of our contemporary logical notation, but also had clearly in mind the fundamental character of the language of logic as basic to all fields of study. As Peano said [14]

> ... I think that the propositions of any science can be expressed by these signs of logic alone, provided we add signs representing the objects of that science.

It should not be thought that the advances in logical understanding being chronicled here were being acclaimed on all sides. In fact, there were those who bitterly attacked the work being done. And since these attacks themselves played an important role in the further development of the subject, it is necessary to discuss some of this criticism. Perhaps the strongest attacks came from the great mathematician Henri Poincaré, who was particularly incensed by Peano's work. The spirit of Poincaré's criticism will be gathered from his remark [15] that:

> It is difficult to admit that the word *if* acquires when written $\supset$, a virtue it did not possess when written if.

One should not infer from this that Poincaré had not realized the relationship between work on symbolic logic and the possibility of mechanizing human reason. Poincaré realized perfectly well that if the claims of the logicians were to be taken seriously, this possibility would be very real. But to Poincaré the very absurdity of such a possibility, which threatened everything creative and beautiful in mathematical thought, showed that in fact the logicians' claims need not be taken seriously. Poincaré's argument by reductio ad absurdum is expressed picturesquely as follows: [16]

> Thus it will be readily understood that in order to demonstrate a theorem, it is not necessary or even useful to know what it means. We might replace geometry by the *reasoning piano* imagined by Stanley Jevons; or, if we prefer, we might imagine a machine where we should put in axioms at one end and take out theorems at the other, like that legendary machine in Chicago where pigs go in alive and come out transformed into hams and sausages. It is not more necessary for the mathematician than it is for these machines to know what he is doing.

In order to understand the intellectual climate in which many of the key technical ideas used in automated deduction developed (e.g. Skolemization of a formula, Herbrand's theorem), it is important to be aware of some of the work in mathematics towards the close of the nineteenth century which contributed to the negative attitude of Poincaré and others. Perhaps most important was Cantor's introduction of a theory of infinite sets into mathematics. Before Cantor it was generally assumed that the actual infinite was to be relegated to the philosophers and had no place in mathematics. "Infinity" was in all cases to be regarded as shorthand for a limiting process. Thus in 1831, Gauss [17] wrote:

> I protest against the use of infinite magnitude as something completed, which in mathematics is never permissible. Infinity is merely a *facon de parler*, the real meaning being a limit which certain ratios approach indefinitely near, while others are permitted to increase without restriction.

When Cantor developed his theory of transfinite cardinal and ordinal numbers, there were those who accepted it as a beautiful and important extension of traditional mathematics. But, a number of first-rate mathematicians reacted negatively, not to say hostilely, to Cantor's innovations. Among these were Kronecker, Poincaré and Brouwer. In fact, Brouwer proceeded to develop a thorough-going critique of the foundations of mathematics, calling his new doctrine, intuitionism. Brouwer's analysis went much further than simply rejecting the Cantorian infinite. Brouwer objected to the use of non-constructive methods in mathematical proofs. In order to understand just what Brouwer was objecting to, let us examine a nonconstructive proof of a simple theorem, familiar to all computer scientists:

> *there is a context-free language* L *such that* $\bar{L}$ (i.e. the complement of L) *is not context-free*. We suppose it known that the union of context-free languages is context-free, but that the intersection of context-free languages need not be context-free. (The familiar example:
>
> $L_1 = \{a^n b^n c^m\}$, $L_2 = \{a^m b^n c^n\}$, $L_1 \cap L_2 = \{a^n b^n c^n\}$ is one
>
> where L_1, L_2 are context-free, but $L_1 \cap L_2$ is not.)
>
> But now, the de Morgan identity:
>
> $$L_1 \cap L_2 = \overline{(\bar{L}_1 \cup \bar{L}_2)}$$
>
> shows that if the class of context-free languages were closed under complementation, it would also be closed under intersection, which contradiction proves the theorem.

What does this proof tell us about the particular context-free languages L_1, L_2 mentioned above? Only that either $\bar{L}_1$ is not context-free or $\bar{L}_2$ is not context-free or $L = \bar{L}_1 \cup \bar{L}_2$ is context-free, but $\bar{L}$ is not. Thus, we have three languages about which we know that one is context-free while its complement is not; but *this proof does not tell us which*. Now, everyone agrees that a constructive proof of the theorem which would actually supply a single example is to preferred. But Brouwer [18] went further: he denied that a proof of an existential statement (such as the above theorem) should be accepted as being valid unless the proof itself supplies a specific example. Merely supplying three concrete objects with a proof that at least one of them satisfies the theorem is not enough for an intuitionist. Of course most mathematicians do not limit themselves to intuitionistic methods, but a surprising number still feel uneasy about nonconstructive proofs, and a few still opt for a thorough-going constructivism. [19]

The great mathematician David Hilbert was powerfully affected by what he saw as Brouwer's attack on classical mathematics, especially when Hilbert's colleague Herman Weyl associated himself at least partially with Brouwer's critique. Hilbert accused Weyl and Brouwer [20] of "trying to establish mathematics by pitching overboard everything that does not suit them and setting up an embargo". He warned that: [21]

> The effect is to dismember our science and run the risk of losing a large part of our most valuable possessions. Weyl and Brouwer condemn the general notions of irrational numbers, of functions — even of such functions as occur in the theory of numbers — Cantor's transfinite numbers, etc., the theorem that an infinite set of positive integers has atleast, and even the law of excluded middle, as for example the assertion: Either there is only a finite number of primes or there are infinitely many. These are examples of forbidden theorems and modes of reasoning. I believe that impotent as Kronecker was to abolish irrational numbers (Weyl and Brouwer do permit us to retain a torso), no less impotent will their efforts prove today. No!

Hilbert concluded: [21]

> Brouwer's program is not a revolution, but merely the repetition of a futile *coup de main* with old methods, but which was then undertaken with greater verve, yet failed utterly. Today the State is thoroughly armed through the labors of Frege, Dedekind, and Cantor. The efforts of Brouwer and Weyl are foredoomed to futility.

It should not be thought that Hilbert proposed to content himself with polemics. Hilbert responded to Brouwer's challenge with a heroic program. Hilbert was going to establish the legitimacy of nonconstructive mathematics by methods that even an intuitionist would be forced to accept. Hilbert proposed to make use of the ideas of Frege and Peano to construct a formal calculus which would encompass the entire edifice of classical mathematics. (In the meanwhile, Whitehead and Russell had carried out the monumental task of showing in detail that classical mathematics could realy be developed in such a calculus.) Then classical mathematics was to be justified by giving a *constructive consistency proof* for this calculus. In fact Hilbert and his followers never succeeded in carrying out their program which indeed, as the later work of Gödel showed, faced obstacles of a subtlety that were not even imagined. And Brouwer made it quite clear that even if Hilbert had succeeded, he would not have been prepared to withdraw his criticism. [22]

> ... nothing of mathematical value will be attained in this manner; a false theory which is not stopped by a contradiction is none the less false, just as a criminal policy unchecked by a reprimanding court is none the less criminal.

Nevertheless, Hilbert's ideas were enormously influential. The most important legacy of Hilbert's program is the idea of *metamathematics*, of formalized systems of logic as the subject of mathematical investigation. However, the fact that Hilbert's ideas developed as part of a defense of classical mathematics against Brouwer's critique, led to the principle that in metamathematical investigation only constructive methods were to be tolerated. (This is not the place to discuss the continuing important work in this field which has become known as *proof theory* that has developed directly from Hilbert's constructivist ideas.) It is only by keeping this circumstance clearly in mind that it is possible to make sense of the work of Skolem, Herbrand, Gödel, and Gentzen in the 1920's which has so heavily influenced work in automated deduction.

It should be mentioned that quite independently of the Hilbert school, the American logician Emil Post had, in his doctoral dissertation of 1920, clearly formulated a metamathematical program (but without constructivist restrictions) and carried out a careful metamathematical investigation of what is now called the propositional calculus. Post [24] began with the statement:

"We here wish to emphasize that the theorems of this paper are *about* the logic of propositions but are *not included* therein."

The key work for automated deduction was that of Skolem. He carried out a systematic study of the problem of the existence of an interpretation which will satisfy a given formula of the predicate calculus, or, as one says, whether the given formula is satisfiable. In 1920, he [25] introduced what have become known as Skolem functions, but only for $\forall\exists$ formulas. The full treatment of Skolem functions came in 1928.[26] This remarkable paper, not only has a treatment of what is usually called Herbrand's theorem in writings on automated deduction, but has a clear and complete definition of what is usually called in this field the *Herbrand universe* for a formula. (It should, it seems, be called the *Skolem universe*,[27] since Herbrand's thesis only appeared two years later.) Let us briefly recapitulate the idea of Skolem functions and their use: the sentence, say,

$$(\forall x)\ (\exists y)\ (\forall u)\ (\exists v)\ \ R(x,y,u,v), \qquad \text{(a)}$$

where R(x,y,u,v) is a formula containing no quantifiers, no variables other than x, y, u, v, and no symbols for constants or functions, is replaced by

$$(\forall x)\ (\forall u)\ \ R(x,\ f(x),\ u,\ g(x,u)) \qquad \text{(b)}$$

and it is argued that the first sentence has a *model* (i.e. an interpretation which makes it true) if and only if the second does. Now, from the point of view of a constructivist, there is a difficulty here: in any interpretation which makes (a) true, for each value of x, there are *one or more* values of y such that

$$(\forall u)\ (\exists v)\ \ R(x,y,u,v)$$

is true. The "Skolem function" f is then defined as associating with each value of x *one of the corresponding* values of y. But for a constructivist such a definition (without an algorithm for obtaining an appropriate value of y, given x) is not admissible. In fact the existence of such a function f depends on the controversial *axiom of choice*. So this part of Skolem's work was not acceptable in the framework of Hilbert's metamathematical program. To continue with our recapitulation, if we set $H_0 = \{a\}$, where *a* is a constant, and $H_{n+1} = \{f(\mu), g(\mu,\nu) \mid \mu,\nu \in H_n\}$, then $H = \bigcup_{n=0}^{\infty} H_n$ is called the Herbrand universe[27] for sentence (a). "Herbrand's theorem" for the sentence (a) then asserts that (a) has no model just in case some finite conjunction of sentences of the form

$$R(\mu,\ f(\mu), \nu,\ g(\mu,\nu)) \qquad \text{(c)}$$

where $\mu,\nu \in H$, is truth-functionally unsatisfiable. As Skolem recognized, this technique furnishes a proof-procedure for the predicate calculus, independent of any particular formal axiomatization. And of course it is this very procedure which is the underlying basis for most computer implemented theorem-provers.

It should not be supposed from Skolem's use of nonconstructive methods just discussed, that Skolem himself remained untouched by the constructivist ideas which were so widespread among logicians in the 1920's. Skolem's work on quantifier-free recursive arithmetic [28] was intended to provide a constructive foundation for certain parts of arithmetic. In fact, Skolem carried the use of recursive definitions of number-theoretic functions far beyond the early work of Dedekind and Peano.

In 1928, a little book [29] on logic appeared under the joint authorship of Hilbert and Ackermann. In this book some of the ideas that are central to work on automated deduction appeared for the first time. A formal axiomatization was given for the predicate calculus, and a pair of key unsolved problems concerning the predicate calculus were posed. The first was the problem of the *completeness* of the proposed axiomatization, i.e. the problem: is every sentence which is *valid*, i.e. true in all interpretations, formally derivable from the axioms. The second, the famous *Entscheidungsproblem*, was the problem of finding an algorithm to test whether or not a given sentence is valid. Hilbert and Ackermann declared that the Entscheidungsproblem "must be regarded as the principal problem of mathematical logic." Their reason for the prominence accorded this one problem was the explicit recognition of the fundamental role played by the predicate calculus in mathematical deduction. Any specific mathematical field is characterized by a set of axioms (which to simplify the discussion, we assume to be finite in number); to say that a certain sentence σ holds in a particular branch of mathematics is then just to say that the sentence obtained by placing the conjunction of the axioms to the left of an implication sign and σ to its right, is valid. An algorithm which would settle the Entscheidungsproblem, would then serve as an algorithm to settle all questions in mathematics.

The problem of completeness of Hilbert-Ackermann's predicate calculus was settled in 1930 by Kurt Gödel in his doctoral dissertation.[30] Gödel's work also contains a form of "Herbrand's theorem". In effect, Gödel showed that sentence (a) above is refutable in Hilbert-Ackermann's system (that is, its negation is provable from the axioms using the rules of inference) if and only if a finite conjunction of sentences of the form (c) is truth-functionally unsatisfiable. When *combined with Gödel's completeness theorem, this form of "Herbrand's theorem" is equivalent to Skolem's*. Moreover, Gödel's proof makes no use of the axiom of choice.

Finally, we come to Herbrand himself. Herbrand's thesis is written from the point of view of strict adherence to Hilbert's program.[31] At the time of writing, Herbrand knew Skolem's work, but not Gödel's, and there is considerable overlap between Herbrand's work and Gödel's. However, Herbrand could not permit himself the nonconstructive step necessary to obtain the completeness theorem itself. In fact the very notion of arbitrary interpretation was well beyond the bounds which Herbrand permitted himself. Unlike Skolem and Gödel Herbrand's formulations are valid for all sentences, not only for socalled *prenex sentences* (those consisting of a string of quantifiers followed by a quantifier-free formula). Moreover Herbrand's work contains a form of what has come to be known as the *unification algorithm*.[32]

Gödel's epoch-making 1931 paper[33] not only proved the existence of undecidable sentences of arithmetic in every reasonable formal system, but also showed what enormous obstacles the Hilbert program faced. In the first place, Gödel showed that no reasonable system is powerful enough to demonstrate its own consistency. Thus Hilbert's hope that the consistency of powerful formal systems would be proved by a restricted class of the methods available in that system was utterly demolished. (Gödel did hold out the hope that "finitary" methods might be found not encompassed by the systems being studied which methods might be powerful enough to yield the desired consistency proof.) Gödel's work also made it seem very doubtful that a solution would be obtained to the Entscheidungsproblem. Namely Gödel made heavy use of what are now called the *primitive recursive functions*, and in particular gave an algorithm by means of which for every primitive recursive scheme defining a primitive recursive function f(x), a sentence of the predicate calculus σ_f is associated such that σ_f is satisfiable (that is has a model) if and only if f(x) is 0 for all x. From this reduction (or as we might say nowadays, simulation), and Gödel's undecidability results, it followed at once, that for every reasonable formal system, there are satisfiable sentences σ_f whose satisfiability could not be demonstrated in the given formal system.

In fact the unsolvability of the Entscheidungsproblem was obtained, independently of one another, in 1936, by Alan Turing and Alonzo Church. Turing had used his now famous "machines" to give a precise explication for the intuitive notion of algorithmic procedure, and inferred the unsolvability of the halting problem. Now, after Hilbert's observation that a solution to the Entscheidungsproblem should, in principle, lead to the algorithmic decidability of all mathematical questions, it was clear that the existence of any algorithmically unsolvable problem should in turn lead, via the expression of that problem in predicate calculus, to the unsolvability of the Entscheidungsproblem. And in fact Turing [34] showed how to associate with any Turing machine T a sentence σ_T of predicate calculus such that σ_T is valid if and only if T beginning with a blank tape eventually prints a particular symbol. Thus, an algorithm which solved the Entscheidungsproblem could be used to solve the latter problem which is of course equivalent to the Halting Problem.

Church similarly reduced an unsolvable problem he had obtained from his proposal to define algorithmic computability as definability in his lambda-calculus [35] to the Entscheidungsproblem. But the problem that Church proves to be unsolvable is that of determining of a given sentence σ of predicate calculus whether or not it can be derived from the Hilbert-Ackermann axioms using their rules of inference. Of course, this is equivalent to the Entscheidungsproblem, but only via Gödel's completeness theorem for the Hilbert-Ackermann formal system. Because Gödel's completeness proof was seen as being tainted by nonconstructivism, Church felt impelled to warn his readers that [36]

> "... the unsolvability of ... the Entscheidungsproblem cannot, therefore, be regarded as established beyond question."

What seems very strange is that no one in 1936 seems to have noticed the relevance of Gödel's construction, referred to above, of a sentence σ_f of predicate calculus associated with every primitive recursive function f. Namely it follows at once from the work of Church's student Kleene [37] that the problem of determining of a given scheme for defining a primitive recursive function whether or not that function vanished identically is unsolvable. And, this leads at once, via Gödel's reduction, to the unsolvability of the Entscheidungsproblem in the form originally stated.

We have now come to the end of the history of the technical ideas from logic which proved important in automating deduction. Before going on to the early history of automated deduction, it is worth noting a curious statement made by Gödel in an essay [38] that appeared in 1944 and which shows how some of the notions discussed by Leibniz, Frege, Peano, and Poincaré continued to cast their shadow forward:

> It seems reasonable to suspect that it is this incomplete understanding of the foundations which is responsible for the fact that Mathematical Logic has up to now remained so far behind the high expectations of Peano and others who (in accordance with Leibniz's claims) had hoped that it would facilitate theoretical mathematics to the same extent as the decimal system of numbers has facilitated numerical computations. For how can one expect to solve mathematical problems systematically by mere analysis of the concepts occurring, if our analysis so far does not even suffice to set up the axioms. [39] But there is no need to give up hope. Leibniz did not in his writings about the *Characteristica universalis* speak of a utopian project; if we are to believe his words he had developed this calculus of reasoning to a large extent, but was waiting with its publication till the seed sould fall on fertile ground. He went even so far as to estimate the time which would be necessary for his calculus to be developed by a few select scientists to such an extent "that humanity would have a new kind of an instrument increasing the powers of reason far more than any optical instrument has ever aided the power of vision." The time he names is five years, and he claims that his method is not any more difficult to learn than the mathematics or philosophy of his time. Furthermore, he said repeatedly that, even in the rudimentary state to which he had developed the theory himself, it was responsible for all his mathematical discoveries; which, one should expect, even Poincaré would acknowledge as a sufficient proof of its fecundity.

With modern digital computers becoming available in the 1950's, it is not surprising that interest began to develop in automating deduction. It is interesting to look back at two early implementations of mathematical theorem provers, both of them extremely crude, produced with a difference in outlook which has been part of work in the field ever since, and has occasioned much controversy.

One was the "logic machine" of Newell, Shaw, and Simon.[40] This was a computer program intended to prove theorems in the propositional calculus using the particular axiomatization of Whitehead and Russell's *Principia Mathematica.* There was no attempt at logical sophistication. Rather the interest was in simulating the behavior of a human "problem-solver" attempting the same task. The most important legacy of this work remains its pinpointing of some of the data structures which have been crucial in automatic theorem-proving. A proof of a conclusion from given premises is conceived as a path in a tree with nodes representing the premises at the "top" and a node representing the conclusion. Edges joining a node to nodes "above" it represent legitimate derivations according to the permitted rules of proof. The crucial importance of subgoals and of seeking substitutions to produce a "match" both first appeared in this seminal work.

The other early theorem-prover referred to was the implementation in 1954 by the present author[41] of a decision procedure for the arithmetic of addition that had been given by Presburger. The computer was the Institute for Advanced Study's "johniac" — a vacuum tube machine with a cathode ray tube memory. Since it is now known that Presburger's procedure has worse than exponential complexity, it is not surprising that this program did not perform very well. Its great triumph was to prove that the sum of two even numbers is even.

The controversy referred to may be succinctly characterized as being between the two slogans: "Simulate people" and "Use mathematical logic". Although this controversy has generated much heat, there has never been much doubt among serious workers in the field that both streams of ideas were important and that a really useful automated deduction system would have to draw upon both of them. Thus as early as 1961 Minsky[41a] remarked:

"... it seems clear that a program to solve real mathematical problems will have to combine the mathematical sophistication of Wang with the [23] heuristic sophistication of Newell, Shaw and Simon."

Continuing in the spirit of Newell, Shaw, and Simon, high school geometry was chosen as an area of study by Gelernter. [42] The most interesting success of his program which he called the "geometry machine", was the rediscovery of the proof (well known to later Greek commentators on Euclid, but quite unknown to the designers of the geometry machine) that the base angles of an isosceles triangle are equal, based on the nontrivial congruence of such a triangle with itself. For substantially more difficult problems, however, the geometry machine had to be permitted to use a limited analytic geometry capability as a guide to plausible proof steps in order to achieve results.

The first suggestion that methods based on "Herbrand's theorem" were appropriate for general purpose theorem-provers seems to have been made by Abraham Robinson in a brief talk [43] delivered at the Summer Institute for Symbolic Logic at Cornell University in 1954. (This five week Institute was a most stimulating meeting at which many logicians as well as computer scientists were present.) Robinson made the suggestive remark that additional points, lines or circles "constructed" as part of the solution to a geometry problem, can be construed as being elements of what is now called the Herbrand universe for the problem (when formalized in predicate calculus).

The earliest implemented theorem-provers for predicate calculus using Herbrand's theorem were based on a totally unguided search through the Herbrand universe. However the format usually did not explicitly use Skolem functions. [44] Instead, all instantiations (for universally quantified as well as for existentially quantified variables) were of parameters, so that dependencies needed to be kept track of "on the side". The necessary tests for truth-functional satisfiability were to be carried out either by direct truth table calculations or by expansion into disjunctive normal form. Not surprisingly, the combination of a wasteful search through the Herbrand universe with inefficient propositional calculus techniques produced programs capable of proving only the simplest theorems. [44]

In [Davis and Putnam 1960], Skolem functions were introduced and the now familiar conjunctive normal form clausal arrangement of the initial data was proposed. The authors naively took "the critical difficulty" to be the lack of a feasible technique for testing for truth functional satisfiability. They dealt with this "difficulty" (which certainly had crippled previous theorem-provers) by proposing a new algorithm for this purpose. This algorithm and its improved version in an implemented version [45] have been much studied. In [Davis, Logemann and Loveland 1962] the authors express disappointment with the results obtained with their program. That effectively eliminating the truth-functional satisfiability obstacle only uncovered the deeper problem of the combinatorial explosion inherent in unstructured search through the Herbrand universe was evidently something of a surprise. The conclusion was now clear: [45]

> "... the most fruitful future results will come from ... excluding ... 'irrelevant' quantifier-free lines [from the Herbrand expansion]."

Meanwhile the logician Hao Wang [46] was bringing to bear on the problem some of the more sophisticated work that had been done in proof theory and on solvable cases of Hilbert's Entscheidungsproblem. He was able to announce a computer program which was able to prove all theorems (about 400) of Whitehead and Russell's "Principia Mathematica" which belonged to the "pure" predicate calculus (with equality). However, although this stunning announcement seemed a great achievement in automating deduction, Wang was quick to point out that it was really a matter of all of the theorems in question belonging to a subclass which is particularly easy to handle. "The most interesting lesson from these results is perhaps that even in a fairly rich domain, the theorems actually proved are mostly ones which call on a very small portion of the available resources of the domain."

It was Prawitz in his influential paper [Prawitz 1960] who taught us that unnecessary terms in the Herbrand expansion could be avoided by arranging our algorithms so that terms are not produced until actually needed. This key insight is the germ of all future progress in the field. In effect Prawitz proposed to obtain expansions into disjunctive normal form *before* substitution of elements from the Herbrand universe. The algorithm thus took the form of producing longer and longer disjunctive normal forms to see whether one could be obtained into which an appropriate substitution from the Herbrand universe would produce a truth-functionally unsatisfiable formula. This led to the fruitful idea of seeking substitutions which would cause two literals to be negatives of one another. With a Skolem function format, this idea leads at once to the unification algorithm. Since Prawitz was still working with parameters and dependencies, instead he was led to systems of equations which needed to be satisfied.

Prawitz' procedure was an enormous improvement over the "Davis-Putnam" procedure in that only "relevant" clauses were generated. But, as was pointed out in [Davis 1963], it was subject to the very inefficiencies in testing for truth-functional satisfiability that the Davis-Putnam procedure had overcome. The present author was led to propose [47] "a new kind of procedure which seeks to combine the virtues of the Prawitz procedure and those of the Davis-Putnam procedure". This "linked conjunct" procedure was based on the simple observation that a truth-functionally unsatisfiable conjunction of disjunctive clauses remains unsatisfiable if all clauses are deleted in which some literal ℓ occurs such that $\sim \ell$ occurs in none of the remaining clauses. (This observation, which was basically just one of the "Davis-Putnam" rules, was made, independently, in [Dunham and North 1962].) Thus a proof procedure can be based on "mating" (or, unifying, as one now says) a literal in one clause with a literal in another clause via appropriate substitutions of terms for variables so that the two literals are negations of one another. Once a completely mated "linked conjunct" is obtained, a truth-functional test for satisfiability, e.g. using the Davis-Putnam rules, is required. Procedures of this kind have the virtue that no more clauses are ever produced than are actually required for a proof, and no search through the entire Herbrand universe is needed. A theorem-proving program based on these ideas was written by D. McIlroy and completed in November 1962 at Bell Telephone Laboratories. The program was improved and corrected by Peter Hinman during the summer of 1963. The "mating" procedure naturally required what is now called unification and the unification algorithm incorporated in the program was exactly the one later given in print in [Robinson 1965a]. A discussion of the implementation with its virtues and limitations was given in [Chinlund et al. 1964]. [48]

J. A. Robinson began his important and fruitful work in the field with his [Robinson 1963]. Using the basic data structures of [Davis and Putnam 1960] and referencing [Davis 1963], Robinson also considered the problem of capturing some of the power of [Prawitz 1960] in a feasible procedure. However rather than a general mating or unifying procedure, Robinson proposed methods that were based on previous knowledge of the elements of the Herbrand universe required to obtain a proof. Robinson proposed that locating the appropriate elements of the Herbrand universe "seems to be the really 'creative' part of the art of proof-construction - ...".

It was Robinson's method of resolution which more than anything else, brought automatic theorem-proving to the attention of a wide audience of computer scientists, and may fairly be said to have revolutionized the subject. In the historical context, Robinson's step was to propose combining the present author's "mating" of literals and their negations with the Davis-Putnam rules [49] into a single step: just as soon as a mating was accomplished, so the set of clauses could be thought of as being of the form $(\ell \vee A) \& (\sim \ell \vee B) \& C$, where ℓ is the mated literal and A, B, C are free of ℓ, one replaces the system of clauses by $(A \vee B) \& C$. Robinson proved [50] that this single "rule of inference" was complete. Thus, no separate testing for truth-functional satisfiability was required following the mating phase (as in the linked conjunct method). A single simple combinatorial principle had been found which was adequate for all logical deduction.

The price which had to be paid for this simplicity was the loss of the guarantee in the linked conjunct method that no more clauses would be present at any stage than were required for a proof. In fact, as was soon seen, without the help of heuristics the resolution method could easily produce many thousands of clauses without reaching a proof. From the point of view of resolution the problem of organizing the search for a proof takes on the special form: in which order shall resolutions be attempted?

One early way to proceed was given by Robinson himself when he observed that by making use of the natural partition of literals into unnegated ("positive") literals and negated ("negative") literals, the search space can be considerably cut down without losing completeness. This elegant method (which has been much generalized) was called hyper-resolution. [51]

Less theoretically elegant, but highly useful, heuristics were given by the team of L. Wos, D. Carson and G. A. Robinson at the Argonne National Laboratory.[52] These heuristics were based on singling out certain clauses for preferential treatment.

Thus, we end our 330 year survey at the beginning of the explosion of interest in mechanical theorem proving following the introduction of Robinson's resolution. Resolution has so dominated the field that discussions have tended to identify all theorem-proving efforts based on deeper logical analysis with resolution. However more recently there has been renewed interest in nonresolution proof procedures based on proof-theoretic considerations. Whatever the future may bring, efforts to increase the logical power of automatic devices will surely continue.

Footnotes

[1] This article is based on an invited address delivered before the Fourth Workshop on Automated Deduction at Austin, Texas in January 1979.

[2] Martin Davis, Courant Institute of Mathematical Sciences, New York University, 251 Mercer Street, New York, N.Y. 10012.
This work was supported by NSF Grant MCS 76-24212.

[3] [Chang & Lee 1973] and [Loveland 1978] are the treatises. See also the bibliographies in these books.

[4] These "four figures" showed how to reduce arithmetic operations to geometric constructions.

[5] p. 105 [Parkinson 1966].

[6] Ibid., p. xvii.

[7] Ibid., pp. 132-133.

[8] [Kneale 1962], p. 328.

[9] [Boole 1847].

[10] [Kneale 1962], p. 421.

[11] [van Heijenoort 1967], p. 2.

[12] Ibid., p. 3.

[13] *Modus ponens* also known as the *rule of detachment* permits one to infer a proposition B from the premises A and $A \rightarrow B$.

[14] Ibid., p. 86.

[15] [Poincaré 1908], p. 156.

[16] Ibid., p. 147. Actually the direct target of the quoted paragraph is Hilbert's abstract version of Euclid's geometry.

[17] Quoted in [Moritz 1914].

[18] [Brouwer 1923]

[19] In [Bishop 1967], readers will find a strongly worded "manifesto" in favor of constructivism.

[20] Quoted in [Bell 1945], p. 569.

[21] Ibid., p. 570.

[22] Ibid., p. 569.

[23] See for example [Takeuti 1975].

[24] [Post 1921], p. 265.

[25] [Skolem 1920].

[26] [Skolem 1928].

[27] The present author is apparently responsible for this historical error. The term *Herbrand universe* seems to have first occurred in [Davis 1963].

[28] [Skolem 1923].

[29] [Hilbert and Ackermann 1928].

[30] [Gödel 1930].

[31] [Herbrand 1930].

[32] I am indebted to Gêrard Huet for calling this to my attention. Herbrand (for reasons already discussed) was not using Skolem functions and so his formulation involved satisfying systems of equations between terms, anticipating a proof procedure later given by Prawitz, [Prawitz 1960].

[33] [Gödel 1931].

[34] [Turing 1936].

[35] [Church 1936].

[36] Ibid., p. 102.

[37] [Kleene 1936], p. 741.

[38] [Gödel 1944].

[39] Gödel here seems to make the point that the best available foundation for classical mathematics, the axioms for set theory of Zermelo-Fraenkel, do not adequately encompass our set-theoretic intuitions.

[40] [Newell, Shaw and Simon 1957].

[41] [Davis 1957].

[41a] [Minsky 1961].

[42] [Gelernter 1959].

[43] [Robinson 1957].

[44] For example, see [Gilmore 1960]. However the discussion in [Robinson 1957] does use Skolem functions. [Prawitz, Prawitz and Voghera 1960] is not based on Herbrand's theorem but suffers from similar limitations.

[45] [Davis, Logemann and Loveland 1962].

[46] [Wang 1960a], [Wang 1960b], [Wang 1963].

[47] [Davis 1963].

[48] [Chinlund et al.] was submitted for publication to the CACM and was accepted subject to some relatively minor changes being made. It is difficult to reconstruct, at this date, why this was never done.

[49] In the original [Davis, Putnam 1960] version, rather than that of the "improved" [Davis, Logemann, Loveland 1962] version.

[50] [Robinson 1965a].

[51] [Robinson 1965b]. An early generalization of hyper-resolution was given in [Meltzer 1966].

[52] [Wos, Carson and Robinson 1964], [Wos, Robinson and Carson 1965].

References

Bell, E. T., *The Development of Mathematics.* Second edition, McGraw-Hill, 1945.

Boole, George, *The Mathematical Analysis of Logic.* Cambridge, 1847. Reprinted, Oxford, 1948.

Chang, Chin-Liang and Richard Char-Tung Lee, *Symbolic Logic and Mechanical Theorem Proving.* Academic Pres, 1973.

Chinlund, T. J., M. Davis, P. G. Hinman, and M. D. McIlroy, *Theorem Proving by Matching.* Bell Laboratories, 1964.

Church, Alonzo, A Note on the Entscheidungsproblem. *The Journal of Symbolic Logic* 1 (1936) 40-41; correction, Ibid., 101-102. Reprinted (with correction incorporated into text) in [Davis 1965].

Davis, Martin, A computer program for Presburger's procedure. *Summaries of Talks Presented at the Summer Institute for Symbolic Logic*, 1957. Second edition, published by Institute for Defense Analysis, 1960.

Davis, Martin, Eliminating the irrelevant from mechanical proofs. *Proc. Symp. Applied Math.* XV (1963), 15-30.

Davis, Martin (editor), *The Undecidable*, Raven Press, New York, 1965.

Davis, Martin, G. Logemann and D. Loveland, A machine program for theorem proving. *C ACM* 5 (1962), 394-397.

Davis, Martin, and Hilary Putnam, A computing procedure for quantification Theory. *J. ACM* 7 (1960), 201-215.

Descartes, René, *Geometry.* (translated by D. E. Smith and M. L. Latham) Dover, 1954. Original publication as appendix to *Discours de la Methode,* Jan Maire, Leyden, 1637.

Dunham, B. and J. H. North, Theorem testing by computer, *Symp. Math. Theory Machines,* Brooklyn Poly. Inst., 172-177, 1962.

Feigenbaum, E. and J. Feldman, *Computers and Thought,* McGraw-Hill, 1963.

Gelernter, H., Realization of a geometry-theorem proving machine. *Proc. Intern. Conf. on Inform. Processing,* UNESCO House, 1959, 273-282. Reprinted in *Computers and Thought* (Feigenbaum and Feldman, editors), McGraw-Hill, 1963.

Gilmore, P. C., A proof method for quantification theory: its justification and realization. *IBM J. Res. Develop.* 4 (1960), 28-35.

Gödel, Kurt, Die Vollständigkeit der Axiome des logischen Funktionenkalküls. *Monatshefte f. Math. u. Phys.* 37 (1930), 349-360. Translated as: The completeness of the axioms of the functional calculus logic, in [van Heijenoort, 1967].

Gödel, Kurt, Über formal unentscheidbare Sätze der Principia mathematica und verwandter Systeme I, *Monatshefte f. Math. u. Phys.* 38 (1931), 173-198. English translation in [Davis 1965], in [van Heijenoort 1967] and as [Gödel 1962].

Gödel, Kurt, Russell's mathematical logic. *The Philosophy of Bertrand Russell* (Paul A. Schlepp, editor), Northwestern University Press, 1944, 123-153.

Gödel, Kurt, *On formally undecidable propositions of Principia mathematica and related systems* (translated by B. Meltzer), Oliver and Boyd, 1962.

Herbrand, Jacques, *Recherches sur la théorie de la démonstration.* Thesis at the University of Paris, 1930. Translated as: Investigations in proof theory, in *Jacques Herbrand - Logical Writings* (Warren D. Goldfarb, editor), Harvard University Press, 1971.

Hilbert, D. and W. Ackermann, *Grundzüge der Theoretischen Logik*, Julius Springer, 1928.

Kleene, S. C., General recursive functions of natural numbers. *Math. Annalen* 112 (1936), 727-742. Reprinted in [Davis 1965].

Kneale, W. and M., *The Development of Logic*, Oxford, 1962.

Loveland, Donald, *Automated Theorem Proving: A Logical Basis*, North-Holland, 1978.

Meltzer, B., Theorem-proving for computers: some results on resolution and renaming. *Comput. J.* 8, 341-343 (1966).

Minsky, Marvin, Steps toward artifical intelligence, *Proc. Inst. Radio Enineers*, 49, 8-30, 1961. Reprinted in [Feigenbaum and Feldman, 1963].

Moritz, Robert, *Memorabilia Mathematica*. Macmillan Co., 1914.

Newell, A., J. C. Shaw, and H. Simon, Empirical explorations with the logic theory machine, *Proc. West. Joint Comp. Conf.*, 1957, 218-239. Reprinted in [Feigenbaum and Feldman 1963].

Parkinson, G. H. R., *Leibniz -- Logical Papers*. Oxford, 1966.

Poincaré, Henri, *Science and Method*. Dover, 1952. (translated by F. Maitland from the original French edition of 1908.)

Post, E. L., Introduction to a general theory of elementary propositions. *Amer. J. Math.* 43 (1921), 163-185. Reprinted in [van Heijenoort 1967].

Prawitz, Dag., An improved proof procedure. *Theoria*. 26 (1960), 102-139.

Prawitz, D., H. Prawitz and N. Voghera, A mechanical proof procedure and its realization in an electronic computer, *JACM*, 102-128, 1960.

Robinson, Abraham, Proving theorems, as done by man, machine and logician. *Summaries of Talks Presented at the Summer Institute for Symbolic Logic*, 1957. Second edition, published by Institute for Defense Analysis, 1960.

Robinson, J. A., Theorem-proving on the computer, *JACM*, 10, 163-174, 1963.

Robinson, J. A., A machine oriented logic based on the resolution principle. *JACM*, 12 (1965a), 23-41.

Robinson, J. A., Automatic deduction with hyperresolution, *Int. J. Comput. Math.* 1, 227-234, 1965b.

Skolem, Thoralf, Logisch-kombinatorische Untersuchungen über die Erfüllbarkeit oder Beweisbarkeit mathematischer Sätze nebst einem Theorem über dichte Mengen. *Videnskopsselskapits skifter*, I. *Matematik-naturvidenskabelig klasse*, No. 4 (1920). (Translated as: Logico-combinatorial investigations in the satisfiability or provability of mathematical propositions: A simplified proof of a theorem by L. Löwenheim and generalizations of the theorem [van Heijenoort 1967].)

Skolem, Thoralf, Begründung der elementaren Arithmetik durch rekurrierende Denkweise ohne Anwendung scheinbarer Veränderlichen mit unendlichem Ausdehnungsbereich. Ibid., No. 6 (1923). (Translated as: The foundation of elementary arithmetic established by means of the recursive mode of thought, without the use of apparent variables ranging over infinite domains. [van Heijenoort 1967].)

Skolem, Thoralf, Über die mathematische Logik. *Norsk matematisk tideskrift*, 10 (1928), 125-142. (Translated as: On mathematical logic. [van Heijenoort 1967].)

Takeuti, G., *Proof Theory*, North-Holland, 1975.

Turing, Alan M., On computable numbers with an application to the Entscheidungsproblem. *Proc. London Math. Soc.* 2nd series, 42 (1937), 230-265; correction, Ibid. 43 (1937), 544-546. Reprinted in [Davis 1965].

van Heijenoort, Jean (editor), *From Frege to Gödel.* Harvard University Press, 1967.

Wang, Hao, Proving theorems by pattern recognition I, *CACM* 3, 220-234 (1960a).

Wang, Hao, Towards mechanical mathematics, *IBM J. Res. Develop.* 4, 224-268 (1960b).

Wang, Hao, The mechanization of mathematical arguments, *Proc. Symp. in Applied Math.* XV (1963), 31-40.

Wos, L., D. Carson and G. A. Robinson, The unit preference strategy in theorem proving. *Proc. IFIPS 1964 Fall Joint Comp. Conf.* 26, 616-621, 1964.

Wos, L., G. A. Robinson and D. Carson, Efficiency and completeness of the set of support strategy in theorem proving, *JACM* 12, 536-541, 1965.

Mechanical Proof-Search and the Theory of Logical Deduction in the USSR

S. Yu. Maslov, G. E. Mints, V. P. Orevkov

In the present survey of the Soviet publications devoted to the problem of the automatic theorem-proving the authors consider in the first place those papers which use the methods of mathematical logic for solving this problem. Thus we shall only superficially touch upon the heuristic direction of the automation of the mathematical proofs. The excellent results (in the first place due to Siberian mathematicians) on the decidable theories [11] will not be mentioned because the corresponding decision algorithms are of pure theoretical nature. That is why we shall deal with the decision problem for the predicate calculus only.

Co, C, $\bar{C}$ will denote respectively classical propositional calculus, pure classical predicate calculus, classical predicate calculus with function symbols. The calculi with equality corresponding to C and $\bar{C}$ are denoted by $C^{=}$ and $\bar{C}^{=}$. Symbols Jo, J, $\bar{J}$, $J^{=}$, $\bar{J}^{=}$ have the similar meaning for intuitionistic calculi.

1. The direction of principal importance in the automation of the theorem-proving is one connected with the construction of logical and applied calculi well fitted for finding for every formula a small number of its possible predecessors (that is the formulas the given one can be obtained from by one application of a deduction rule). Such calculi are well adapted for the proof-search « from bottom to top ». The construction of such calculi is based first of all on the development of the Gentzen's ideas about Sequenzen-calculi with the « subformula-property ». Some methods similar to the Sequenzen-methods are also suggested by the famous Herbrand theorem (an analog of the theorem for $\bar{J}^{=}$ was proposed by G. E.

Mints [28]) and by the apparatus of the regular formulas introduced by P. S. Novikov [31, chapter VI] for the construction of a variant of the arithmetic without induction (1).

Some interesting Sequenzen-variants of C with the subformula-property were proposed by V. A. Matulis [26]. In these variants the deduction tree of every formula is unique. A convenient Sequenzen-variant of J was constructed by R. A. Pluškevičius [38]; he also constructed (in [39], [40]) a number of cut-free variants of the constructive logic (in the sense of N. A. Shanin and A.A. Markov) for the normal formulas. Similar variants of the applied calculi with equality and of some types of axiomatic theories were constructed by M. G. Rogava [43] and A. J. Pluškevičene [41].

When proving the completeness of proof-procedures one often uses the possibility of some specializations of the form of deduction. Various specializations of such kind for $C^{=}$ and $J^{=}$ were established by V. A. Lifshits [17], G. E. Mints [29] and V. P. Orevkov [36].

The property of a calculus called the invertibility of its rules is very essential for the deduction-search « from bottom to top » (the rule of inference is invertible if the derivability of the conclusion implies the derivability of all premises). All variants of classical calculi mentioned above are invertible. S. Ju. Maslov [21] proposed the invertible Sequenzen-variant for $J^{=}$ and a simple method for constructing an invertible variant of an arbitrary calculus starting from an uninvertible variant.

To the Gentzen-methods there adjoin some results of A. Tauts [44] and the papers where good decision algorithms for Jo are constructed; such are for example the algorithms by N. N. Vorobjev [2] and Ya. Ya. Golota [3], [4] (2).

2. The basic difficulty of the proof-searching « from bottom to top » lies in the « inability » to find the terms to be substituted for the variables in the applications of quantifier-rules. In this connection there arises the idea of « metavariables » (3) which are to be replaced

(1) L. L. Tsinman [46] considered arithmetical systems with various restrictions of the induction rule.

(2) B. Ju. Pil'čak [37] gives a proof of the decidability of J_0 in terms of Jaskowski truth tables.

(3) S. Kanger (Comput. Progr. and Formal Syst., Studies in Logic and Found of Math., 1968, 87-93) and independently N. A. Shanin.

by terms only after the process of the proof-search from bottom to top is finished. Following this idea it is not the proof we get at first but some intermediate object — prededuction ; after that we verify whether some substitutions (concretizations) can be found for the metavariables turning this object into a correct deduction. The direct application of the method of metavariables is inefficient because much work is needed to verify the possibility of turning the sufficiently complicated predeductions into correct deductions (and this amount of work is generally speaking useless for the further steps if the given prededuction cannot still be turned into a correct deduction and have to be built over). This is connected first of all with the « globality » of work with the prededuction. That is why more perspective are apparently the methods enabling us to combine the unpreciseness of the values of variables and « local independence » of work (i.e. the using of a relatively small part of deduction at the given stage of work and independence of this work from the remaining parts of the deduction).

The resolution method of J. Robinson for $\tilde{C}$ proposed in 1965 is the most known method of this kind. At the same time S. Ju. Maslov independently proposed another method of the similar kind — the so-called « inverse method » [18]. The initial idea of this method differs from that of the resolution method and turns out to be applicable for the establishing of deducibility of formulas of arbitrary structure in every Sequenzen-calculus with the subformula-property [20], [22]. The inverse method for a fixed calculus K is a method of construction of a special calculus K_F (the calculus of favourable sets) for every formula F ; the derivability F in K is equivalent to derivability in K_F of a special object (the empty set). The inverse method enables one to use expediently the individual peculiarities of the trial formula in constructing its deduction. The deduction is obtained successively : first one determines the structure of the upper formulas of the deduction-tree, then the structure of the formulas lying under them and so on. Combining the qualities of the method of metavariables with the « local independence » the inverse method turned out to be an effective tool of the construction of the machine-oriented proof-procedures (a computer program of such a procedure is accomplished in Leningrad [9]).

For further increasing of the efficiency of the machine-oriented proof procedures (based either on the inverse method or on the resolution principle) it is of great interest to study the strategies restricting the process of the construction of the favourable sets (respectively, clauses). Such strategies for $\bar{C}$ are studied in [23], [24]. There was found a close connection between the resolution method (in terms of « clash » (4)) and the inverse method for the formulas of the same standard form as for the resolution method. This connection leads to the useful interaction of two methods, using the strategies of one of them in the other. The most interesting results concern the strategies of restriction of the order of members in electrons (if one speaks in clash-terms). In [12] a general method of proving the completeness of the great number of strategies of the resolution method is proposed.

3. It seems essential to devise efficient proof procedures that would be at the same time decision algorithms for as broad classes of formulas as possible. So it is important to study solvable classes of undecidable calculi. By solvable or reduction classes we understand those relative to deducibility (validity). Completing the investigations of a number of authors (including [13]). Ju. Sh. Gurevich has shown [5] that the only (5) decidable classes of $C^{=}$ are $\forall^m \exists^2 \forall^n$, $\forall^m \exists^n$ and the singular predicate calculus. The principal point of his proof is the existence of such an r that the class of closed formulas of the form $\exists \forall \exists M_1 \forall \forall^n M_2$, where M_1, M_2 are quantifier-free and contain at most one binary and r monadic predicate letters, is a reduction class for C.

The inverse method enables us to organize an efficient deduction search in such a way as to decide at the same time the unprovability of the formulas from broad classes. In fact the inverse method gives us a convenient tool for the theory of deduction and for finding of solvable classes and reduction classes. Proofs of solvability of almost all solvable classes in literature are covered by a single scheme, also it turns out to be possible to extend well-known decidable classes and to obtain new ones [19], [20]. For example the solvability of two classes (K and M) described below is proved by means of the inverse method. Let us consider only formulas of $\bar{C}$ satisfying the conditions :

(4) For terminology consult J. R. Slagle — *J. Assoc. Comput. Mach*, 1967, 14, 4, pp. 687-697.

(5) If the classes are formulated in terms of prefix and signature and the finite sets of formulas are disregarded.

1) the variables of different occurences of quantifiers are different and distinct from the free variables of the formula;

2) the only propositional connectives that may occur in the formula are &, V, $\rceil$ and only elementary formulas can be negated. Let F be a formula and G be its elementary subformula. By F-prefix of G we shall mean a quantifier prefix obtained in the following way: F is read from left to right and those quantifiers which bind the variables in G are written down successively. We shall say that a formula F (being possibly open) of C belongs to K if we can find quantifiers $\exists x_1, \ldots, \exists x_k$ not within the scope of any universal quantifier such that the only F-prefixes are F-prefixes of length $\leqslant 1$, F-prefixes ending with a universal quantifier and the F-prefix $\exists x_1 \ldots \exists x_k$. The class of arbitrary disjunctions of the formulas from K is solvable. This implies the solvability of cases III, IV, VI, VII, VIII, IX, XII and XIII from the list of § 46 of the monography by A. Church.

We shall say that the formula F of $\bar{C}$ belongs to M if in every F-prefix every quantifier except possibly the last one is a universal quantifier. The class M is solvable. The immediate consequence is the solvability for $\bar{C}$ of the class $V^m \exists V^n$ (V. P. Orevkov [34]) and of the class of formulas with at most singular predicate and function symbols. Ju. Sh. Gurevich has shown that these are only (6) solvable classes of $\bar{C}$ (Symp. on Math. Logic, Alma-Ata, 1969). As to $\bar{C}^=$ the following class is a reduction class: $\exists x D$ where D is a disjunction of (possibly negated) elementary formulas with at most 2 singular function symbols and without predicate letters (V. A. Lifshits, [14]).

It was shown that the class of formulas containing exactly I singular predicate letter [25] and the class $\rceil\rceil V^n \exists^m M$ (M—quantifier-free) [33] are reduction classes for **J**. Two decidable classes were found by G. E. Mints: the analog of the classical $V^m \exists^n$ for $\mathcal{J}^=$ [27] and the class of formulas without negative quantifiers for $\mathcal{J}$ [30].

A new approach to the analysis of reduction procedures that enables one to prove constructively reductions of classical calculi and to carry on these reductions to intuitionistic case was proposed in [15]. Many other solvable and reduction classes are contained in [14], [16], [20], [22], [32], [35].

(6) See preceeding footnote.

4. N. A. Shanin suggested a formulation of the problem of deduction search that calls for algorithms which, instead of the mere testing of deducibility would construct deductions of « high quality ». By this are meant those deductions that are as compact as possible (avoiding, in particular, the unnecessary applications of rules of inference), as « joint » as possible (avoiding the repetitions of similar parts of deduction), as « natural » as possible (having a usual form, habitual for mathematicians) and such like.

Under the direction of N. A. Shanin there was developed in Leningrad an algorithm for C_0 partially satisfying the above mentioned demands [47]. For every propositional formulas $Q_1, \ldots, Q_n$, R this algorithm not only decides if R is implied by $Q_1, \ldots, Q_n$ but constructs (if the answer is positive) a natural deduction of R from $Q_1, \ldots, Q_n$. This algorithm was programmed for computer and the outputs were presented in the form of the sequences of sentences of the Russian mathematical language forming a deduction in a calculus of natural inferences (some extension of NJ-calculus of G. Gentzen — « Math. Zeitschr. », 1934, 39, 2, 176-210).

The theoretical base of the algorithm is formed by a group of invertible Sequenzen-calculi proposed by N. A. Shanin in 1961. Deductions in these calculi are easy to transform in deductions in NJ.

The methods useful for the solution of the problem of the searching of natural deductions in $\bar{C}$ are developed in the Davydov's paper [8]; he also considered [6], [7] the problem of the automatic generation of new theorems by means of correcting given hypotheses (that is by transforming the underivable formulas in derivable ones by means of the least possible modifications).

5. The heuristic direction in the automation of mathematical proofs is represented in USSR by the Kiev school. Algorithms of the logicians of this school are based on the analysis of reasoning usual for a common mathematician. The resulting algorithms are incomplete because they are aimed at the obtaining of the proofs constructed in correspondence with some fixed schemata. In [1] the computer proof procedure for the group theory is described. The proof procedure for analisis described in [42] allow one to find proofs of some elementary theorems in the theory of limits. In [10] a program for the algorithm of A. Tarski is mentioned. A proof procedure for primitive recursive arithmetic was programmed by M. Šiltere [48].

6. The question of the obtaining of exact mathematical estimations of the complexity of deductions is very urgent now. The corresponding mathematical apparatus has not yet been developed and al discussions on the comparing of the proof procedures are only qualitative, intuitive and not sufficiently convincing. In this connection the work of G. S. Tseytin [45] is of great importance. In his paper estimations of the complexity are given for deductions in C_0. The estimations are made for several variants of C_0 close to the calculus of resolutions. In these estimations an influence of the cut rule on the length of deductions is revealed. (The author considers deductions of two forms : the tree form and the line-form).

Some preliminary results about the comparing of the strategies of the proof procedures are given in [24].

7. In conclusion the authors would like to list the most essential (in their opinion) directions of the investigations to be developed next : 1) the studying of the possibility of using the individual features of mathematical theories for the increasing of the efficiency of proof procedures for these theories ; 2) the development of the mathematical apparatus of estimations of the qualify of logical algorithms ; 3) the creating of a general theory including various partial results about the strategies of deduction-search (at least for $\bar{C}$).

REFERENCES

Abbreviations :

PSIM = ТМИ = «Труды Математического института им. В. А. Стеклова АН СССР» (Trudy Matem. Inst. Steklov ; Translation : Proc. Steklov Inst. Math.).

SIM = ЗНС = «Загивки Научных Семинаров ЛОМИ» (Translation : Seminars in mathematics V. A. Steklov Mathem. Institute, Leningrad, Consultants Bureau, New York-London).

SMD = ДАН = «Доклады АН СССР» (Translation : Soviet Math. Dokl.).

1. Anufriev, F. V., Fedjurko, V. V., Letičevskii, A. A., Asel'derov, Z. M., Diduh, J. J., *On a Certain Algorithm for Search of Proofs of Theorems in the Theory of Groups.* Kibernetika (Kiev), 1966, N° I, 23-29.
2. Vorob'ev, N. N., *A New Decision Algorithm in the Constructive Propositional Calculus.* ТМИ, 1958, 2, 52, 193-225.
3. Golota, Ya. Ya., *Nets of Marks and Deducibility in the Intuitionistic Propositional Calculus.* SIM, 1971, 16, 11-19 (= ЗНС, 1969, 16, 28-43).
4. Golota, Ya. Ya., *Some Techniques for Simplifying the Construction of Nets of Marks.* SIM, 1971, 16, 20-25 (= ЗНС, 1969, 16, 44-53).

5. Gurevič, Yu. Š., *Effective Recognition of Realizability of Formulae of the Restricted Predicate Calculus.* Algebra i Logika Sem., 1966, 5, N° 2, 25-55.
6. Davydov, G. V., *Method of Establishing Deducibility in Classical Predicate Calculus.* SIM, 1969, 5, 1-4 (= ЗНС, 1967, 4, 8-17).
7. Davydov, G. V., *On the Correction of Unprovable Formulas,* SIM, 1969, 4, 5-8 (= ЗНС, 1967, 4, 18-29).
8. Davydov, G. V., *Some Remarks on Proof Search in the Predicate Calculus.* SIM, 1970, 4, 1-6 (= ЗНС, 1968, 8, 8-20).
9. Davydov, G. V., Maslov, S. Yu., Mints, G. E., Orevkov, V. P., Slisenko, G. O., *A Computer Algorithm for the Determination of Deducibility on the Basis of the Inverse Method,* SIM, 1971, 16, 1-6 (= ЗНС, 1969, 16, 8-19).
10. Džafarova, N. N., *Machine Realization of Tarski Algorithm.* Trudy Vyčisl. Centra Akad. Nauk Azerb. SSR, 1965, 3, 3-9.
11. Eršov, Yu. L., Lavrov, J. A., Taimanov, A. D., Taiclin, M. A., *Elementary theories.* Uspehi Mat. Nauk, 1965, 20, 4 (124), 37-108.
12. Zamov, N. K., Sharonov, V. J., *A Class of Strategies for the Determination of Provability by the Resolution Method.* SIM, 1971, 16, 26-31 (= ЗНС, 1969, 16, 54-64).
13. Kostyrko, V. F., *The reduction class* $\forall\exists^n\forall$. Algebra i Logika Sem., 1964, 3, N° 5-6, 45-55.
14. Lifshits, V. A., *Some Reduction Classes and Undecidable Theories.* SIM, 1969, 4, 24-25 (= ЗНС, 1967, 4, 65-68).
15. Lifshits, V. A., *Deductive Validity and Reduction Classes.* SIM, 1969, 4, 26-28 (= ЗНС, 1967, 4, 69-77).
16. Lifshits, V. A., *Problem of Decidability for Some Constructive Theories of Equalities.* SIM, 1969, 4, 29-31 (= ЗНС, 1967, 4, 78-85).
17. Lifshits, V. A., *Specialization of the Form of Deduction in the Predicate Calculus with Equality and Functional Symbols,* I, PSIM (= ТМИ, 1968, 98, 5-25).
18. Maslov, S. Yu., *An Inverse Method of Establishing Deducibility in the Classical Predicate Calculus.* SMD, 1964, 5, 1420-1424 (= ДАН, 1964, 159, N° I, 17-20).
19. Maslov, S. Yu., *Application of the Inverse Method for Establishing Deducibility to the Theory of Decidable Fragments in the Classical Predicate Calculus.* SMD, 1966, 7, N° 6, 1653-1657 (=ДАН, 1966, 171, N° 6, 1282-1285).
20. Maslov, S. Yu., *An Inverse Method of Establishing Deducibility of Nonprenex Formulas of the Predicate Calculus.* SMD, 1967, 8, N° I, 16-19 (= ДАН, 1967, 172, N° I, 22-25).
21. Maslov, S. Yu., *Invertible Sequential Variant of Constructive Predicate Calculus.* SIM, 1969, 4, 36-42 (= ЗНС, 1967, 4, 96-111).
22. Maslov, S. Yu., *The Inverse Method for Establishing Deducibility for Logical Calculi.* PSIM (= ТМИ, 1968, 98, 26-87).
23. Maslov, S. Yu., *Deduction-Search Tactics Based on Unification of the Order of Members in a Favourable Set.* SIM, 1971, 16, 64-68 (= ЗНС, 1969, 16, 126-136).

24. Maslov, S. Yu., *Relationship between Tactics of the Inverse Method and the Resolution Method.* SIM, 1971, 16, 69-73 (= ЗНС, 1969, 16, 137-146).
25. Maslov, S. Yu., Mints, G. E., Orevkov, V. P., *Unsolvability in the Constructive Predicate Calculus of Certain Classes of Formulas Containing only Monadic Predicate Variables.* SMD (=ДАН, 1965, 163, N° 2, 295-297).
26. Matulis, V. A., *Variants of the Classical Predicate Calculus with Unique Deduction Tree.* ДАН, 1963, 148, N° 4, 768-770.
27. Mints, G. E., *Choice of Terms in Quantifier Rules of Constructive Predicate Calculus.* SIM, 1969, 4, 43-46 (= ЗНС, 1967, 112-122).
28. Mints, G. E., *Analog of Herbrand's Theorem for Non-Prenex Formulas of Constructive Predicate Calculus,* SIM, 1969, 4, 47-51 (= ЗНС, 1967, 4, 123-233).
29. Mints, G. E., *Variation in the Deduction Search Tactics in Sequential Calculi,* SIM, 1969, 4, 52-59 (= ЗНС, 1967, 4, 134-151).
30. Mints, G. E., *Solvability of the Problem of Deducibility in for a Class of Formulas Not Containing Negative Occurences of Quantifiers.* PSIM (= ТМИ, 1968, 98, 121-130).
31. Novikov, P. S., *Elements of mathematical logic.* Gosudarstv. Izdat. Fiz.-Mat. Lit., Moscow, 1959.
32. Orevkov, V. P., *Certain Reduction Classes and Solvable Classes of Sequents for the Constructive Predicate Calculus,* SMD, 1965, 6, 888-891 (= ДАН, 1965, 163, N° I, 30-32).
33. Orevkov, V. P., *Unsolvability in the Constructive Predicate Calculus of the Class of the Formulas of the Type* $\neg\neg\forall\exists$ SMD, 1965, (= ДАН, 1965, 163, N° 3, 581-583).
34. Orevkov, V. P., *A Decidable Class of Formulas of Classical Predicate Calculus with Function Symbols.* « *II Cybernetic Symposium (Theses)* ». Tbilisi, 1965, 176.
35. Orevkov, V. P., *Two Undecidable Classes of Formulas in Classical Predicate Calculus.* SIM, 1970, 8, 98-102 (= ЗНС, 1968, 8, 202-310).
36. Orevkov, V. P., *On Nonlengthening Applications of Equality Rules,* SIM, 1971, 16, 77-79 (= ЗНС, 1969, 16, 152-156).
37. Pil'čak, B. Yu., *On the Calculus of Problems.* Ukrain. Mat. Ž., 1952, 4, 174-194.
38. Plyushkevichus, R. A., *On a Variant of the Constructive Predicate Calculus without Structural Deduction Rules.* ДАН, 1965, 161, N° 2, 292-295.
39. Plyushkevichus, R. A., *Sequent-Variant of the Calculus of Constructive Logic for Normal Formulas.* PSIM (= ТМИ, 1968, 98, 155-202).
40. Plyushkevichus, R. A. (Pluskevicus, R. A.), *Kanger's Variant of Predicate Calculus with Symbols for Functions That Are Not Everywhere Defined.* SIM 1970, 8, 103-109 (= ЗНС, 1968, 8, 211-224).
41. Phyushkevichene, A. Yo. (Pluškevičene, A. Yo.), *Elimination of Cut-Type Rules in Axiomatic Theories with Equality.* SIM, 1971, 16, 90-94 (= ЗНС, 1969, 16, 175-184).
42. Pšeničnikova, S. V., *On an Algorithm for the Automatic Proof of Certain Theorems in Analysis.* Izv. Akad. Nauk Azerbaidžan. SSR. Ser. Fiz-Tehn. Mat. Nauk, 1964, N° 4, 65-71.

43. Rogava, M. G., *On Sequential Modifications of Applied Predicate Calculi.* SIM, 1969, 4, 77-81 (= ЗНС, 1967, 4, 189-200).
44. Tauts, A., *Solution of Logical Equations in the First Order Predicate Calculus by the Iteration Method.* Proceedings of the Inst. Phys. Astr. of the Acad. Sci. Est., 1964, N° 24, 17-24.
45. Tseitin, G. S., *On the Complexity of Derivation in Propositional Calculus.* SIM 1970, 8, 115-125 (= ЗНС, 1968, 8, 234-259).
46. Cinman, L. L., *The Role of the Principle of Induction in a Formal Arithmetic System.* Mat. Sb. (N,S.), 1968, 77 (119), 71-104.
47. Sanin, N. A., Davydov, G. V., Maslov, S. Yu., Mints, G. E., Orevkov, V. P., Slisenko, A. O., *An algorithm for a machine search of a natural logical deduction in a propositional calculus.* Izdat. « Nauka », Moscow, 1965.
48. Šiltere, M. Ya., *Mechanical Deduction of Arithmetical Identities.* « Automatics and Telemechanics », 1969, N° 6, 110-114.

Institut de mathématiques de l'Académie des Sciences d'URSS.

1957

A Computer Program for Presburger's Algorithm

M. Davis

1. Introduction

Presburger has shown that for the fragment of elementary number theory which involves addition only (not multiplication) a decision procedure exists. (Cf. Presburger, M.: Über die Vollständigkeit eines gewissen Systems der Arithmetik ganzer Zahlen, in welchem die Addition als einzige Operation hervortritt, Sprawozdanie z I Kongresu Matematyków Krajów Slowiańskich Warszawa, 1929, pp. 92-101.)

This paper reports on the author's program, written for the Institute for Advanced Study electronic digital computer (hereafter abbreviated IASC), for Presburger's decision procedure. This program enables the IASC to furnish the truth value of any proposition (i.e., "truth", if the proposition is true, "falsehood", if it is false) from elementary additive number theory, (subject, of course, to the usual limitations of time and memory space). The program was written during the summer of 1954 under contract number DA-36-034-ORD-1645 with the Office of Ordnance Research.

2. Notation

We use the notation "$\sim$" for negation, "&" for conjunction, "$\vee$" for (inclusive) alternation, and "(Ex_i)", where x_i is an individual variable for existential quantification. Following the practice of the Polish school we write binary connectives preceding the symbols on which they operate. Thus, & A B renders "A and B". As is well known, the other operations of elementary logic are representable in terms of these. Thus, "A implies B" is rendered by $\vee \sim AB$, and universal quantification by $\sim(Ex_i)\sim$.

The variables $x_1, x_2, x_3, \ldots,$ are to be thought of as ranging over the integers positive, negative, or zero. A proposition belongs to <u>additive elementary number theory</u> if it can be written in terms of 0, 1, +, =, the variables x_i, and the operations of elementary logic.

As usual, if α, β are any expressions, we write $\alpha \neq \beta$ for $\sim(\alpha = \beta)$. We note that congruence modulo any definite integer n given metalogically is representable in additive number theory. For,

$$|| \; \alpha \equiv \beta \pmod{n}, \text{ if and only } (Ex_i)(\alpha = \beta + \underbrace{x_i + x_i + \ldots + x_i}_{n \text{ terms}}).$$

(Here, if α and β are thought of as symbolic expressions, rather than definite integers, i must be chosen so that x_i does not occur in either α or β.)

Presburger's algorithm is an inductive one which, beginning with a proposition written in terms of the original notions of additive elementary number theory, introduces congruences and negations of congruences (or as we shall say, <u>incongruences</u>). Hence, we permit $\equiv n$ and $\not\equiv n$ among our notions. Also, if $n > 0$ is a definite integer, we write nx_i for the expression $\underbrace{x_i + x_i + \ldots + x_i}_{n \text{ terms}}$ and n for the expression $\underbrace{1 + 1 + \ldots + 1}_{n \text{ terms}}$. We call an expression of the form:

$$n_o + n_1 x_1 + n_2 x_2 + \ldots + n_k x_k$$

a <u>linear polynomial</u>, and write it

$$n_o + \sum_{i=1}^{k} n_i x_i.$$

An elementary relation is understood to be an expression having one of the forms:

$$P = 0, \; P \neq 0, \; P \underset{m}{\equiv} 0, \; P \underset{m}{\not\equiv} 0,$$

where P is a linear polynomial and m is an integer. It is easy to see that the propositions of additive elementary number theory are precisely those which can be formed by applying the operations of elementary logic to elementary relations.

3. Presburger's decision procedure

Let P be a sentence of additive elementary number theory. The following steps will yield a sentence P' such that P is true if and only if P' is and, P' contains no quantifiers. We take P as given in terms of the operations of elementary logic, operating on elementary relations.

I. Obtain a Disjunctive Normal Form

A. Locate right-most existential quantifier if any. If none, we are through.

B. Locate the right-most negation sign which follows the right-most quantifier. If none, go to D.

1. $\sim$ is followed by $=$, $\neq$, $\equiv_n$, or $\not\equiv_n$. Then, erase $\sim$, and replace $=$ by $\neq$, $\neq$ by $=$, $\equiv_n$ by $\not\equiv_n$, or $\not\equiv_n$ by $\equiv_n$, whichever is appropriate.
2. $\sim$ is follwed by $\sim$. Then erase both $\sim$'s.
3. $\sim$ is followed by & or $\vee$. Then, & (or $\vee$) occurs in a context & A B (or $\vee$AB) where A, B are w.f.f's. Replace $\sim$& A B by $\vee\sim$ A $\sim$B (or $\sim\vee$ A B by &$\sim$ A $\sim$B).

C. Go back to B.

D. Distribute any & which precedes an $\vee$. That is, &$\vee$ABC becomes $\vee$ & A C & B C .

II. Obtain Test Equality

A. Begin with left-most conjunction.

B. In this conjunction is there an equality?

If not, go to IV.

C. Let $mx_i + \alpha = 0$ be the right-most, so called test, equality where (Ex_i) is the quantifier in question.

III. Presburger's Procedure for a Test Equality

A. Begin with the right-most elementary relation.

B. Let this relation be $m'x_i + \alpha' = 0$, $\neq 0$, $\equiv_r 0$, or $\not\equiv_r 0$. Replace relation by $(-p)\alpha + q\alpha' = 0$, $\neq 0$, $\equiv_s 0$, $\not\equiv_s 0$, respectively, where $pm = qm' = $ l. c. m. (m,m'), and $s = $ g. c. d. (r,q).

C. Choose next relation in conjunction if any and go back to B. If none, go on to D.

D. Is there another conjunction preceding quantifier? If so go to II, B. If none, erase quantifier and go back to I, A.

IV. Presburger's Procedure for no Equality Present

A. Erase all inequalities in the conjunction.

B. Set, as test congruence, $x_i \equiv_1 0$.

C. Begin with right-most elementary relation.

D. Operate on this relation:

1. If relation is $m'x_i + \alpha' \equiv_{p'} 0$, with test congruence $mx_i + \alpha \equiv_p 0$, first replace relation with $n'x_i + \beta' \equiv_q 0$, and test congruence by $nx_i + \beta_q\ 0$, where $q = \text{l.c.m}(p,p') = pr = p'r'$, $n' = m'r'$, $\beta' = \alpha'r'$, $n = mr$, $\beta = \alpha\ r$. Then subtract one from the other and replace the one with greater n by the difference. Continue until the coefficient of x_i in the relation is 0.

2. If relation is $m'x_i + \alpha' \not\equiv_{p'} 0$, proceed as in 1. However, whenever, it is the test congruence which is to be replaced, reject this method, and go to V.

E. Choose next relation in conjunction if any and go back to D. If none, go to III D.

V. <u>Expansion of Incongruence</u>

A. If relation is $m'x_i + \alpha' \not\equiv_{p'} 0$, replace it by:

$$\underbrace{\vee \vee \vee \ldots \vee}_{p'-2} \equiv_{p'} m'x_i + \alpha + 1 \equiv_{p'} m'x_i + \alpha + 2 \ldots \equiv_{p'} m'x_i + \alpha + p-1$$

and go back to I D.

This completes our description of Presburger's procedure. Our version represents a modification of Presburger's in several minor respects. In particular, Presburger uses the expansion of V for all incongruences. This represents a prohibitive drain on the memory capacity of digital computers of the size of the IASC. Therefore, we have had incongruences handled by other methods where practical, i.e., in III, B and IV, D2.

To complete the decision procedure, that is to obtain the truth value of the sentence P, we need only determine the truth value of P'. This is accomplished by using the truth functional interpretation of the propositional connectives in the obvious manner.

4. <u>The IASC</u>

The IASC has an electrostatic storage consisting of 1024 40 bit words. Each word may contain a pair of orders. We shall use <u>pentad</u> notation for representation of addresses and 40 bit words, current among the IAS computer group.

In essence the pentad system uses the base 32. Thus, an address requires 2 and a 40 bit word, 8 digits. However, as the 32 digits, the decimal digit pairs 00, 01, 02, ..., 31 are used.

The sentence being tested by our routine is stored in consecutive memory locations beginning at 28, 01. 96 is the maximum length permitted. Linear polynomials in at most 5 variables with coefficients <512 in absolute value are considered. The sentence is written backwards, i.e. from right to left. The initial symbol is preceded by the termination symbol:

15,00 00,00 00,00 00,00.

(Ex_i), i = 1, 2, 3, 4, 5 is represented by:

12,00 00,00 00,00 00,0i.

∿ is represented by:

08,00 00,00 00,00 00,00.

& is represented by:

04,00 00,00 00,00 00,00.

∨ is represented by:

28,00 00,00 00,00 00,00.

= is represented by:

02,00 00,00 00,00 00,00.

≠ is represented by:

30,00 00,00 00,00 00,00.

$\equiv_n$, 0 < n < 512 is represented by:

01,00 00,00 00,00 n

where the "n" is the pentad representation of n.

$\neq_n$, 0 < n < 512 is represented by the machine negative of the representation of ≡n.

The linear polynomial

$$n_1 x_1 + n_2 x_2 + n_3 x_3 + n_4 x_4 + n_5 x_5 + n$$

is represented by the pair of words

00,01	n_1	n_2	n_3,
00,03	n_4	n_5	n.

5. Some Coding Problems and Technique

The reader will probably have noted that the machine representations of the pairs &, $\vee$; =, $\neq$; $\equiv_n$, $\not\equiv_n$ are negatives of one another. This fact is quite useful in the coding. Thus, in applying the de Morgan Laws (cf. I, B3, above),

$$\sim \& A B \equiv \vee \sim A \sim B$$

$$\sim \vee A B \equiv \& \sim A \sim B ,$$

it is unnecessary to know whether we are using & or $\vee$. The "add clear absolute value" order is, of course, invaluable here. Also, in replacing $\sim =$ by $\neq$, $\sim\neq$ by =, $\sim\equiv_n$ by $\not\equiv_n$, and $\sim\not\equiv_n$ by $\equiv_n$, it is simply necessary to use the "clear Subtract" order; it is unnecessary to know which of the four relations is involved.

In computing truth values, the addition order is helpful. Namely, we take $T = \frac{1}{4}$, $F = -\frac{1}{4}$. Then, the truth values of p & q and p $\vee$ q are given by the signs of $p + q - \frac{1}{4}$ and $p + q$ respectively.

The above example gives one case in which an arithmetical operation is useful in a purely logical problem. Actually, of course, the only property of the "clear subtract" order used, is that it is an involution. As to the multiplication and division orders - they are scarcely used. In computing g.c.d.'s and l.c.m.'s it is usually most convenient to go back to the addition order. The automatic semi-roundoff on the division order is a particular nuisance.

Finally, we should like to mention two relatively time-consuming subroutines. Namely, in applying, e.g.

$\sim\vee A\,B \equiv \&\sim A \sim B$, or worse,

$\&\vee A\,B\,C \equiv \vee\ \&\,A\,C\,\&\,B\,C$,

it is necessary to know where A (and in the second case, B) end. With the Polish notation this is reasonably easy to do. (Otherwise, it would nevertheless not be too-difficult - the end would occur when the number of left parentheses was equal to the number of right parentheses.) Moreover, it is necessary to move an entire block over, which requires yet another subroutine.

This last subroutine involves increasing the length of the sentence being tested. The same is true of the expansion of VA. However, once this length reaches 97 memory locations, the routine breaks down. Hence, the routine is provided with a special print-out indicating failure because of lack of memory space.

6. Other Decision Procedures

The writer's experience would indicate that with equipment presently available, it will prove impracticable to code any decision procedures considerably more complicated than Presburger's. Tarski's procedure for elementary algebra falls under this head.

Empirical Explorations with the Logic Theory Machine: A Case Study in Heuristics

A. Newell, J. C. Shaw, H. A. Simon

This is a case study in problem-solving, representing part of a program of research on complex information-processing systems. We have specified a system for finding proofs of theorems in elementary symbolic logic, and by programming a computer to these specifications, have obtained empirical data on the problem-solving process in elementary logic. The program is called the Logic Theory Machine (LT); it was devised to learn how it is possible to solve difficult problems such as proving mathematical theorems, discovering scientific laws from data, playing chess, or understanding the meaning of English prose.

The research reported here is aimed at understanding the complex processes (heuristics) that are effective in problem-solving. Hence, we are not interested in methods that guarantee solutions, but which require vast amounts of computation. Rather, we wish to understand how a mathematician, for example, is able to prove a theorem even though he does not know when he starts how, or if, he is going to succeed.

This focuses on the pure theory of problem-research solving (Newell and Simon, 1956*a*). Previously we specified in detail a program for the Logic Theory Machine; and we shall repeat here only as much of that specification as is needed so that the reader can understand our data. In a companion study (Newell and Shaw, 1957) we consider how computers can be programmed to execute processes of the kinds called for by LT, a problem that is interesting in its own right. Similarly, we postpone to later papers a discussion of the implications of our work for the psychological theory of human thinking and problem-solving. Other areas of application will readily occur to the reader, but here we will limit our attention to the nature of the problem-solving process itself.

Our research strategy in studying complex systems is to specify them in detail, program them for digital computers, and study their behavior empirically by running them with a number of variations and under a variety of conditions. This appears at present the only adequate means to obtain a thorough understanding of their behavior. Although the problem area with which the present system, LT, deals is fairly elementary, it provides a good example of a difficult problem—logic is a subject taught in college courses, and is difficult enough for most humans.

Our data come from a series of programs run on the JOHNNIAC, one of RAND's high-speed digital computers. We will describe the results of these runs, and analyze and interpret their implications for the problem-solving process.

The Logic Theory Machine in Operation

We shall first give a concrete picture of the Logic Theory Machine in operation. LT, ot course, is a program, written for the JOHNNIAC, represented by marks on paper or holes in cards. However, we can think of LT as an actual physical machine and the operation of the program as the behavior of the machine. One can identify LT with JOHNNIAC after the latter has been loaded with the basic program, but before the input of data.

LT's task is to prove theorems in elementary symbolic logic, or more precisely, in the sentential calculus. The sentential calculus is a formalized system of mathematics, consisting of expressions built from combinations of basic symbols. Five of these expressions are taken as axioms, and there are rules of inference for generating new theorems from the axioms and from other theorems. In flavor and form elementary symbolic logic is much like abstract algebra. Normally the variables of the system are interpreted as sentences, and the axioms and rules of inference as formalizations of logical operations, *e.g.,* deduction. However, LT deals with the system as a purely formal mathematics, and we will have no further need of the interpretation. We need to introduce a smattering of the sentential calculus to understand LT's task.

There is postulated a set of *variables p, q, r, . . . A, B, C, . . . ,* with which the sentential calculus deals. These variables can be combined into expressions by means of *connectives.* Given any variable *p,* we can form the expression "not-*p*." Given any two variables *p* and *q,* we can form the expression "*p* or *q,*" or the expression "*p* implies *q,*" where "or" and "implies" are the connectives. There are other connectives, for example "and," but we will not need them here. Once we have formed expressions, these can be further combined into more complicated expressions. For example, we can form:[1]

"(*p* implies not-*p*) implies not-*p*." (2.01)

There is also given a set of expressions that are axioms. These are taken to be the universally true expressions from which theorems are to be derived by means of various rules of inference. For the sake of definiteness in our work with LT, we have employed the system of axioms, definitions, and rules that is used in the *Principia Mathematica,* which lists five axioms:

(*p* or *p*) implies *p* (1.2)
p implies (*q* or *p*) (1.3)
(*p* or *q*) implies (*q* or *p*) (1.4)
[*p* or (*q* or *r*)] implies [*q* or (*p* or *r*)] (1.5)
(*p* implies *q*) implies [(*r* or *p*) implies (*r* or *q*)]. (1.6)

Given some true theorems one can derive new theorems by means of three rules of inference: *substitution, replacement, and detachment.*

1. By the rule of substitution, any expression may be substituted for any variable in any theorem, provided the substitution is made throughout the theorem wherever that variable appears. For example, by substitution of "*p* or *q*" for "*p*," in the second axiom we get the new theorem:

(*p* or *q*) implies [*q* or (*p* or *q*)].

2. By the rule of replacement, a connective can be replaced by its definition, and *vice versa,* in any of its occurrences. By definition "*p* implies *q*" means the same as "not-*p* or *q*." Hence the former expression can always be replaced by the latter and *vice versa.* For example from axiom 1.3, by replacing "implies" with "or," we get the new theorem:

not-*p* or (*q* or *p*).

3. By the rule of detachment, if "*A*" and "*A* implies *B*" are theorems, then "*B*" is a theorem. For example, from:

(*p* or *p*) implies *p*,

and [(*p* or *p*) implies *p*] implies (*p* implies *p*),

we get the new theorem:

p implies *p*.

Given an expression to prove, one starts from the set of axioms and theorems already proved, and applies the various rules successively until

[1] For easy reference we have numbered axioms and theorems to correspond to their numbers in *Principia Mathematica,* 2nd ed., vol. 1, New York: by A. N. Whitehead and B. Russell, 1935.

the desired expression is produced. The proof is the sequence of expressions, each one validly derived from the previous ones, that leads from the axioms and known theorems to the desired expression.

This is all the background in symbolic logic needed to observe LT in operation. LT "understands" expressions in symbolic logic—that is, there is a simple code for punching expressions on cards so they can be fed into the machine. We give LT the five axioms, instructing it that these are theorems it can assume to be true. LT already knows the rules of inference and the definitions—how to substitute, replace, and detach. Next we give LT a single expression, say expression 2.01, and ask LT to find a proof for it. LT works for about 10 seconds and then prints out the following proof:

(p implies not-p) implies not-p	(theorem 2.01, to be proved)
1. (A or A) implies A	(axiom 1.2)
2. (not-A or not-A) implies not-A	(subs. of not-A for A)
3. (A implies not-A) implies not-A	(repl. of "or" with "implies")
4. (p implies not-p) implies not-p	(subs. of p for A; *QED*).

Next we ask LT to prove a fairly advanced theorem (Whitehead and Russell, 1935), theorem 2.45; allowing it to use all 38 theorems proved prior to 2.45. After about 12 minutes, LT produces the following proof:

not (p or q) implies not-p	(theorem 2.45, to be proved)
1. A implies (A or B)	(theorem 2.2)
2. p implies (p or q)	(subs. p for A, q for B in 1)
3. (A implies B) implies (not-B implies not-A)	(theorem 2.16)
4. [p implies (p or q)] implies [not (p or q) implies not-p]	[subs. p for A, (p or q) for B in 3]
5. not (p or q) implies not-p	(detach right side of 4, using 2; *QED*).

Finally, all the theorems prior to (2.31) are given to LT (a total of 28); and then LT is asked to prove:

[p or (q or r)] implies [(p or q) or r]. (2.31)

LT works for about 23 minutes and then reports that it cannot prove (2.31), that it has exhausted its resources.

Now, what is there in this behavior of LT that needs to be explained? The specific examples given are difficult problems for most humans, and most humans do not know what processes they use to find proofs, if they find them. There is no known simple procedure that will produce such proofs. Various methods exist for verifying whether any given expression is true or false; the best known procedure is the method of truth tables. But these procedures do not produce a proof in the meaning of Whitehead and Russell. One can invent "automatic" procedures for producing proofs. We will look at one briefly later, but these turn out to require computing times of the orders of thousands of years for the proof of (2.45).

We must clarify why such problems are difficult in the first place, and then show what features of LT account for its successes and failures. These questions will occupy the rest of this study.

Problems, Algorithms, and Heuristics

In describing LT, its environment, and its behavior we will make repeated use of three concepts. The first of these is the concept of *problem*. Abstractly, a person is given a problem if he is given a set of possible solutions, and a test for verifying whether a given element of this set is in fact a solution to his problem.

The reason why problems are problems is that the original set of possible solutions given to the problem-solver can be very large, the actual solutions can be dispersed very widely and rarely throughout it, and the cost of obtaining each new element and of testing it can be very expensive. Thus the problem-solver is not really "given" the set of possible solutions; instead he is given some process for generating the elements of that set in some order. This generator has properties of its own, not usually specified in stating the problem; *e.g.*, there is associated with it a certain cost per element produced, it may be possible to change the order in which it produces the elements, and so on. Likewise the verification test has costs and times associated with it. The problem can be solved if these costs are not too large in relation to the time and computing power available for solution.

One very special and valuable property that a generator of solutions sometimes has is a guarantee that if the problem has a solution, the generator will, sooner or later, produce it. We will call a process that has this property for some problem an *algorithm* for that problem. The guarantee provided by an algorithm is not an unmixed blessing, of course, since nothing has been specified about the cost or time required to produce the solutions. For example, a simple algorithm for opening a combination safe is to try all combinations, testing each one to see if it opens the safe. This algorithm is a typical problem-solving process: there is a generator that produces new combinations in some order, and there is a verifier that determines whether each new combination is in fact a solution to the problem. This search process is an algorithm because it is known that *some* combination will open the safe, and because the generator will exhaust all combinations in a finite interval of time. The algorithm is sufficiently expensive, however, that a combination safe can be used to protect valuables even from people who know the algorithm.

A process that *may* solve a given problem, but offers no guarantees of doing so, is called a *heuristic*[2] for that problem. This lack of a guarantee is not an unmixed evil. The cost inflicted by the lack of guarantee depends on what the process costs and what algorithms are available as alternatives. For most run-of-the-mill problems we have only heuristics, but occasionally we have both algorithms and heuristics as alternatives for solving the same problem. Sometimes, as in the problem of finding maxima for simple differentiable functions, everyone uses the algorithm of setting the first derivative equal to zero; no one sets out to examine all the points on the line one by one as if it were possible. Sometimes, as in chess, everyone plays by heuristic, since no one is able to carry out the algorithm of examining all continuations of the game to termination.

The Problem of Proving Theorems in Logic

Finding a proof for a theorem in symbolic logic can be described as selecting an element from a generated set, as shown by Fig. 1. Consider the *set of all possible sequences of logic expressions*—call it E. Certain of these sequences, a very small minority, will be proofs. A proof sequence satisfies the following test:

Each expression in the sequence is either

1. One of the accepted theorems or axioms, or
2. Obtainable from one or two previous expressions in the sequence by application of one of the three rules of inference.

Call the *set of sequences that are proofs* P. Certain of the sequences in E have the *expression to be proved*—call it X, as their final expression. Call this set of sequences T_X. Then, to find a proof of a given theorem X means to select an element of E that belongs to the intersection of P and T_X. The set E is given implicitly by rules for generating new sequences of logic expressions.

The difficulty of proving theorems depends on the scarcity of elements in the intersection of P and T_X, relative to the number of elements in E. Hence, it depends on the cost and speed of the available generators that produce elements of E, and on the cost and speed of making tests that determine whether an element belongs to T_X or P. The difficulty also de-

[2] As a noun, "heuristic" is rare and generally means the art of discovery. The adjective "heuristic" is defined by Webster as: serving to discover or find out. It is in this sense that it is used in the phrase "heuristic process" or "heuristic method." For conciseness, we will use "heuristic" as a noun synonymous with "heuristic process." No other English word appears to have this meaning.

pends on whether generators can be found that guarantee that any element they produce automatically satisfies some of the conditions. Finally, as we shall see, the difficulty depends heavily on what heuristics can be found to guide the selection.

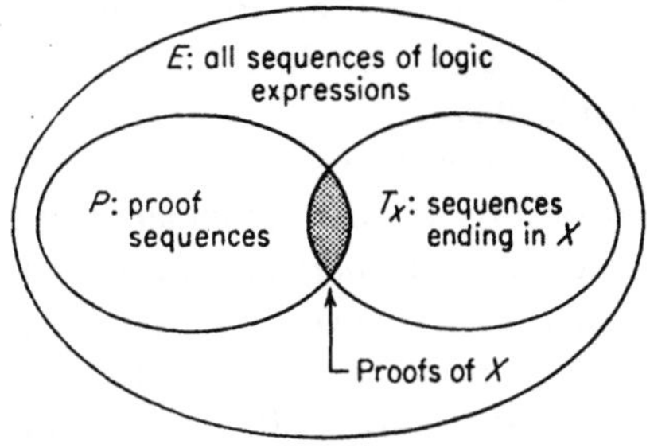

Figure 1. Relationships between E, P, and T_X.

A little reflection, and experience in trying to prove theorems, make it clear that proof sequences for specified theorems are rare indeed. To reveal more precisely why proving theorems is difficult, we will construct an algorithm for doing this. The algorithm will be based only on the tests and definitions given above, and not on any "deep" inferred properties of symbolic logic. Thus it will reflect the basic nature of theorem proving; that is, its nature prior to building up sophisticated proof techniques. We will call this algorithm the British Museum algorithm, in recognition of the supposed originators of procedures of this type.

The British Museum Algorithm

The algorithm constructs all possible proofs in a systematic manner, checking each time (1) to eliminate duplicates, and (2) to see if the final theorem in the proof coincides with the expression to be proved. With this algorithm the set of one-step proofs is identical with the set of axioms (*i.e.*, each axiom is a one-step proof of itself). The set of n-step proofs is obtained from the set of $(n-1)$-step proofs by making all the permissible substitutions and replacements in the expressions of the $(n-1)$-step proofs, and by making all the permissible detachments of pairs of expressions as permitted by the recursive definition of proof.[3]

Figure 2 shows how the set of n-step proofs increases with n at the very start of the proof-generating process. This enumeration only extends to replacements of "or" with "implies," "implies" with "or," and negation of variables (*e.g.*, "not-*p*" for "*p*"). No detachments and no complex substitutions (*e.g.*, "*q* or *r*" for "*p*") are included. No specializations have been made (*e.g.*, substitution of *p* for *q* in "*p* or *q*"). If we include the specializations, which take three more steps, the algorithm will generate

[3] A number of fussy but not fundamental points must be taken care of in constructing the algorithm. The phrase "all permissible substitutions" needs to be qualified, for there is an infinity of these. Care must be taken not to duplicate expressions that differ only in the names of their variables. We will not go into details here, but simply state that these difficulties can be removed. The essential feature in constructing the algorithm is to allow only one thing to happen in generating each new expression, i.e., one replacement, substitution of "not-*p*" for "*p*," etc.

an (estimated) additional 600 theorems, thus providing a set of proofs of 11 steps or less containing almost 1000 theorems, none of them duplicates.

In order to see how this algorithm would provide proofs of specified theorems, we can consider its performance on the sixty-odd theorems of chap. 2 of *Principia.* One theorem (2.01) is obtained in step (4) of the generation, hence is among the first 42 theorems proved. Three more (2.02, 2.03, and 2.04) are obtained in step (6), hence among the first 115. One more (2.05) is obtained in step (8), hence in the first 246. Only one more is included in the first 1000, theorem 2.07. The proofs of all the remainder require complex substitutions or detachment.

We have no way at present to estimate how many proofs must be generated to include proofs of all theorems of chap. 2 of *Principia.* Our best guess is that it might be a hundred million. Moreover, apart from the six theorems listed, there is no reason to suppose that the proofs of these theorems would occur early in the list.

Our information is too poor to estimate more than very roughly the times required to produce such proofs by the algorithm; but we can estimate times of about 16 minutes to do the first 250 theorems of Fig. 2 [*i.e.,* through step (8)] assuming processing times comparable with those in LT. The first part of the algorithm has an additional special property, which holds only to the point where detachment is first used; that no check for duplication is necessary. Thus the time of computing the first few thousand proofs only increases linearly with the number of theorems generated. For the theorems requiring detachments, duplication checks must be made, and the total computing time increases as the square of the number of expressions generated. At this rate it would take hundreds of thousands of years of computation to generate proofs for the theorems in chap. 2.

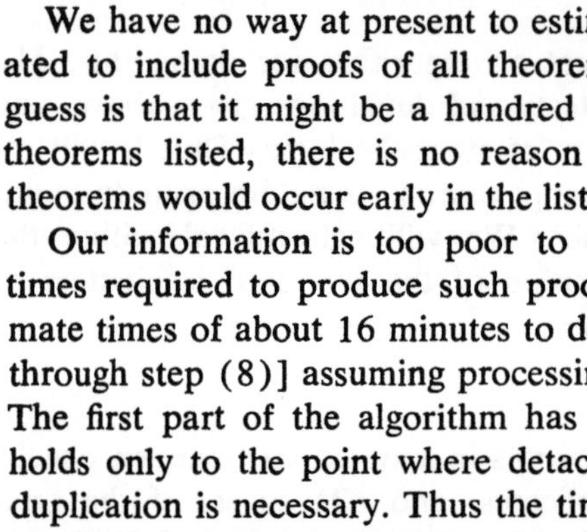

Figure 2. Number of proofs generated by first few steps of British Museum algorithm.

The nature of the problem of proving theorems is now reasonably clear. When sequences of expressions are produced by a simple and cheap (per element produced) generator, the chance that any particular sequence is the desired proof is exceedingly small. This is true even if the generator produces sequences that always satisfy the most complicated and restrictive of the solution conditions: that each is a proof of something. The set of sequences is so large, and the desired proof so rare, that no practical amount of computation suffices to find proofs by means of such an algorithm.

The Logic Theory Machine

If LT is to prove any theorems at all it must employ some devices that alter radically the order in which possible proofs are generated, and the way in which they are tested. To accomplish this, LT gives up almost all the guarantees enjoyed by the British Museum algorithm. Its procedures guarantee neither that its proposed sequences are proofs of something, nor that LT will ever find the proof, no matter how much effort is spent. However, they *often* generate the desired proof in a reasonable computing time.

Methods

The major type of heuristic that LT uses we call a *method*. As yet we have no precise definition of a method that distinguishes it from all the other types of routines in LT. Roughly, a method is a reasonably self-contained operation that, if it works, makes a major and permanent contribution toward finding a proof. It is the largest unit of organization in LT, subordinated only to the executive routines necessary to coordinate and select the methods.

THE SUBSTITUTION METHOD

This method seeks a proof for the problem expression by finding an axiom or previously proved theorem that can be transformed, by a series of substitutions for variables and replacements of connectives, into the problem expression.

THE DETACHMENT METHOD

This method attempts, using the rule of detachment, to substitute for the problem expression a new subproblem which, if solved, will provide a proof for the problem expression. Thus, if the problem expression is *B*, the method of detachment searches for an axiom or theorem of the form "*A* implies *B*." If one is found, *A* is set up as a new subproblem. If *A* can be proved, then, since "*A* implies *B*" is a theorem, *B* will also be proved.

THE CHAINING METHODS

These methods use the transitivity of the relation of implication to create a new subproblem which, if solved, will provide a proof for the problem expression. Thus, if the problem expression is "*a* implies *c*," the method of forward chaining searches for an axiom or theorem of the form "*a* implies *b*." If one is found, "*b* implies *c*" is set up as a new subproblem. Chaining backward works analogously: it seeks a theorem of the form "*b* implies *c*," and if one is found, "*a* implies *b*" is set up as a new subproblem.

Each of these methods is an independent unit. They are alternatives to one another, and can be used in sequence, one working on the subproblems generated by another. Each of them produces a major part of a proof. Substitution actually proves theorems, and the other three generate subproblems, which can become the intermediate expressions in a proof sequence.

These methods give no guarantee that they will work. There is no guarantee that a theorem can be found that can be used to carry out a proof by the substitution method, or a theorem that will produce a subproblem by any of the other three methods. Even if a subproblem is generated, there is no guarantee that it is part of the desired proof sequence, or even that it is part of any proof sequence (*e.g.*, it can be false). On the other hand, the generated methods do guarantee that any subproblem generated is part of a sequence of expressions that ends in the desired theorem (this is one of the conditions that a sequence be a proof). The methods also guarantee that each expression of the sequence is derived by the rules of inference from the preceding ones (a second condition of proof). What is not guaranteed is that the beginning of the sequence can be completed with axioms or previously proved theorems.

There is also no guarantee that the combination of the four methods, used in any fashion whatsoever and with unlimited computing effort, comprises a sufficient set of methods to prove all theorems. In fact, we have discovered a theorem [(2.13), "*p* or not-not-not-*p*"] which the four methods of LT cannot prove. All the subproblems generated for (2.13) after a certain point are false, and therefore cannot lead to a proof.

We have yet no general theory to explain why the methods transform LT into an effective problem-solver. That they do, in conjunction with the other mechanisms to be described shortly, will be demonstrated amply in the remainder of this study. Several factors may be involved. First, the methods organize the sequences of individual processing steps into larger units that can be handled as such. Each processing step can be oriented toward the special function it performs in the unit as a whole, and the units can be manipulated and organized as entities by the higher-level routines.

Apart from their "unitizing" effect, the methods that generate subproblems work "backward" from the desired theorem to axioms or known theorems rather than "forward" as did the British Museum algorithm. Since there is only one theorem to be proved, but a number of known true theorems, the efficacy of working backward may be analogous to the ease with which a needle can find its way out of a haystack, compared with the difficulty of someone finding the lone needle in the haystack.

The Executive Routine

In LT the four methods are organized by an executive routine, whose flow diagram is shown in Fig. 3.

1. When a new problem is presented to LT, the substitution method is tried first, using all the axioms and theorems that LT has been told to assume, and that are now stored in a *theorem list.*

2. If substitution fails, the detachment method is tried, and as each new subproblem is created by a successful detachment, an attempt is made to prove the new subproblem by the substitution method. If substitution fails again, the subproblem is added to a *subproblem list.*

3. If detachment fails for all the theorems in the theorem list, the same cycle is repeated with forward chaining, and then with backward chaining: try to create a subproblem; try to prove it by the substitution method; if unsuccessful, put the new subproblem on the list. By the nature of the methods, if the substitution method ever succeeds with a single subproblem, the original theorem is proved.

4. If all the methods have been tried on the original problem and no proof has been produced, the executive routine selects the next untried subproblem from the subproblem list, and makes the same sequence of attempts with it. This process continues until (1) a proof is found, (2) the time allotted for finding a proof is used up, (3) there is no more available memory space in the machine, or (4) no untried problems remain on the subproblem list.

In the three examples cited earlier, the proof of (2.01) [(*p* implies not-*p*) implies not-*p*] was obtained by the substitution method directly, hence did not involve use of the subproblem list.

The proof of (2.45) [not (*p* or *q*) implies not-*p*] was achieved by an application of the detachment method followed by a substitution. This proof required LT to create a subproblem, and to use the substitution method on it. It did not require LT ever to select any sub-

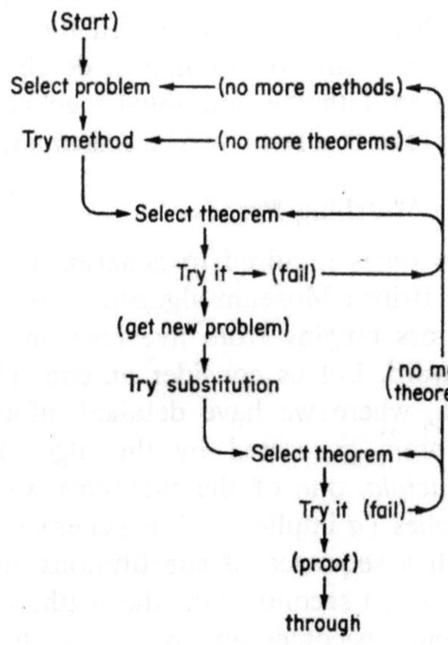

Figure 3. General flow diagram of LT.

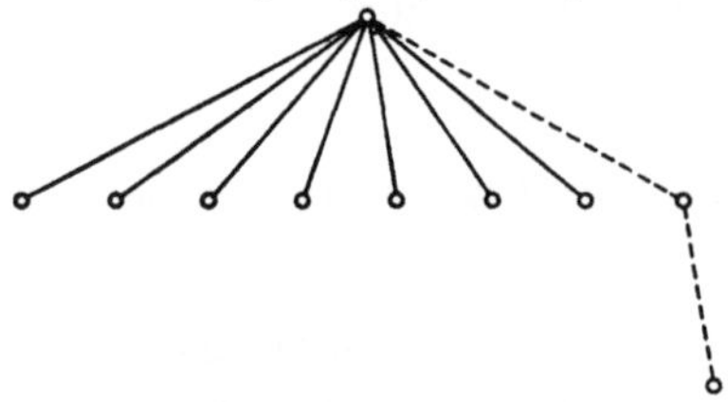

Figure 4. Subproblem tree of proof by LT of (2.45) (all previous theorems available).

problem from the subproblem list, since the substitution was successful. Figure 4 shows the *tree of subproblems* corresponding to the proof of (2.45). The subproblems are given in the form of a downward branching tree. Each node is a subproblem, the original problem being the single node at the top. The lines radiating down from a node lead to the new subproblems generated from the subproblem corresponding to the node. The proof sequence is given by the dashed line; the top link was constructed by the detachment method, and the bottom link by the substitution method. The other links extending down from the original problem lead to other subproblems generated by the detachment method (but not provable by direct substitution) prior to the time LT tried the theorem that leads to the final proof.

LT did not prove theorem 2.31, also mentioned earlier, and gave as its reason that it could think of nothing more to do. This means that LT had considered all subproblems on the subproblem list (there were six in this case) and had no new subproblems to work on. In none of the examples mentioned did LT terminate because of time or space limitations; however, this is the most common result in the cases where LT does not find a proof. Only rarely does LT run out of things to do.

This section has described the organization of LT in terms of methods. We have still to examine in detail why it is that this organization, in connection with the additional mechanisms to be described below, allows LT to prove theorems with a reasonable amount of computing effort.

The Matching Process

The times required to generate proofs for even the simplest theorems by the British Museum algorithm are larger than the times required by LT by factors ranging from five (for one particular theorem) to a hundred and upward. Let us consider an example from the earliest part of the generation, where we have detailed information about the algorithm. The 79th theorem generated by the algorithm (see Fig. 2) is theorem 2.02 of *Principia,* one of the theorems we asked LT to prove. This theorem, "p implies (q implies p)," is generated by the algorithm in about 158 seconds with a sequence of substitutions and replacements; it is proved by LT in about 10 seconds with the method of substitution. The reason for the difference becomes apparent if we focus attention on axiom 1.3, "p implies (q or p)," from which the theorem is derived in either scheme.

Figure 5 shows the tree of proofs of the first twelve theorems obtained from (1.3) by the algorithm. The theorem 2.02 is node (9) on the tree and is obtained by substitution of "not-*q*" for "*q*" in axiom 1.3 to reach node (5); and then by replacing the "(not-*q* or *p*)" by "(*q* implies *p*)" in (5) to get (9). The 9th theorem generated from axiom 1.3 is the 79th generated from the five axioms considered together.

This proof is obtained directly by LT using the following *matching* procedure. We compare the axiom with (9), the expression to be proved:

$$p \text{ implies } (q \text{ or } p) \qquad (1.3)$$
$$p \text{ implies } (q \text{ implies } p). \qquad (9)$$

First, by a direct comparison, LT determines that the main connectives are identical. Second, LT determines that the variables to the left of the main connectives are identical. Third, LT determines that the connectives within parentheses on the right-hand sides are different. It is necessary to replace the "or" with "implies," but in order to do this (in accordance with the definition of implies) there must be a negation sign before the variable that precedes the "or." Hence, LT first replaces the "*q*" on the right-hand side with "not-*q*" to get the required negation sign, obtaining (5). Now LT can change the "or" to "implies," and determines that the resulting expression is identical with (9).

The matching process allowed LT to proceed directly down the branch from (1) through (5) to (9) without even exploring the other branches. Quantitatively, it looked at only two expressions instead of eight, thus reducing the work of comparison by a factor of four. Actually, the saving is even greater, since the matching procedure does not deal with whole expressions, but with a single pair of elements at a time.

An important source of efficiency in the matching process is that it proceeds componentwise, obtaining at each step a feedback of the results of a substitution or replacement that can be used to guide the next step. This feedback keeps the search on the right branch of the tree of possible ex-

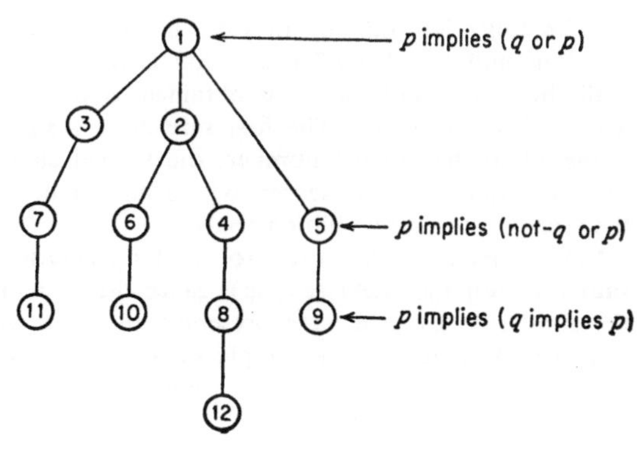

Figure 5. Proof tree of proof 2.02 by British Museum algorithm (using axiom 1.3).

pressions. It is not important for an efficient search that the goal be known from the beginning; it is crucial that hints of "warmer" or "colder" occur as the search proceeds.[4] Closely related to this feedback is the fact that where LT is called on to make a substitution or replacement at any step, it can determine immediately what variable or connective to substitute or replace by direct comparison with the problem expression, and without search.

Thus far we have assumed that LT knows at the beginning that (1.3) is the appropriate axiom to use. Without this information, it would begin matching with each axiom in turn, abandoning it for the next one if the matching should prove impossible. For example, if it tries to match the theorem against axiom 1.2, it determines almost immediately (on the second test) that "*p* or *p*" cannot be made into "*p*" by substitution. Thus, the matching process permits LT to abandon unprofitable lines of search as well as guiding it to correct substitutions and replacements.

MATCHING IN THE SUBSTITUTION METHODS

The matching process is an essential part of the substitution method. Without it, the substitution method is just that part of the British Museum algorithm that uses only replacements and substitutions. With it, LT is able, either directly or in combination with the other methods, to prove many theorems with reasonable effort.

To obtain data on its performance, LT was given the task of proving in sequence the first 52 theorems of *Principia*. In each case, LT was given the axioms plus all the theorems previously proved in chap. 2 as the material from which to work (regardless of whether LT had proved the theorems itself).[5]

Of the 52 theorems, proofs were found for a total 38 (73 per cent). These proofs were obtained by various combinations of methods, but the substitution method was an essential component of all of them. Seventeen of these proofs, almost a half, were accomplished by the substitution method alone. Subjectively evaluated, the theorems that were proved by

[4] The following analogy may be instructive. Changing the symbols in a logic expression until the "right" expression is obtained is like turning the dials on a safe until the right combination is obtained. Suppose two safes, each with ten dials and ten numbers on a dial. The first safe gives a signal (a "click") when any given dial is turned to the correct number; the second safe clicks only when all ten dials are correct. Trial-and-error search will open the first safe, on the average, in 50 trials; the second safe, in five billion trials.

[5] The version of LT used for seeking solutions of the 52 problems included a similarity test (see next section). Since the matching process is more important than the similarity test, we have presented the facts about matching first, using adjusted statistics. A notion of the sample sizes can be gained from Table 1. The sample was limited to the first 52 of the 67 theorems in chap. 2 of *Principia* because of memory limitations of JOHNNIAC.

the substitution method alone have the appearance of "corollaries" of the theorems they are derived from; they occur fairly close to them in the chapter, generally requiring three or fewer attempts at matching per theorem proved (54 attempts for 17 theorems).

The performance of the substitution method on the subproblems is somewhat different, due, we think, to the kind of selectivity implicit in the order of theorems in *Principia.* In 338 attempts at solving subproblems by substitution, there were 21 successes (6.2 per cent). Thus, there was about one chance in three of proving an original problem directly by the substitution method, but only about one chance in 16 of so proving a subproblem generated from the original problem.

MATCHING IN DETACHMENT AND CHAINING

So far the matching process has been considered only as a part of the substitution method, but it is also an essential component of the other three methods. In detachment, for example, a theorem of form "*A* implies *B*" is sought, where *B* is identical with the expression to be proved. The chances of finding such a theorem are negligible unless we allow some modification of *B* to make it match the theorem to be proved. Hence, once a theorem is selected from the theorem list, its right-hand subexpression is matched against the expression to be proved. An analogous procedure is used in the chaining methods.

We can evaluate the performance of the detachment and chaining methods with the same sample of problems used for evaluating the substitution method. However, a successful match with the former three methods generates a subproblem and does not directly prove the theorem. With the detachment method, an average of three new subproblems were generated for each application of the method; with forward chaining the average was 2.7; and with backward chaining the average was 2.2. For all the methods, this represents about one subproblem per 7½ theorems tested (the number of theorems available varied slightly).

As in the case of substitution, when these three methods were applied to the original problem, the chances of success were higher than when they were applied to subproblems. When applied to the original problem, the number of subproblems generated averaged eight to nine; when applied to subproblems derived from the original, the number of subproblems generated fell to an average of two or three.

In handling the first 52 problems in chap. 2 of *Principia,* 17 theorems were proved in one step—that is, in one application of substitution. Nineteen theorems were proved in two steps, 12 by detachment followed by substitution, and seven by chaining forward followed by substitution. Two others were proved in three steps. Hence, 38 theorems were proved in all. There are no two-step proofs by backward chaining, since, for two-step proofs only, if there is a proof by backward chaining, there is also one by forward chaining. In 14 cases LT failed to find a proof. Most of these unsuccessful attempts were terminated by time or space limitations. One of these 14 theorems we know LT cannot prove, and one other we believe it cannot prove. Of the remaining twelve, most of them can be proved by LT if it has sufficient time and memory (see section on subproblems, however).

Similarity Tests and Descriptions

Matching eliminates enough of the trial and error in substitutions and replacements to make LT into a successful problem solver. Matching permeates all of the methods, and without it none of them would be useful within practical amounts of computing effort. However, a large amount of search is still used in finding the correct theorems with which matching works. Returning to the performance of LT in chap. 2, we find that the over-all chances of a particular match being successful are 0.3 per cent for substitution, 13.4 per cent for detachment, 13.8 per cent for forward chaining, and 9.4 per cent for backward chaining.

The amount of search through the theorem list can be reduced by interposing a screening process that will reject any theorem for matching that has low likelihood of success. LT has such a screening device, called the *similarity test*. Two logic expressions are defined to be similar if both their left-hand and right-hand sides are equal, with respect to, (1) the maximum number of *levels* from the main connective to any variable; (2) the number of *distinct* variables; and (3) the number of *variable places*. Speaking intuitively, two logic expressions are "similar" if they look alike, and look alike if they are similar. Consider for example:

$$(p \text{ or } q) \text{ implies } (q \text{ or } p) \qquad (1)$$

$$p \text{ implies } (q \text{ or } p) \qquad (2)$$

$$r \text{ implies } (m \text{ implies } r). \qquad (3)$$

By the definition of similarity, (2) and (3) are similar, but (1) is not similar to either (2) or (3).

In all of the methods LT applies the similarity tests to all expressions to be matched, and only applies the matching routine if the expressions are similar; otherwise it passes on to the next theorem in the theorem list. The similarity test reduces substantially the number of matchings attempted, as the numbers in Table 1 show, and correspondingly raises the probability of a match if the matching is attempted. The effect is particularly strong in substitution, where the similarity test reduces the matchings attempted by a factor of ten, and increases the probability of a successful match by a factor of ten. For the other methods attempted matchings were

TABLE 1 Statistics of Similarity Tests and Matching

Method	Theorems considered	Theorems similar	Theorems matched	Per cent similar of theorems considered	Per cent matched of theorems similar
Substitution	11,298	993	37	8.8	3.7
Detachment	1,591	406	210	25.5	51.7
Chain. forward	869	200	120	23.0	60.0
Chain. backward	673	146	63	21.7	43.2

reduced by a factor of four or five, and the probability of a match increased by the same factor.

These figures reveal a gross, but not necessarily a net, gain in performance through the use of the similarity test. There are two reasons why all the gross gain may not be realized. First, the similarity test is only a heuristic. It offers no guarantee that it will let through only expressions that will subsequently match. The similarity test also offers no guarantee that it will not reject expressions that would match if attempted. The similarity test does not often commit this type of error (corresponding to a type II statistical error), as will be shown later. However, even rare occurrences of such errors can be costly. One example occurs in the proof of theorem 2.07:

$$p \text{ implies } (p \text{ or } p). \tag{2.07}$$

This theorem is proved simply by substituting p for q in axiom 1.3:

$$p \text{ implies } (q \text{ or } p). \tag{1.3}$$

However, the similarity test, because it demands equality in the number of distinct variables on the right-hand side, calls (2.07) and (1.3) dissimilar because (2.07) contains only p while (1.3) contains p and q. LT discovers the proof through chaining forward, where it checks for a direct match before creating the new subproblem, but the proof is about five times as expensive as when the similarity test is omitted.

The second reason why the gross gain will not all be realized is that the similarity test is not costless, and in fact for those theorems which pass the test the cost of the similarity test must be paid in addition to the cost of the matching. We will examine these costs in the next section when we consider the effort LT expends.

Experiments have been carried out with a weaker similarity test, which compares only the number of variable places on both sides of the expression. This test will not commit the particular type II error cited above, and (2.07) is proved by substitution using it. Apart from this, the modification had remarkably little effect on performance. On a sample of ten problems it admitted only 10 per cent more similar theorems and about 10 per cent more subproblems. The reason why the two tests do not differ more radically is that there is a high correlation among the descriptive measures.

Effort in LT

So far we have focused entirely on the performance characteristics of the heuristics in LT, except to point out the tremendous difference between the computing effort required by LT and by the British Museum algorithm. However, it is clear that each additional test, search, description, and the like, has its costs in computing effort as well as its gains in performance. The costs must always be balanced against the performance gains, since there are always alternative heuristics which could be added to the system in place of those being used. In this section we will analyze the computing effort used by LT. The memory space used by the various processes also constitutes a cost, but one that will not be discussed in this study.

MEASURING EFFORTS

LT is written in an interpretive language or pseudocode, which is described in the companion paper to this one. LT is defined in terms of a set of primitive operations, which, in turn, are defined by subroutines in JOHNNIAC machine language. These primitives provide a convenient unit of effort, and all effort measurements will be given in terms of total number of primitives executed. The relative frequencies of the different primitives are reasonably constant, and, therefore, the total number of primitives is an adequate index of effort. The average time per primitive is quite constant at about 30 milliseconds, although for very low totals (less than 1000 primitives) a figure of about 20 milliseconds seems better.

COMPUTING EFFORT AND PERFORMANCE

On *a priori* grounds we would expect the amount of computing effort required to solve a logic problem to be roughly proportional to the total number of theorems examined (*i.e.,* tested for similarity, if there is a similarity routine; or tested for matching, if there is not) by the various methods in the course of solving the problem. In fact, this turns out to be a reasonably good predictor of effort; but the fit to data is much improved if we assign greater weight to theorems considered for detachment and chaining than to theorems considered for substitution.

Actual and predicted efforts are compared below (with the full similarity test included, and excluding theorems proved by substitution) on the assumption that the number of primitives per theorem considered is twice as great for chaining as for substitution, and three times as great for detachment. About 45 primitives are executed per theorem considered with the substitution method (hence 135 with detachment and 90 with chaining). As Table 2 shows, the estimates are generally accurate within a few per cent, except for theorem 2.06, for which the estimate is too low.

TABLE 2 Effort Statistics with "Precompute Description" Routine

	Total primitives, thousands	
Theorem	Actual	Estimate
2.06	3.2	0.8
2.07	4.3	4.4
2.08	3.5	3.3
2.11	2.2	2.2
2.13	24.5	24.6
2.14	3.3	3.2
2.15	15.8	13.6
2.18	34.1	35.8
2.25	11.1	11.5

There is an additional source of variation not shown in the theorems selected for Table 2. The descriptions used in the similarity test must be computed from the logic expressions. Since the descriptions of the theorems are used over and over again, LT computes these at the start of a problem and stores the values with the theorems, so they do not have to be computed again. However, as the number of theorems increases, the space devoted to storing the precomputed descriptions becomes prohibitive, and LT switches to recomputing them each time it needs them. With recomputation, the problem effort is still roughly proportional to the total number of theorems considered, but now the number of primitives per theorem is around 70 for the substitution method, 210 for detachment, and 140 for chaining.

Our analysis of the effort statistics shows, then, that in the first approximation the effort required to prove a theorem is proportional to the number of theorems that have to be considered before a proof is found; the number of theorems considered is an effort measure for evaluating a heuristic. A good heuristic, by securing the consideration of the "right" theorems early in the proof, reduces the expected number of theorems to be considered before a proof is found.

EVALUATION OF THE SIMILARITY TEST

As we noted in the previous section, to evaluate an improved heuristic, account must be taken of any additional computation that the improvement introduces The net advantage may be less than the gross advantage, or the extra computing effort may actually cancel out the gross gain in selectivity. We are now in a position to evaluate the similarity routines as preselectors of theorems for matching.

A number of theorems were run, first with the full similarity routine, then with the modified similarity routine (which tests only the number of variable places), and finally with no similarity test at all. We also made some comparisons with both precomputed and recomputed descriptions.

When descriptions are precomputed, the computing effort is less with the full similarity test than without it; the factor of saving ranged from 10 to 60 per cent (*e.g.*, 3534/5206 for theorem 2.08). However, if LT must recompute the descriptions every time, the full similarity test is actually more expensive than no similarity test at all (*e.g.*, 26,739/22,914 for theorem 2.45).

The modified similarity test fares somewhat better. For example, in proving (2.45) it requires only 18,035 primitives compared to the 22,914 for no similarity test (see the paragraph above). These comparisons involve recomputed descriptions; we have no figures for precomputed descriptions, but the additional saving appears small since there is much less to compute with the abridged than with the full test.

Thus the similarity test is rather marginal, and does not provide anything like the factors of improvement achieved by the matching process, although we have seen that the performance figures seem to indicate much more substantial gains. The reason for the discrepancy is not difficult to find. In a sense, the matching process consists of two parts. One is a testing part that locates the differences between elements and diagnoses the corrective action to be taken. The other part comprises the processes of substituting and replacing. The latter part is the major expense in a matching that works, but most of this effort is saved when the matching fails. Thus matching turns out to be inexpensive for precisely those expressions that the similarity test excludes.

Subproblems

LT can prove a great many theorems in symbolic logic. However, there are numerous theorems that LT cannot prove, and we may describe LT as having reached a plateau in its problem solving ability.

Figure 6 shows the amount of effort required for the problems LT solved out of the sample of 52. Almost all the proofs that LT found took less than 30,000 primitives of effort. Among the numerous attempts at proofs that went beyond this effort limit, only a few succeeded, and these required a total effort that was very much greater.

The predominance of short proofs is even more striking than the approximate upper limit of 30,000 primitives suggests. The proofs by substitution

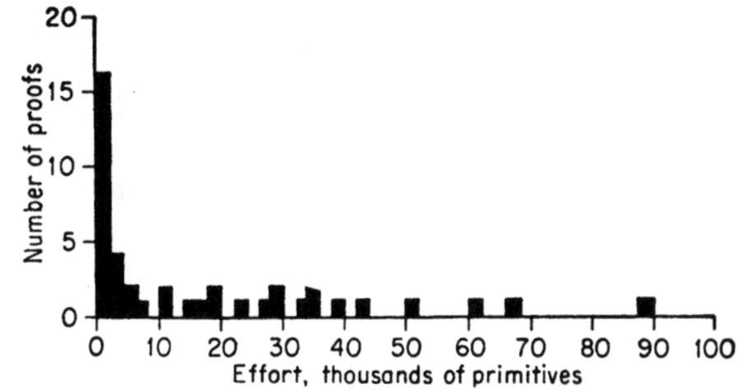

Figure 6. Distribution of LT's proofs by effort. Data include all proofs from attempts on the first 52 theorems in chap. 2 of *Principia*.

—almost half of the total—required about 1000 primitives or less each. The effort required for the longest proof—89,000 primitives—is some 250 times the effort required for the short proofs. We estimate that to prove the 12 additional theorems that we believe LT can prove requires the effort limit to be extended to about a million primitives.

From these data we infer that LT's power as a problem solver is largely restricted to problems of a certain class. While it is logically possible for LT to solve others by large expenditures of effort, major adjustments are needed in the program to extend LT's powers to essentially new classes of problems. We believe that this situation is typical: good heuristics produce differences in performance of large orders of magnitude, but invariably a "plateau" is reached that can be surpassed only with quite different heuristics. These new heuristics will again make differences of orders of magnitude. In this section we shall analyze LT's difficulties with those theorems it cannot prove, with a view to indicating the general type of heuristic that might extend its range of effectiveness.

The Subproblem Tree

Let us examine the proof of theorem 2.17 when all the preceding theorems are available. This is the proof that cost LT 89,000 primitives. It is reproduced below, using chaining as a rule of inference (each chaining could be expanded into two detachments, to conform strictly to the system of *Principia*).

(not-*q* implies not-*p*) implies (*p*-implies *q*)	(theorem 2.17, to be proved)
1. *A* implies not-not-*A*	(theorem 2.12)
2. *p* implies not-not-*p*	(subs. *p* for *A* in 1)
3. (*A* implies *B*) implies [(*B* implies *C*) implies (*A* implies *C*)]	(theorem 2.06)
4. (*p* implies not-not-*p*) implies [(not-not-*p* implies *q*) implies (*p* implies *q*)]	(subs. *p* for *A*, not-not-*p* for *B*, *q* for *C* in 3)

5. (not-not-*p* implies *q*) implies (*p* implies *q*) — (det. 4 from 3)
6. (not-*A* implies *B*) implies (not-*B* implies *A*) — (theorem 2.15)
7. (not-*q* implies not-*p*) implies (not-not-*p* implies *q*) — (subs. *q* for *A*, not-*p* for *B*)
8. (not-*q* implies not-*p*) implies (*p* implies *q*) — (chain 7 and 5; *QED*)

The proof is longer than either of the two given earlier. In terms of LT's methods it takes three steps instead of two or one: a forward chaining, a detachment, and a substitution. This leads to the not surprising notion, given human experience, that length of proof is an important variable in determining total effort: short proofs will be easy and long proofs difficult, and difficulty will increase more than proportionately with length of proof. Indeed, all the one-step proofs require 500 to 1500 primitives, while the number of primitives for two-step proofs ranges from 3000 to 50,000. Further, LT has obtained only six proofs longer than two steps, and these require from 10,000 to 90,000 primitives.

The significance of length of proof can be seen by comparing Fig. 7, which gives the proof tree for (2.17), with Fig. 4, which gives the proof tree for (2.45), a two-step proof. In going one step deeper in the case of (2.17), LT had to generate and examine many more subproblems. A comparison of the various statistics of the proofs confirms this statement: the problems are roughly similar in other respects (*e.g.,* in effort per theorem considered); hence the difference in total effort can be attributed largely to the difference in number of subproblems generated.

Let us examine some more evidence for this conclusion. Figure 8 shows the subproblem tree for the proof of (2.27) from the axioms, which is the only four-step proof LT has achieved to date. The tree reveals immediately

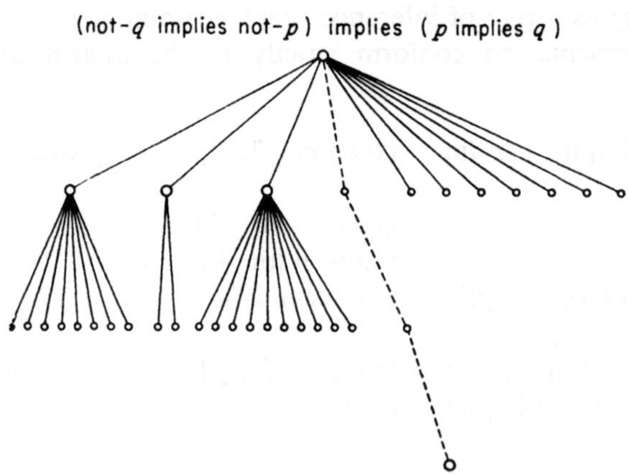

Figure 7. Subproblem tree of proof by LT of (2.17) (all previous theorems available).

why LT was able to find the proof. Instead of branching widely at each point, multiplying rapidly the number of subproblems to be looked at, LT in this case only generates a few subproblems at each point. It thus manages to penetrate to a depth of four steps with a reasonable amount of effort (38,367 primitives). If this tree had branched as the other two did, LT would have had to process about 250 subproblems before arriving at a proof, and the total effort would have been at least 250,000 primitives. The statistics quoted earlier on the effectiveness of subproblem generation support the general hypothesis that the number of subproblems to be examined increases more or less exponentially with the depth of the proof.

The difficulty is that LT uses an algorithmic procedure to govern its generation of subproblems. Apart from a few subproblems excluded by the type II errors of the similarity test, the procedure guarantees that all subproblems that can be generated by detachment and chaining will in fact be obtained (duplications are eliminated). LT also uses an algorithm to determine the order in which it will try to solve subproblems. The subproblems are considered in order of generation, so that a proof will not be missed through failure to consider a subproblem that has been generated.

Because of these systematic principles incorporated in the executive program, and because the methods, applied to a theorem list averaging 30 expressions in length, generate a large number of subproblems, LT must find a rare sequence that leads to a proof by searching through a very large set of such sequences. For proofs of one step, this is no problem at all; for proofs of two steps, the set to be examined is still of reasonable size in relation to the computing power available. For proofs of three steps, the size of the search already presses LT against its computing limits; and if one or two additional steps are added the amount of search required to

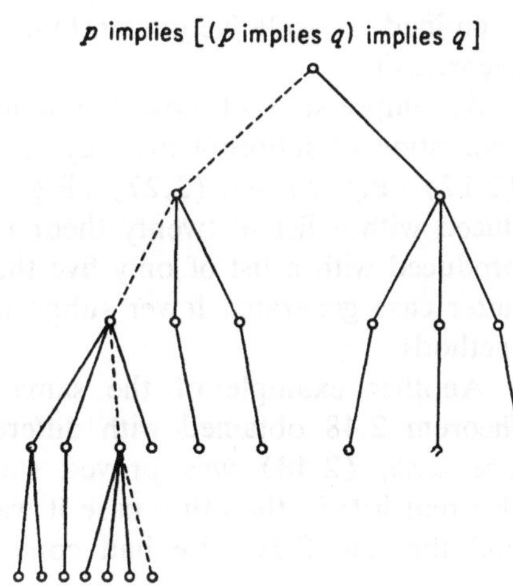

Figure 8. Subproblem tree of proof by LT of (2.27) (using the axioms).

find a proof exceeds any amount of computing power that could practically be made available.

The set of subproblems generated by the Logic Theory Machine, however large it may seem, is exceedingly selective and rich in proofs compared with the set through which the British Museum algorithm searches. Hence, the latter algorithm could find proofs in a reasonable time for only the simplest theorems, while proofs for a much larger number are accessible with LT. The line dividing the possible from the impossible for any given problem-solving procedure is relatively sharp; hence a further increase in problem-solving power, comparable to that obtained in passing from the British Museum algorithm to LT, will require a corresponding enrichment of the heuristic.

Modification of the Logic Theory Machine

There are many possible ways to modify LT so that it can find proofs of more than two steps in a way which has reason and insight, instead of by brute force. First, the unit cost of processing subproblems can be substantially reduced so that a given computing effort will handle many more subproblems. (This does not, perhaps, change the "brute force" character of the process, but makes it feasible in terms of effort.) Second, LT can be modified so that it will select for processing only subproblems that have a high probability of leading to a proof. One way to do this is to screen subproblems before they are put on the subproblem list, and eliminate the unlikely ones altogether. Another way is to reduce selectively the number of subproblems generated.

For example, to reduce the number of subproblems generated, we may limit the lists of theorems available for generating them. That this approach may be effective is suggested by the statistics we have already cited, which show that the number of subproblems generated by a method per theorem examined is relatively constant (about one subproblem per seven theorems).

An impression of how the number of available theorems affects the generation of subproblems may be gained by comparing the proof trees of (2.17) (Fig. 7) and (2.27) (Fig. 8). The broad tree for (2.17) was produced with a list of twenty theorems, while the deep tree for (2.27) was produced with a list of only five theorems. The smaller theorem list in the latter case generated fewer subproblems at each application of one of the methods.

Another example of the same point is provided by two proofs of theorem 2.48 obtained with different lists of available theorems. In the one case, (2.48) was proved starting with all prior theorems on the theorem list; in the other case it was proved starting only with the axioms and theorem 2.16. We had conjectured that the proof would be more difficult to obtain under the latter conditions, since a longer proof chain would have to be constructed than under the former. In this we were wrong: with the longer theorem list, LT proved theorem 2.48 in two steps, employing 51,450 primitives of effort. With the shorter list, LT proved the theorem in three steps, but with only 18,558 primitives, one-third as many as before. Examination of the first proof shows that the many "irrelevant" theorems on the list took a great deal of processing effort. The comparison provides a dramatic demonstration of the fact that a problem solver may be encumbered by too much information, just as he may be handicapped by too little.

We have only touched on the possibilities for modifying LT, and have seen some hints in LT's current behavior about their potential effectiveness. All of the avenues mentioned earlier appear to offer worthwhile modifications of the program. We hope to report on these explorations at a later time.

Conclusion

We have provided data on the performance of a complex information processing system that is capable of finding proofs for theorems in elementary symbolic logic. We have used these data to analyze and illustrate the difference between systematic, algorithmic processes, on the one hand, and heuristic, problem-solving processes, on the other. We have shown how heuristics give the program power to solve problems in a reasonable computing time that could be solved algorithmically only in large numbers of years. Finally, we have assessed the limitations of the present program of the Logic Theory Machine and have indicated some of the directions that improvement would have to take to extend its powers to problems at new levels of difficulty.

Our explorations of the Logic Theory Machine represent a step in a program of research on complex information processing systems that is aimed at developing a theory of such systems and applying that theory to such fields as computer programming and human learning and problem-solving.

Proving a Theorem (as Done by Man, Logician, or Machine)

A. Robinson

1. This talk is related to H. Gelernter's interesting description (given in the course of this Institute) of a programme designed to enable a machine to prove theorems of Elementary Geometry with the help of some built-in "artificial intelligence" rather than by an automatic decision procedure. The author owes much to discussions with E. Beth, P.C. Gilmore and G. Kreisel. Our main purpose is to provide some background and to encourage further discussion of this problem but we shall also make some constructive suggestions.

There exist some interesting books on the strategy of mathematical proofs (Heuristics) which make no reference to Mathematical Logic. Can Mathematical Logic do more than provide a notation for the detailed formulation of a proof on a computer? Believing that the answer to this question is in the affirmative, we place at the centre of our discussion Herbrand's procedure, which is explained briefly in the following section.

2. Suppose that we wish to deduce the conclusion

(2.1) $$(\exists u)(t)G(u,t)$$

from the premiss

(2.2) $$(x)(\exists y)(z)(\exists w)F(x,y,z,w)$$

where both F and G are quantifier-free expressions in the lower predicate calculus. This is impossible if and only if there exist functions $\phi_1(x)$, $\phi_2(x,z)$, $\phi_3(u)$ in a model M such that both

(2.3) $$F(x,\phi_1(x),z,\ \phi_2(,x,z))$$

and

(2.4) $$\sim G(u,\phi_3(u))$$

hold for arbitrary x,z,u in M. And if we consider the sequence of all terms obtained by the repeated application of ϕ_1, ϕ_2, ϕ_3, to initial constants a,b,c, i.e.

(2.5) $\quad a,b,c,\phi_1(a),\ \phi_2(a),\phi_3(a)\ldots\phi_3(c),\ldots,\phi_2(\phi_1(a),\phi_2(a,b)),\ldots$

then we can obtain such a model M if and only if the "infinite conjunction," A, obtained by substituting the terms of 2.5 in all possible ways for the free variables of 2.3 and 2.4 is consistent. Moreover, a well known mathematical argument shows that if A is not consistent, and hence, if 2.1 is derivable from 2.2, then a finite sub-conjunction A' of A is already contradictory. This is to say, in order to deduce 2.1 from 2.2 we only have to find a finite sub-conjunction A' of A such that ~A' is valid. The validity of ~A' can be decided by the propositional calculus alone, by regarding atomic relations whose places are occupied by specific elements of 2.5 as propositional constants.

Moreover, the generation of 2.5 and of the infinite sequence of particular instances of 2.3 and 2.4 can be easily systematized (and hence programmed). Once the contradictory character of a sub-conjunction A' has been established by means of some mechanized version of the propositional calculus, we may claim that we have deduced 2.1 from 2.2. Note that this is a perfectly good formal method of proof and that there is no need to go to the more familiar deductive method of the predicate calculus in order to establish that the machine can fulfill its purpose.

3. However, when a proof is found by the above procedure it may well be unnecessarily long and quite beyond the capacity of a contemporary computer. The question therefore arises how to replace the systematic method sketched in section 2 by a less systematic method which, however, may give us a reasonable chance of finding a relatively short sub-conjunction of A which is contradictory, provided 2.1 is at all deducible from 2.2. Now some reflection shows that this is precisely what Man (i.e. the working mathematician) has been doing intuitively for a long time when trying to prove a mathematical theorem. For example, to prove that the sum of two sides of a triangle is greater than the third side, $AB+BC \geq AC$, we construct a single auxiliary point C' (on AB produced, so that BC' = BC) and we then make use of a previously established theorem. Now this point C' may be regarded as a Herbrand functor of the initial A,B,C, and hence, as a particular element of the sequence 2.5. Having selected it, we are in a position to write down a suitable sub-conjunction A'. Thus, instead of producing the elements of 2.5 in large numbers, we tend to select only a small number of them and to see whether the instances of 2.3 and 2.4 which are formulated in

terms of these, already yield a contradictory sub-conjunction. It may be unwarranted to say that all mathematical proofs adopt this procedure (in fact, in some cases, Herbrand's method itself suggests a reduction in the number of functors ϕ_k), but nevertheless it seems to be far more general than might be thought at first sight.

It is clear that in the course of developing a mathematical theory one gradually has to increase the number of functors used. To do this efficiently, the mathematician adds the previously proved theorems to his list of premisses and this creates a new pattern for formulae which may be used in the selection of sub-conjunctions for the solution of subsequent problems. But he also adds to the number of original relations (e.g. when he introduces the six-point relation of perspective triangles in Projective Geometry) and introduces new functors which are considered to be on a par with the original functors (e.g. the "centroid of a triangle"). These, together with the theorems concerning them which were proved previously, will help to guide the mathematician in the selection of a suitable sub-conjunction.

5. The following conclusions may be drawn from the above analysis. Without wishing to exclude other possibilities, it is suggested that a theorem-proving programme might be based on Herbrand's procedure rather than on the standard predicate calculus; that it instruct the computer to select, in turn, small numbers of elements of 2.5 in order to find sub-conjunctions, as indicated above; and that the number of sentences in the premiss as well as of relations and functors, be increased gradually, thus providing new patterns for the construction of sub-conjunctions of the required character.

We have confined our discussion to sentences in prenex normal form. Modifications can be introduced to cope with sentences of a more general type but it may well be that in this respect E. Beth's method of semantic tableaux is superior. On the other hand, it would appear that the explicit introduction of functors, which is basic in Herbrand's procedure, is also very important from the heuristic point of view. It might be included equally well in programmes which are not based directly on our present approach.

1958

On Machines Which Prove Theorems

E. W. Beth

1. Introduction — Computation and Formal Deduction.— The invention of computation goes far back in the history of the human mind. And, even though for many centuries it remained technically inefficient and theoretically crude, it fascinated mankind already in this underdeveloped stage. Perhaps the very defects in the art of computation offer the best explanation of the fact that, for a long time, it was considered as a kind of black magic.

At any rate, the fascination inspired by the art of computation explains why, at a certain moment, a more scientific interest in numbers developed. At first, this interest naturally focused on the study of numbers as individuals. There still remain clear traces of a stage in the development of arithmetic where the factorisation of a large number or the discovery of a large prime were considered as brilliant achievements. Such investigations, however, could in themselves hardly lead to any important progress in the art of computation. Two further steps were necessary. In the first place, scientific interest had to shift from numbers as individuals to general properties of numbers; this step eventually led to a considerable deepening of theoretic insight. Secondly, a deeper insight into the general properties of numbers made it possible to invent more convenient systems of notation. Finally, the combination of a deeper insight with a better notation gave rise to the development of more and more powerful computational techniques.

It has been observed long ago that there is a striking analogy between computation and formal deduction. On the basis of this analogy, we may expect the development of arithmetic and of logic to be quite similar. This expectation is corroborated by the facts, with this understanding, however, that there is an enormous lag in time. In fact, the present stage of development in logic may well be compared to a rather primitive stage in the development of arithmetic. Formal logic is still often looked upon as a kind of esoteric doctrine; and a man, who succeeds in deducing a certain theorem from axioms by which apparently it was not entailed, creates a similar stir as was formerly elicited by a calculator factorising some large number previously believed to be prime.

2. Formal Deduction and Computing Machines. — On the other hand, in view of recent developments in the design and the construction of very powerful automatic computers and in the art of programming their operation, it is only natural to consider the possibility of using such computers in solving certain problems in formal deduction. This possibility is also strongly suggested by the fact that actually sentential logic plays a certain role in the design of computers and that GÖDEL's method of arithmetisation enables us, in principle, to convert every logical problem into an arithmetical one.

This matter was first taken up by NEWELL and SIMON [1], and it was recently discussed by H. GELERNTER and by A. ROBINSON [2]. Its importance seems to justify still another discussion, even though at present I am not in a position to offer concrete results.

I intend to show that further progress in the domain under discussion strongly depends on the development of a deeper insight into the general properties of formal deductions but that, as in the case of arithmetic, we must anticipate that a deeper insight on the theoretical level will not, in itself, be sufficient. We need, in addition, a suitable notation. And what we need is not only a convenient notation for the sentences which appear in formal deductions; we need a concise notation for formal deductions as well.

3. The Subformula Principle. — NEWELL and SIMON discuss the construction of a machine which is able to deduce theses of the two-valued (or classical) sentential logic on an axiomatic basis. Such a basis usually consists of a certain (finite) number of axioms, combined with two rules of deduction, namely:

(i) the *rule of substitution*, which allows us to substitute arbitrary formulas U, V, W, ... for the atoms p, q, r, ... which appear in a given thesis T, and (ij) the *rule of detachment* (or *modus ponens*), which allows us, from two given theses S and $S \rightarrow T$, to deduce the thesis T.

It is argued by NEWELL and SIMON that the success of an attempt to deduce a certain thesis T on a given axiomatic basis depends upon the selection of the right substituends $U, V, W, \ldots$. These substituends must be selected from the infinite stock of all formulas and, hence, either the machine has to exhaust all possible substitutions (the «*British Musem method*», which is found to be too much time-consuming), or it must be equipped with certain «*heuristic*» devices, which enable it to select in advance those substitutions which are most likely to prove successful. These heuristic devices are to be copied from the manner in which a human being would use its intelligence in a similar situation.

However, interesting and, perhaps, enlightening though such an «*anthropomorphic*» approach may be, there is no real need for the introduction of heuristic devices. In fact, suppose we wish to deduce a certain thesis, for instance:

(1) $$[p \rightarrow (q \vee r)] \rightarrow [(p \ \& \ q) \rightarrow r],$$

on a certain given axiomatic basis. Then it is possible to enumerate, in advance, all formulas which ought to be taken into account as substituends; they are:

(i) the formula (1) itself;

(ij) the *proper subformulas* of formula (1), namely: $p \rightarrow (q \vee r)$, $(p \ \& \ \bar{q}) \rightarrow r$, $q \vee r$, $p \ \& \ q$, $\bar{q}$, p, q, and r;

(iij) certain simple combinations of the formulas under (i) and (ij), depending upon the choice of an axiomatic basis.

So the machine can first construct an exhaustive list of all possible substituends, then carry out all possible substitutions in the axioms, and finally go through all possible applications of *modus ponens*; the formula (1) must appear among the formulas which result. If instead of the formula (1) we consider a certain formula U which happens not to be a thesis, then the machine will go through all the above operations, but U will not appear among the formulas which result. — We shall give later on a summary explanation of the above facts; it will then become clear that a further reduction is possible of the substitutions to be carried out by the machine.

4. Semantic Tableaux. — The discussion in Section 3 was

meant to show the importance of taking into account all relevant theoretic insight which is available at present. I now turn to my second point, namely, the need for a concise notation for formal deductions. With a view to this requirement I may in the first place explain the *method of semantic tableaux*, which I have developed and investigated since 1955.

As a concrete example, let us first consider the semantic tableau for the above formula (1).

True		False	
(2) $p \to (q \vee r)$		(1) as above	
(4) $p \;\&\; \bar{q}$		(3) $(p \;\&\; \bar{q}) \to r$	
(6) p		(5) r	
(7) $\bar{q}$		(8) q	
———	(10) $q \vee r$	(9) p	
(11) q	(12) r		

This tableau may be considered in the first place as a more convenient presentation of the familiar *truth-table analysis*, which results in the conclusion that formula (1) is a *logical identity*.

However, the tableau may also be considered under a different angle, in which case it is construed as a *formal deduction*; this aspect will be more clearly seen if the formulas are rearranged, as follows:

(2) $p \to (q \vee r)$	(+ hyp 1)
(4) $p \;\&\; \bar{q}$	(+ hyp 2)
(6) p	(4)
(7) $\bar{q}$	(4)
(10) $q \vee r$	(6), (2)
(11) q | (12) r	(alt)
(5) r	
(3) $(p \;\&\; \bar{q}) \to r$	(— hyp 2)
(1) as above	(— hyp 1)

We have clearly obtained a formal deduction which, however, does not fit intc an axiomatic development as outlined in Section **3**. It belongs to a certain formal system F which is closely related to GENTZEN's system NK[3]). In such a system, no axioms are introduced, but the set of rules of deduction is expanded, so as to make allowance, among others, for the introduction and elimination of hypotheses. It is one of the attractive features in these systems that all formulas which appear in the deduction of a thesis U are subformulas of U.

If, instead of the system F, we wish to use a conventional (or «Hilbert-type») axiomatic basis B, this advantage will be lost, but not completely. For each step in a deduction in F (or in the construction of the corresponding semantic tableau) corresponds to a certain procedure in connection with B. Suppose that the step involves the formulas X, Y, and Z (which, as we found, were subformulas of the formula U to be deduced); then this procedure will consists in substituting, in certain axioms of B, the formulas X, Y, Z, and, say, $X \rightarrow Y$ and $Y \rightarrow Z$. Therefore, every deduction in the system F can be duplicated on a given axiomatic basis.

5. Elementary Predicate Logic. — It is an interesting fact that the method of semantic tableaux, although apparently adapted exclusively to classical sentential logic, can be extended so as to apply to several more intricate domains of logic. Let us consider the semantic tableau for the formula:

(1) $(x)\,(\mathrm{E}y)\,[A(y) \rightarrow B(x)] \rightarrow \{(x)A(x) \rightarrow (y)B(y)\}$;

which is a thesis of elementary predicate logic.

<table>
<tr><th colspan="2">True</th><th colspan="2">False</th></tr>
<tr><td colspan="2">(2) $(x)\,(\mathrm{E}y)\,[A(y) \rightarrow B(x)]$
(4) $(x)A(x)$
(7) $(\mathrm{E}y)\,[A(y) \rightarrow B(a)]$
(8) $A(a)$
(9) $A(b) \rightarrow B(a)$
(10) $A(b)$</td><td colspan="2">(1) as above
(3) $(x)A(x) \rightarrow (y)B(y)$
(5) $(y)B(y)$
(6) $B(a)$</td></tr>
<tr><td></td><td>(12) $B(a)$</td><td>(11) $A(b)$</td><td></td></tr>
</table>

The construction and the «*closure*» of the tableau are again clearly suggested by the semantic rules (or truth conditions), as applied to formula (1) and to its proper subformulas (2)-(12). As a whole, the closed tableau may be construed to express the fact that it is impossible to find a non-empty domain S and two predicates A and B, defined in S, for which formula (1) would become false.

The method of semantic tableaux again offers the following advantages: (i) all formulas which appear are subformulas of the formula to be deduced; (ij) the construction of a tableau does not require a reduction of the formulas involved to some normal form; (iij) every closed tableau can be rewritten so as to obtain a deduction in a certain formal system F (which is closely related to GENTZEN's system NK), and this deduction can be duplicated on any given conventional axiomatic basis.

It would carry us to far to substantiate these statements, but it is relatively easy to prove that the transformations under (iij) can be carried out in a purely mechanical manner. A further study of semantic tableaux provides relatively simple proofs of various deeper metamathematical results concerning elementary predicate logic, such as HERBRAND's Theorem and GENTZEN's Subformula Theorem. These results are of importance with a view to the problems which I should like to discuss.

6. Complications. — In principle, the method of semantic tableaux can be said to provide a theoretical basis for the construction of a «*logic theory machine*», up to the problem of setting up the program. However, there are certain complications, which are connected with CHURCH's Theorem concerning the impossibility of an effective decision procedure for deducibility in elementary predicate logic, and which, therefore, present themselves also in all other adequate formalisations of this logical domain. Again, a study of certain semantic tableaux may help us in understanding the mechanism behind these complications.

Suppose we are asked to deduce the formula :

(1) $$(x)\,(\mathbf{E}y)A(x, y) \rightarrow (\mathbf{E}z)A(z, z)$$

as a thesis of elementary logic. We construct tho semantic tableau:

True	False
(2) $(x)\,(\mathbf{E}y)A(x, y)$	(1) as above
(5) $(Ey)A(a, y)$	(3) $(\mathbf{E}z)A(z, z)$
(6) $A(a, b)$	(4) $A(a, a)$
(8) $(\mathbf{E}y)A(b, y)$	(7) $A(b, b)$
(9) $A(b, c)$	(10) $A(c, c)$
(11) $(\mathbf{E}y)A(c, y)$	
(12) $A(c, d)$	*etc.*

We may clearly go on indefinitely, without ever reaching a closure. It follows that formula (1) cannot be deduced by the method of semantic tableaux; hence, the formula cannot be a thesis and so cannot be deduced in any adequate formalisation of elementary predicate logic. In fact, it is easy to see that on the basis of the above tableau we can even find a domain of individuals S and a predicate A, defined in S, for which formula (1) will be false.

Our next example illustrates a more involved situation. Let us ask the same question as before in connection with the formula:

(1) $(x)\,(\mathbf{E}y)\,[A(x, y) \vee A(y, y)] \rightarrow (\mathbf{E}z)A(z, z)$;

we obtain the following tableau:

True		False
(2) $(x)\,(\mathbf{E}y)\,[A(x, y) \vee A(y, y)]$		(1) as above
(5) $(\mathbf{E}y)\,[A(a, y) \vee A(y, y)]$		(3) $(\mathbf{E}z)A(z, z)$
(6) $A(a, b) \vee A(b, b)$		(4) $A(a, a)$
		(7) $A(b, b)$
(8) $A(a, b)$	(9) $A(b, b)$	
(10) $(\mathbf{E}y)\,[A(b, y) \vee A(y, y)]$		(12) $A(c, c)$
(11) $A(b, c) \vee A(c, c)$		
(13) $A(b, c)$	(14) $A(c, c)$	*etc.*
(15) $(\mathbf{E}y)\,[A(c, y) \vee A(y, y)]$		

which clearly shows that formula (1) is not deducible and hence cannot be a thesis.

Of course, the examples discussed so far were extremely simple ones. Nevertheless, the last example shows that we must be prepared, in general, to meet with a quite irregular succession of introductions of new individuals *a*, *b*, *c*, ..., of splittings into subtableaux, and of closures of some of the subtableaux obtained. Very soon the tableau may become to involved as to make the operation of the machine too much time-consuming.

It will be clear that the number $n(U)$ of the individuals *a*. *b*. *c*, ..., which must be introduced in a tentative deduction of a given formula U, is the decisive factor here. Once $n(U)$ has been fixed, the construction of a semantic tableau reduces to a finite procedure. Hence, the problem of simplifying this construction (or, at least, of avoiding unnecessary complications) breaks up into two partial problems:

(i) minimising or fixing the value of $n(U)$;

(ij) for given $n(U)$, simplifying the tableau construction.

Now it would be very nice if we could fix $n(U)$ in such a manner that, whenever a formula U is at all deducible, its deduction can be achieved by introducing at most $n(U)$ individuals. However, this would imply the existence of an effective decision procedure, which is excluded by CHURCH's Theorem.

Before going more deeply into point (i), it will be good to devote a few words to point (ij). As an illustration, we consider the tableau for the formula:

(1) $(x)\,(y)\,[A(x) \vee B(y)] \rightarrow \{(x)A(x) \vee (y)B(y)\}$.

True	False
(2) $(x)\,(y)\,[A(x) \vee B(y)]$	(1) as above
(8) $(y)\,[A(a) \vee B(y)]$	(3) $(x)A(x) \vee (y)B(y)$
(9) $A(a) \vee B(a)$	(4) $(x)A(x)$
(10) $A(a) \vee B(b)$	(5) $(y)B(y)$
(11) $(y)\,[A(b) \vee B(y)]$	(6) $A(a)$
(12) $A(b) \vee B(a)$	(7) $B(b)$
(13) $A(b) \vee B(b)$	

Of the four formulas (9), (10), (12), and (13), it is clearly sufficient to consider formula (10). This leads to one splitting of the tableau, immediately followed by the closure of each of the resulting subtableaux. However, if we overlook this fact, we are compelled to carry out 4 successive splittings, followed by the closure of each of the 16 subtableaux which result. This example shows that, by an intelligent selection of the relative order in which the operations are to be carried out, we are able to simplify the construction. It remains to be seen, inhowfar it would prove convenient to imitate this intelligent way of proceeding in a logical theory machine.

7. Introduction of New Individuals. — Returning to point (i), we first ask the following question: to what operation in, say, geometrical proof does the introduction of individuals *a*, *b*, *c*, ... correspond? The answer is rather simple: it corresponds to the construction of auxiliary points, lines, circles, and so on. This insight has been stressed by G. KREISEL on the basis of HERBRAND's Theorem; the same logician also made the observation that in all known mathematical proofs the number of individuals involved is surprisingly small.

This observation might be construed as pointing to a rather profound difference between proof procedures, as studied in formal logic, and methods of proof as actually applied in mathematics. However, this would be a mistaken notion. The methods, applied in mathematical proof, differ from those which we study in logic only in this respect that, in mathematics, if we wish to prove a certain conclusion X on the basis of certain assumptions $A_1, A_2, \ldots, A_m$, we usually intercalate a number of lemmas $L_1, L_2, \ldots, L_n$. So we prove L_1 on the basis of the A's alone, L_2, on the basis of the A's plus L_1, and so on; finally, we prove X on the basis of all A's and all L's. Now it will be clear that we have to take into account the total number of all individuals involved in the respective proofs of these lemmas; roughly speaking, this will add up to the number of individuals involved in a straightforward proof of X on the basis of the A's alone.

Nevertheless, a mechanical application of the method of semantic tableaux will naturally lead to the introduction of individuals which do not contribute to the closure of the tableau, and thus to a premature exhaustion of the capacity of the machine.

8. Types of Logical Problems. — Therefore, we now return to point (ij) and we ask for devices which may prevent the machine from going into useless operations. It seems to me that, before this point can be discussed, we have to make up our mind as to the types of problems which a logical theory machine is meant to handle. As far as I see, three types of problems can be distinguished.

(I) To check a given proof;

(II) To prove a given theorem on the basis of given assumptions;

(III) To discover theorems deducible from given assumptions.

Type (I) is clearly the simplest one. In this case we have at least an estimate of the number and the kind of the individuals which must be introduced. By giving the machine suitable instructions, we may compel it to follow the given proof as closely as possible. If at a certain point the machine fails to reproduce the given proof, this will presumably point to a gap in the argument.

In the case of type (II), we have hardly any sound basis for an estimate of $n(U)$ and, if we simply make a guess as to its value, we still have to select those operations which are most likely to be successful; in other words, we have to provide the machine with certain «heuristic» devices. In this connection, I should like te offer the following suggestion. There are a number of solvable cases of the decision problem; this means, essentially, that for certain classes of formulas U a number $n(U)$ can be effectively computed. Now run a large number of formulas U through the machine, and make a statistical analysis of those operations which prove successful and of those which are not. Then provide the machine with such instructions as to give preference to the more successful operations. This might considerably enhance the efficiency of the machine.

This statistical approach to heuristics might even be extended to the case of type (III). The danger to be avoided here is clearly the production of an endless sequence of trivialities. Again we might try to find out, by way of statistical analysis, what operations are most likely to lead to interesting results; this would enable us to provide the machine with suitable instructions.

But it seems wiser at present not to go into problems of type (III). In my opinion, the most promising direction at this moment would be to study case (I). To mention one point, it would not be necessary for the machine to follow the given proof in every detail.

We know at present various metamathematical results of the following kind: if a formula U is at all provable, then it has a proof of such and such a form (in fact, the method of semantic tableaux is based on such a theorem). So we may give the machine such instructions as to compel it to produce, if possible, a proof of that specific form. Nevertheless, the machine might at the same time take advantage of the results previously obtained by an analysis of the given proof.

9. Concluding Remarks. — The heuristic devices, discussed in Section 8, are meant to be adapted to the specific operations of a given automatic computer. NEWELL and SIMON [1] and GELERNTER [2], however, rather seem to think of using those heuristic devices which are often applied by human beings in finding a proof or solving a problem. It seems not unfair to draw a comparison between such devices and certain tricks which one resorts to in mental computation. A number of these tricks may be of sufficient importance to be taken into account in designing a computer, or, perhaps, rather in programming a specific problem. But many tricks, helpful though they may be in mental computation, will be devoid of any value with a view to computational machinery.

Similarly, as I have shown in a recent book [3], certain heuristic devices used in proving theorems reappear as mere corollaries, so to speak, to the method of semantic tableaux. Other heuristic devices of this kind may turn up again as a result of a statistical analysis as suggested in Section 8. But many others will be found to be devoid of any value in connection with the problem of constructing an efficient logic theory machine.

NOTES

[1] NEWELL and SIMON, in : *Proceedings of the Western Joint Computer Conference*, 1957; *IRE Transactions on Information Theory*, vol. IT-2, No. 3, Sept. 1956. I take this reference from the paper by GELERNTER [2].

[2] H. GELERNTER, Theorem proving by machine; A. ROBINSON, Proving a theorem (as done by man, logician, or machine), in : *Summaries of talks presented at the Summer Institute of Symbolic Logic in 1957 at Cornell University* (mimeographed), vols II and III. — I had an opportunity to discuss these matters with Dr. GELERNTER, Professor ROBINSON, and several other scholars during the Summer Institute and during a visit to the IBM Research Center in Yorktown, N. Y. The present article is

a revised version of a talk which I presented in a seminar at the Research Center.

[3] E. W. Beth, *La crise de la raison et la logique*, Paris-Louvain 1957. —, Ueber Lockes «Allgemeines Dreieck», *Kant-Studien*, Bd. 48 (1956/57).

1959

A Non-heuristic Program for Proving Elementary Logical Theorems

B. Dunham, R. Fridshal, G. L. Sward

MOTIVATION: If our basic objective is to obtain greater problem-solving capacity, we need not only better computers, but also a better understanding of how to use computers. With the latter in mind, we have undertaken a long-range study which is first of all an experiment in problem-solving techniques and secondly an attempt to fabricate very powerful problem-solving programs.

One area especially where we are in the dark as to the computer's potentialities is symbolic logic. More specifically, we do not know how best to use the machine in deriving logical theorems nor can we guess how effective the ultimate theorem-proving programs will be.

There is much loose talk in this connection. It is sometimes bluntly asserted, for example, that algorithmic techniques are inefficient in proving even elementary logical theorems; and questionable statistics are cited. On purely formal grounds it can be seen certain classes of problems do not admit of a general systematic solution, so that a problem-solver for such systems would presumably involve "strategy." Nevertheless, in seeking after new powerful theorem-proving techniques, we must first clarify the extent to which systematic, non-heuristic procedures can be found suitable for machine use.

Our first modest objective is, therefore, a systematic, efficient machine program for validating truth-functional formulae. In pursuing this, not only can we learn more about problem solving, but also provide ourselves with a necessary subroutine for later theorem-proving mechanisms. Our objective is not actually modest, however, nor, as we shall see, yet perfectly realized. Nevertheless, results do indicate the general workability of the algorithmic approach in truth-functional validation.

LOGIC: The two most familiar algorithms for testing the validity of truth-functional expressions are (a) conversion to conjunctional normal form and (b) the truth table. Both have their drawbacks, in that we can easily devise examples which neither can manage in decent time; but both serve reasonably well for the great majority of expressions normally encountered. Working entirely by rote with pencil and paper, the average clerical worker, using either method, could validate in a matter of hours all the truth-functional theorems in Principia Mathematica. This does not mean, however, that the drawbacks in question can be dismissed lightly.

The chief difficulty with the truth-table method is, of course, that the size of table doubles with each new variable. The conversion to conjunctional normal form, on the other hand, has two major limitations. First, when IF-AND-ONLY-IF and EXCLUSIVE-OR (symbolically, $\equiv$ and $\not\vee$) are eliminated in favor of AND, OR, and NOT (symbolically, $\cdot$, V , and $\overline{}$), we must double the relevant part of the expression. Second, when we distribute alternations through conjunctions in normalization, we often very much extend the length of the formula. Further complications might result, of course, in eliminating propositional connectives having more than two variables; but there is no real need to admit such connectives in the formulae with which we shall be concerned.

In view of the difficulties with the usual methods, we wish to suggest an alternative validation procedure. The general technique is one of progressive variable elimination and simplification. Suppose for the time being we need deal only with formulae in basic form, that is, formulae composed from AND, OR, and NOT exclusively with denials reduced. Under such conditions, a variable will occur in full state if it occurs both denied and undenied; otherwise, in partial state. Simplification is to be accomplished by the following equivalences, easily adapted to machine use:

$$\phi \vee 1 \ \underline{eq} \ 1$$
$$\phi \vee O \ \underline{eq} \ \phi$$
$$\phi \cdot 1 \ \underline{eq} \ \phi$$
$$\phi \cdot O \ \underline{eq} \ O$$

First, all literals representing variables in partial state are replaced by O and the resulting expression simplified. Although not commonly recognized, it is easily verified that in this way expressions can be reduced to shorter expressions not necessarily equivalent but valid if and only if the original expression is valid. When no further reductions can be made, a variable in full state is eliminated by branching. Two new expressions are formed and in turn simplified. In one, the variable (not literal) to be eliminated is replaced by 1; in the other, O. The process continues until the question of validity is settled. Simplification of a single branch to O assures non-validity. Since variables often drop out during intervening simplifications or are left to be dealt with in partial state, the simplification of branches to 1 or O is generally easier than one might expect. There is also computer advantage in the fact that branches can be dealt with serially.

To illustrate the technique, let us test the expression which follows (some parentheses are omitted because AND and OR are associative):

$$(X_1 \cdot (\overline{X}_2 \vee X_3 \vee (X_4 \cdot (X_5 \vee (X_6 \cdot \overline{X}_1 \cdot X_7))))) \vee (((\overline{X}_6 \cdot ((X_8 \cdot X_9 \cdot \overline{X}_5)$$

$$\vee (\overline{X}_7 \cdot \overline{X}_8 \cdot \overline{X}_9))) \vee \overline{X}_1) \cdot X_2 \cdot \overline{X}_3) .$$

We replace X_4, the only variable in partial state, by O. After simplification, we are left with the following:

$$(X_1 \cdot (\overline{X}_2 \vee X_3)) \vee (((\overline{X}_6 \cdot ((X_8 \cdot X_9 \cdot \overline{X}_5) \vee (\overline{X}_7 \cdot \overline{X}_8 \cdot \overline{X}_9)))$$

$$\vee \overline{X}_1) \cdot X_2 \cdot \overline{X}_3) .$$

$\overline{X}_6$ now represents a variable in partial state. We replace it by O and simplify to the following:

$$(X_1 \cdot (\overline{X}_2 \vee X_3)) \vee (\overline{X}_1 \cdot X_2 \cdot \overline{X}_3) .$$

All variables are now in full state, so that we must branch to eliminate any one of them. We first replace X_1 by 1 and simplify. We are left with

$$\overline{X}_2 \vee X_3 .$$

Both $\overline{X}_2$ and X_3, now representing variables in partial state, are replaced by O. The non-validity of the original expression is determined, and we do not need to continue the problem further.

Thus far, we have presented a technique for validating expressions in basic form. The partitioning into branches speeds up the overall test, since we are more likely to encounter variables in partial state. A similar strategy should be adopted for conjunctions. A conjunction is valid if and only if each of its full components is valid, but a variable is more likely to occur in partial state within a single component than in the entire expression. Hence, there is advantage in testing such components one at a time. The question remains, however, what we are to do with expressions not in basic form.

The procedure for converting formulae into basic form is well known. All logical connectives except AND, OR, and NOT are eliminated, and denials reduced. Hence, a machine program for such conversion, combined with a program for testing basic formulae, would constitute a validity check for truth-functional expressions. This method, which we have in fact programmed, is not unappealing; but it does have certain drawbacks. Foremost among these is the fact, already mentioned, that elimination of IF-AND-ONLY-IF or EXCLUSIVE-OR doubles the relevant part of the expression. We might, of course, restrict our family of expressions to those not containing the two connectives; but this does not seem reasonable, since both are in such common use by logicians. A formula with only fifteen IF-AND-ONLY-IF's would overflow the fast memory of almost all machines upon conversion to basic form. As a result, we have found it necessary to modify the validation procedure just described.

Let a formula be in <u>standard</u> form which is composed from AND, OR, IF-AND-ONLY-IF, and NOT, with denials reduced. If we do not admit propositional connectives having more than two variables, it is trivial to convert a truth-functional expression into standard form; and, as we shall see, a slightly modified version of the validation test already worked out for basic formulae can be directly applied.

There are two advantages in converting expressions to basic form, which we should like to retain in dealing with expressions in standard form. First, the rules for simplification are uncomplicated and readily adapted to machine use. Second, variables in partial state can be immediately determined. Let us deal with the second problem first. It is easily verified that any variable falling within the overall scope of an IF-AND-ONLY-IF will occur in full state in a new expression where the IF-AND-ONLY-IF is eliminated in favor of AND, OR, and NOT. Hence we can tell by direct inspection of a formula in standard form which of its variables would be in partial state were it converted to basic form. A variable is, thus, in full state which occurs both denied and undenied or which falls within the overall scope of an IF-AND-ONLY-IF; otherwise, it is in partial state and can be dealt with accordingly. Finally, to the rules of simplification, we add the two which follow:

$$\phi \equiv 1 \;\underline{\text{eq}}\; \phi$$
$$\phi \equiv O \;\underline{\text{eq}}\; -(\phi)$$

Unreduced denials can be reduced upon appearance.

Although the method indicated is reasonably efficient, further improvements are possible. Let a <u>prime</u> conjunction, alternation, or biconditional be a formula containing only one kind of two-variable connective. A prime formula is <u>stark</u> if it does not contain 1 or O, and no variable within it occurs twice. Without appreciable complication, a routine can be added which will eliminate or render stark all prime conjunctions, alternations, or biconditionals within an expression before each operation of branching. No stark expression is, of course, valid, so that, apart from other advantages, reduction of any branch to a stark formula will terminate the problem.

Without dwelling upon details, we need only comment that the "stark" routine is readily available from the commutativity and associativity of conjunctions, alternations, and biconditionals. The following equivalences are also used:

$$P \equiv P \ \underline{eq}\ 1 \qquad P \vee P \ \underline{eq}\ P \qquad P \cdot P \ \underline{eq}\ P$$
$$P \equiv \bar{P} \ \underline{eq}\ O \qquad P \vee \bar{P} \ \underline{eq}\ 1 \qquad P \cdot \bar{P} \ \underline{eq}\ O$$

Although the "stark" routine will slow up but not prevent the solution of some problems, in other cases solutions are rendered easy which would be impossible without it. It is less important to deal quickly with simple cases than to expand the domain of manageable expressions. Further sophistications can, of course, be introduced; but some compromise is also required with other needs.

It is true examples can be devised in a straightforward way which the method outlined above will not resolve. Nevertheless, it should handle easily the vast majority of expressions we shall have occasion to test; and we do not, in fact, know of a comparable method, systematic or not. The technique is in certain ways a blend of the truth-table and conversion approaches mentioned earlier with the major drawbacks of each minimized so far as possible.

COMPUTATION: Although details of programs are generally without interest, it may be pertinent to comment briefly upon the way in which we have represented logical expressions in the IBM 704. How this is best accomplished is not obvious from the computer itself; but the matter is important, since the structure of a program is largely determined by the representation of its computational data.

In this case, we set ourselves two immediate objectives: (1) an easy way of expanding and contracting expressions, and (2) an easy way of determining how the various connectives operate upon one another. Logical notations usually represent formulae by a serial arrangement of symbols with grouping indicated by parentheses, dots, special rules, or position. To expand or contract a formula, we must shift the position of symbols. Also, the status of any given connective within an expression can be determined only by scanning a substantial part of the expression. Hence, none of the familiar notations seem appropriate to our purpose.

The actual scheme of representation adopted is roughly as follows. Five numbers are associated with each occurrence of a logical connective. One of these designates the type. Another specifies whether the connective is main, intermediate, or terminal. A terminal connective will take as an argument either O, 1, or a variable. The other three numbers specify the locations of the three possible connectives with which the particular occurrence can be associated. The representation thus resembles a tree with one main connective and at most 2^n terminal connectives, where n is the number of connective levels in the expression. Every set of connectives which, in the usual notation, would be grouped together, form a subtree. The ordering of connectives in the computer is immaterial.

The advantages of this approach are several. Because no ordering is assumed, expansion of an expression is achieved merely by adjoining additional connectives; and contraction, simply by modifying appropriate numbers. The size of expression is limited only by the amount of storage available. Also, we are provided with an immediate mechanism for determining the interrelationship of connectives. And, finally, the scheme can be easily extended to embrace quantificational formulae.

Although we are still seeking to improve the program's coding, a number of runs have been made on a variety of problems. Parenthetically, it might be noted that all of the truth-functional theorems in Principia were disposed of in approximately two minutes. This is actually somewhat slower than we hope for; but, as remarked earlier, we do not wish to sacrifice problem-solving power for added speed in dealing with simple cases.

Apart from its implications as to the useful range of systematic computational techniques, the experiment has also provided a good illustration of the value of partitioning problems. Where possible, it is usually an effective strategy to tackle problems one section at a time.

Realization of a Geometry-Theorem Proving Machine

H. Gelernter

Introduction

Few of those who have seen a modern high-speed digital computer digest and transform a mass of data in less time than it takes to follow the process in the mind can suppress a certain amount of speculation concerning the future of such machines. Under the assumption that the computer is operating at the mere threshhold of its capacity in performing the tasks we have thus far delegated to it, a long-range program directed at the problem of "intelligent" behavior and learning in machines has been established at the IBM Research Center in New York (Gelernter and Rochester, 1958). In particular the technique of heuristic programming is under detailed investigation as a means to the end of applying large-scale digital computers to the solution of a difficult class of problems currently considered to be beyond their capabilities; namely those problems that seem to require the agent of human intelligence and ingenuity for their solution. It is difficult to characterize such problems further, except, perhaps, to remark rather vaguely that they generally involve complex decision processes in a potentially infinite and uncontrollable environment.

If, however, we should restrict the universe of problems to those that amount to the discovery of a proof for a theorem in some well-defined formal system, then the distinguishing characteristics of those problems of special interest to us are brought clearly into focus. We should like our machine to be able to prove many of the theorems presented to it in a formal system that is manifestly undecidable. Further, as the machine gains "experience" in proving theorems, we should expect it to be able to solve problems that were earlier beyond its capabilities.

The requirement that a machine should deal with undecidable systems places a fundamental restriction on its modus operandi. Finding a suitable algorithm, the obvious technique for the solution of problems on a digital computer, is no longer acceptable for the simple reason that no such algorithm exists. An exhaustive search for the initial axioms and theorems of the proof, combined with exhaustive development of the proof sequence by systematically applying the rules of transformation until the required proof has been produced, has been shown to be much too time-consuming for so simple a logic as propositional calculus (Newell, Shaw and Simon, 1957*a*). It is a fortiori out of the question for any of the more interesting logics. A remaining alternative is to have the machine rely upon heuristic methods, as people usually do under similar circumstances.

Heuristic Methods

A heuristic method is a provisional and plausible procedure whose purpose is to discover the solution of a particular problem at hand. The use of heuristic methods by the human mathematician is quite well understood, at least in its less subtle forms. The reader is referred to the excellent two-volume treatise by Prof. G. Polya (1954) for a definitive treatment of heuristics and mathematical discovery. A machine that functioned under the full set of principles indicated by Polya would be a formidable problem-solver in mathematics, and would be well on the way toward satisfying Turing's requirements for a machine able to compete successfully in the "imitation game" (1950). Such a machine, however, lies in the indefinite future, for the art of instructing a computer is yet in too primitive a state to consider translating Polya into machine language. As a representative problem more in keeping with the present state of computer technology, we have selected the discovery of proofs for theorems in elementary euclidean plane geometry in the manner, let us say, of a high-school sophomore. This problem contains in relatively pure form the difficulties we must surmount in order to attain our stated goal. It must be emphasized that although plane geometry will yield to a decision algorithm, the proofs offered by the machine will not be of this nature. The methods developed will be no less valid for problem-solving in systems where no such decision algorithm exists.

Although we have narrowed the scope of our study to include only those machines that deal with formal systems, there is ample justification for such a restriction. First, the concept of a problem is now well defined, as is the concept of a solution for that problem. Second, our ultimate goal stands clearly before us; it is the design of an efficient theorem-prover in some undecidable system. And, finally, just as manipulation of numbers in arithmetic is the fundamental mode of operation in contemporary computers, manipulation of symbols in formal systems is likely to be the fundamental operating mode of the more sophisticated problem-solving computers of the future. It seems clear that while the problems of greatest concern to lay society will be, for the most part, not completely formalizable, they will have to be expressed in some sort of formal system before they can be dealt with by machine.[1]

Our problem, then, is a statement (or string) in some formal logistic system. A solution for the problem will be a sequence of statements, each of which comprises a string of symbols of the alphabet of the system. The last string of the solution will be the problem itself; the first will always be an axiom or previously established theorem of the system. Every other string will be immediately inferable from some set preceding it or will itself be an axiom or previously established theorem.[2] It is the task of the machine to choose from its stock of axioms and theorems the appropriate ones for the base of the proof, and to generate from these the remaining strings necessary to complete the proof.

The problem of theorem-proving is, in a sense, of a particularly simple nature. Once a sequence of expressions is found that passes the test for a proof of the theorem (such a test always exists), one may, so to speak, "close the book" on that problem, provided that no stipulations have been made concerning the elegance required of the proof. But, basing our estimate on the work of Newell, Shaw, and Simon (1957), any computer extant would require times of the order of a thousand years to prove a not uncommon ten-step geometry theorem by exhaustively developing sequences until one emerged that passed the test for a proof. What is clearly called for is a technique for generating sequences with a much higher *a priori* probability of being the solution to the problem than those generated by an exhaustion algorithm.

As did the *Logic Theorist* of Newell, Shaw, and Simon, the geometry machine relies upon the well-known analytic method to achieve this end. By working backward, the machine is assured that every sequence it considers does indeed terminate in the required theorem. This in itself, however, represents no striking improvement over exhaustion without additional heuristics, for the advantages of working backward are purchased at a steep price; each sequence generated, while terminating properly, is no longer guaranteed to be a proof of anything at all. Indeed, most of the strings generated in this way will be false! But it is here that the great

[1] For a critique of some attempts to formalize scientific, but nonmathematical theories, see Dunham, Fridshal, and Sward (1959).

[2] The machine will use the deduction theorem to get $\vdash\{\text{premises}\}\supset\{\text{conclusions}\}$ from $\{\text{premises}\}\vdash\{\text{conclusions}\}$.

power of the analytic method lies, for if one could find a way of making their falseness manifest, such sequences could be immediately rejected, allowing most of the deadwood to be pruned away from the highly branched problem-solving tree. The set of sequences generated under such a process would contain fewer members by many orders of magnitude by the time the search reached any depth, and the density of possible proofs for the theorem among them would be proportionately greater. It is here, too, that the geometry machine finds the additional theorem-proving power necessary for the complex formal system assigned to it; theorem-proving power that was not necessary, and therefore not sought for in the propositional calculus machine of Newell, Shaw, and Simon (Polya, 1954). Like the human mathematician, the geometry machine makes use of the potent heuristic properties of a diagram to help it distinguish the true from the false sequences. Although the diagram is useful to the machine in other ways as well, the single heuristic "Reject as false any statement that is not valid in the diagram" is sufficient to enable the machine to prove a large class of interesting theorems, some of which contain a certain trivial kind of construction.

Before examining the internal structure of the geometry machine in some detail, we remark on two fundamental, if obvious, principles that must guide the choice of heuristics for any problem-solving machine. A heuristic is, in a very real sense, a filter that is interposed between the solution generator and the solution evaluator for a given class of problems. The first requirement for such a filter is a consequence of the fact that its introduction into the system is never costless. It must, therefore, be sufficiently "nonporous" to result in a net gain in problem-solving efficiency. Secondly, a heuristic will generally remove from consideration a certain number of sequences that are quick and elegant solutions, if not indeed all solutions, to some potential problems within the domain of the problem-solving machine. The filter must, then, be carefully matched to that subclass of problems in the domain containing those that are considered "interesting," and are therefore likely to be posed to the machine. For a given class of heuristics, the balance between these essentially opposing requirements is largely a function of the organization and computing power of the machine, and can under certain rather easily attainable conditions be quite critical. In the case of the *Logic Theorist*[3] experiments with varying "strengths" of a particular heuristic (the similarity test) indicated that the optimum porosity of that heuristic varied markedly with the length of the

[3] The designers of the *Logic Theorist* were not unaware of this heuristic device. In a later version of that machine, they did, in fact, include some syntactic heuristics to reject false subgoals. To use a semantic interpretation of the propositional calculus (a truth table, for example) for this purpose would have reduced the *Logic Theorist* to triviality.

problem and the number of theorems already established in the theorem memory, a consequence of the limited storage capacity of the computer.

The Geometry Machine

With the object of our research program clearly determined, there were a number of specific alternatives to theorem-proving in Euclidean geometry that might have been adopted as a test problem; the evaluation of indefinite integrals, for example, or theorem-proving in the pure functional calculus. The decisive point in favor of geometry was the great heuristic value of the diagram. The creative scientist generally finds his most valuable insights into a problem by considering a model of the formal system in which the problem is couched. In the case of Euclidean geometry, the semantic interpretation is so useful that virtually no one would attempt the proof of a theorem in that system without first drawing a diagram; if not physically, then in the mind's eye. If a calculated effort is made to avoid spurious coincidences in the figure, one is usually safe in generalizing any statement in the formal system that correctly describes the diagram, with the notable exception of those statements concerning inequalities. Further geometry provides illustrative material in treatises and experiments in human problem-solving. It was felt that we could exchange valuable insights with behavioral scientists during the course of our research. In any event, elementary Euclidean geometry is comprehensible to every segment of the scientific community to which we should wish to communicate our results. Finally, it should not be a difficult task to generalize our machine to include the more interesting case of the non-Euclidean geometrics. A program of the same theorem-proving power as our Euclidean theorem-prover should be sufficient to prove a large class of non-obvious theorems in non-Euclidean geometry. A machine furnished with a non-Euclidean diagram (no more difficult to supply than the Euclidean one in suitable analytic form) encounters none of the assault on rationality experienced by a human mathematician searching from some heuristic insight into a theorem by considering a non-Euclidean diagram.

The formalization of geometry must be carried out within the framework of the lower functional calculus. Since we are interested in having the machine produce proofs comparable to those of a high-school student, we have preferred to construct a more or less *ad hoc* system following the scheme of most elementary texts, rather than to adopt as a primitive basis the fundamental axiomatization of Tarski, Hilbert, or Forder. No attempt has been made to provide a formalization that is either complete or non-redundant. If at some later time, the machine is able to prove one axiom from the others, that axiom will be discarded and we shall applaud the elegance displayed by our automaton. With regard to completeness, the

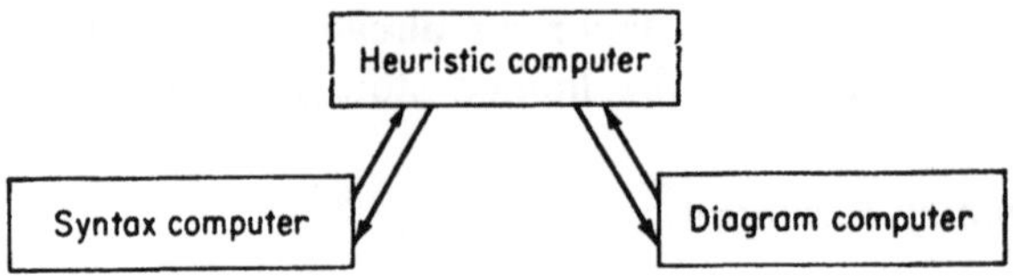

Figure 1.

machine is granted the same privileges enjoyed by the high-school student who is always assuming (*i.e.*, introducing as additional axioms) the truth of a plethora of "obviously self-evident" statements concerning, for example, the ordering properties of points on a line and the intersection properties of lines in a plane. Some of these statements are indeed independent of his original axioms, and must be introduced to complete the system. Most could be derived (but usually with some difficulty) from what he already has. There is nothing essentially wrong with this procedure of extracting assumptions from the model, provided that one is fully aware that this is being done (of course, this is rarely the case for the average student), and it simplifies the proof considerably without invalidating it. The geometry machine explicitly records its assumptions for a given proof. It could, if necessary, minimize the danger that it is proving a specific instance of a given theorem by drawing alternate diagrams to test the generality of its assumptions.

The geometry machine is in reality a particular state configuration of the IBM 704 electronic Data Processing Machine specified by a rather long and complex program written for the computer. Its organization falls naturally into three parts: a *syntax computer* and a *diagram computer* both embedded in an executive routine, which is a *heuristic computer*. The flow of control is indicated in Fig. 1.

Manipulation of the formal system is relegated to the syntax computer, which has within it the equivalent of most of the syntactic heuristics used by the *Logic Theorist*.[4] The diagram computer contains a coordinate representation of the theorem to be established together with a series of routines that produce a qualitative description of the diagram. It is important to point out that although the procedures of analytic geometry are used to generate the description, the only information transmitted to the heuristic computer (there is no direct link between the diagram and the formal system) is of the form: "Segment AB appears to be equal to segment CD in the diagram," or "Triangle ABC does not contain a right angle in the diagram." The behavior of the system would not be changed if the diagram computer were replaced by a device that could draw figures on paper and scan them.

[4] The process of *chaining* as defined by Newell et al. is under the control of the heuristic computer.

The major function of the heuristic computer for our first system, the subject of this report, is to compare strings generated by the syntax computer (working backward) with their interpretation in the diagram, rejecting those sequences that are not supported by the model. In addition to the above, the heuristic computer performs several other tasks. Among these are the organization of the proof-search process and the recognition of the syntactic symmetry of certain classes of strings. The latter function produces behavior equivalent to that of the human mathematician who, when A and B are syntactically symmetric, and both must be established, will merely prove A, and say "Similarly, B." It is an important feature, and is described in detail in an earlier report (Gelernter, 1959*a*). The procedures above are clearly independent of geometry; they are applicable to any formal system with its corresponding interpretation. The heuristic computer applies some additional semantic heuristics that are not indepedent of geometry. These may be "switched off" so that the behavior of the machine can be observed with and without specific geometry heuristics.

The character of the theorem-proving machine is determined largely by the heuristic computer. Modifications and improvements in the system (the introduction of learning processes, for example) will be made by modifying this part of the program.

Our first system does not "draw" its own initial figure, but is, instead, supplied with the diagram in the form of a list of possible coordinates for the points named in the theorem. This point list is accompanied by another list specifying the points joined by segments. Coordinates are chosen to

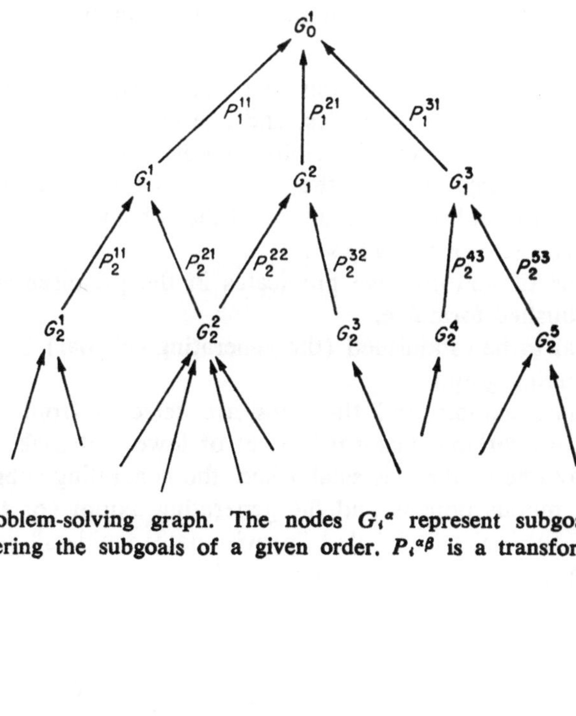

Figure 2. Problem-solving graph. The nodes G_i^α represent subgoals of order i, with α numbering the subgoals of a given order. $P_i^{\alpha\beta}$ is a transformation on G_i^α into G_{i-1}^β.

reflect the greatest possible generality in the figures. Later systems will construct their own interpretation of the premises, but since most problems for high school students are accompanied by a diagram, it was felt that we could dispense with this additional spate of programming at the current stage. When the machine is drawing its own figures, points will be chosen at random, subject to the constraints of the premises.

In working backward, the system generates a problem-solving graph, defined in the following way: Let G_0 be the formal statement to be established by the proof. It will be called the problem goal. If G_i is a formal statement with the property that G_{i-1} may be immediately inferred from G_i, then G_i is said to be a subgoal of order i for the problem. All G_j such that $j < i$ are higher subgoals than G_i, where G_0 is considered to be a subgoal of order zero. The problem-solving graph (Fig. 2) has as nodes the G_i, with each G_i joined to at least one G_{i-1} by a directed link. Each link represents a given transformation from G_i to G_{i-1}. The problem is solved when any G_i can be immediately inferred from the premises and axioms. If, as is generally the case in geometry, a given subgoal is a conjunction of statements, the graph splits at that point, and each parallel subgoal must be separately established. At any given time, the problem-solving graph is a complete representation of the current status of the proof-search process.

The organization of the heuristic computer (which is also the organization of the entire system) is displayed in greatly simplified form in Fig. 3. The diagram and syntax computers are accessible as subroutines to the heuristic computer. In operation, the machine executes the following processes (numbered below to correspond with like-numbered blocks in the flow chart).

1. The diagram is scanned to construct three lists, one containing every segment in the figure, one the angles, and one the triangles. Each element on a list is followed by a sublist describing that element.

2. The initial configuration of the system is set up, with the premises placed on a list of established formulas, and the conclusion on the problem-solving graph as a zero-order subgoal.

3. Definitions of nonprimitive predicates in the premises are added to the list of established formulae.

4. A subgoal to be established (the generating subgoal) is chosen from the problem-solving graph.

5. Appropriate axioms and theorems are selected from the theorem memory and, by working backward, a set of lower subgoals is generated such that if any one of these is established, the generating subgoal may be established by modus ponens and the generating axiom (or theorem). If the generating subgoal was labeled "provisionally fruitless" (see step 8), constructions are attempted (see below, p. 144).

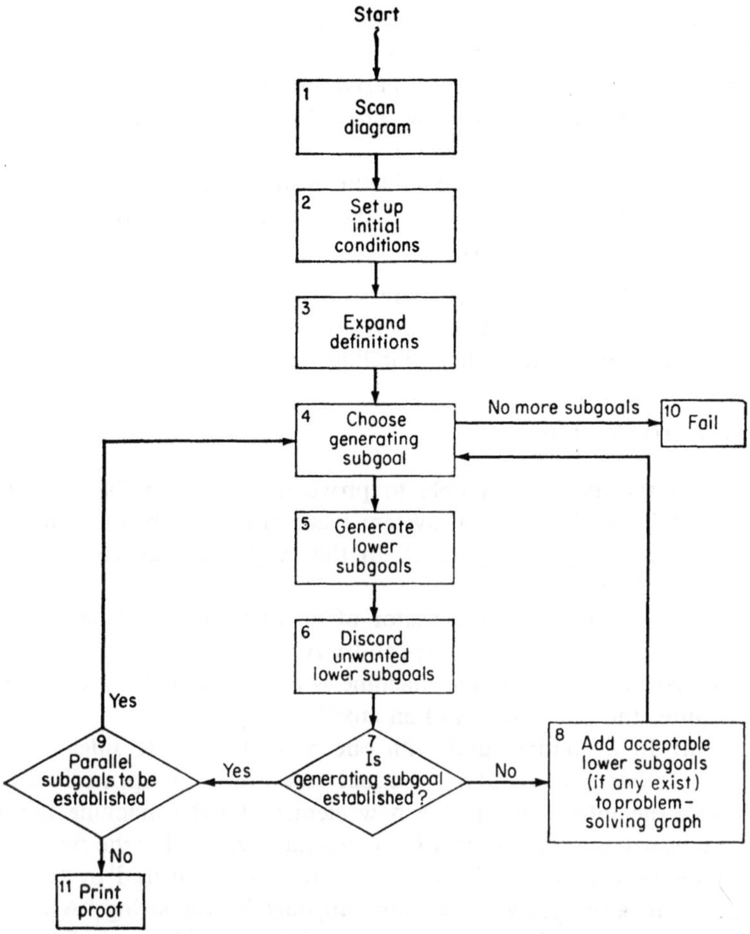

Figure 3. Simplified flow chart for the geometry-theorem proving machine.

6. Subgoals that are not valid in the diagram are rejected, as are those that appear as higher subgoals on the graph (or are syntactically symmetric to some higher subgoal).

7. If any lower subgoal is valid by virtue of its instance on the list of established formulas or if it may be assumed from the diagram, the generating subgoal is established; otherwise—

8. Acceptable nonredundant lower subgoals are added to the graph, and a new subgoal generator is chosen (4). If there are no acceptable lower subgoals and a construction is possible at this point, the generating subgoal is designated as provisionally fruitless. If a construction is not possible, or if the machine has tried and failed to find one, the generating subgoal is designated as fruitless.

9. If the generating subgoal is established, it is added to the list of established formulas, together with all of its higher consequences as determined by the graph. If there are no parallel subgoals remaining to be established, the machine reconstructs the proof from the problem-solving graph and prints it (11).

10. If at any time, every free subgoal on the graph is fruitless, the machine fails, providing it has not previously exhausted its available storage or the patience of the operator.

It is within blocks 4, 5 and 6, where subgoals are chosen, developed, and discarded, that the major heuristics reside. These subprograms represent, if you like, the seat of our artificial intelligence.

Some Early Results

The geometry machine is able to prove many of the theorems within the scope of its ad hoc formal system using the diagram only to indicate which subgoals are probably valid. In this way, the following theorem is proved in less than a minute.[5]

Theorem: A point on the bisector of an angle is equidistant from the sides of the angle (see Fig. 4 in Appendix A).

In less than five minutes, the machine is able to find the attached proof, which requires the construction of an auxiliary segment.

Theorem: In a quadrilateral with one pair of opposite sides equal and parallel, the other pair of sides are equal (see Fig. 5 in Appendix B).

Although the introduction of a new element by the machine is impressive, the construction in this proof is essentially trivial, for the new segment merely joins two already existing points. It was discovered by the following process. In attempting to develop subgoals for the string "AB = CD," the machine could find none that were valid in the diagram. The normal procedure at this point is to seek an alternative path on the problem-solving graph. But when none is available (as was the case here, since the offending string is a zero-order subgoal), the machine reexamines those of the previously rejected subgoals containing instances of predicates for which there was no representation in the diagram. The machine then considers for each one an augmented set of premises such that the

[5] In the proofs displayed herein, the nonobvious predicates have the following interpretations:

OPP-SIDE XYUV	Points X and Y are on opposite sides of the line through points U and V.
SAME-SIDE XYUV	Points X and Y are on the same side of the line through points U and V.
PRECEDES XYZ	Points X, Y, and Z are collinear in that order.
COLLINEAR XYZ	Points X, Y, and Z are collinear.

interpretation does contain a representation of the predicate. If the string is valid in the augmented system, and there exist theorems permitting the required additional premises to be derived from the original set, then the string becomes a subgoal in the augmented system. The added premises specify a construction in the diagram that is permitted by virtue of the theorems through which they were derived. Returning to our example, the subgoal "$\Delta ABD \cong \Delta CDB$" is generated for the string "AB = CD," but the required triangles are not represented in the diagram until the premise "Segment BD exists" is added. The axiom "Two distinct points determine a segment" justifies the construction.[6] The entire process is a variant of the major heuristic above, and is clearly independent of the particular formal system under consideration. Note, too, that the process is finite, since no new points are introduced into the predicates; the old ones are merely reconsidered.

Our second example illustrates one further point. Although it is clear in the diagram (Fig. 5) that the transversal BD makes alternate interior angles with sides BC and AD, this is a consequence of the theorem "Opposite vertices of a convex quadrilateral fall on opposite sides of the diagonal through the other vertices." That this is not true of a general quadrilateral becomes clear when one considers the outside diagonal of a reflex quadrilateral. A completely rigorous solution, then, requires that one prove the lemma above if it is not already available, and that one demonstrate that the quadrilateral ABCD can only be convex. Rather than do this, the machine makes the usual assumption that the diagonal forms alternate interior angles with the opposite sides of the quadrilateral. Unlike the usual high-school text, however, the assumption is made explicit in the proof.

The theorem-proving system described thus far is adequate for many problems of greater complexity than the ones cited above. However, with a linear increase in the number of individual points mentioned in the premises, the rate of growth of the problem-solving graph increases exponentially and the time required to explore the graph increases correspondingly. If the machine were able to select those among a given set of subgoals that were more likely to lead to a solution, much of the wasted search time could be eliminated. Two specific geometry heuristics have been introduced to enable the machine to do this. The first is a routine that recognizes certain of the subgoals that are usually established in just one step. Identities are in this category, for example, as are equalities between angles that are observed to be vertical angles in the diagram. Such subgoals

[6] Our ad hoc formal system requires that the segments joining the vertices of a triangle be specified, as well as the vertices themselves, to define the triangle. This is necessary in order to avoid the difficulties that would otherwise arise when the theorem names a large number of noncollinear points. If our formal system were a true point geometry, all such constructions would be implicit in the diagram.

are placed on a priority list and developed before any of the others are considered. The second specific heuristic is a routine that assigns a "distance" between each subgoal string and the set of premise strings in some vaguely defined formula space. After those on the priority list have been developed, the next subgoal chosen is that which is "closest" to the premise set in formula space.

It is instructive to examine the machine's behavior in proving complex theorems both with and without the expanded set of semantic heuristics. For the theorem "Two vertices of a triangle are equidistant from the median to the side determined by those vertices," the machine finds a proof in about eight minutes with the basic heuristics alone (see Fig. 6 in Appendix C). The expanded set of heuristics produces a proof in one minute. In addition, the second proof is quite short and to the point, while the first proof meanders blindly about the direct path to the goal before reaching it.

Reflecting the greater efficiency with which the machine attacked the problem in the second trial, only four circuits of the subgoal-generating loop were required compared with twenty-four circuits required without the extended heuristics. Twenty-one intermediate subgoals were generated, compared with sixty-one in the first case, and the problem-solving graph extended to a depth of only three levels, rather than twelve levels for the proof with basic heuristics alone.

For a particular case of a problem taken from a Brooklyn technical high school final examination in plane geometry a solution was found with the extended heuristics in less than five minutes. With the basic heuristics alone, the machine exhausted its working storage in half an hour without having completed the problem. On the other hand, there are problems for which the machine achieves no net gain by applying the additional heuristics. The theorem: "Diagonals of a parallelogram bisect one another" was proved in about three minutes in either mode. The proofs produced in each trial were equivalent, though not the same. A Brooklyn technical high school final examination supplied an example of an intermediate case, where the machine found identical proofs in both modes, but took almost three times as long with the basic heuristics alone (eight minutes, compared with three minutes with extended heuristics). We shall undoubtedly encounter cases for which the application of the extended set will result in a net loss of efficiency, although none has appeared yet in our limited tests.

Conclusion

It is well at this point in our discussion to reemphasize the fact that the object of this research has not been the design of a machine capable of proving theorems in Euclidean plane geometry, or even one able to prove theorems in some undecidable system such as number theory. We are, rather, interested in understanding the use of heuristic methods (or strategies) by machines for the solution of problems that would otherwise be inaccessible to them. Theorem-proving machines in themselves are objects of much interest to mathematicians and logicians, and important work at IBM is being done on this approach by Wang and by Gilmore. Wang (1960*a*) has written a program for the IBM 704 that is able to prove all theorems in propositional logic offered by Russell and Whitehead in the *Principia Mathematica,* whereas the Logic Theorist could master only about 38 of the 52 theorems appearing in chap. 2 of that volume. Also, the time required by the latter machine was far in excess of that used by the former. Newell, Shaw, and Simon, however, were interested in heuristic methods, whereas Wang, and also Gilmore, whose machine deals with the first order predicate calculus, are searching for algorithms, which, though less than a decision procedure, will produce "interesting" proofs within a reasonable amount of time. Both Wang and Gilmore find that for more complex formal systems, heuristics are required (they prefer the word "strategies") to make their algorithms sufficiently selective to produce, within acceptable bounds on space and time, proofs of any great interest.

The work of Wang and Gilmore is most relevant to a new branch of applied logic first characterized by Wang. He names this discipline "inferential analysis," and defines it to include "treatment of proofs as numerical analysis does calculations" (1960*a*). The results of inferential analysis are expected to "lead to mechanical checks of new mathematical results," and ultimately "lead to proofs of difficult new theorems by machine." The present author feels that inferential analysis is relevant, too, to the problem of intelligent behavior in machines. An automaton confronted with the real world, however, will certainly have to rely heavily on heuristics, for the unorthodox formal systems describing its environment will probably be far from amenable to the traditional methods of mathematical logic.

In conclusion, we should like to specify the course of this research for the immediate future. The machine described above is purely a problem-solving system. Except for the annexation of new theorems to the list of axioms, its structure is static. A sequence of practice problems given to the machine will not improve its performance unless a usable theorem is among them. Because it is incapable of developing its own structure, the machine will always be limited in the class of problems it can solve by the initial intent of the designer. It seems that the problem of designing a more general problem-solving machine will be enormously greater than that of designing one not so intelligent but with the capacity to learn.

An immediately obvious approach to the problem of introducing learning into the geometry machine is to allow the machine to adjust all of the parameters that determine its specific semantic heuristics, maximizing the predicted utility of those subgoals that prove to be useful in practice. The machine will thus improve the match of its heuristic filters to the class of problems considered interesting enough to be presented to it for solution. Of greater significance would be the introduction of routines enabling the machine to recognize recurring patterns in its proof-search procedure. Once discovered, such a pattern would enable the machine to construct its own heuristics designed to induce a repetition of the pattern in later proofs. For example, the machine might notice that certain classes of premise strings are regularly followed by the same first step in a proof. The heuristic derived from this pattern would search the premises for such strings and perform the first deduction before starting on the problem-solving graph. The difficult subject of abstract pattern recognition must be understood first, however, and the transformation of pattern to effective heuristic is by no means trivial. But whatever approach to learning is considered most worthwhile to explore, the geometry machine should serve as an excellent framework within which the explorations may be pursued.

Appendix A

Premises

Angle ABD equals angle DBC
Segment AD perpendicular segment AB
Segment DC perpendicular segment BC

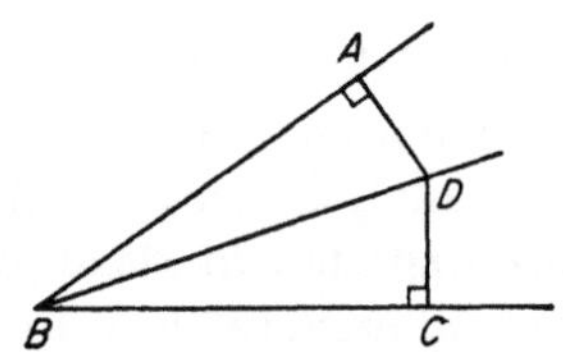

Figure 4.

Definition

Right angle DAB
Right angle DCB

Syntactic Symmetries

CA, BB, AC, DD

Goals

Segment AD equals segment CD

Solution

Angle ABD equals angle DBC
  Premise
Right angle DAB
  Definition of perpendicular
Right angle DCB
  Definition of perpendicular
Angle BAD equals angle BCD
  All right angles are equal

Segment DB
  Assumption based on diagram
Segment BD equals segment BD
  Identity
Triangle BCD
  Assumption based on diagram
Triangle BAD
  Assumption based on diagram
Triangle ADB congruent triangle CDB
  Side-angle-angle
Segment AD equals segment CD
  Corresponding elements of congruent triangles are equal

Total elapsed time = 0.3200 minute

Appendix B

Premises

Quad-lateral ABCD
Segment BC parallel segment AD
Segment BC equals segment AD

CA	BA	DA
DB	AB	CB
AC	DC	BC
BD	CD	AD

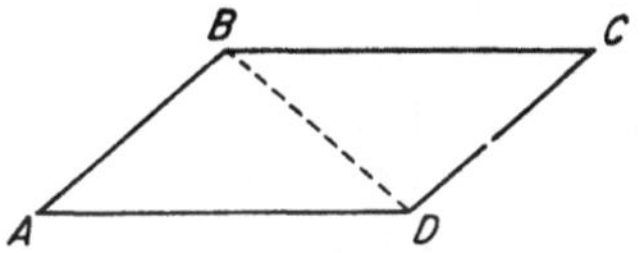

Figure 5.

Goals

Segment AB equals segment CD

I am stuck, elapsed time = 0.88 minute

Construct segment DB
Add premise segment DB
Restart problem

Solution

Segment BC parallel segment AD
  Premise
Opp-side CADB
  Assumption based on diagram
Segment DB
  Premise
Angle ADB equals angle CBD
  Alternate interior angles of parallel lines

Segment BC equals segment AD
  Premise
Segment BD equals segment DB
  Identity
Triangle CDB
  Assumption based on diagram
Triangle ABD
  Assumption based on diagram
Triangle ABD congruent triangle CDB
  Side-angle-side
Segment AB equals segment CD
  Corresponding elements of congruent triangles

Total elapsed time = 4.06 minutes

Appendix C

Premises

Triangle ABC
Precedes BMC
Segment BM equals segment MC
Precedes ADM
Precedes DME
Segment BD perpendicular segment AM
Segment CE perpendicular segment ME

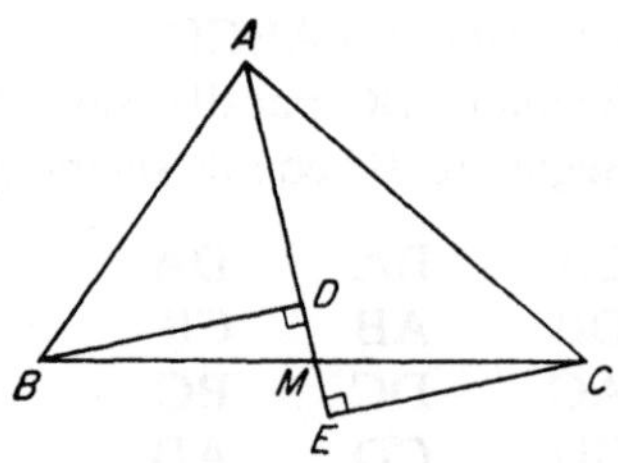

Figure 6.

Definition

Right angle MEC
Right angle BDM
Right angle BDA
Angle BDA equals angle BDM

No syntactic symmetries

Goals

Segment BD equals segment EC

Solution

Precedes BMC
  Premise
Segment EC
  Assumption based on diagram

Angle ECM equals angle BCE
Same angle
Precedes DME
Premise
Angle CED equals angle MEC
Same angle
Right angle BDM
Definition of perpendicular
Right angle MEC
Definition of perpendicular
Angle BDM equals angle CEM
Right angles are equal
Angle CED equals angle BDM
Angles equal to the same angle are equal
Same side MEDB
Assumption based on diagram
Same side DACE
Assumption based on diagram
Collinear EDM
Ordered collinearity implies collinearity
Precedes ADM
Premise
Precedes EDA
Combinatorial properties of ordered collinearity
Collinear EDA
Ordered collinearity implies collinearity
Angle BDE equals angle AEC
Different names for equal angles
Precedes EMA
Combinatorial properties of ordered collinearity
Angle CEM equals angle AEC
Same angle
Angle BDE equals angle MEC
Angles equal to the same angle are equal
Angle CED equals angle EDB
Angles equal to the same angle are equal
Opp side CBED
Assumption based on diagram
Segment ED
Assumption based on diagram
Segment EC parallel segment BD
Segments are parallel if alternate interior angles are equal
Opp side EDCB
Assumption based on diagram

Segment CB
  Assumption based on diagram
Angle BCE equals angle DBC
  Alternate interior angles of parallel lines
Angle ECM equals angle DBC
  Angles equal to the same angle are equal
Same side CMBD
  Assumption based on diagram
Same side MBEC
  Assumption based on diagram
Collinear CMB
  Ordered collinearity implies collinearity
Angle DBM equals angle BCE
  Different names for equal angles
Angle MBD equals angle MCE
  Angles equal to the same angle are equal
Angle DMB equals angle EMC
  Vertical angles
Segment BM equals segment MC
  Premise
Triangle BDM
  Assumption based on diagram
Triangle CEM
  Assumption based on diagram
Triangle BDM congruent triangle CEM
  Angle-side-angle
Segment BD equals segment EC
  Corresponding elements of congruent triangles

Total elapsed time = 8.08 minutes

WITH BASIC HEURISTICS

Solution

Precedes DME
  Premise
Precedes BMC
  Premise
Angle DMB equals angle EMC
  Vertical angles
Right angle BDM
  Definition of perpendicular
Right angle MEC
  Definition of perpendicular

Angle BDM equals angle CEM
  Right angles are equal
Segment BM equals segment MC
  Premise
Triangle CEM
  Assumption based on diagram
Triangle BDM
  Assumption based on diagram
Triangle BDM congruent triangle CEM
  Side-angle-angle
Segment BD equals segment EC
  Corresponding elements of congruent triangles

Total elapsed time = 1.06 minutes

WITH EXTENDED HEURISTICS

Commentary by the Author

H. Gelernter - Comments on the geometry papers.

The reprinting of these early papers on theorem-proving in Euclidean plane geometry provides me with a welcome opportunity to dispel a small mythology that has grown to the extent that it has supplanted the facts concerning this work. One aspect of the myth, which has begun to appear with some regularity in uncritical reviews of the history of artificial intelligence research, describes the Geometry Theorem Machine as if it were a software realization of the proposal first put forth by Marvin Minsky at the 1956 Dartmouth Summer Research Project on Artificial Intelligence. Minsky's ideas have been well documented[1], and they did in fact strongly influence my first approach to the problem[2]. However, even a cursory examination of the paper "Machine-generated Problem-solving Graphs" (vide seq.), which contains a detailed exposition of the programmed theorem-proving paradigm, will reveal that aside from the extremely important (though hardly theopneustic) notion that an interpretation of the formal system would provide crucially valuable guidance in searching the problem space, the geometry machine that emerged from my work had little in common with the one envisaged by Minsky. Indeed, it differed markedly from my own early conception of the ultimate form that the problem-solving system would assume. One has merely to compare the proposed interaction between syntax and semantics suggested by Minsky with that realized in the working program to conclude that the Geometry Machine was not the theorem-prover conceived by Minsky at Dartmouth. To my knowledge, the latter has never gone beyond hand simulation.

A second aspect of the myth concerns the "rediscovery" of the fourth century Pappus proof for the Isosceles Triangle Base Angles theorem, the so-called pons asinorum. This branch originated with an offhand and inaccurate comment by W.R. Ashby in a review of Feigenbaum and Feldman's book Computers and Thought[3], namely that "Gelernter's theorem-proving program has discovered a new proof of the pons asinorum that demands no construction....(one) which the greatest mathematicians of 2000 years have failed to notice....(and) which would have evoked the highest praise had it occurred." The remark was picked up by H.L. Dreyfus in his article "Alchemy and Artificial Intelligence"[4], where he observed, "The theorem sounds important, and the naive reader cannot help sharing Ashby's enthusiasm. A little research, however, reveals that the pons asinorum, or ass's bridge, is (an) elementary theorem.... Moreover, the first announcement of the "new" proof "discovered" by the machine is attributed to Pappus(A.D. 300). There is a striking disparity between Ashby's excitement and the antiquity and simplicity of this proof."

The simplicity of Dreyfus's excitement over Ashby's error (which was undoubtedly a consequence of Ashby's confusion of the Geometry Machine with Minsky's hand simulation) would probably have sufficed to bring the discussion to an end then and there had it not spilled over into the mass media. In 1966, the editors of The New Yorker magazine, profoundly relieved to learn that Dreyfus had vitiated the threat to the natural intelligensia posed by this line of research, popularized the issue with an item in their widely-read feature, "Talk of the Town"[5].

Dreyfus's misapprehension that a simple proof is somehow an unimportant or an unworthy one is unimportant and unworthy of further comment. However, had he done just a little more research (beyond the literature of the fourth century, say), he might have discovered that in no published paper nor at any formal presentation have I ever claimed that the Geometry Machine "discovered" the Pappus proof. I was well aware that the proof had emerged in Minsky's hand simulation of his own conception of a geometry paradigm, and I could not fail to know of its fourth century origin, since I had steeped myself in the history and lore of Euclid in preparing for my research. I knew, too, that in my formulation of the syntax of Euclid to reflect a high school student's approach to theorem-proving in geometry, the Pappus proof lay so close to the primitive basis and so comfortably below the threshold for combinatorial explosion that it would turn up in the most unsophisticated exploration of the problem space. After all, the combinatorics of three points is not overwhelming, even for so modest a force as the IBM 704. I therefore never commented in public on the Pappus proof until forced to do so by Ashby's error. And even then, my comments were confined to the remarks above, and to brief mention that the Pappus proof had indeed emerged as a lemma embedded in the problem-solving graph for a rather more complex problem. The lemma was not, as a matter of fact, used by the machine in constructing the final proof sequence for that problem. It was unearthed during our routine analysis of the output trace of the search process.

In retrospect, I would guess that the confusion has been a consequence of the fact that the last and definitive paper describing the geometry machine, namely "Machine-generated Problem-solving Graphs", has not heretofore been generally available, and because the title gives no hint of its relevance to the subject, it was rarely consulted in that connection. I thank the editors of this collection for providing the opportunity to correct the record.

References

1. See, for example, Minsky, M.L., "Heuristic Aspects of the Artificial Intelligence Problem", Lincoln Laboratory Report 34-55, Dec. 1956.

2. Gelernter, H.L. and Rochester, N. "Intelligent Behavior in Problem-solving Machines", IBM Jour. R. and D., 2 (1958).

3. Ashby, W.R., "Review of Feigenbaum's Computers and Thought", J. Nervous and Mental Diseases. This reference is taken from Dreyfus, H.L., "What Computers Can't Do" (Harper, 1972). I have not been able to track down the original to verify the quotation.

4. Dreyfus, H.L., "Alchemy and Artificial Intelligence", RAND Corporation Report (December 1965). The report was expanded into the book mentioned in reference 3.

5. The New Yorker, 11 June 1966, pp. 27-28.

1960

A Computing Procedure for Quantification Theory

M. Davis, H. Putnam

The hope that mathematical methods employed in the investigation of formal logic would lead to purely computational methods for obtaining mathematical theorems goes back to Leibniz and has been revived by Peano around the turn of the century and by Hilbert's school in the 1920's. Hilbert, noting that all of classical mathematics could be formalized within quantification theory, declared that the problem of finding an algorithm for determining whether or not a given formula of quantification theory is valid was the central problem of mathematical logic. And indeed, at one time it seemed as if investigations of this "decision" problem were on the verge of success. However, it was shown by Church and by Turing that such an algorithm can not exist. This result led to considerable pessimism regarding the possibility of using modern digital computers in deciding significant mathematical questions. However, recently there has been a revival of interest in the whole question. Specifically, it has been realized that while no *decision procedure* exists for quantification theory there are many proof procedures available—that is, uniform procedures which will ultimately locate a proof for any formula of quantification theory which is valid but which will usually involve seeking "forever" in the case of a formula which is not valid—and that some of these proof procedures could well turn out to be feasible for use with modern computing machinery.

Hao Wang [9] and P. C. Gilmore [3] have each produced working programs which employ proof procedures in quantification theory. Gilmore's program employs a form of a basic theorem of mathematical logic due to Herbrand, and Wang's makes use of a formulation of quantification theory related to those studied by Gentzen. However, both programs encounter decisive difficulties with any but the simplest formulas of quantification theory, in connection with methods of doing propositional calculus. Wang's program, because of its use of Gentzen-like methods, involves exponentiation on the total number of truth-functional connectives, whereas Gilmore's program, using normal forms, involves exponentiation on the number of clauses present. Both methods are superior in many cases to truth table methods which involve exponentiation on the

Received September, 1959. This research was supported by the United States Air Force through the Air Force Office of Scientific Research of the Air Research and Development Command, under Contract No. AF 49(638)-527. Reproduction in whole or in part is permitted for any purpose of the United States Government.

Article from: Journal of the Association for Computing Machinery, Volume 7, Number 3, July 1960.

total number of variables present, and represent important initial contributions, but both run into difficulty with some fairly simple examples.

In the present paper, a uniform proof procedure for quantification theory is given which is feasible for use with some rather complicated formulas and which does not ordinarily lead to exponentiation. The superiority of the present procedure over those previously available is indicated in part by the fact that a formula on which Gilmore's routine for the IBM 704 causes the machine to compute for 21 minutes without obtaining a result was worked successfully by *hand computation* using the present method in 30 minutes. Cf. §6, below.

It should be mentioned that, before it can be hoped to employ proof procedures for quantification theory in obtaining proofs of theorems belonging to "genuine" mathematics, finite axiomatizations, which are "short," must be obtained for various branches of mathematics. This last question will not be pursued further here; cf., however, Davis and Putnam [2], where one solution to this problem is given for elementary number theory.

1. *General Remarks*

We shall describe a computational procedure, or algorithm, which when applied to a logically valid formula written in the notation described below will terminate and yield a proof of the validity of that formula; for formulas which are not logically valid, the computation will continue indefinitely without giving a result.[1]

The symbols of which our formulas are constructed are divided into the classes: punctuation marks, logical symbols, (individual) variables, predicate symbols, and function symbols. The punctuation marks are:

$$, \quad (\quad)$$

The logical symbols are:

$$\sim \quad \& \quad \vee \quad \rightarrow \quad \leftrightarrow \quad E$$

We shall take as the variables the terms of the following infinite sequence:

$$x_1 \quad x_2 \quad x_3 \quad x_4 \quad \cdots$$

The predicate symbols will be the letters F, G, H, with or without subscripts, and the function symbols[2] will be the terms of the infinite sequence:

$$f_1 \quad f_2 \quad f_3 \quad f_4 \quad \cdots$$

Among all of the expressions (e.g. $\rightarrow \vee Fx_5E$) which can be formed using these symbols, we distinguish three classes: the *terms*, the *atomic formulas*, and the *well-formed formulas* (abbreviated *w.f.f.*).

[1] Since by results of Church and Turing the set of formulas involved is a recursively enumerable set which is not recursive (for terminology, and a proof of this fact, cf. [1] or [6]), this kind of algorithm is the best one can hope to obtain.

[2] We intend to use function symbols not only to stand for functions of one or more arguments but also for individuals. In the latter use they may be thought of as standing for functions of zero argument.

The notion *term* will be defined inductively:

(1) *The expressions* f_i *and* x_i *are terms for each* $i = 1, 2, 3, \cdots$.

(2) *If* $p_1, p_2, \cdots, p_n$ *are terms,*[3] *then so is* $f_i(p_1, p_2, \cdots, p_n)$, *and* $p_1, p_2, \cdots, p_n$ *are called the arguments of* f_i.

(3) *The terms consist exactly of the expressions generated by* (1) *and* (2).

Next:

The expression $p(p_1, p_2, \cdots, p_n)$ *is an atomic formula if* p *is a predicate symbol and* $p_1, p_2, \cdots, p_n$ *are terms.* $p_1, p_2, \cdots, p_n$ *are called the arguments of* p.

Finally:

(1) *An atomic formula is a w.f.f.*

(2) *If* R *is a w.f.f., then so are* $\sim R$, $(x_i)R$, *and* $(Ex_i)R$.

(3) *If* R *and* S *are w.f.f.'s, then so are* $(R \mathbin{\&} S)$, $(R \vee S)$, $(R \rightarrow S)$, *and* $(R \leftrightarrow S)$.

We introduce the following abbreviative conventions:

a stands for f_1.

f stands for f_2.

pq stands for $p(q)$ if p is a function symbol and q is a term.

$\bar{p}(p_1, p_2, \cdots, p_n)$ stands for $\sim p(p_1, \cdots, p_n)$, where p is a predicate symbol and $p_1, \cdots, p_n$ are terms.

An occurrence of x_i in a w.f.f. R is a *bound occurrence* if it is in a w.f. part of R of the form $(x_i)P$ or $(Ex_i)P$. An occurrence of x_i which is not bound is called a *free occurrence.* x_i is *free* in R if it has at least one free occurrence in R.

If $x_{i_1}, x_{i_2}, \cdots, x_{i_n}$ are all of the free variables in R, we sometimes write $R(x_{i_1}, x_{i_2}, \cdots, x_{i_n})$ for R. If $p_1, p_2, \cdots, p_n$ are terms, we write $R(p_1, p_2, \cdots, p_n)$ for the result of replacing x_{i_k} by p_k, $k = 1, 2, \cdots, n$, at all free occurrences of x_{i_k} in R.

Parentheses will be omitted wherever their omission can cause no confusion.

Our next step is to single out from the class of w.f.f.'s those which are *logically valid.* This can be done either by specifying axioms and rules of inference or by referring to "interpretations" of the w.f.f.'s of the system, and by a basic result due to Gödel[4] both of these procedures will lead to the same class of formulas. For our present purposes it is most convenient to use the latter formulation employing "interpretations."

An *interpretation* for a formula R consists of a nonempty set of elements U called a *universe* and an assignment of "values" to each function symbol and predicate symbol as follows:

To each function symbol which occurs in R with n arguments,[5] we assign a function of n variables ranging over U, whose values are in U.[6]

To each predicate symbol which occurs in R with n arguments, we assign a

[3] Note that the symbols p_1, p_2, etc. occur here as "syntactic variables." That is, they stand for expressions made up of our symbols.

[4] The Gödel completeness theorem. Cf. [5], [6], or [7].

[5] Thus, if $n = 0$, f_i is assigned an element of U.

[6] Note that if f_i occurs in R both with m arguments and with n arguments, $m \neq n$, it is assigned different functions in each case. In practice this will not happen in examples considered below. (However, two occurrences in R of f_i with the same number of arguments are, of course, to be assigned the same value.)

function of n variables ranging over U, whose values are the truth values, 0 (*falsehood*) and 1 (*truth*).[7]

Let $R(x_{n_1}, x_{n_2}, \cdots, x_{n_k})$ be a w.f.f. Then, given an interpretation of R over universe U, the value 0 or 1 will be assigned to $R(t_1, t_2, \cdots, t_k)$ for each ordered k-tuplet $(t_1, t_2, \cdots, t_k)$ of elements of U. This value may be obtained simply by interpreting 0 as falsehood and 1 as truth, using the usual truth tables for $\sim$ & $\vee$ $\rightarrow$ and $\leftrightarrow$, interpreting $(x_i)P(x_i)$ as 0 unless $P(t)$ has the value 1 for all t in U, and interpreting $(Ex_i)P(x_i)$ as 1 unless $P(t)$ has the value 0 for all t in U.

A w.f.f. R is called *valid* if under every interpretation and for every set of arguments from U, R is assigned the value 1.

A w.f.f. R is called *consistent* (or *satisfiable*) if there is some interpretation under which R is assigned the value 1, for *some choice* of *arguments from* U. R is *inconsistent* if it is not consistent.

We shall make use of the obvious fact that:

R is valid if and only if $\sim R$ is inconsistent.

That is, to "prove" R it suffices to "refute" $\sim R$, and indeed our *proof procedure* for validity will be couched in the form of a *refutation procedure.*

R is called *logically equivalent* to S if the *w.f.f.* $(R \leftrightarrow S)$ is valid.

A w.f.f. is called *quantifier-free* if it contains no occurrence of (x_i) or (Ex_i). A w.f.f. is a *prenex formula*, or in *prenex normal form*, if it begins with a sequence of quantifiers (x_i) and (Ex_i) in which no variable occurs more than once (called the *prefix*) and if the sequence is followed by a quantifier free w.f.f. (called the *matrix*). An example of a prenex formula is:

$$(x_1)(Ex_3)(x_7)(Ex_2)F(f(x_3), f_3(x_1, x_2), x_5)$$

S is called *a prenex normal form of R* if S is a prenex formula which is logically equivalent to R. There is a simple algorithm (cf. [5], [7]), for obtaining a prenex normal form of a given w.f.f. *Thus, for the purpose of our refutation procedure it suffices to consider prenex formulas.*

The *disjunction of* $R_1, \cdots, R_n$, $n \geqq 1$, is the w.f.f. $R_1 \vee R_2 \vee \cdots \vee R_n$; their *conjunction* is the w.f.f. $R_1 \& R_2 \& \cdots \& R_n$. A *literal* is a w.f.f. which is either an atomic formula or $\sim R$, where R is atomic. A *clause* is a disjunction $R_1 \vee R_2 \vee \cdots \vee R_n$ in which each R_i is a literal and in which no atomic formula occurs twice. (E.g., $F(x_1) \vee \bar{G}(x_2, x_3)$ is a clause, but $F(x_1) \vee \bar{F}(x_1)$ is not.)

A conjunction of clauses is said to be *a formula in conjunctive normal form.*

EXAMPLE: $(p \vee q \vee \bar{r}) \& (s \vee \bar{t})$ is a formula in conjunctive normal form if p, q, r, s, t are atomic formulas.

If a w.f.f. A is in conjunctive normal form and A is logically equivalent to B, then A is called a *conjunctive normal form of B.*

EXAMPLE: $(p \vee \bar{q}) \& (q \vee \bar{p})$ is a conjunctive normal form of $p \leftrightarrow q$ if p and q are any atomic formulas.

For further discussion of conjunctive normal form the reader may consult Hilbert and Ackermann [5]. In particular, there is a simple algorithm by which

[7] The comment in footnote 6 regarding function symbols applies also to predicate symbols.

a conjunctive normal form is obtainable for any quantifier-free formula which is not valid; if the formula is valid the same algorithm will establish that fact. (Cf. [5].) Hence, we may assume that the w.f.f. which is offered for refutation is a prenex formula whose matrix is in conjunctive normal form. Later we shall see why this is a useful and practical assumption.

2. *Replacement of Existential Quantifiers by Function Symbols*

The refutation algorithm to be presented will exploit the following idea (which, in essence, goes back to Lowenheim): that existential quantifiers in a prenex formula can be replaced by function symbols without affecting consistency. The notion may be clarified by an example: Suppose the given prenex formula is

$$(x_1)(Ex_2)(Ex_3)(x_4)(Ex_5)R(x_1, x_2, x_3, x_4, x_5), \quad \text{(i)}$$

where the matrix $R(x_1, x_2, x_3, x_4, x_5)$ is supposed to be quantifier-free and to contain ony the free variables indicated. Then the formula (i) is consistent only if the formula

$$(x_1)(x_4)R(x_1, f_2(x_1), f_3(x_1), x_4, f_5(x_1, x_4)) \quad \text{(ii)}$$

is, where f_2 and f_3 are one-place function symbols and f_5 is a two-place function symbol. To verify this, observe that (ii) logically implies (i), so if (ii) is consistent, so is (i). On the other hand, if (i) is true in some universe U (under some interpretation of the predicate letters in R), then there are functions[8] f_2, f_3 and f_5 over U such that (ii) is true in U under the same interpretation of the predicate letters in R. Thus if (i) is consistent, so is (ii).

Throughout the present paper, accordingly, the instruction "replace the existential quantifiers in F by function symbols" (where F is a prenex formula) will have the following meaning: Let the variables in the prefix of F (in order of occurrence) be $x_1, x_2, \cdots, x_N$. Let the *existentially* quantified variables in the prefix be $x_{i_1}, x_{i_2}, \cdots, x_{i_M}$. Then, (1) the quantifier (Ex_{i_j}) (for $j = 1, 2, \cdots, M$) is to be deleted from the prefix, and (2) each occurrence of x_{i_j} in the matrix is to be replaced by an occurrence of the term $f_{i_j}(x_{q_1}, x_{q_2}, \cdots, x_{q_p}$ where $(x_{q_1}), (x_{q_2}), \cdots, (x_{q_p})$ are all the universal quantifiers that precede (Ex_{i_j}) in the prefix of F.

In the above example, following the instruction "replace the existential quantifiers in (i) by function symbols," as just explained, would lead to formula (ii). Finally, (recalling that 0-place function symbols are interpreted simply as individual constants) replacing the existential quantifiers by function symbols in

$$(Ex_1)(x_2)(Ex_3)(x_4)M(x_1, x_2, x_3, x_4)$$

[8] This agreement tacitly employs a nonconstructive principle known as the Axiom of Choice. Alternatively, one can use the theorem that if (*i*) is consistent then (*i*) has a true interpretation in some denumerable universe U (Skolem-Löwenheim theorem; cf. [7], pp. 253–260), and then explicitly define the functions f_2, f_3 and f_5 in terms of some fixed ordering of the elements of U.

would lead to the formula

$$(x_2)(x_4)M(f_1, x_2, f_3(x_2), x_4).$$

3. *The Sequence of Quantifier-Free Lines*

The way our whole refutation-algorithm will "look" may now be indicated in a general way. Suppose the given formula is

$$(x_1)(Ex_2)(x_3)R(x_1, x_2, x_3),$$

where R is quantifier-free and contains only the indicated variables. Then the first step will be to replace the existential quantifier(s) by function symbols, which will lead in this case to

$$(x_1)(x_3)R(x_1, f(x_1), x_3)$$

(recall that "f" abbreviates f_2 and that "a" abbreviates f_1). Next we will form a sequence of *quantifier-free lines* as follow (certain parentheses are omitted for brevity):

$R(a, fa, a)$	(Observe that the variables x_1, x_3 are replaced in all possible ways with terms from the sequence $a, fa, ffa, \cdots$.)	(1)
$R(a, fa, fa)$		
$R(fa, ffa, a)$		
$R(fa, ffa, fa)$		
$R(a, fa, ffa)$		
$\vdots$		

As these quantifier-free lines are generated, we will test the conjunction of the first n lines (for $n = 1, 2, 3, \cdots$) for consistency (by methods described in the next section). If the conjunction of the first n lines is inconsistent, for any n, then the formula $(x_1)(x_3)R(x_1, f(x_1), x_3)$ is inconsistent (since it implies all of the quantifier free lines), and hence the given formula was inconsistent. On the other hand, if the conjunctive of the first n lines is consistent for every n, then the algorithm never terminates, and the given formula was consistent.[9]

We now state the general *rule for forming the sequence of quantifier-free lines.* Let F be the given formula after the existential quantifiers have been replaced by function symbols. Let $f_{i_1}, \cdots, f_{i_M}$ be all the function symbols in F, and let f_{i_k} be an n_k-place function symbol (for $k = 1, 2, \cdots, M$). Let D be the following set: the smallest set containing the individual constant a and having the property that whenever it contains $t_1, \cdots, t_{n_k}$ then it contains the expression $f_{i_k}(t_1, \cdots, t_{n_k})$, for $k = 1, 2, \cdots, M$. Let L be the number of *universal* quantifiers in F, and let S be the sequence of all ordered L-tuplets of members of D,

[9] For the proof of this statement see [7], pp. 253–260. The key point in the proof is that an infinite set of quantifier-free formulas is inconsistent if and only if some *finite* subset is inconsistent.

in lexicographic order.[10] Then the nth quantifier free line (for $n = 1, 2, 3, \cdots$) is the result of substituting[11] t_{n1} for the first universally quantified variable (in F), t_{n2} for the second universally quantified variable, $\cdots$, t_{nL} for the Lth universally quantified variable, where $t_{n1}; \cdots; t_{nL}$ is the nth L-tuplet in the sequence S.

REMARKS:

(A) One may, if one desires, abbreviate the expressions in the set D by numbers according to some convenient scheme. If one adopted this policy, the quantifier-free lines (1) above might look like this:

$$\begin{array}{c} R(1,2,1) \\ R(1,2,2) \\ R(2,3,1) \\ R(2,3,2) \\ R(1,2,3) \\ \vdots \end{array} \tag{2}$$

Such a scheme of numerical abbreviation is extremely worthwhile from the standpoint of *hand* computation (because it cuts down the length of the formulas). On the other hand, there may be little or no advantage to adopting such a scheme if the algorithm is going to be programmed for a computer.

(B) Instead of testing the conjunction of the first n quantifier-free lines for consistency when $n = 1, 2, 3, \cdots$, one might test "intermittently," e.g., when $n = 10, 20, 30, \cdots$. The relative advantages and disadvantages of such "intermittent" applications of the testing for consistency should be investigated if the algorithm we are describing is to be actually programmed for a computer.

4. *Feasible Methods in the Propositional Calculus*

The idea of a refutation-algorithm, of the sort described in general terms in the preceding section, is not new. In essence, it goes back to Herbrand[12], and formulations of the kind we have given (based on the idea of generating a sequence of quantifier-free lines, and then testing the conjunction of the first n lines for consistency as $n = 1, 2, 3, \cdots$) have been previously given by Quine[13], Gilmore[14], and others. However, the crucial difficulty, to which little attention ap-

[10] For the purposes of defining "lexicographic order," subscripts are to be thought of as if they were written on the line (e.g., $f_{12}(a)$ is to be treated as if it were "$f12(a)$"). Then our alphabet consists of the symbols: ( ) f 0 1 2 3 4 5 6 7 8 9 , ; (the latter symbol being used to separate the members of an L-tuplet thus: "$f2(f1);f6(f1,f2(f1))$"), and the "lexicographic ordering" of the L-tuplets is the ordering in which they are arranged like words in a dictionary.

[11] As indicated in the example, a universal quantifier is deleted whenever something is substituted for the variable it contains. This sort of "substitution" is technically known as *universal instantiation* (cf. [7], p. 147).

[12] Cf. [4].

[13] Cf. [8].

[14] Cf. [3].

pears to have been given in this connection, is that of finding a *feasible* technique for testing the conjunction of the first n lines for consistency when n is large. Quine's "uniform proof procedure" is described with hand computation in mind, and thus Quine limits himself to truth-tables as a method in the propositional calculus. However, the number of lines in a truth table, when k propositional variables are involved, is 2^k and so truth-tables quickly become unfeasible for our purposes. Gilmore's procedure is to put the conjunction of the first n lines into disjunctive normal form, but this too leads to exponentiation (on the number of clauses in the matrix of the given formula), and so this method too is unfeasible in general (although fortuitous cancellations may keep the formulas involved down to manageable length in special cases). Still another procedure has been proposed by Wang in [9]. Wang's procedure is less easy to compare with ours because it does not use prenex normal form; however his routine employs a "Gentzentype" formal system in which proofs have a "tree" structure[15] (as opposed to the usual "linear" structure) with "branching" possible at any line. As far as the propositional calculus is concerned, the difficulty with Wang's technique is that the number of branches tends to increase exponentially with the number of logical connectives involved. Thus, none of the three methods just described—truth-tables, disjunctive normal forms, or Gentzen-type systems—is satisfactory as a method for testing the conjunction of the first n lines (in our sequence of quantifier-free lines) for truth-functional consistency when n becomes at all large (e.g., $n > 10$).

By contrast, the method to be described always terminates in at most $2(R-1)$ steps, where R is the number of variables (i.e., the number of steps increases *linearly*, not exponentially, in the number of variables). Moreover, the process will rarely lead to formulas which are much more complicated than those with which one started in examples of the sort likely to arise in practice. Actually it has been found possible to work quite complicated formulas by this method even by *hand* computation.

The method to be described depends on putting the conjunction of the first n lines into *conjunctive* normal form. Since putting a formula into conjunctive normal form does not of itself enable one to tell whether or not the formula is consistent, it is necessary to make one or two remarks explaining our choice of this normal form. Briefly, the reasons are as follows: although normal forms may in certain cases be used as decision-methods (e.g., putting a formula into *disjunctive* normal form automatically reveals whether or not the formula is inconsistent[16]), they have also another function, as the term "normal form" indicates, namely, their use serves to regularize formulas and to cut down structural complexity. For instance, every formula F in conjunctive normal form has the structure A & B & R where A is the conjunction of the clauses containing a given atomic formula (say, p), B is the conjunction of the clauses containing the negation of that formula (say, $\bar{p}$), and R is the conjunction of the remaining clauses. Moreover, it can be shown that F is inconsistent if and only if A' & R

[15] For an explanation of "tree structure" cf. [6], pp. 106–107.

[16] Cf. [7], pp. 52–59.

and B' & R are both inconsistent, where A' is obtained from A by deleting occurrences of p, and B' is obtained from B by deleting occurrences of $\bar{p}$. Such regularities are hardly to be hoped for in the case of arbitrary formulas not in normal form.

Our problem, as indicated above, is how to deal with cases in which the number of quantifier-free lines is too large to make it feasible to put the whole system of lines into disjunctive normal form. In such cases there is one normal form we can use: namely, the *conjunctive normal form.*

That the conjunctive normal form can be employed follows from the remark that to put a whole system of formulas into conjunctive normal form we have only to put the individual formulas into conjunctive normal form. Thus, even if a system has hundreds or thousands of formulas, it can be put into conjunctive normal form "piece by piece", without any "multiplying out." This is a feasible (if laborious) task even for *hand* computation: thus no specialization is introduced here beyond supposing that the individual formulas in the system are "manageable" (i.e., short enough to be put into conjunctive normal form by hand) and that the whole system can be written down by a human being.

In the case of our "sequences of quantifier-free lines" (generated according to the rule in the preceding section), the situation is even more pleasant than in the general case of testing some "big" system of formulas for consistency: namely, it suffices to put the *matrix* of the given formula (after the existential quantifiers have been replaced by function symbols) into conjunctive normal form, and then the "quantifier-free lines" will be automatically generated in conjunctive normal form!

In stating our method for testing the conjunction of the first n "quantifier-free lines" for consistency, we shall assume that the matrix of the given formula was in conjunctive normal form (so that the conjunction of the first n lines will likewise automatically be in conjunctive normal form), and we shall speak of the entire conjunction as a single formula F.

Our method consists of the following three rules, in which p, q, r, s are atomic formulas:

I. *Rule for the Elimination of One-Literal Clauses:*

(a) If a formula F in conjunctive normal form contains an atomic formula p as a one-literal clause and also contains $\bar{p}$ as a one-literal clause, then F may be replaced by 0. (I.e., F is self-contradictory).

(b) If case (a) does not apply, and if an atomic formula p appears as a clause in a formula F in conjunctive normal form, then one may modify F by striking out all clauses that contain p affirmatively[17] and deleting all occurrences of $\bar{p}$ from the remaining clauses, thus obtaining a formula F' which is inconsistent if and only if F is.

(c) If case (a) does not apply and $\bar{p}$ appears as a clause in a formula F in conjunctive normal form, then one may modify F by striking out all clauses that con-

[17] An occurrence of p without a negation bar is called an *affirmative* occurrence; one with a negation bar is called a *negative* occurrence.

tain $\bar{p}$ and deleting all occurrences of p from the remaining clauses, again obtaining a formula F' which is inconsistent if and only if F is.

(d) In cases (b) and (c), if F' is empty, then F is consistent.

II. *Affirmative-Negative Rule.* If an atomic formula p occurs in a formula F in conjunctive normal form only affirmatively, or if p occurs only negatively, then all clauses which contain p may be deleted. The resulting formula F' is inconsistent if and only if F is. (If F' is empty, then F is consistent).

III. *Rule for Eliminating Atomic Formulas.* Let the given formula be put into the form $(A \vee p)$ & $(B \vee \bar{p})$ & R where A, B, and R are free of p. (This can be done simply by grouping together the clauses containing p and "factoring out" occurrences of p to obtain A, grouping the clauses containing $\bar{p}$ and "factoring out" $\bar{p}$ to obtain B, and grouping the remaining clauses to obtain R.) Then F is inconsistent if and only if $(A \vee B)$ & R is inconsistent.

Justification. For Rule I: The justification of case (a) of the rule is obvious. For case (b), let the formula F be p & A. Then F is clearly false when $p = 0$; hence F is inconsistent, provided F is false when $p = 1$. Substituting 1 for p in F and simplifying has the following effect: All clauses that contain p affirmatively reduce to 1 and may be deleted. All clauses that contain p negatively reduce to 0 (in case the whole clause was $\bar{p}$) or to $0 \vee B$, where B is the remainder of the clause. But there cannot be any clauses which consist of *just* $\bar{p}$ (otherwise case (a) would apply); and $0 \vee B = B$. Hence the effect of substituting 1 for p in F and simplifying is to strike out all the clauses that contain p affirmatively and delete all occurrences of $\bar{p}$ from the remaining clauses. Thus

$$F' \text{ is inconsistent} \leftrightarrow F \text{ is false whenever } p = 1$$
$$\leftrightarrow F \text{ is inconsistent.}$$

Case (c) is symmetrical to case (b). Case (d) reduces to the observation that if p occurs in every clause, then $F = 1$ when $p = 1$.

For Rule II: Let p occur in F only affirmatively, and let F be A & R where A is the conjunction of all the clauses containing p. Then if F is inconsistent, F is false when $p = 1$. But when $p = 1$ we have $A = 1$, and therefore $(A \& R) \leftrightarrow R$ when $p = 1$. Hence, if F is inconsistent, so is R. But, since $(A \& R) \rightarrow R$, if R is inconsistent, so is $(A \& R)$. (If R is empty, $F = 1$ when $p = 1$, and therefore F is consistent.) The argument is similar when p occurs only negatively, using $p = 0$ instead of $p = 1$.

For Rule III: F is inconsistent if and only if F is false when $p = 0$ and false when $p = 1$. But in the first case, F reduces to $(A \& R)$ and in the second case to $(B \& R)$. So F is inconsistent if and only if $(A \& R)$ and $(B \& R)$ are both inconsistent, and $(A \& R) \vee (B \& R) \leftrightarrow (A \vee B) \& R$.

Examples. (1) Consider the formula:

$$(p \vee q \vee \bar{r}) \& (p \vee \bar{q}) \& \bar{p} \& r$$

There are two one-literal clauses. Elimination of these leads immediately to $q \& \bar{q} = 0$.

(2) Consider the formula

$$(p \vee q) \,\&\, \bar{q} \,\&\, (\bar{p} \vee q \vee \bar{r}).$$

Elimination of the one-literal clause yields $p \,\&\, (\bar{p} \vee \bar{r})$, which in turn yields $\bar{r}$. By Rule I or Rule II, this formula is consistent.

(3) The formula

$$(p \vee \bar{q}) \,\&\, (\bar{p} \vee q) \,\&\, (q \vee \bar{r}) \,\&\, (\bar{q} \vee \bar{r})$$

contains r only negatively. By Rule II, it is inconsistent if and only if $(p \vee \bar{q}) \,\&\, (\bar{p} \vee q)$ is. By Rule III (eliminating p), this is inconsistent if and only if $q \vee \bar{q}$ is. But $q \vee \bar{q} = 1$, so this is consistent.

(4) The following example is worked using only Rule III. (Note that it is necessary to put the formula back into conjunctive normal form after each elimination).

$(p \vee r) \,\&\, (p \vee \bar{s}) \,\&\, (\bar{p} \vee s) \,\&\, (\bar{p} \vee \bar{r}) \,\&\, (s \vee \bar{r}) \,\&\, (\bar{s} \vee r)$
$[(r \,\&\, \bar{s}) \vee p] \,\&\, [(\bar{r} \,\&\, s) \vee \bar{p}] \,\&\, (s \vee \bar{r}) \,\&\, (\bar{s} \vee r)$
$(s \vee r) \,\&\, (\bar{s} \vee \bar{r}) \,\&\, (s \vee \bar{r}) \,\&\, (\bar{s} \vee r)$ (p eliminated)
$[(s \,\&\, \bar{s}) \vee r] \,\&\, [(s \,\&\, \bar{s}) \vee \bar{r}]$
$s \,\&\, \bar{s}$ (r eliminated)

To complete the refutation, it suffices to note that $s \,\&\, \bar{s}$ is inconsistent by Rule I.

5. *The Complete Algorithm*

In the preceding sections we have stated the various rules which make up our refutation-algorithm. It remains to "put the pieces together." The following is the complete sequence of steps to be followed in employing the algorithm (we adopt the policy of alluding to rules which have been completely stated in earlier sections of this paper, rather than restating them in full; also we assume the given formula to be prenex, and to have a matrix in conjunctive normal form):

Step 1. Generate one more quantifier-free line (if none have previously been generated, this means: generate a first quantifier-free line). Then test the conjunction of all the so-far-generated quantifier-free lines for consistency by the following steps:

Step 2. Apply the rule for eliminating one-literal clauses (Rule I) to the conjunction obtained at step 1 if it contains any one-literal clauses, and continue applying this rule until the resulting formula has no one-literal clauses. If the empty formula results, the conjunction obtained at step 1 was consistent. If a formula results which is inconsistent by Rule I, the conjunction obtained at step 1 was inconsistent. If a nonempty formula with no one-literal clauses results, go on to—

Step 3. Apply the affirmative-negative rule (Rule II) to the formula obtained at step 2 (or to the conjunction obtained at step 1, if step 2 did not apply) unless that formula had the property that every atomic formula that occurred in it occurred both affirmatively and negatively. Then go back to step 2 if the result contains any one-literal clauses. Otherwise, repeat step 3 if the result contained some literal which occurred only affirmatively or only negatively If the result is the empty formula, the conjunction obtained at step 1 was consistent. If a nonempty formula with no one-literal clauses and with the property that every atomic formula that occurs in it occurs both affirmatively and negatively results, go on to—

Step 4. Using Rule III, eliminate the first atomic formula from the first clause of minimal length in the formula that has resulted from the preceding steps (or from the conjunction obtained at step 1, if steps 2 and 3 did not apply). If the resulting formula cannot be put back into conjunctive normal form (because every clause would contain an atomic formula both negated and not-negated), the conjunction obtained at step 1 was consistent. Otherwise, put the resulting formula back into conjunctive normal form, and go back to step 2.

Continue in this way (i.e., going through the "cycle" steps 2–3–4) until either (a) it has been decided at some application of steps 2, 3, or 4 that the conjunction obtained at step 1 was consistent; or (b) it has been decided that the conjunction obtained at step 1 was inconsistent. (This can only happen at an application of step 2.)

If it is decided that the conjunction obtained at step 1 was inconsistent, then the algorithm terminates, and the given formula was inconsistent (i.e., "refutation" has been accomplished). If it is decided that the conjunction obtained at the preceding application of step 1 was consistent, go back to step 1, and continue.

6. *An Example*

P. C. Gilmore[18] tested his refutation-procedure on a number of formulas, including the following one:

$$(Ex)(Ey)(z)\{(F(x, y) \rightarrow (F(y, z) \mathbin{\&} F(z, z))) \mathbin{\&} ((F(x, y) \mathbin{\&} G(x, y)) \rightarrow (G(x, z) \mathbin{\&} G(z, z)))\} \tag{1}$$

We have selected this example for purposes of comparison because (a) it is not so long as to make hand computation immediately impractical (e.g., it is already in prenex form, and the matrix can easily be put into conjunctive normal form); yet (b) Gilmore's procedure did *not* lead to a refutation although an IBM 704 was employed for 21 minutes.

Our procedure, on the other hand, *did* lead to a refutation in under a half-hour of *hand* computation! For the purposes of hand computation, one modification was made in the algorithm: instead of testing the conjunction of the first n-lines for consistency when $n = 1, 2, 3, \cdots$, we adopted the scheme of "intermittent" testing alluded to at the end of section 3, and tested at $n = 10, 20, 30$. The conjunction of the first n lines was *consistent* when $n = 10$ and $n = 20$ and inconsistent when $n = 30$. Inspection later revealed that the smallest n for which the conjunction of the first n lines was inconsistent was $n = 25$. That the difficulty

[18] Cf. [3].

with Gilmore's procedure lies in the propositional calculus method employed is confirmed by the fact that in the 21 minutes the IBM 704 was running, only 7 "substitutions" were made; only what amounts to 7 quantifier-free lines were generated. *We adopt the abbreviation, here and below, of omitting the symbol* $\vee$, writing, e.g.,

$$\bar{F}(y, z)\bar{F}(z, z)G(x, y) \quad \text{for} \quad (\bar{F}(y, z) \vee \bar{F}(z, z) \vee G(x, y)).$$

The following is the negation of formula (1) with matrix in conjunctive normal form:

$$(x)(y)(Ez)(F(x, y) \ \& \ \bar{F}(y, z)\bar{F}(z, z)G(x, y) \\ \& \ \bar{F}(y, z)\bar{F}(z, z)\bar{G}(x, z)\bar{G}(z, z)) \tag{2}$$

Replacing the existential quantifier by a function symbol gives:

$$(x)(y)[F(x, y) \ \& \ \bar{F}(y, f(x, y))\bar{F}(f(x, y), f(x, y))G(x, y) \\ \& \ \bar{F}(y, f(x, y))\bar{F}(f(x, y), f(x, y))\bar{G}(x, f(x, y))\bar{G}(f(x, y), f(x, y))]. \tag{3}$$

In writing the first 25 quantifier-free lines generated we have used numbers up to 25 instead of "$f(a, a)$", "$f(f(a, a), a)$", etc, in order to make the formulas shorter and the over-all pattern more clear. Also we have omitted parentheses between predicate symbols and their arguments. The lines are as follows:

Quantifier-Free Lines:

1.	Fa, a & $\bar{F}a, 1$	$\bar{F}1, 1$	Ga, a & $\bar{F}a, 1$	$\bar{F}1, 1$	$\bar{G}a, 1$	$\bar{G}1, 1$
2.	$F1, a$ & $\bar{F}a, 2$	$\bar{F}2, 2$	$G1, a$ & $\bar{F}a, 2$	$\bar{F}2, 2$	$\bar{G}1, 2$	$\bar{G}2, 2$
3.	$F1, 1$ & $\bar{F}1, 3$	$\bar{F}3, 3$	$G1, 1$ & $\bar{F}1, 3$	$\bar{F}3, 3$	$\bar{G}1, 3$	$\bar{G}3, 3$
4.	$Fa, 1$ & $\bar{F}1, 4$	$\bar{F}4, 4$	$Ga, 1$ & $\bar{F}1, 4$	$\bar{F}4, 4$	$\bar{G}a, 4$	$\bar{G}4, 4$
5.	$Fa, 2$ & $\bar{F}2, 5$	$\bar{F}5, 5$	$Ga, 2$ & $\bar{F}2, 5$	$\bar{F}5, 5$	$\bar{G}a, 5$	$\bar{G}5, 5$
6.	$Fa, 3$ & $\bar{F}3, 6$	$\bar{F}6, 6$	$Ga, 3$ & $\bar{F}3, 6$	$\bar{F}6, 6$	$\bar{G}a, 6$	$\bar{G}6, 6$
7.	$Fa, 4$ & $\bar{F}4, 7$	$\bar{F}7, 7$	$Ga, 4$ & $\bar{F}4, 7$	$\bar{F}7, 7$	$\bar{G}a, 7$	$\bar{G}7, 7$
8.	$F1, 2$ & $\bar{F}2, 8$	$\bar{F}8, 8$	$G1, 2$ & $\bar{F}2, 8$	$\bar{F}8, 8$	$\bar{G}1, 8$	$\bar{G}8, 8$
9.	$F1, 3$ & $\bar{F}3, 9$	$\bar{F}9, 9$	$G1, 3$ & $\bar{F}3, 9$	$\bar{F}9, 9$	$\bar{G}1, 9$	$\bar{G}9, 9$
10.	$F1, 4$ & $\bar{F}4, 10$	$\bar{F}10, 10$	$G1, 4$ & $\bar{F}4, 10$	$\bar{F}10, 10$	$\bar{G}1, 10$	$\bar{G}10, 10$
11.	$F2, a$ & $\bar{F}a, 11$	$\bar{F}11, 11$	$G2, a$ & $\bar{F}a, 11$	$\bar{F}11, 11$	$\bar{G}2, 11$	$\bar{G}11, 11$
12.	$F2, 1$ & $\bar{F}1, 12$	$\bar{F}12, 12$	$G2, 1$ & $\bar{F}1, 12$	$\bar{F}12, 12$	$\bar{G}2, 12$	$\bar{G}12, 12$
13.	$F2, 2$ & $\bar{F}2, 13$	$\bar{F}13, 13$	$G2, 2$ & $\bar{F}2, 13$	$\bar{F}13, 13$	$\bar{G}2, 13$	$\bar{G}13, 13$
14.	$F2, 3$ & $\bar{F}3, 14$	$\bar{F}14, 14$	$G2, 3$ & $\bar{F}3, 14$	$\bar{F}14, 14$	$\bar{G}2, 14$	$\bar{G}14, 14$
15.	$F2, 4$ & $\bar{F}4, 15$	$\bar{F}15, 15$	$G2, 4$ & $\bar{F}4, 15$	$\bar{F}15, 15$	$\bar{G}2, 15$	$\bar{G}15, 15$
16.	$F3, a$ & $\bar{F}a, 16$	$\bar{F}16, 16$	$G3, a$ & $\bar{F}a, 16$	$\bar{F}16, 16$	$\bar{G}3, 16$	$\bar{G}16, 16$
17.	$F3, 1$ & $\bar{F}1, 17$	$\bar{F}17, 17$	$G3, 1$ & $\bar{F}1, 17$	$\bar{F}17, 17$	$\bar{G}3, 17$	$\bar{G}17, 17$
18.	$F3, 2$ & $\bar{F}2, 18$	$\bar{F}18, 18$	$G3, 2$ & $\bar{F}2, 18$	$\bar{F}18, 18$	$\bar{G}3, 18$	$\bar{G}18, 18$
19.	$F3, 3$ & $\bar{F}3, 19$	$\bar{F}19, 19$	$G3, 3$ & $\bar{F}3, 19$	$\bar{F}19, 19$	$\bar{G}3, 19$	$\bar{G}19, 19$
20.	$F3, 4$ & $\bar{F}4, 20$	$\bar{F}20, 20$	$G3, 4$ & $\bar{F}4, 20$	$\bar{F}20, 20$	$\bar{G}3, 20$	$\bar{G}20, 20$
21.	$F4, a$ & $\bar{F}a, 21$	$\bar{F}21, 21$	$G4, a$ & $\bar{F}a, 21$	$\bar{F}21, 21$	$\bar{G}4, 21$	$\bar{G}21, 21$
22.	$F4, 1$ & $\bar{F}1, 22$	$\bar{F}22, 22$	$G4, 1$ & $\bar{F}1, 22$	$\bar{F}22, 22$	$\bar{G}4, 22$	$\bar{G}22, 22$
23.	$F4, 2$ & $\bar{F}2, 23$	$\bar{F}23, 23$	$G4, 2$ & $\bar{F}2, 23$	$\bar{F}23, 23$	$\bar{G}4, 23$	$\bar{G}23, 23$
24.	$F4, 3$ & $\bar{F}3, 24$	$\bar{F}24, 24$	$G4, 3$ & $\bar{F}3, 24$	$\bar{F}24, 24$	$\bar{G}4, 24$	$\bar{G}24, 24$
25.	$F4, 4$ & $\bar{F}4, 25$	$\bar{F}25, 25$	$G4, 4$ & $\bar{F}4, 25$	$\bar{F}25, 25$	$\bar{G}4, 25$	$\bar{G}25, 25$

Applying our "one-literal clause rule," we obtain:

$Ga, a \ \& \ \bar{G}a, 1 \quad \bar{G}1, 1 \ \&$
$G1, a \ \& \ \bar{G}1, 2 \quad \bar{G}2, 2 \ \&$
$G1, 1 \ \& \ \bar{G}1, 3 \quad \bar{G}3, 3 \ \&$
$Ga, 1 \ \& \ \bar{G}a, 4 \quad \bar{G}4, 4 \ \&$
$\bar{F}2, 5 \quad \bar{F}5, 5 \quad Ga, 2 \ \& \ \bar{F}2, 5 \quad \bar{F}5, 5 \quad \bar{G}a, 5 \quad \bar{G}5, 5 \ \&$
$\bar{F}3, 6 \quad \bar{F}6, 6 \quad Ga, 3 \ \& \ \bar{F}3, 6 \quad \bar{F}6, 6 \quad \bar{G}a, 6 \quad \bar{G}6, 6 \ \&$
$\bar{F}4, 7 \quad \bar{F}7, 7 \quad Ga, 4 \ \& \ \bar{F}4, 7 \quad \bar{F}7, 7 \quad \bar{G}a, 7 \quad \bar{G}7, 7 \ \&$
$\bar{F}2, 8 \quad \bar{F}8, 8 \quad G1, 2 \ \& \ \bar{F}2, 8 \quad \bar{F}8, 8 \quad \bar{G}1, 8 \quad \bar{G}8, 8 \ \&$
$\bar{F}3, 9 \quad \bar{F}9, 9 \quad G1, 3 \ \& \ \bar{F}3, 9 \quad \bar{F}9, 9 \quad \bar{G}1, 9 \quad \bar{G}9, 9 \ \&$
$\bar{F}4, 10 \quad \bar{F}10, 10 \quad G1, 4 \ \& \ \bar{F}4, 10 \quad \bar{F}10, 10 \quad \bar{G}1, 10 \quad \bar{G}10, 10 \ \&$
$\bar{F}a, 11 \quad \bar{F}11, 11 \quad G2, a \ \& \ \bar{F}a, 11 \quad \bar{F}11, 11 \quad \bar{G}2, 11 \quad \bar{G}11, 11 \ \&$
$\bar{F}1, 12 \quad \bar{F}12, 12 \quad G2, 1 \ \& \ \bar{F}1, 12 \quad \bar{F}12, 12 \quad \bar{G}2, 12 \quad \bar{G}12, 12 \ \&$
$\bar{F}2, 13 \quad \bar{F}13, 13 \quad G2, 2 \ \& \ \bar{F}2, 13 \quad \bar{F}13, 13 \quad \bar{G}2, 13 \quad \bar{G}13, 13 \ \&$
$\bar{F}3, 14 \quad \bar{F}14, 14 \quad G2, 3 \ \& \ \bar{F}3, 14 \quad \bar{F}14, 14 \quad \bar{G}2, 14 \quad \bar{G}14, 14 \ \&$
$\bar{F}4, 15 \quad \bar{F}15, 15 \quad G2, 4 \ \& \ \bar{F}4, 15 \quad \bar{F}15, 15 \quad \bar{G}2, 15 \quad \bar{G}15, 15 \ \&$
$\bar{F}a, 16 \quad \bar{F}16, 16 \quad G3, a \ \& \ \bar{F}a, 16 \quad \bar{F}16, 16 \quad \bar{G}3, 16 \quad \bar{G}16, 16 \ \&$
$\bar{F}1, 17 \quad \bar{F}17, 17 \quad G3, 1 \ \& \ \bar{F}1, 17 \quad \bar{F}17, 17 \quad \bar{G}3, 17 \quad \bar{G}17, 17 \ \&$
$\bar{F}2, 18 \quad \bar{F}18, 18 \quad G3, 2 \ \& \ \bar{F}2, 18 \quad \bar{F}18, 18 \quad \bar{G}3, 18 \quad \bar{G}18, 18 \ \&$
$\bar{F}3, 19 \quad \bar{F}19, 19 \quad G3, 3 \ \& \ \bar{F}3, 19 \quad \bar{F}19, 19 \quad \bar{G}3, 19 \quad \bar{G}19, 19 \ \&$
$\bar{F}4, 20 \quad \bar{F}20, 20 \quad G3, 4 \ \& \ \bar{F}4, 20 \quad \bar{F}20, 20 \quad \bar{G}3, 20 \quad \bar{G}20, 20 \ \&$
$\bar{F}a, 21 \quad \bar{F}21, 21 \quad G4, a \ \& \ \bar{F}a, 21 \quad \bar{F}21, 21 \quad \bar{G}4, 21 \quad \bar{G}21, 21 \ \&$
$\bar{F}1, 22 \quad \bar{F}22, 22 \quad G4, 1 \ \& \ \bar{F}1, 22 \quad \bar{F}22, 22 \quad \bar{G}4, 22 \quad \bar{G}22, 22 \ \&$
$\bar{F}2, 23 \quad \bar{F}23, 23 \quad G4, 2 \ \& \ \bar{F}2, 23 \quad \bar{F}23, 23 \quad \bar{G}4, 23 \quad \bar{G}23, 23 \ \&$
$\bar{F}3, 24 \quad \bar{F}24, 24 \quad G4, 3 \ \& \ \bar{F}3, 24 \quad \bar{F}24, 24 \quad \bar{G}4, 24 \quad \bar{G}24, 24 \ \&$
$\bar{F}4, 25 \quad \bar{F}25, 25 \quad G4, 4 \ \& \ \bar{F}4, 25 \quad \bar{F}25, 25 \quad \bar{G}4, 25 \quad \bar{G}25, 25.$

Now applying the one-literal clause rule again to eliminate Ga, a, $G1, a$, and $G1, 1$ yields a formula containing $Ga, 1$ and $\bar{G}a, 1$ as clauses, which is inconsistent by Rule I.

The reader may be interested to see how the method works when the conjunction of quantifier-free lines being tested is *not* truth-functionally inconsistent. To illustrate this, let us test the conjunction of the first 10 quantifier-free lines listed above for consistency. Applying the one-literal clause rule yields:

1. $[Ga, a \ \&] \ \bar{G}a, 1 \quad \bar{G}1, 1$
2. $\bar{F}2, 2 \quad G1, a \ \& \ \bar{F}2, 2 \quad \bar{G}1, 2 \quad \bar{G}3, 3$
3. $\bar{F}3, 3 \quad G1, 1 \ \& \ \bar{F}3, 3 \quad \bar{G}1, 3 \quad \bar{G}4, 4$
4. $\bar{F}4, 4 \quad Ga, 1 \ \& \ \bar{F}4, 4 \quad \bar{G}a, 4 \quad \bar{G}5, 5$

6.–10. Same as in above list of "quantifier-free lines" except with first clause omitted.

A second application of the one-literal clause rule deletes the clause "Ga, a" (which was bracketted above in anticipation of this deletion). Now all the clauses containing an atomic formula beginning "$\bar{F}$" can be deleted by the affirmative-negative rule, and we obtain $\bar{G}a, 1 \vee \bar{G}1, 1$, which reduces to the empty formula by one more application of the affirmative-negative rule. Thus the conjunction of the first 10 quantifier-free lines was *consistent*. A similar result is obtained on testing the result of the first 20 quantifier-free lines.

NOTE ADDED IN PROOF: The "affirmative-negative rule" has also been employed, independently of our work, for testing propositional-calculus formulas by B. Dunham, R. Fridshal, and G. L. Sward: "A nonheuristic program for proving elementary logical theorems," *Proceedings of the First International Conference on Information Processing, Paris, 1959.*

To the list of reports of working proof procedure programs should be added: Dag Prawitz, Hakan Prawitz, and Neri Vogera, "A mechanical proof procedure and its realization in an electronic computer," *J. Assoc. Comput. Mach.*, *7* (1960), 102–128.

REFERENCES

1. MARTIN DAVIS, *Computability and Unsolvability*, New York, Toronto, and London, McGraw-Hill, 1958, xxv + 210 pp.
2. MARTIN DAVIS AND HILARY PUTNAM, A finitely axiomatizable system for elementary number theory. Submitted to the *Journal of Symbolic Logic.*
3. PAUL C. GILMORE, A proof method for quantification theory. *IBM J. Research Dev.* **4** (1960), 28–35.
4. JACQUES HERBRAND, *Recherches sur la theorie de la demonstration.* Travaux de la Societe des Sciences et des Lettres de Varsovie, Classe III science mathematiques et physiques, no. 33, 128 pp.
5. DAVID HILBERT AND WILHELM ACKERMANN, *Principles of Mathematical Logic.* New York, Chelsea, 1950, xii + 172 pp.
6. STEPHEN C. KLEENE, *Introduction to Metamathematics.* New York and Toronto, D. Van Nostrand, 1952, x + 550 pp.
7. WILLARD V. O. QUINE, *Methods of Logic.* New York, Henry Holt, revised 1959, xx + 272 pp.
8. WILLARD V. O. QUINE, A proof procedure for quantification theory. *J. Symbolic Logic* *20* (1955), 141–149.
9. HAO WANG, Towards mechanical mathematics. *IBM J. Research Dev.* **4** (1960) 2–22.

Empirical Explorations of the Geometry-Theorem Proving Machine

H. Gelernter, J. R. Hansen, D. W. Loveland

Introduction

In early spring, 1959, an IBM 704 computer, with the assistance of a program comprising some 20,000 individual instructions, proved its first theorem in elementary Euclidean plane geometry (Gelernter, 1959*b*). Since that time, the geometry-theorem proving machine (a particular state configuration of the IBM 704 specified by the afore mentioned machine code) has found solutions to a large number of problems[1] taken from high-school textbooks and final examinations in plane geometry. Some of these problems would be considered quite difficult by the average high-school student. In fact, it is doubtful whether any but the brightest students could have produced a solution for any of the latter group when granted the same amount of prior "training" afforded the geometry machine (*i.e.,* the same vocabulary of geometric concepts and the same stock of previously proved theorems).

The research project which had as its consequence the geometry-theorem proving machine was motivated by the desire to learn ways to use modern high-speed digital computers for the solution of a new and difficult class of problems; a class heretofore considered to be beyond the capabilities of a finite-state automaton. In particular, we wished to make our computer perform tasks which are generally considered to require the intervention of human intelligence and ingenuity for their successful completion. The reasons behind our choice of theorem proving in geometry as a representative task are set forth in detail in an earlier study (1958). We

[1] More than fifty proofs are on file at the present time.

only remark here that problem-solving in geometry satisfies our definition of an intellectual activity, while being at the same time especially well suited to the approach we wished to explore. The fact that geometry is decidable is irrelevant for the purpose of our investigation. The methods employed by the machine are suitable as well for the proof of theorems in systems for which no decision algorithm can exist.

We shall not labor the question as to whether our machine is indeed behaving intelligently in performing a task for which humans are credited with intelligence. The psychologists offer us neither aid nor comfort here; they have yet to satisfactorily characterize such behavior in humans, and have rarely considered the abstract concept of intelligence independent of its agent. In the final analysis, people are occasionally observed to do things that may best be described as intelligent, however vague the connotations of the word. These are, in general, tasks involving highly complex decision processes in a potentially infinite and uncontrollable environment. We should be most happy to have our machine duplicate this kind of behavior, whatever label is affixed to it.

Heuristic Programming and the Geometry Machine

The geometry machine is able to discover proofs for a significant number of interesting theorems within the domain of its *ad hoc* formal system (comprising theorems on parallel lines, congruence, and equality and inequality of segments and angles) without resorting to a decision algorithm or exhaustive enumeration of possible proof sequences. Instead, the theorem-proving program relies upon heuristic methods to restrain it from generating proof sequences that do not have a high *a priori* probability of leading to a proof for the theorem in question.

The general problem of heuristic programming has been discussed by Minsky (1959*a*) and Newell, Shaw, and Simon (1959*a*). The particular approach pursued by the authors has been described at length in the papers to which we have already referred (Gelernter et al., 1958, 1959*b*). We shall therefore defer to the presentation of the machine's detailed results in the full study summarized here for a description of how these results were achieved. It should be recorded here, however, that the geometry machine operates principally in the analytic mode (reasoning backward). At each stage of the search for a proof, a goal exists which must be "connected" with the premises for the problem by a bridge of axioms and previously established theorems or lemmas. If the connection cannot be made directly, then a set of "subgoals" is generated and the process is repeated for one of the subgoals. Heuristic rules are used to reject subgoals that are not likely to prove useful, to select one from those remaining to work on, and to choose particular axioms and theorems to use in generating new subgoals. The machine does depart from this procedure in a number of circumstances (in setting up an indirect proof, for example), but these cases account for only a small fraction of the total search time.

The computer program itself was written within the framework of the so-called Newell-Shaw-Simon list memory (1957*b*). In order to ease the task of writing so massive and complex a machine code, a convenient special-purpose list processing language was designed to be compiled by the already available FORTRAN system for the IBM 704 computer (Gelernter et al., 1960*b*). The authors feel that had they not made free use of an intermediate programming language, it is likely that the geometry program could not have been completed.

Summary of Results

Since its initial solo performance, the geometry machine has existed in several different configurations. In its earliest and most primitive form, the system was equipped with a single major semantic heuristic.[2] That first system was, however, able to prove a large number of interesting, though admittedly simple theorems in elementary plane geometry.[3] The heuristic rule in question, which is independent of the particular formal system under consideration, may be described in the following way. All subgoal formulas that are generated at a given stage of the proof search are interpreted in a model of the formal system; in our case, the model is a diagram, a formal semantic interpretation. If the interpreted subgoal is valid in the diagram, it is accepted as a possible step in the proof, provided that it is noncircular (Gelernter, 1959*a*). Otherwise, it is rejected.

As an experiment, a number of attempts were made to prove extremely simple theorems with the latter heuristic "disconnected" from the system (*i.e.,* all noncircular subgoals generated were accepted). In each case, the computer's entire stock of available storage space was quickly exhausted by the initial several hundreds of first level subgoals generated, and, in fact, the machine never finished generating a complete set of first level subgoals. We estimate conservatively that on the average, a number of the order of 1000 subgoals are generated per stage by the decoupled system. If one compares the latter figure with the average of 5 subgoals per stage accepted when the diagram is consulted by the machine, it is easy to see that the use of a diagram is crucial for our system. (Note that the total number of subgoals appearing on the problem-solving graph grows exponentially with the number accepted per stage.)

Since the procedure described above is a heuristic one, errors are oc-

[2] A semantic heuristic is one based on an interpretation of the formal system rather than on the structure of the strings within that system.

[3] A number of these proofs are reproduced in Gelernter, 1959*b*.

casionally made in the selection or rejection of formulas as subgoals. The diagram is made available to the machine in coordinate representation to finite precision. Formulas are interpreted by transforming them into an appropriate calculation on the numerical coordinates representing the point variables. For example, to check the validity of a statement concerning the equality of two segments, the length of each segment in the figure is calculated, and they are then compared to a certain preassigned number of decimal places. If, instead, the statement concerned parallel segments, the slopes would be calculated and compared. In a small number of cases, round-off error has propagated beyond the allowed value, so that valid subgoals were rejected, or invalid ones accepted. It is important to point out, however, that in no case could this effect result in a false proof. Where valid subgoals were rejected, the machine found alternate paths to the solution. Where invalid ones were accepted, the machine failed, of course, to establish them within the formal system. In the worst possible case, the interpretation error could prevent the computer from finding any solution at all, but never could it lead to an invalid proof.

It should be clear at this point that the diagram is used only to guide the search for a proof by supplying yes or no answers to questions of the form: "Is segment AB equal to segment CD in the figure?", or "Is angle ABC a right angle in the figure?". There is no direct link between the diagram and the formal system in the geometry machine. The behavior of the machine would not be changed if the coordinate representation were replaced by a device capable of drawing figures on paper and scanning them.

In the basic theorem-proving system described above, after a set of subgoals has been generated, each member of the set is explored in order. The next subgoal in line is not examined until the one preceding it has been followed down to a dead end. Too, in generating the next level for a given subgoal, every applicable theorem available is pressed into service.

This system was soon extended by the introduction of selection heuristics for both subgoals and subgoal-generating theorems. The subgoal selection heuristic assigns a "distance" between each subgoal string and the set of premises in a vaguely defined *ad hoc* formula space. At each stage, the next subgoal selected is that which is "closest" to the premises in formula space. The generator selection routine recognizes certain classes of subgoals that are usually established in one step. For such "urgent" subgoals, the appropriate generator is withdrawn immediately, and an attempt is made for a one-step proof (of that particular subgoal) before generating the full set for that formula.

The extended system is able to prove a number of somewhat more difficult theorems that are beyond the capacity of the basic machine. For those problems within the range of both systems, the former is, on the

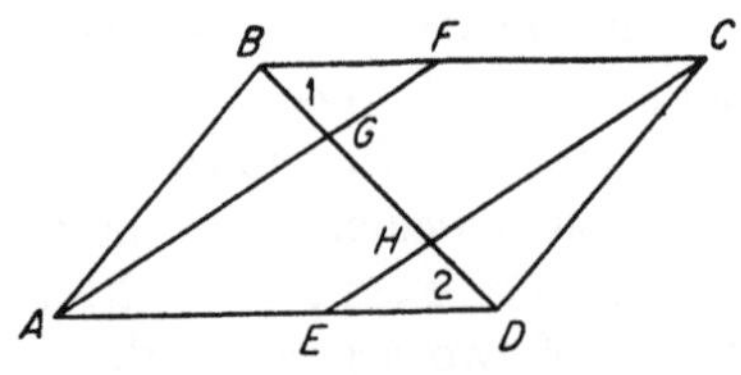

Figure 1.

average, about three times faster, and generates about two-thirds the total number of subgoals in half as many subgoal generation cycles as required by the basic system. The average depth of the problem-solving graph for the refined system, about seven to nine levels, is two-thirds the average depth for the basic system.

By the addition of a simple construction routine, the theorem-proving power of the machine is expanded to include an entirely new class of problem, hitherto logically unattainable. The routine, called upon only when all other attempts have failed, allows the machine to join two previously unconnected points in the diagram, and extends the newly created segment to its intersections with all other segments in the figure. The new segment, when it intersects previously given ones, introduces new points into the problem which are named by the machine and become part of the problem system.

At this stage in its development, the geometry machine was capable of producing proofs that were quite impressive (Appendix 1).[4] Its performance, however, fell off rapidly as the number of points in the diagram increased. This effect was due largely to the fact that unlike humans, who generally identify angles visually by their vertices and rays, the computer specifies an angle by a predicate on three variables, the vertex and a point on each ray. Consequently, the equality of angles 1 and 2 in Fig. 1 may be represented in thirty-six different ways, since each angle has six different names. Formal rigor demands, too, that the equality of angles ADH and EDG, for example, be proved rather than taken for granted. It should be clear that where the condition above exists, the search for a proof quickly bogs down in a mass of uninteresting detail.

In the current system, the angle problem is solved by allowing the machine to use the diagram to identify a given angle with its full set of names, and to assume the equality relationship between different names for the same angle, as does its human counterpart. The geometry machine in its present configuration is able to find proofs for theorems of the order of difficulty represented by the following:

Theorem: If the segment joining the midpoints of the diagonals of a

[4] In the proofs appended to this paper, the nonobvious predicates have the following interpretations:

OPP-SIDE XYUV	Points X and Y are on opposite sides of the line through points U and V.
SAME-SIDE XYUV	Points X and Y are on the same side of the line through points U and V.
PRECEDES XYZ	Points X, Y, and Z are collinear in that order,
COLLINEAR XYZ	Points X, Y, and Z are collinear,

trapezoid is extended to intersect a side of the trapezoid, it bisects that side (Appendix 2).

Limitations of the System

It will be immediately evident to those familiar with the properties of formal logistic systems that unless a construction which generates a new point is introduced by the machine, all problems are solved within the framework of a propositional calculus, however complex its structure. Although the machine's present construction routine can and does generate new points, we could not expect our results to be of great interest to logicians until a full set of possible constructions (corresponding to a complete set of existentially quantified axioms) is made available to the system to abet its search for a proof.

An equally serious limitation on the formal generality of the theorem-proving machine is imposed by our method for determining the well-formedness of strings within the logical system. In order to attain the necessary speed and efficiency in processing, well-formed formulas are defined by schema rather than recursively. The kind of statement that can be made in the system is then determined by the schema available to the machine. The practical effect of this loss in generality is to restrict rather severely the freedom with which algebraic statements in geometry may be manipulated.

In addition to the above, there are a number of nonessential bounds on the theorem-proving ability of the machine. These are a consequence of the limited speed and memory capacity of the computer for problems of such highly combinatorial character. Improvements in either of the above will be immediately effective in extending the class of machine-solvable problems in both quantity and difficulty.

Conclusion

The initial goal of our research program in machine intelligence has been attained. If the interrogator were to restrict his probing to the area of theorem-proving in elementary Euclidean plane geometry, our machine could be expected to give an excellent account of itself in competition with a human in Turing's well-known "imitation game" (1950). Of course there are many other problem areas (solving arithmetic problems, for example) where computers have always been able to compete successfully with humans. The significant point is that a knowledgeable interrogator would certainly avoid such areas in his questioning, while he might well (until now, at any rate) introduce a plane geometry problem in a calculated attempt to separate the men from the machines.[5] Although the stage is now set for the argument that any distinct area of human intellectual activity will in the same way succumb to the inexorable logic of electrons, switches, and gates, we defer to our philosopher colleagues for debate on the implications of that contention, at least until the time that computers have been programmed to consider such issues.

There are a number of consequences of our work that are, fortunately, more concrete than that alluded to above. Perhaps the most important are those relating to inferential analysis, a new branch of applied logic first characterized by Wang (1960*a*). Inferential analysis "treats proofs as numerical analysis does calculations," and is expected to "lead to mechanical checks of new mathematical results" and, more important, "lead to proofs of difficult new theorems by machine." It is expected that our techniques for the manipulation and efficient search of problem-solving trees and our results concerning syntactic symmetry will prove to be useful tools in pursuing the goals of inferential analysis.

Contributions have been made, too, in the area of techniques for computer implementation of complex information processes. Results pertaining to the design and use of intermediate languages for the specification of list manipulation processes have been reported elsewhere (Gelernter et al., 1960*b*). The latter work indicates clearly the requirements of a digital computer system designed for optimum execution of such list processes. In brief, a list processing computer should possess hardware facilities for:

1. Generalized indirect addressing; specified in the indirectly addressed instruction to arbitrary depth and in arbitrary order from either the left or the right field of a two-address register,
2. Effective address recovery; making available the terminal content of the address register (the final address in a long and complex indirect address chain, for example) as the address field for a subsequent operation,
3. Field logic; a greatly expanded set of interfield operations within a full register sectioned according to some previously established convention, and
4. List search operations; a list equivalent of the conventional table look-up instruction.

The bulk storage input-output requirements for a list processing computer are severe, and are not included in the enumeration above. The system

[5] It may be argued (and undoubtedly, it *will* be argued) that the truly knowledgeable interrogator, cognizant of the decidability of geometry, would certainly avoid this area as well, perhaps preferring the manifestly undecidable parts of the predicate calculus or number theory to effect the distinction between man and machine. We recall here that our methods are independent of the decidability of the formal system, and, in fact, Wang (1960*a*) and Gilmore (1960) have developed proofs for theorems in the undecidable area of the predicate calculus.

design of a digital computer for the manipulation of list structures will be described in detail in a subsequent paper.

Finally, we consider the implications of our work for the basic problem of machine intelligence. The geometry machine, we feel, offers convincing evidence of the power and fruitfulness of heuristic programming for the solution of problems of a certain class by computer. In our experience, the theorem-proving power of the machine has often been extended by the addition of a single heuristic to a degree equivalent to a three-to-fivefold increase in the speed or storage capacity of the computer.

Our program has proved to be disappointing as a tool for the study of the more elementary trial-and-error types of machine learning, largely because of the rather low rate at which it accumulates experience. It is reasonable to expect, however, that the geometry machine might yet be pressed into service in an investigation of the higher, conceptual types of machine learning, providing that one will someday know how to formulate the problem.

If nothing else, our work offers some qualitative indication of the order of magnitude of difficulty for problems that could be expected to yield to contemporary computer technology. Three years ago, the dominant opinion was that the geometry machine would not exist today. And today, hardly an expert will contest the assertion that machines will be proving interesting theorems in number theory three years hence.

Appendix 1

Premises

Quad-lateral ABCD
Point E midpoint segment AB
Point F midpoint segment AC
Point G midpoint segment CD
Point H midpoint segment BD

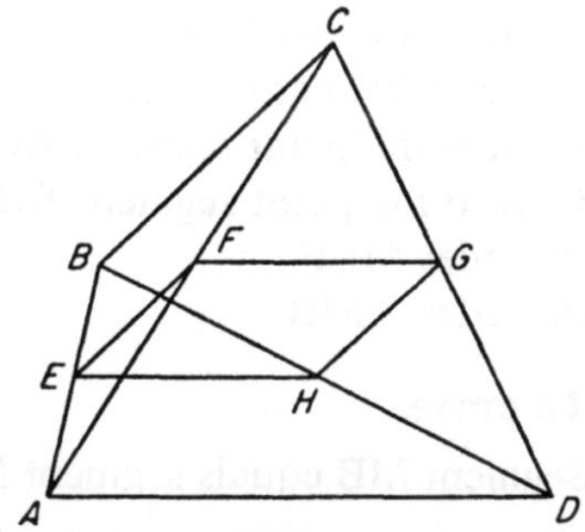

Figure 2.

To Prove

Parallelogram EFGH

Syntactic Symmetries

BA, AB, DC, CD, EE, HF, GG, FH, CA, DB, AC, BD, GE, FF, EG, HH, DA, CB, BC, AD, GE, HF, EG, FH

Proof

Segment DG equals segment GC
Definition of midpoint

Segment CF equals segment FA
  Definition of midpoint
Triangle DCA
  Assumption based on diagram
Precedes DGC
  Definition of midpoint
Precedes CFA
  Definition of midpoint
Segment GF parallel segment AD
  Segment joining midpoints of sides of triangle is parallel to base
Segment HE parallel segment AD
  Syntactic conjugate
Segment GF parallel segment EH
  Segments parallel to the same segment are parallel
Segment HG parallel segment FE
  Syntactic conjugate
Quad-lateral HGFE
  Assumption based on diagram
Parallelogram EFGH
  Quadrilateral with opposite sides parallel is a parallelogram

Total elapsed time = 1.03 minutes

Appendix 2

Premises

Quad-lateral ABCD
Segment BC parallel segment AD
Point E midpoint segment AC
Point F midpoint segment BD
Precedes MEF
Precedes AMB

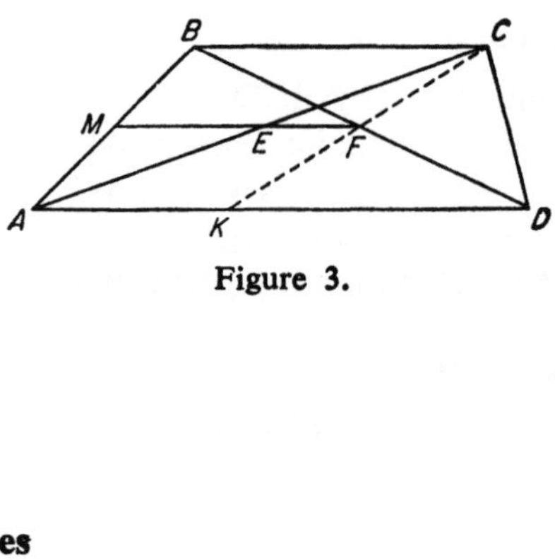

Figure 3.

To prove

Segment MB equals segment MA

No Syntactic Symmetries

I Am Stuck, Elapsed Time = 8.12 Minutes

Construct segment CF
Extend segment CF to intersect segment AD in point K

Add to Premises the Following Statements

Precedes CFK
Collinear AKD

Proof

Segment BC parallel segment AD
  Premise
Collinear AKD
  Premise
Segment KD parallel segment BC
  Segments collinear with parallel segments are parallel
Opp-side KCDB
  Assumption based on diagram
Segment DB
  Assumption based on diagram
Angle KDB equals angle CBD
  Alternate interior angles of parallel lines are equal
Precedes CFK
  Premise
Precedes DFB
  Definition of midpoint
Angle KFD equals angle CFB
  Vertical angles are equal
Segment DF equals segment FB
  Definition of midpoint
Triangle FDK
  Assumption based on diagram
Triangle FBC
  Assumption based on diagram
Triangle FDK congruent triangle FBC
  Two triangles are congruent if angle-side-angle equals angle-side-angle
Segment KF equals segment CF
  Corresponding segments of congruent triangles are equal
Segment CE equals segment EA
  Definition of midpoint
Triangle AKC
  Assumption based on diagram
Precedes CEA
  Definition of midpoint
Segment EF parallel segment AK
  Segment joining midpoints of sides of triangle is parallel to base
Segment EF parallel segment KD
  Segments collinear with parallel segments are parallel
Segment FE parallel segment BC
  Segments parallel to the same segment are parallel

Precedes MEF
  Premise
Collinear MEF
  Ordered collinear points are collinear
Segment FM parallel segment BC
  Segments collinear with parallel segments are parallel
Segment FM parallel segment DA
  Segments parallel to the same segment are parallel
Triangle DBA
  Assumption based on diagram
Precedes AMB
  Premise
Segment MB equals segment MA
  Line parallel to base of triangle bisecting one side bisects other side

Total elapsed time = 30.68 minutes

A Proof Method for Quantification Theory: Its Justification and Realization

P. C. Gilmore

Abstract: A program is described which can provide a computer with quick logical facility for syllogisms and moderately more complicated sentences. The program realizes a method for proving that a sentence of quantification theory is logically true. The program, furthermore, provides a decision procedure over a subclass of the sentences of quantification theory. The subclass of sentences for which the program provides a decision procedure includes all syllogisms. Full justification of the method is given.

A program for the IBM 704 Data Processing Machine is outlined which realizes the method. Production runs of the program indicate that for a class of moderately complicated sentences the program can produce proofs in intervals ranging up to two minutes.

Introduction

Before a rigorous mathematical proof of any theorem can be given, it is necessary that the theorem, and the axioms from which the theorem is to be deduced, be precisely stated in an unambiguous language. The formal language variously called the first-order predicate calculus or logic, the first-order functional calculus or logic, or quantification theory, is adequate for the expression of any theorem or axiom. Further, although the sentences of quantification theory are ambiguous in the sense of being capable of possessing many different meanings, it is an easy matter to remove this ambiguity simultaneously from all sentences of the language by assigning meaning to the primitive symbols from which the sentences are formed. But most important of all, it is possible to give for sentences $T, S_1, S_2, \ldots$ of quantification theory a precise definition of "T is a logical consequence of the sentences $S_1, S_2, \ldots$" to replace the intuitive but vague notion in natural languages that one sentence is a logical consequence of other sentences. Thus quantification theory not only provides a language for mathematics but also permits a description of what constitutes a rigorous mathematical proof.

The primitive symbols of quantification theory are brackets (,), the logical connectives $-$, &, v, $\supset$, $\equiv$, the predicate letters $A, B, C, \ldots$ the individual names, $a, b, c, \ldots$ the individual variables $x, y, z, \ldots$ and the quantifiers E and A. Any finite sequence of these primitive symbols is a formula of the theory. From the formulae of the theory are selected some which in some sense are meaningful and are therefore called *well-formed formulae*, or briefly wff. From the wff of the theory will be selected the sentences of the theory.

Inductive definitions of wff and of "the (individual) variable X occurs free in the wff S" are given simultaneously:

1. A formula consisting of a predicate letter followed by any number of individual names or variables is an *atomic* wff and a wff and the variables occurring in it occur free.

2. If S is a wff then so is $-S$, and the variables occurring free in S occur free in $-S$.

3. If S and T are wff then so are $(S \& T)$, $(S \vee T)$, $(S \supset T)$ and $(S \equiv T)$, and the variables occurring free in either S or T occur free in these wff.

4. If S is a wff and X is any variable then $(EX)S$ and $(AX)S$ are wff and the variables other than X occurring free in S occur free in these wff.

A variable X which occurs in a wff but does not occur free is said to be bound in the wff and must, therefore, by part (4) of the definition, occur with quantifier symbols as in (EX) or (AX). A *sentence* of the theory is a wff in which no variable occurs free. Calling such formulae of the theory sentences is reasonable if the primitive symbols of the theory are interpreted as follows:

Atomic sentences (atomic wff which are sentences) can be understood as abbreviations for sentences such as "1+3=5", "3<7", or "John is tall"; these could be expressed in quantification theory as $A135$, $B37$, or Ca, allowing numerals also to be names in quantification theory. Atomic wff which are not sentences can be understood as abbreviations for sentence forms such as "$1+x=5$", or "$x+y=5$", or "$x+y=z$", or "$3<y$" or "$x<y$" or "x is tall", where if variables in a sentence form are replaced by names the result is a sentence. The logical connectives $-$, &, v, $\supset$ and $\equiv$ are to be understood as expressing negation, conjunction, disjunction, implication and equivalence respectively in the sense that, for example, $(S \,\&\, T)$ is a sentence which is true if and only if both S and T are true, and $(S \supset T)$ is a sentence which is true if and only if S is false or T is true, and $-S$ is a sentence which is true if and only if S is false. When the logical connectives are attached to sentence forms rather than sentences the result is to be understood as a new sentence form with the expected properties. For example, if $S(x)$ is a sentence form with one free variable x, then $-S(x)$ is a sentence form with one free variable x and such that for any name α, the sentence $-S(\alpha)$ formed from $-S(x)$ by replacing x everywhere by α is true if and only if $S(\alpha)$ is false. Finally a sentence such as $(Ex)S(x)$ can be understood to be a sentence which is true if and only if for some name α, $S(\alpha)$ is true, while a sentence $(Ax)S(x)$ can be understood to be a sentence which is true if and only if for every possible name α, $S(\alpha)$ is true. For sentence forms $S(x, y)$ with two free variables x and y, $(Ey)S(x, y)$ and $(Ay)S(x, y)$ are sentence forms with one free variable x and are related in the obvious manner to the sentence form $S(x, y)$.

A sentence is generally thought to be either true or false. But a sentence of quantification theory may be true or false depending upon the interpretations given to the predicate letters and names which occur in it. Thus "$(Ey)(Bay \,\&\, Cy)$" is true when "Bay" is understood as "a is the brother of y", "Cy" is understood as "y is tall" and "a" is the name of someone with a tall brother, but can be false under other interpretations.

If atomic sentences or sentences formed from atomic sentences by attaching a negation sign "$-$" are called *ground sentences*, then by an interpretation can be understood a set I of ground sentences of quantification theory with the following properties:

1. No atomic sentence and its negation are members of I (I is *consistent*);

2. For every possible atomic sentence which can be formed from names and predicate letters occurring in ground sentences in I, either the sentence or its negation is a member of I (I is *complete*).

A predicate letter or name occurring in a member of an interpretation I is said to be a predicate letter or name of I.

The members of an interpretation I are the only ground sentences of the language which are true for the interpretation. A ground sentence not in I, but formed from a predicate letter and names of I, is false for I. Hence a sentence in which occurs a predicate letter or name not of an interpretation is neither true nor false for the interpretation. For sentences which are not ground sentences and in which occur only predicate letters and names of I, "true for the interpretation" and "false for the interpretation" can be readily defined considering the meanings which have been given to the logical connectives and quantifiers. Thus $(Ey)(Bay \,\&\, Cy)$ is true in the interpretation $\{Ca, Cb, Bab, -Bba, -Baa, -Bbb\}$ because Bab and Cb are true in the interpretation and therefore also $(Bab \,\&\, Cb)$ and hence the above sentence. But the same sentence is false in the interpretation $\{Ca, -Cb, Bab, -Bba, -Baa, -Bbb\}$ because there is no name α such that both $Ba\alpha$ and $C\alpha$ are true in the interpretation, and hence no name α for which $(Ba\alpha \,\&\, C\alpha)$ is true for the interpretation. Clearly any consistent set I of ground sentences can be completed to an interpretation by adding to I, for any atomic sentence formed from a predicate letter and names of I, either the atomic sentence or its negation, should neither already appear in I. Such an interpretation will be called a *completion* of I.

This definition of interpretation will appear a little strange because it does not require the specification of any set of objects which is to be the range of the variables $x, y, z, \ldots$, to which the names $a, b, c \ldots$ are assigned or the properties and relations of which are to be assigned to the predicate letters $A, B, C, \ldots$. But since it is possible for every object of a set to be assigned a single name, it is irrelevant whether one uses names to discuss the objects or simply discusses the names. However, this has as a consequence that two apparently quite different interpretations may be, for all practical purposes, the same interpretation. An object given a single name in one interpretation may be given a different name, or many different names in another interpretation. For this reason the notion of a homomorphism for interpretations is convenient.

Let ψ be a single valued mapping of the names of an interpretation I_1 onto the names of an interpretation I_2, and for any sentence S in which only names of I_1 occur, let $\psi(S)$ be the sentence obtained from S by replacing each name α occurring in S by $\psi(\alpha)$. If for any ground sentence G in which only names of I_1 occur, G is a member of I_1 if and only if $\psi(G)$ is a member of I_2, ψ is said to be a homomorphism of I_1 onto I_2. If there exists a homomorphism of I_1 onto I_2 then I_2 is said to be a homomorphic image of I_1. It can be readily seen that if ψ is a homomorphism of I_1 onto I_2 then for any sentence S, S is true or false for I_1 if and only if $\psi(S)$ is true or false respectively for I_2.

A counterexample for a conjectured theorem T in a mathematical theory with axioms $S_1, S_2, \ldots$ consists in a mathematical structure for which $S_1, S_2, \ldots$ are all true and for which T is false. Assuming that $T, S_1, S_2, \ldots$ are all sentences of quantification theory, this can be more precisely expressed: The counterexample consists in an interpretation for which $S_1, S_2, \ldots$ are all true and T is false. Thus, a conjectured theorem is actually a theorem if it is impossible to find a counterexample for it. Hence one can say that a sentence T is a *logical consequence* of sentences $S_1, S_2, \ldots$ if and only if it is not possible to find an interpretation for which $S_1, S_2, \ldots$ are all true and T is false. If T is a logical consequence of an empty set of sentences, that is, if it is impossible to find an interpretation in which T is false, then T is *logically* true.

A program for the IBM 704 Data Processing Machine will be described which will, for any logically true sentence T, construct a proof that T is logically true. For a special class M of sentences the program will do more; for any member T of M the program can decide whether or not T is logically true, produce a proof of T should it be logically true and provide essentially a counterexample to T should T be not logically true. The class M of sentences is sufficiently wide to include, for example, all syllogisms. Results of production runs will be given. In addition it will be shown how the program can be adapted for producing a proof that a sentence T is a logical consequence of sentences $S_1, S_2, \ldots$.

The theoretical basis of the program is a process which, for any sentence S of quantification theory, generates at the nth step of the process a finite number k_n, $k_n \geq 0$, of finite sets $I_{n1}, I_{n2}, \ldots, I_{nk_n}$ of ground sentences. Each set formed at the $n+1^{\text{st}}$ step is either one of the sets formed at the nth step or is obtained from such a set by adding new members to it. Hence if $k_n > 0$ for all n then it is possible to find a function ϕ such that $I_{n\phi(n)} \subseteq I_{n+1\phi(n+1)}$ for all n. It is then the case that any completion of the set $\bigcup_{n=1}^{\infty} I_{n\phi(n)}$ is an interpretation for which S is true. Conversely if there is any interpretation whatsoever for which S is true, then there is also one which is a completion of $\bigcup_{n=1}^{\infty} I_{n\phi(n)}$ for some ϕ. It follows that there is an interpretation for which S is true if and only if $k_n > 0$ for all n.

The process can be applied to proving that a sentence T is logically true by taking as input to the process the sentence $-T$. Should $k_n = 0$ for some n, then T is true for every interpretation, and therefore logically true, since it is not possible for there to be an interpretation for which $-T$ is true and thus for which T is false. On the other hand, however, the process cannot, in general, show that a sentence T is not logically true; that is, that $-T$ has an interpretation, since it would be necessary to show that $k_n > 0$, for all n, while only a finite number of k_n can ever be computed.[1] Nevertheless, for one special class M of sentences of quantification theory the process does actually provide a decision procedure; that is, for any member T of M, if $-T$ is used as input for the process, it will be possible to decide after a finite number of steps whether or not T is logically true. For, for any member T of M, it is possible to compute a number N such that if $-T$ is used as input to the process then there is an interpretation for which $-T$ is true if and only if $k_n > 0$ for all n, $n \leqslant N$. Further if ϕ is a function such that $I_{n\phi(n)} \subseteq I_{n+1\phi(n+1)}$ for $n \leqslant N$, then any completion of $\bigcup_{n=1}^{N} I_{n\phi(n)}$ is an interpretation for which $-T$ is true. Thus for T in M, if the process is carried out with $-T$ as input and $k_n = 0$ at the nth step, $n \leqslant N$, then T has been proven to be logically true; while if $k_n > 0$ for each of the first N steps of the process, T is known to be not logically true and a counterexample to T can actually be exhibited.

In the description of the process some well-known concepts and theorems from logic will be used. Two sentences are said to be *logically equivalent* if and only if there exists no interpretation for which one is true and the other is false. A sentence in which no quantifiers appear, and which is in disjunctive normal form, is called a *matrix*. Since a matrix consists of one or more disjunctions of conjunctions of one or more ground sentences, from two matrices M_1 and M_2 can be formed a product matrix as follows: For each conjunction of M_1 and each conjunction of M_2, form a conjunction of the product matrix by conjoining the two conjunctions. Contradictory conjunctions, i.e., ones in which both an atomic sentence and its negation appear, can be dropped from the product matrix.

A wff which is not a sentence but is in disjunctive normal form and does not contain any quantifiers is called a *matrix form*. A sentence which is in prenex form, that is, one in which all of the quantifiers occur initially, and which consists of a sequence of quantifiers attached to a wff which is a matrix form, will be said to be in *standard form*. For any sentence of quantification theory, there exists a sentence in standard form which is logically equivalent to the given sentence. A sentence which is in standard form and which is logically equivalent to a given sentence will be said to be a standard form of the given sentence.

The original motivation for the proof method of this paper was the method of semantic tableaux of Beth, although in its present form it is closer to the work of Hintikka.[2] The proof method is related in form, if not entirely in motivation, to the methods of Herbrand and Gentzen.[3]

Although much previous work has been done in proving theorems by machine, the present work is the first working program for quantification theory. The work of Newell, Shaw and Simon did not have as its primary aim the proof of theorems, but did result in a program for the propositional logic.[4]

Gelernter's and Rochester's program[5] for proving theorems in elementary geometry is applied to proving theorems which can be put into the standard form:

$$(AX_1) \ldots (AX_n)(P(X_1, \ldots, X_n) \supset Q(X_1, \ldots, X_n)),$$

where $P(X_1, \ldots, X_n)$ and $Q(X_1, \ldots, X_n)$ are conjunctions of atomic wff, from axioms which can be put into the same standard form. Although there exists a simple decision procedure for theorems in such an axiomatic theory, the Gelernter-Rochester program does not make use of it. Their program, with motivations similar to those of Newell and Simon, instead of exhaustively generating substitutions for the axioms, chooses substitutions which are expected to lead more directly to a proof of the theorem. Dunham, Fridshal and Sward have developed a program which uses an efficient decision procedure for theorems in propositional logic.[6] The work of Wang is most directly related to the present paper.[7] Wang has written two programs and has proposed a third. His first program is a decision procedure for theorems of the propositional logic, his second a decision procedure for theorems of quantification theory which are members of the Class M, while the proposed program is a proof method for theorems of quantification theory which incorporates the decision procedure of the first two programs. Thus the intention of Wang's third program is exactly the same as the program outlined in this paper; the two programs differ, however, in several ways. For example, it is based on a method of proof more directly related to that of Gentzen and Herbrand rather than that of Beth and Hintikka. Also, at the same time, Wang's program is to accomplish more, for it is to accept as input data the sentence to be proven, while the input data for the program of this paper is a standard form of the negation of the sentence to be proven. A further comparison of the two programs will have to wait upon production runs from Wang's program.[8]

The process which has been described as the theoretical basis of the program of this paper is given in detail in the next section. The following section describes the program for the IBM 704 Data Processing Machine for proving sentences to be logically true, and its extension to a program for proving that a given sentence is a logical consequence of other sentences. Results of production runs are given in the subsequent section.

The process

Given any sentence, choose a standard form S for it. Then S can be written:

$$(Q_1X_1)(Q_2X_2)\dots(Q_mX_m)M(X_1,\dots,X_m),$$

where $X_1, X_2, \dots, X_n$ are all the variables which occur free in the matrix form $M(X_1, \dots, X_m)$ and each Q_j is either E or A. If Q_j is $E(A)$, then X_j is said to be existentially (universally) quantified.

The matrix form $M(X_1, \dots, X_m)$ consists of the disjunction of one or more conjunctions of one or more basic sentence forms or basic sentences. Two conjunctions are said to be linked by given variables if there exists a finite sequence of conjunctions of the matrix form, beginning with one of the conjunctions and ending with the other, such that for each adjoining pair of conjunctions in the sequence, one of the given variables occurs free in both members of the pair. An existentially quantified variable X_j is said to depend upon all universally quantified variables X_i for which $i<j$ and which occur free in conjunctions where X_j occurs free, or which are linked to the conjunctions in which X_j occurs free by the universally quantified variables X_k for which $j<k$.

Let there be r universally quantified variables $X_{u_1}, X_{u_2}, \dots, X_{u_r}$. Then a sequence of r-tuples $(p_{n1}, p_{n2}, \dots, p_{nr})$, $n=1, 2, \dots$, of positive non-zero integers is generated as follows:

(1) For all j, $p_{1j}=1$;

(2) Let μ_n be the maximum of $p_{n1}, p_{n2}, \dots, p_{nr}$, then (**a**) if for all j, $1\leq j\leq r$, $p_{nj}=\mu_n$, then $p_{n+1r}=\mu_n+1$ and $p_{n+1j}=1$ for $j<r$; (**b**) if there is a k such that $p_{nj}=\mu_n$ for $k<j\leq r$ and $p_{nk}=\mu_n-1$, then $p_{n+1j}=p_{nj}$ for $1\leq j<k$, $p_{n+1k}=\mu_n$, and $p_{n+1j}=1$ for $k<j\leq r$; (**c**) if $p_{nr}<\mu_n$, then $p_{n+1r}=p_{nr}+1$ and $p_{n+1j}=p_{nj}$ for $1\leq j<r$; (**d**) if there is a k such that $p_{nj}=\mu_n$ for $k<j\leq r$ and $p_{nk}<\mu_n-1$, then $p_{n+1j}=p_{nj}$ for $1\leq j<k$, $p_{n+1k}=p_{nk}+1$, $p_{n+1,r}=\mu_n$ and $p_{n+1j}=1$ for $k<j<r$.

The sequence of r-tuples so generated is without duplications and is such that for any k, $1\leqslant k\leqslant r$, and any positive non-zero integer μ, every k-tuple of numbers m, $1\leqslant m\leqslant \mu$, occurs in the sequence $(p_{n(r-k)+1}, \dots, p_{nr})$, $n=1, 2, \dots$, before any k-tuple with a number $\mu+1$ appears.

A sentence S in standard form is a member of the set D if and only if there is no (existentially quantified) variable of S dependent upon any (universally quantified) variable of S. A sentence T is then a member of the set M if and only if the sentence $-T$ has a standard form which is a member of D.

Let there occur s names in S. Let S not be in D. A sequence $(q_{n1}, q_{n2}, \dots, q_{nm})$, $n=1, 2, \dots$, of m-tuples of positive non-zero integers is generated as follows:

(1) For all k and n, if X_k is universally quantified and $k=u_j$, then $q_{nk}=p_{nj}$;

(2) For all k, if X_k is existentially quantified and is dependent upon the variables $X_{uj_1}, \dots, X_{uj_t}$, then (**a**) q_{1k} is the maximum of $s+1$, $q_{11}+1, \dots, q_{1k-1}+1$; (**b**) for all $n>1$, if h is the smallest positive integer for which $(p_{nj_1}, \dots, p_{nj_t})$ is identical with $p_{hj_1}, \dots, p_{hj_t})$, then if $h<n$, q_{nk} is q_{hk}; while if $h=n$, q_{nk} is the maximum of $s+1$, $q_{11}+1, \dots, q_{1m}+1, \dots, q_{n-11}+1, \dots, q_{n-1m}+1$, $q_{n1}+1, \dots, q_{nk-1}+1$; in particular if $t=0$, then $q_{nk}=q_{1k}$.

Should S be in D so that it can be assumed that in the sequence $Q_1, Q_2, \dots, Q_m$ no A precedes an E, then a finite sequence $(q_{n1}, \dots, q_{nm})$, $n=1,2,\dots,\max(1,(m-r+s)^r)$, of m-tuples of positive non-zero integers is generated as follows:

(1) For all k and n, if X_k is universally quantified and $k=u_j$, then $q_{nk}=p_{nj}$;

(2) For all k and n, if X_k is existentially quantified, then q_{nk} is the maximum of $s+1$, $q_{11}+1, \dots, q_{1k-1}+1$. In particular, therefore, if either $r=0$ or $r=m$, then only one term is generated.

Let $a_1, a_2, \dots, a_s, a_{s+1}, \dots$ be a list of names of quantification theory in which the names $a_1, a_2, \dots, a_s$ which occur in S are all listed first. Let P_1 be the matrix $M(a_{q11}, \dots, a_{q1m})$, and for $n>1$, let P_n be the product matrix of P_{n-1} with the matrix $M(a_{q_{n1}}, \dots, a_{q_{nm}})$. For S in D, P_n is defined for $n\leqslant \max(1, (m-r+s)^r)$, while for S not in D, it is defined for all $n\geq 1$. Then for any n and j, I_{nj} is the set of ground sentences conjoined together to form the jth conjunction of P_n. Let k_n be the number of conjunctions of P_n.

- *Theorem 1*

For S not in D the sets I_{nj} possess the following properties:

(1) For any n for which $k_{n+1}>0$ and any j, $1\leq j\leq k_{n+1}$, there exists an i such that $I_{ni}\subseteq I_{n+1j}$;

(2) For any function ϕ for which $I_{n\phi(n)}\subseteq I_{n+1\phi(n+1)}$ for

$n \geq 1$, S is true for any completion of $\bigcup_{n=1}^{\infty} I_{n\phi(n)}$;

(3) For any interpretation I'' for which S is true there exists an interpretation I', $I' \subseteq I''$, for which S is true and a function ϕ for which $I_{n\phi(n)} \subseteq I_{n+1\phi(n+1)}$ for $n \geq 1$, such that I' is the homomorphic image of some completion of $\bigcup_{n=1}^{\infty} I_{n\phi(n)}$.[9]

• *Proof*

That they possess property (1) is immediate. A completion I of a set $\bigcup_{n=1}^{\infty} I_{n\phi(n)}$ is formed from the latter set by adding to it ground sentences consistent with it formed from names and predicate letters occurring in members of the latter set. Hence, all of the sentences $M(a_{q_{n1}}, \ldots, a_{q_{nm}})$, $n=1, 2, \ldots$, are true for I and therefore also S is true for I.

Let S be true for I''. Choose a sequence $b_1, b_2, \ldots$ of names of I'' as follows:

(1) If names occur in S, list these first in the order in which they are listed in $a_1, a_2, \ldots, a_s$;

(2) If no names occur in S and Q_1 is A, choose any name of I'' as b_1;

(3) Assume that the sequence of names has been completed for t members (t may be 0 if $s=0$ and Q_1 is E). Let n be the smallest integer for which $q_{nj}=t+1$ for some j. It then follows that for only one k is $q_{nk}=t+1$, so that one can assume that $b_{q_{n1}}, \ldots, b_{q_{nk-1}}$ have already been chosen, and further that Q_k is E. Hence, choose $b_{q_{nk}}$ to be such that $(Q_{k+1}X_{k+1}) \ldots (Q_m X_m) M(b_{q_{n1}}, \ldots, b_{q_{nk}}, X_{k+1}, \ldots, X_m)$ is true for I''.

For any n, $M(b_{q_{n1}}, \ldots, b_{q_{nm}})$ is true for I''. Let P'_1 be $M(b_{q_{11}}, \ldots, b_{q_{1m}})$ and let P'_n be the product matrix of P'_{n-1} and $M(b_{q_{n1}}, \ldots, b_{q_{nm}})$. Let I'_{nj} be the set of ground sentences in the jth conjunction of P'_n. Then since P'_n is true for I'' for any n, there must be a ϕ' such that the $\phi'(n)^{\text{th}}$ conjunction of P'_n is true for I'' and $I'_{n\phi'(n)} \subseteq I'_{n+1\phi'(n+1)}$ for any n. Since I'' is complete and $\bigcup_{n=1}^{\infty} I'_{n\phi'(n)} \subseteq I''$, there is a unique completion I' of $\bigcup_{n=1}^{\infty} I'_{n\phi'(n)}$ such that $I' \subseteq I''$. Since for any n, $M(b_{q_{n1}}, \ldots, b_{q_{nm}})$ is true for I', S also is true for I'.

From the manner in which the sets $I'_{n\phi'(n)}$ have been defined, it follows that if J'_{nj} is the set of ground sentences in the jth conjunction of $M(b_{q_{n1}}, \ldots, b_{q_{nm}})$ then there exists a function γ such that $\bigcup_{k=1}^{n} J'_{k\gamma(k)} = I'_{n\phi'(n)}$ for all n. Let J_{nj} be the set of ground sentences in the jth conjunction of $M(a_{q_{n1}}, \ldots, a_{q_{nm}})$ and let ϕ be a function such that $\bigcup_{k=1}^{n} J_{k\gamma(k)} = I_{n\phi(n)}$ for all n. Let the map ψ be the single-valued map of the names $a_1, a_2, \ldots$ onto the names $b_1, b_2, \ldots$ defined by: for all i, $\psi(a_i)=b_i$. Let I^* be the interpretation with ground sentences G as members for which $\psi(G)$ is in I'. Then since $\bigcup_{n=1}^{\infty} I_{n\phi(n)} \subseteq I^*$, there exists a unique completion I of $\bigcup_{n=1}^{\infty} I_{n\phi(n)}$ such that $I \subseteq I^*$. ψ is a homomorphism of I onto I'.

Corollary. For S not in D there exists an interpretation for which S is true if and only if $k_n > 0$ for all n.

Since by (1) if $k_n > 0$ for all n then there exists a function ϕ for which $I_{n\phi(n)} \subseteq I_{n+1\phi(n+1)}$ for all n, and hence by (2) there exists an interpretation for which S is true. Conversely if there is an interpretation for which S is true then by (3) there is such a ϕ and hence $k_n > 0$ for all n.

• *Theorem 2*

When S is in D and $N = \max(1, (m-r+s)^r)$, the sets I_{nj} possess the following properties:

(1) For any n, $1 \leq n < N$, for which $k_{n+1} > 0$ and any j, $1 \leq j \leq k_{n+1}$, there exists an i such that $I_{ni} \subseteq I_{n+1j}$.

(2) For any function ϕ for which $I_{n\phi(n)} \subseteq I_{n+1\phi(n+1)}$ for $1 \leq n < N$, S is true for any completion of $\bigcup_{n=1}^{\infty} I_{n\phi(n)}$.

(3) For any interpretation I'' for which S is true there exists an interpretation I', $I' \subseteq I''$, for which S is true and a function ϕ for which $I_{n\phi(n)} \subseteq I_{n+1\phi(n+1)}$, for $1 \leq n < N$, such that I' is the homomorphic image of some completion of $\bigcup_{n=1}^{N} I_{n\phi(n)}$.

Properties (1) and (2) follow in exactly the same manner as in Theorem 1. The proof of (3) is similar to that in Theorem 1 except that the sequence $b_1, b_2, \ldots$ of names of I'' is finite, the sequence being terminated after all the names in S have been listed and either one member has been chosen for each existentially quantified variable in S, should one exist, or otherwise after only one member has been chosen. Then $(b_{q_{n1}}, \ldots, b_{q_{nm}})$ is defined for $1 \leq n \leq N$. Similarly one can prove:

Corollary. For S in D there exists an interpretation for which S is true if and only if $k_n > 0$ for $n \leq N$.

The program to produce proofs for logically true sentences

The input for the program is a standard form of the negation of the sentence T to be proven. The input data provides, therefore, a matrix form as well as a list of quantifiers with the dependencies of the existential quantifiers indicated. The matrix form data consists of a list of 36-bit computer words each member in the list being an atomic wff, negated or not, or a dividing word to indicate the occurrence of the connective "v". The negated and unnegated atomic wff are expressed within the computer words as follows: The sign bit is used to indicate the occurrence of the negation sign, a negative word being a negated atomic wff. The next five bits are used for the predicate letter while the remaining 30 bits are used for the variables in a manner dependent upon the number of variables. The 30 bits are broken up, as nearly as possible, into equal fields, the number of which is the same as the maximum number of variables attached to predicate letters in the sentence to be proven. Variables are expressed as non-zero binary numbers occupying the fields of the 30 bits. The names which replace the variables are also binary numbers and occupy the same field as the variable they replace. Thus positive and negative ground sentences and unnegated and negated atomic wff are represented in the program within a single computer word in exactly the same manner. There results a limitation on the maximum number of names that can be introduced which is determined by the maximum number of variables attached to predicate letters; for example, if this number is 6 then the maximum number of names that can be introduced is 2^5-1, or 31.

In outline the program is very simple. A substitution generator generates the m-tuples $(a_{q_{n1}}, \ldots, a_{q_{nm}})$, $n=1, 2, \ldots$, and as each is produced it is substituted into the matrix form to produce the matrices $M(a_{q_{n1}}, \ldots, a_{q_{nm}})$. As each matrix $M(a_{q_{n1}}, \ldots, a_{q_{nm}})$, $n>1$, is produced it is multiplied by the previous product matrix P_{n-1} and the resulting product matrix P_n is tested to determine whether or not it has any conjunctions. Should P_n have no conjunctions then a proof has been produced for the sentence to be proven. Should the sentence T to be proven be a member of M and should $n=N$ for T, then if P_n has a conjunction, T is not provable. Should $n<N$ or should T not be a member of M then if P_n has a conjunction the program goes on to produce P_{n+1}.

This outline of the program is not complete, however, since in order to conserve computing time and space, the matrices are coded before being multiplied. A matrix is coded by expressing each of its conjunctions as a pair of 36-bit computer words as follows: As the ground sentences are produced by substitution into the negated and unnegated atomic wff of the matrix form, they receive in turn a code number from 1 to 36, a ground sentence with a negation sign attached receiving the same number as the ground sentence without the negation sign. In the pair of words representing the conjunction of a matrix, a zero in the ith place of the first word indicates that the negative ground sentence with that code number is a member of the conjunction, and a zero in the ith place of the second word indicates that the positive ground sentence with that code number is a member of the conjunction.

The limitation of 36 on the number of distinct positive ground sentences that may appear is severe. Therefore, immediate consideration is being given to writing a program in which each conjunction of a matrix is expressed by two, four, six, et cetera, words as needed, allowing the limitation to be raised to 72, 108, 144, et cetera. However, two features in the program for conserving code numbers make the limitation less severe than would first be apparent.

A ground sentence which appears in every conjunction of some P_n must necessarily appear in every conjunction of P_{n+j} for $j>0$. Hence such common factors can be removed from all of the conjunctions of P_n and put onto a special list called the *truth list*. Since a ground sentence on the truth list does not require a code number, the inclusion in the program of a routine to remove common factors results in a saving of code numbers.

If $I_{nj} \subseteq I_{nk}$, for some j and k, then it is clear from Theorems 1 and 2 that there is no loss in ignoring the set I_{nk} and considering only further the set I_{nj}. For the program this amounts to discarding redundant conjunctions from P_n, a redundant conjunction being one for which another conjunction of P_n exists, of which each ground sentence occurs in the redundant conjunction. After the removal of redundant conjunctions, fewer distinct atomic sentences may occur in a product matrix than before the removal so that code numbers may be freed for reuse.

The main steps in the program can now be fully described.

(1) Should T not be in M or should T be in M and n be not greater than N for T, then generate a new substitution $(a_{q_{n1}}, \ldots, a_{q_{nm}})$ according to the dependencies of existentially quantified variables on universally quantified variables. Should T be in M and should n be greater than N, then T is not logically true and the program stops.

(2) Generate the jth conjunction of $M(a_{q_{n1}}, \ldots, a_{q_{nm}})$ by substituting $a_{q_{n1}}$ for $X_1, \ldots, a_{q_{nm}}$ for X_m in the jth conjunction of the matrix form of the input. If a member of the conjunction contradicts any ground sentence on the truth list, or another ground sentence in the conjunction, discard the conjunction and go to (5).

(3) Determine the code numbers that have been previously assigned to the ground sentences of the conjunction, and assign new numbers to those ground sentences which have not previously been assigned numbers. Should a code number larger than 36 be required, the program stops. Express the conjunction as a pair of words.

(4) Form the product of the coded j^{th} conjunction with the previous product matrix, dropping from the resulting new product any contradictory conjunctions. As each conjunction of the new product matrix is formed, test to see if it is redundant or if it makes an already appearing conjunction of the new product matrix redundant, and store it according to the results of this test.

(5) Check to see if the new product matrix is complete; that is check to see if j is the number of conjunctions in the matrix form. If the product matrix is not complete go to (2) to generate the $j+1^{\text{th}}$ conjunction; if it is, go to (6).

(6) Check to see if the product matrix is empty. If it is, the proof is complete and the program stops. If it is not, remove common factors from the product matrix and put them onto the truth list. Determine and record all code numbers that have been freed for reuse.

The only change that would be necessary in order to use the program for proving theorems in an axiomatic theory with axioms $S_1, S_2, \ldots$, would be in (1) and (2). Instead of only generating substitutions for a matrix form obtained from the negation of the theorem T to be proven, the program would also have to generate substitutions for matrix forms obtained from the axioms. The remainder of the program would be unchanged.

Table 1 **The sentences to be proven.**[11]

(1) $(Ex)(Ay)(Az)\{[((Fy \supset Gy) \equiv Fx) \& ((Fy \supset Hy) \equiv Gx) \& (((Fy \supset Gy) \supset Hy) \equiv Hx)] \supset (Fz \& Gz \& Hz)\}$.

(2) $(Ex)(Ey)(Az)\{[(Fxz \equiv Fzy) \& (Fzy \equiv Fzz) \& (Fxy \equiv Fyx)] \supset (Fxy \equiv Fxz)\}$.

(3) $(Ex)(Ay)(Az)\{[((Fyz \supset (Gy \supset Hx)) \supset Fxx) \& ((Fzx \supset Gx) \supset Hz) \& Fxy] \supset Fzz\}$.

(4) $(Ex)(Ey)(Az)\{(Fxy \supset (Fyz \& Fzz)) \& ((Fxy \& Gxy) \supset (Gxz \& Gzz))\}$.

(5) $\{[(Ax)(Ey)(Fxy \vee Fyx) \& (Ax)(Ay)(Fxy \supset Fyy) \supset (Ez)Fzz\}$.

(6) $(Ax)(Ey)(Px \supset (Py \vee Qy))$, where the atomic wff "$Px$" is replaced by: $(Eu)(Av)(Fux \supset (Gvu \& Gux))$, the atomic wff "$Py$" is replaced by a corresponding wff, and the atomic wff "Qy" is replaced by: $(Au)(Av)(Ew)((Gvu \vee Hwyv) \supset Guw)$.

(7) $\{[(Ax)(Kx \supset (Ey)(Ly \& (Fxy \supset Gxy))) \& (Ez)(Kz \& (Au)(Lu \supset Fzu))] \supset (Ev)(Ew)(Kv \& Lw \& Gvw)\}$.

(8) (3) in which the atomic wff "Hx" is replaced by: $(Au)(Ev)Huvx$, and the atomic wff "Hz" is replaced by a corresponding wff.

(9) $(Ax)(Ey)(Az)\{(Pyx \supset (Pxz \supset Pxy)) \& (Pxy \supset (-Pxz \supset (Pyx \& Pzy)))\}$, where the atomic wff "$Pxy$" is replaced by: $(Au)(Ev)(Fxuv \& Gyu \& -Hxy)$, and the other atomic wffs are replaced by corresponding wffs.

Table 2 **The input.**

	Quantifier List	*No. of Conjunctions*
(1)	$(Ez)(Ax)(Ey)$	18
(2)	$(Ax, y)(Ez)$	2
(3)	$(Ax)(Ey, z)$	4
(4)	$(Ax, y)(Ez)$	4
(5)	$(Ax)(Ey)(Az)$	4
(6)	$(Ex)(Ay)(Ez, u, v)(Aw)(Ex')(Ay', z')$	8
(7)	$(Ex)(Ay)(Ez)(Av, w)$	9
(8)	$(Ax)(Ey, z, u)(Av)(Ew)$	6
(9a)	$(Ex)(Ay)(Ez, u)(Av)(Ew)(Ax')(Ey')(Az', u', v')$	21
(9b)	$(Ex)(Ay)(Ez, u, v, w)(Ax')(Ey', z')(Au', v', w')$	21

Results from production runs[10]

The sentences for which proofs were attempted by the program are given in Table 1. Information about the quantifier list and matrix function provided by the sentences of Table 1 are given in Table 2. Finally in Table 3 the results of the production runs are given. No names appear in any of the sentences used as inputs. None of the inputs to the program are in the class *D* although Example (1) is a member of class *M*.

In evaluating the difficulty of Examples (6), (8), and (9), it is important to recognize that although they are described as substitutions into easily proven theorems (the sentence of (9) into which the substitution is made can be proven by the program in less than one-hundredth of a minute), they are themselves only easily proven when

Table 3 **The results.**

	Status	*Time*	*Proof*	*No. of Subst.*	*No. of Conj.*	*Truth L.*	*Code No.*
(1)	yes	0.01	yes	4	4	10	0
(2)	no	0.01	no	11	2	0	36+
(3)	yes	1.42	yes	13	590	53	33
(4)	yes	21*	no	7	2900+	7	30
(5)	yes	0.01	yes	3	2	4	5
(6)	yes	0.12	yes	27	24	16	14
(7)	yes	0.01	yes	5	6	4	12
(8)	yes	0.74	no	10	150	6	36+
(9a)	yes	15.06	no	3	1900	3	36+
(9b)	yes	21*	no	6	1850	5	35

Status indicates whether or not the sentence is logically true. *Time* is given in minutes. *Proof* indicates whether or not a proof was produced. *No. of Subst.* is the number of times a new matrix was generated. *No. of Conj.* is the maximum number of conjunctions in any product matrix. *Truth L.* is the number of entries on the truth list. *Code No.* is the maximum number of code numbers used. The two starred examples were manually stopped.

the substitutions are recognized. When the substitutions are disguised, as they are in the input data for the program, the difficulty of the example is considerably increased.

One run was used to evaluate that part of the program which removes redundant conjunctions from the product matrix, as this portion of the program consumes a large proportion of the running time when the product matrix is large. The result was to prove conclusively the value of this portion of the program.

A number of examples other than those used in production runs were considered but rejected because they are all too easy for the program; that is, the program can produce a proof for them in less than one-hundredth of a minute. Included in such examples are all of the syllogisms. Indeed, as the syllogisms belong to that special class of sentences which are decidable by the program, the program can decide of any syllogism whether or not it is valid within one-hundredth of a minute.

In Table 2, the dependencies indicated by the order of the quantifiers in the quantifier list were the only ones indicated. Two different equivalent quantifier lists for Example (9) were tried. The number of conjunctions is the number in the matrix function. Variables in a group with *E* or *A* indicate a sequence of quantifiers.

Conclusions

Without considering the results of the production runs, it might be concluded that the program had little chance of success. With each multiplication the number of conjunctions in the product matrix can increase rapidly. But the removal of contradictory conjunctions and redundant conjunctions, in some cases at any rate, keeps the number of conjunctions down to a manageable size. The limitation on the number of coding numbers, although quite stringent, is not as serious in practice as would first appear. The results certainly encourage the writing of programs with double or triple the present number of coding numbers. Pessimism regarding the program is confirmed, however, in one respect. If the quantifier list of the input contains m universal quantifiers, then in order to consider all possible substitutions of up to k names into the universally quantified variables it is necessary to generate k^m matrices and form their product. Thus, for example, in (9a) if all possible substitutions of up to only three individuals are to be considered then 3^6 or 729 matrices must be generated and multiplied, the product matrix of which can have up to 21^{729} conjunctions! Considering the number of multiplications performed by the program in production runs, it is clear that success can only be had with such a problem by a program which is more discriminating in the matrices that it generates and multiplies. Example (4) is a good case in point. A very short proof of this sentence can be produced by hand simulation if a very obvious refinement is added to the routine for generating matrices. Nevertheless, the program has succeeded in producing proofs for moderately complicated sentences. For simple logical deductions such as the syllogisms, the program has very fast logical facility.

References and footnotes

1. Actually, from a theorem of Church in "An Unsolvable Problem of Elementary Number Theory," *American Journal of Mathematics,* **58,** 345-363, it can be concluded that there can exist no effective process for the construction of interpretations for sentences. In particular, therefore, there can exist no effective process for determining, in general, whether or not it is true that $k_n>0$ for all n.
2. E. W. Beth, "Semantic Entailment and Formal Derivability," Amsterdam, North Holland Publishing Co., 1955. K. J. J. Hintikka, "Form and Content in Quantification Theory," appearing in "Two Papers on Symbolic Logic," Helsinki, *Acta Philosophica Fennica,* Fasc. VIII, 1955.
3. G. Gentzen, "Untersuchungen über das logische Schliessen," *Mathematische Zeitschrift* **39,** 176-210, 405-431 (1934-5). J. Herbrand, *Recherches sur la théorie de la démonstration,* Travaux de la Société des Sciences et des Lettres de Varsovie, Classe III sciences mathématiques et physiques, No. 33 (1930).
4. A. Newell, J. C. Shaw and H. A. Simon, "Empirical Explorations of the Logic Theory Machine: A Case Study in Heuristics," *Proceedings of The Western Joint Computer Conference,* 218-230 (1957). Further references can be obtained from this paper.
5. H. L. Gelernter and N. Rochester, "Intelligent Behavior in Problem-Solving Machines," *IBM Journal,* **2,** 336-345 (1958). See also, H. Gelernter, "Realization of a Geometry Theorem Proving Machine," *Proceedings of the International Conference on Information Processing, Paris,* 1959.
6. B. Dunham, R. Fridshal and G. L. Sward, "A Non-Heuristic Program for Proving Elementary Logical Theorems," *Proceedings of the International Conference on Information Processing, Paris,* 1959.
7. Hao Wang, "Toward Mechanical Mathematics," see p. 2, this journal.
8. There has come to my attention the work of D. Prawitz, H. Prawitz, and Neri Voghera of Stockholm. A brief outline of their work appears in the *Proceedings of the International Conference on Information Processing, Paris,* 1959, in the discussion of the session on theorem proving.
9. A stronger result can be proven but is not needed for the purposes of this paper. For the stronger result I' is such that any sentence is true for I' if and only if it is true for I''.
10. First announced in a paper "A Program for the Production of Proofs for Theorems Derivable within the First Order Predicate Calculus from Axioms," *Proceedings of the International Conference on Information Processing, Paris,* 1959.
11. Alonzo Church, *Introduction to Mathematical Logic,* Princeton University Press, 1956, is the source of Example 1 to 4 and 9. They are, respectively, **Ex. 3,** p. 262; **2,** p. 265; **1,** p. 262; **5,** p. 265; and **2,** p. 262. Example 7 appears in Rosser, J. B., *Logic for Mathematicians,* McGraw-Hill Book Co., New York, p. 150 **Ex. (e).** Example 6 and the modification for Example 9 are due to J. D. Rutledge.

Commentary by the Author

Preface to 'An improved proof procedure'

As explained in the preface to the paper in this volume by myself, H. Prawitz, and N. Voghera, it had soon become obvious that the main weakness of the first programs for theorem proving was the manner of generating substitution instances of universal formulas that were assumed as hypotheses and of existential formulas that were to be proved. The generation was restricted by considering only those constants that had already been introduced, but otherwise it was essentially by random. The aim of the present paper was to overcome this predicament by introducing a method for calculating the values that were to be substitued for the quantified variables of the problematic kind. The way in which I conceived the situation may be illustrated by the problem of finding integral solutions to, say, second order equations. We may of course run through the integers in some order, testing whether they are roots, and in some lucky cases this may be the fastest way, but fortunately, we know a method that allows us to determine the roots by reasoning or indeed by direct calculation. Similarly, it should be possible to calculate which substitutions, if any, give rise to tautological formulas.

Since there is no method to decide in general whether suitable substitutions exist, i.e. to find the negative answer, what the proposed method does is firstly to calculate which substitutions, if any, create a tautology when <u>one</u> instance of each given formula is generated; secondly, if there is no such substitution, to calculate which substitutions, if any, create a tautology when <u>two</u> instances of each given formula are generated; and so on. The calculation proceeds by the use of so-called dummies, which may be understood as syntactical variables (meta-variables) that take symbols for constants as values, and to which are attached different indices to keep control of what constantsymbols the dummies can stand for. Instead of using indices,one can replace the constants by Skolem functions, which would make the restrictions on p. 117 automatic and would simplify the exposition, although, in essence, it comes to the same.

A part of the idea of the method could be summarized by saying that no substitution should be made that cannot contribute to the proof; or more precisely, that the values of the dummies should be determined only to the extent needed to make one atomic formula the negation of another one in the same clause. It has been generally agreed upon that this part of the method, which was also furthered by Stig Kanger as stated in the paper (n. 11; see also Kanger's paper in this volume), is essential, and it has come to general use in automated theorem proving as an integral part of Robinson's resolution method.

But there is another part of the method that does much more than so, and that has received relatively little attention. A set of substitution values that makes a pair of atomic formulas complementary in one clause may be inconsistent with a set that makes a pair complementary in another clause, and may therefore, after all, not be wanted. The method discards substitutions that cannot be part of a system of substitutions that makes <u>every</u> clause tautological, and thus determines such a complete system of substitutions, if any exists for the number of instances of the given formulas tried at the moment.

This part of the method was not built into the resolution method, and as far as I know, one has not seriously studied the question whether the cost of this sorting out the substitution values arising in a first round in order to find a small system of substitution values sufficient for the desired proof does not more than balance the alternative expenses from making non productive substitutions.

When I now reread my paper, I find the explanation of the method in part I quite readable, while I cannot say this about part II. The soundness and completeness proofs in § 6 are not satisfactory, and for the theoretical justification of the method, I want to refer the reader to my more condensed but more mature paper, <u>Advances and Problems in Mechanical Proof Procedures</u>, which is included in the second volume of this collection of papers, and in which also one obvious weakness of the present method on the side of propositional logic is overcome. The rather unwieldy § 7 is concerned with several smaller refinements of the method which could be of importance if a method like this one was implemented on a machine.

A minor classification and a misprint: The claim in the third paragraph of § 1.2 that the number of saf: s of A is finite for a fixed number of applications of TR3 are to be understood with the provisos that the constants substituted at application of TR2 are restricted to those already occurring in the sequent in question or, if there is no such constant, to the alphabetically first constant, and that differences because of different choices of constants at applications of TR1 are disregarded (or the choice is fixed by reference to the alphabetic order). - On p. 125, line 5 from the bottom, read $\underline{a}_n$, for $\underline{a}_{\underline{n}}$.

An Improved Proof Procedure

D. Prawitz

PART I

Introduction

I will here deal with effective procedures for finding a proof of any valid sequent

(1) $$\Gamma \rightarrow \Theta$$

where Γ and Θ are finite sequences of closed well-formed formulas in the predicate (functional) calculus of first order.[2] The existence of such an effective procedure was first shown by Skolem [20] and [21]. Here, we will be concerned with the problem of constructing procedures that could be used in practice.[3]

[1] This work is partly included in a project sponsored by Statens tekniska forskningsråd (Sweden). – I am indebted to professor Wedberg, docent Kanger, fil. lic. Berg, and Mr Voghera for reading the manuscript and making valuable suggestions (see also n. 11).

[2] (1) can be read: if all the formulas of Γ hold, then some formula of Θ holds. Let $\Gamma_{\&}$ be the conjunction of the formulas of Γ and Θ_{v}, the disjunction of the formulas of Θ. Then, (1) is equivalent to the formula $\Gamma_{\&} \supset \Theta_{v}$, or the formula $\sim\Gamma_{\&}$ if Θ is empty, or the formula Θ_{v} if Γ is empty (I will always suppose that one of Γ and Θ is not empty). – I will usually use expressions autonymously in the sequel.

[3] The existence of an effective procedure for proving formulas in the predicate calculus also follows from Gödel's completeness theorem for the predicate calculus (which states that the theorems and the valid formulas of the predicate calculus are the same; the proof of this theorem, however, is built on the Skolem's ideas), since it is always possible to enumerate the expressions that are derivations effectively. (That kind of enumeration procedures is known as "the British museum method" or the "fifty million monkeys technique" from someone's remark that, given long enough, an army of monkeys would be able to type out all the books in the British museum. The

Skolem also showed that the demonstration proving that a sequent is valid can be given in a certain *normal form,* often called Herbrand's normal form; the statement of this fact has later been known as *Herbrand's theorem.* The different effective proof procedures for the predicate calculus of first order described in the literature are all built on variants of Herbrand's theorem and consist simply in generating certain expressions (namely, the saf:s of the formula to be proved as defined below) until one is found which is a proof (in normal form) of the formula to be proved. However, if such a procedure should be usuable in practice, it has to be very discriminating in generating expressions. Rather than generating expressions until the desired one is found, the procedure should by calculations determine the desired proof. An attempt towards constructing such a procedure will be made here.

I will start with discussing Herbrand's theorem and some generalizations of it in § 1. The proof method proposed in this paper is introduced by some examples in § 2 and is stated more exactly in § 3. In §§ 4 and 5 the merits of the method and the possibility of having a machine to execute the procedure is discussed. These sections (1–5) constitute the first part of the paper. Part II contains a demonstration (built on Herbrand's theorem) of the soundness and completeness of the method (§ 6), some possible modifications of the method (§ 7), and an example of how the method sometimes can be used to show that no proof of a formula is possible (§ 8) (the example is essentially Herbrand's case of the decision problem).

§ 1.1. *Herbrand's theorem*

I will use the following version of *Herbrand's theorem* as starting point:[4]

---

first to perform such procedures mechanically seems to have been R. Lull with his machine Ars Magna in the fourteenth century; see Bowden [4] p. 317.)

[4] Herbrand's theorem can be found e.g. in Hilbert and Bernays [9] and Dreben [6] in somewhat different versions from the one given above (cf. § 7.5 n. 24). For other bibliographic references concerning theorems related to Herbrand's see § 1.3. (In its original formulation, Herbrand's theorem speaks about provability instead of validity.)

If A is a formula in prenex normal form, then the sequent $\rightarrow A$ *is valid if and only if it is demonstrable from a valid (tautological) sequent of the propositional calculus by applications of the following three rules of proof.*

PR 1.
$$\frac{\rightarrow \Theta, F(c/x), \Theta'}{\rightarrow \Theta, \mathrm{U}xF, \Theta'}$$ [5]

where c does not occur in the sequent below the line.

PR 2.
$$\frac{\rightarrow \Theta, F(c/x), \Theta'}{\rightarrow \Theta, \mathrm{E}xF, \Theta'}$$

PR 3.
$$\frac{\rightarrow \Theta, \mathrm{E}xF, \Theta', \mathrm{E}xF, \Theta''}{\rightarrow \Theta, \mathrm{E}xF, \Theta', \Theta''}$$ [6]

Let us introduce three *rules of transformation*, TR 1–3, where TR i (i=1, 2, 3) is the rule of transforming the sequent below the line into the sequent above the line in PR i (observing the restriction on c in PR 1); thus, a reverse application of PR i. Let us then define a *sequent of associated formulas of A* (abbreviated: *saf of A*) as a sequent of quantifier-free formulas that can be obtained from $\rightarrow A$ by a series of applications of TR 1–3. A saf of A can thus be written in the form

$$\rightarrow A_1, A_2, \ldots, A_n$$

where A_i (i=1, . . ., n) – which we call an associated formula of A – is obtainable from A by deleting all the quantifiers and substituting constants for the variables.

From Herbrand's theorem we can derive:

(2) *If A is in prenex normal form, then the sequent* $\rightarrow A$ *is valid if and only if there exists a valid (tautological) saf of A.*

---

[5] c and x are used to denote individual constants and variables respectively. The notation $F(a/b)$ where a and b are individual variables or constants, is used to denote the result of substituting an occurrence of a for every free occurrence of b in F. The rule is to be understood as stating that the sequent below the line follows from the one above the line.

[6] PR 1 and 3, as distinguished from PR 2, also hold in the reverse direction, i.e. the premiss follows from the conclusion. This fact is essential for the proof procedures we are to consider.

It is clear that (2) above affords an effective proof method, because if A is in prenex normal form, the following holds

(a) we can enumerate the saf:s of A,

(b) for every saf of A we can decide by propositional calculus whether it is valid or not,

(c) thus, by searching through the saf:s of A testing whether they are valid, we will sooner or later find a valid one, if A is valid,

(d) from a valid saf of A we easily give a proof of $\rightarrow A$ by using the rules of proof PR 1–3 (we may of course omit this last step by simply referring to (2) saying that A is valid if there is a valid saf of A).

To prove a valid sequent $\rightarrow A$, we thus simply generate the saf:s of A in some appropriate order by making different series of applications of the transformation rules TR 1–3 until a valid saf of A is found. In this paper we will consider the problem of so applying the transformation rules that a valid saf of A is found in an expeditious manner.

There are three aspects of this problem corresponding to the following three respects in which the saf:s of A – or rather the series of applications of the transformation rules made in order to obtain the saf:s of A – may differ: They may differ in respect to

(i) the constants substituted at the applications of TR 1 and TR 2,

(ii) which formulas TR 3 has been applied to and the number of such applications, and

(iii) the relative order in which the transformation rules have been applied.

The method to be proposed will determine how the transformation rules shall be applied to a formula A in respect to (i)–(iii) above in order to give a valid saf of A (if such a saf of A exists). Previous proof procedures have found the valid saf of A by trying different combinations of the transformation rules in some arbitrarily defined order until one eventually is found that gives a valid saf of A.

Let us examine the points (i)–(iii) above. First we note that the validity is preserved at the applications of TR 1 and TR 3 (cf. n. 6). The applications of TR 1 that have to be made in order to obtain a saf of A are determined by the quantifiers of A and the applications of TR 3, and can therefore be considered unproblematic as long as we take care to introduce a new constant when replacing the variable.

However, when applying TR 2, it is essential what constant is substituted if validity is to be preserved. In previous proof methods the appropriate constants are found by trying them in some arbitrarily defined order (e.g. an alphabetic one $c_1, c_2, \ldots$) until an appropriate constant is found. The constants tried in this manner may be restricted to the constants already occurring in the sequent to which TR 2 is to be applied.[8] Suppose that there are m constants occurring in a sequent $\rightarrow A$ and that we have to make n applications of TR 2 to obtain a saf of A. Then there are m^n different combinations of applications of TR 2 giving different saf:s of A, one or a few of which may be valid. As m and n grow with increasing complexity of A, the number m^n of different combinations of applications of TR 2 among which we have to search in order to find an appropriate one soon gets intolerably great. Therefore, previous proof methods can in practice be used to prove only rather simple theorems (cf. § 4 and § 5).[9]

It may not be possible to preserve the validity when applying TR 2 without first applying TR 3. Hence, there is also the problem (ii) above of determining to which formulas TR 3 has to be applied and the number of such applications necessary to obtain a valid saf of A. This problem will further be discussed in § 1.2.

Finally, because of the fact that we have to introduce a new constant when making substitutions at an application of TR 1, the saf:s of A are also dependent on the relative order in which the applications of the transformation rules are made. The order between the applications of TR 1 and 2 made to *one* associated for-

---

[8] This restriction is made in the proof procedures developed by Beth [2], Hintikka [10], Kanger [11], Quine [18], and Schütte [19].

[9] The intention of the scholars who have developed proof methods has not only been to give a method for finding derivations but has often primarily been to facilitate the presentation of the completeness proof for the predicate calculus.

mula in a saf of A is of course determined by the order in which the quantifiers stand in A. In the proof method to be proposed the order between the different applications of the transformation rules will conveniently be determined when the order matters and is not determined in the way just described.

§ 1.2. *Applications of TR 3*

To obtain a valid saf of A, we usually have to apply TR 3 to some formulas a certain number of times. Let us consider two methods for finding the appropriate applications of TR 3 in order to obtain a valid saf of A.

The *first method* is the one used in all previous proof procedures that contain a method for enumerating the saf:s of A. The saf:s of A are formed in an order $S_1, S_2, \ldots, S_i, \ldots$, where S_1 contains one formula and S_{i+1} is similar to S_i but is obtained by one more applications of TR 3 (and necessary applications of TR 1 and 2 following this application of TR 3). If A is valid, we finally obtain a valid sequent S_n in this way provided that we vary the constants substituted at applications of TR 2 and the formulas to which we apply TR 3. The sequents in the list $S_1, S_2, \ldots$ that we have to test for validity, will however usually be unmanageably long before we reach the valid sequent S_n (because of all the associated formulas obtained by inappropriate applications of TR 2 not contributing to making S_n tautological).

The number of saf:s of A that differ in respect to the properties (i)–(iii) listed above in § 1 is finite for every fixed number of applications of TR 3. Thus there is a *second method* in which we first form all the saf:s of A obtainable by no application of TR 3, then all the saf:s of A obtainable by one application of TR 3, then all the saf:s of A obtainable by two applications of TR 3, and so on. If A is valid, we finally reach the smallest valid saf of A in this way (i.e. the valid saf of A that contains the least number of formulas; the number of formulas of a saf of A is obviously one greater than the number of applications of TR 3 made in order to obtain the saf).

The method to be proposed here will like the second method

find the smallest valid saf of A, but it will not be necessary to consider all the saf:s of A obtainable by a given number of applications of TR 3. Furthermore, it will also be possible to determine what formulas TR 3 is to be applied to without first having to try the different possibilities.

§ 1.3. *Existing proof methods*

The proof method obtained from Herbrand's theorem has been improved in some respects (besides the improvement of restricting the constants substituted at applications of TR 2, see n. 8). Gentzen [7] extended it to hold for not only one but a number of formulas written in the sequent form like (1) (the same result was later also obtained by Quine [18]). We cover this extension in *Gentzen's theorem* by slightly modifying Herbrand's theorem as formulated in § 1.1:

If the formulas of Γ and Θ are in prenex normal form, then the sequent $\Gamma \to \Theta$ is valid if and only if it is provable from a tautological sequent by using the following six rules of proof.

Three rules, PR ia ($i=1, 2, 3$), are like PR i in Herbrand's theorem except that we add Γ to the antecedents of the sequents that occur in these rules. The other three rules, PR ib ($i=1, 2, 3$) are the duals of PR ia; namely

PR 1b. $$\frac{\Gamma, F(c/x), \Gamma' \to \Theta}{\Gamma, \mathrm{E}xF, \Gamma' \to \Theta}$$

where c does not occur in the sequent below the line.

PR 2b. $$\frac{\Gamma, F(c/x), \Gamma' \to \Theta}{\Gamma, \mathrm{U}xF, \Gamma' \to \Theta}$$

PR 3b. $$\frac{\Gamma, \mathrm{U}xF, \Gamma', \mathrm{U}xF, \Gamma'' \to \Theta}{\Gamma, \mathrm{U}xF, \Gamma', \Gamma'' \to \Theta}$$

Let TR ia and b ($i=1, 2, 3$) be the reverse application of PR ia and b like before.

Gentzen's result, in turn, was generalized to hold for formulas also not in prenex normal form by Beth [2], Hintikka [10],

Kanger [11], and Schütte [19]. To cover this extension, we have to generalize the rules of proof to cover also formulas with quantifiers not initially placed (cf. § 7.2). If we introduce rules of transformation corresponding to such rules of proof and speak about a saf of a sequent $\Gamma \rightarrow \Theta$ accordingly, we obtain the following more general version of (2):

The sequent $\Gamma \rightarrow \Theta$ is valid if and only if there exist a valid (tautological) saf of $\Gamma \rightarrow \Theta$.

§ 2. *A first outline of the proof method*

Before getting into technical details in § 3 let us consider some examples to get an intuitive picture of the proof method that is to be developed. The proof of the following sequent was discussed by Beth [3]:

$$\mathrm{U}x\mathrm{U}y(Px \vee Qy) \rightarrow (\mathrm{U}xPx \vee \mathrm{U}yQy)$$

Let us write this sequent in the equivalent form

(4) $$\mathrm{U}x\mathrm{U}y(Px \vee Qy) \rightarrow \mathrm{U}xPx, \mathrm{U}yQy$$

To prove (4) we now try to find a tautological saf of (4). First we apply TR 1a twice and obtain the sequent

(5) $$\mathrm{U}x\mathrm{U}y(Px \vee Qy) \rightarrow Pc_1, Qc_2$$

We have now to make two applications of TR 2b. If we only consider substitutions of c_1 and c_2, there are four different ways in which the two applications of TR 2b can be made. If we substitute c_1 for both x and y we obtain

$$(Pc_1 \vee Qc_1) \rightarrow Pc_1, Qc_2,$$

if we substitute c_1 for x and c_2 for y we obtain

$$(Pc_1 \vee Qc_2) \rightarrow Pc_1, Qc_2,$$

if we substitute c_2 for x and c_1 for y we obtain

$$(Pc_2 \vee Qc_1) \rightarrow Pc_1, Qc_2,$$

and if we substitute c_2 for both x and y we obtain

$$(Pc_2 \vee Qc_2) \rightarrow Pc_1, Qc_2.$$

Only one of these substitutions, namely the one of substituting c_1 for x and c_2 for y, gives a tautological saf of (4).[10] In order to find the appropriate substitutions in a convenient manner, we may first substitute dummies – syntactical variables, say d_1 and d_2, standing for arbitrary constants – for the variables when we are to apply TR 2a or 2b.[11]

Applying this device to (5), we get

(6) $$(Pd_1 \vee Qd_2) \rightarrow Pc_1, Qc_2$$

The problem is now to determine the values of d_1 and d_2 so that (6) becomes tautological. A solution is found most easily by rewriting (6) in conjunctive normal form:

(7) $$\rightarrow (\sim Pd_1 \vee Pc_1 \vee Qc_2) \;\&\; (\sim Qd_2 \vee Pc_1 \vee Qc_2)$$

or if we also compound sequents with connectives, we can write

(8) $$(Pd_1 \rightarrow Pc_1, Qc_2) \;\&\; (Qd_2 \rightarrow Pc_1, Qc_2).$$

By just comparing the atomic formulas of each conjunction clause of (7) or (8), we now find that the value $d_1 = c_1$ makes the first cluse tautological and that the value $d_2 = c_2$ makes the second tautological.[12] This system of the values of the dummies makes

[10] Beth [3] remarks, "This example shows that an intelligent selection ... (of the substitutions) ... simplifies the construction. It remains to be seen, in how far it would prove convenient to imitate this intelligent way of proceeding in a logical theory machine." He seems to think of finding a solution on the lines of a statistical analysis of which substitutions prove successful on formulas that are decidable. The solution proposed here is quite different but well suited for mechanical use. Cf. the remarks on heuristic methods in n. 22.

[11] Docent Kanger has much promoted the idea of using dummies in this connection. The use of dummies (in the form of free variables standing for arbitrary constants) was suggested in Prawitz [15], an essay for the seminar for theoretical philosophy at the University of Stockholm in 1957, and was discussed at the seminar. At these discussions docent Kanger advocated and furthered the idea, which will also be included in Kanger [12]. I thank him for stimulating talks on these topics.

[12] This is of course the point: We do not need to try the different values of the dummies to find the values that make the conjunction clauses of (8) (e.g) tautological. Instead we can just put identity sign between the arguments of two occurrences of the same predicate sign that stand on different side of the arrow.

thus (7) and (8) tautological and hence also the sequent (6), which becomes a valid saf of (4).

To see how the outlined technique works on a more complicated example, let us consider the proof of the following sequent

(9) $$\rightarrow \mathrm{U}x\mathrm{E}y\mathrm{U}z((Px \supset \sim Pz) \supset \sim Py).$$

Applying TR 1 and substituting c_1 for x, we obtain

$$\rightarrow \mathrm{E}y\mathrm{U}z((Pc_1 \supset \sim Pz) \supset \sim Py).$$

Substituting a dummy d_1 for y when applying TR 2, we obtain

$$\rightarrow \mathrm{U}z((Pc_1 \supset \sim Pz) \supset \sim Pd_1)$$

to which we again apply TR 1 substituting c_2 for z and obtaining

(10) $$\rightarrow ((Pc_1 \supset \sim Pc_2) \supset \sim Pd_1).$$

We rewrite this sequent in conjunctive normal form and obtain

(11) $$(Pd_1 \rightarrow Pc_1) \ \& \ (Pd_1 \rightarrow Pc_2)$$

As in the first example, we now find that the first sequent of the conjunction (11) becomes tautological for the value $d_1 = c_1$ and that the second sequent becomes tautological for the value $d_1 = c_2$. This, however, does not solve our problem to prove (9), because, in the first place, d_1 could clearly not be the constants c_1 and c_2 simultaneously; c_1 and c_2 are supposed to be two distinct constants as they have to be if the second application of TR 1 above shall be correct (*restriction 1*, see below).

In the second place, the value $d_1 = c_2$ is not possible. c_2 is substituted at an application of TR 1 that succeeded the application of TR 2 at which d_1 was substituted. Hence, if d_1 was c_2, then c_2 would already have occurred in the sequent to which TR 1 was applied and the application of TR 1 would thus have been incorrect. (We will express this fact by the inequality $d_1 < c_2$, which could be read "c_2 is introduced later than d_1" and which thus excludes $d_1 = c_2$; *restriction 3*, see below.)

Thus we see that all values of the dummies of a sequent that are obtained by the technique described above, can not be accepted, because – although the values make the sequent tautological – some of them may not make the sequent a saf of the sequent to be proved. We will therefore put some *restrictions* on the values of the dummies. Two of them, called restriction 1 and restriction 3, are illustrated by the example above.

Continuing with the example above, we find that the only system of dummy values that make (11), and hence also (10), tautological contradicts the restrictions (i.e. does not make (10) a saf of the sequent (9) to be proved). A valid saf of (9) must thus contain at least two formulas, thus requiring at least one application of TR 3.

If we had applied TR 3 before applying TR 2, we would have obtained

$$\rightarrow \mathrm{EyUz}((Pc_1 \supset \sim Pz) \supset \sim Py), \mathrm{EyUz}((Pc_1 \supset \sim Pz) \supset \sim Py)$$

Applying TR 2 and TR 1 to the first formula of this sequent as before and TR 2 and TR 1 to the second formula in a similar manner now substituting a dummy d_2 for y and a constant c_3 for z, we obtain the sequent

$$(12) \qquad \rightarrow ((Pc_1 \supset \sim Pc_2) \supset \sim Pd_1), ((Pc_1 \supset \sim Pc_3) \supset \sim Pd_2)$$

Rewriting this sequent in conjunctive normal form we obtain

$$(13) \qquad (Pd_1, Pd_2 \rightarrow Pc_1) \ \& \ (Pd_1, Pd_2 \rightarrow Pc_1, Pc_3) \ \& \\ (Pd_1, Pd_2 \rightarrow Pc_2, Pc_1) \ \& \ (Pd_1, Pd_2 \rightarrow Pc_2, Pc_3)$$

Every sequent of (13) now becomes tautological for a number of values of the dummies. The first sequent e.g. becomes tautological if $d_1 = c_1$ or $d_2 = c_1$. The expression $(d_1 = c_1 \vee d_2 = c_1)$, which states the necessary and sufficient condition of the first sequent being a tautology, is called the *identity condition* associated with the sequent in question. When forming the identity conditions associated with the different sequents of (13), we immediately omit values contradicting restriction 3 (i.e. $d_1 = c_2$ and $d_2 = c_3$). The conjunction of the identity conditions is then as follows:

$$(d_1 = c_1 \vee d_2 = c_1) \ \& \ (d_1 = c_1 \vee d_1 = c_3 \vee d_2 = c_1) \ \& \\ (d_1 = c_1 \vee d_2 = c_2 \vee d_2 = c_1) \ \& \ (d_1 = c_3 \vee d_2 = c_2)$$

Rewriting this conjunction in disjunctive normal form and omitting redundant clauses we obtain

$$(d_1=c_1 \mathbin{\&} d_1=c_3) \vee (d_1=c_1 \mathbin{\&} d_2=c_2) \vee$$
$$\vee (d_2=c_1 \mathbin{\&} d_1=c_3) \vee (d_2=c_1 \mathbin{\&} d_2=c_2).$$

A conjunction of the so obtained expression represents a system of values of the dummies that makes (13), and hence also (12), tautological. The first and fourth system, however, contradict restriction 1. Only the systems $(d_1=c_1 \mathbin{\&} d_2=c_2)$ and $(d_2=c_1 \mathbin{\&} d_1=c_3)$ make (12) both a tautology and a saf of (9).

To illustrate *restriction 2* we consider the proof of a sequent S of the form

$$S: \qquad \rightarrow \mathrm{E}x\mathrm{U}y\mathrm{E}zF$$

where F is a quantifier-free formula. Substituting the dummies d_1 and d_2 for x and z and a constant c_1 for y at the applications of TR 2 and TR 1, we obtain the sequent S_1

$$S_1: \qquad \rightarrow F(d_1/x)(c_1/y)(d_2/z).$$

Suppose now that S_1 becomes a tautology for no values of the dummies. Then we have to make an application of TR 3, applying it either to $\rightarrow \mathrm{E}x\mathrm{U}y\mathrm{E}zF$ or to $\rightarrow \mathrm{E}zF(d_1/x)(c_1/y)$. To determine which application is the appropriate one, our method will work as follows. We generate a new formula from F by substituting dummies d_3 and d_4 for x and z and a constant c_2 for y, and add it to the sequent S_1, obtaining the sequent S_2

$$S_2: \qquad \rightarrow \mathrm{F}(d_1/x)(c_1/y)(d_2/z), \mathrm{F}(d_3/x)(c_2/y)(d_4/z).$$

We now associate identity conditions with the conjunction clauses of S_2 transformed to conjunctive normal form as before, and if we obtain a system of dummy values which makes S_2 valid and which contains the identity $c_1=c_2$, then TR 3 was to have been applied to the sequent $\rightarrow \mathrm{E}zF(d_1/x)(c_1/y)$. In that case, d_1 has to be equal to d_3 if S_2 is to be a saf of S. This is the content of restriction 2, in this case stating: if $c_1=c_2$, then $d_1=d_3$.

§ 3. *The proof method*

The proof procedure is carried out in steps. First there are four *preparatory steps*. Then, there are four steps that may be repeated in the same order any number of times. The procedure is thus naturally divided into *cycles*, and the steps are called the *phases* of respective cycle.

We now suppose that we are to prove a sequent $\Gamma \rightarrow \Theta$ comprising the formulas

$$F_1, \ldots, F_f, F_{f+1}, \ldots, F_g \tag{14}$$

where F_i, $i \leq f$, is a Γ-formula and F_j, $f < j \leq g$, is a Θ-formula. I will speak about the number of a formula thinking of the enumeration (14). Also the results of transforming these formulas will be numbered and divided into Γ- and Θ-formulas in the same way. If individual constants occur in the formulas in (14), we will suppose they are the first n ones in an enumeration $c_1, c_2, \ldots$

§ 3.1. *The preparatory steps of the procedure*

Preparatory step I. Write every formula in prenex normal form.[18]

Definition I. In the formulas resulting from step I, existential quantifiers (E) in Γ-formulas and universal quantifiers (U) in Θ-formulas are called *C-quantifiers* (constant-generating quantifiers; cf. the condition on c in PR 1 a and b). The other quantifiers, U in Γ-formulas and E in Θ-formulas, are called *D-quantifiers* (dummy-generating quantifiers; cf. the preliminary explanations in § 2).

Preparatory step II. Make the following transformation of every formula resulting from step I: Delete every C-quantifier (with its attached variable) not preceded by a D-quantifier and substitute a constant for all (remaining) occurrences of the quantified variable. We use c_{n+1}, c_{n+2}, ... in this order as constants; c_1

[18] The proof procedure will be shortened if we follow certain rules when carrying out this step see § 7.1. I will also consider an alternative method not requiring step I; see § 7.2.

$c_2, \ldots, c_n$ are supposed to be the constants that already occur in the formulas (see § 3). We call the constants c_k ($k=1, \ldots, n, n+1, \ldots$) *simple constants* to distinguish them from other constants introduced below.

Preparatory step III. Make the following transformation of every formula resulting from step II: Delete every C-quantifier (with its attached variable) and substitute a constant $c(p,q,1)$ for all (remaining) occurrences of the variable. The constant $c(p,q,1)$ is called a complex constant and is to be such that

p denotes the number of the formula in which the quantifier occurs (cf. (14)),

q denotes the number of the quantifier in the formula, counting the quantifiers from the left to the right.

In the same way delete every D-quantifier (with its attached variable) and substitute a *dummy* $d(p,q,1)$ for all (remaining) occurrences of the quantified variable.

Preparatory step IV. We now have g quantifier-free formulas

(15) $$F'_1, \ldots, F'_f, F'_{f+1}, \ldots, F'_g.$$

Form the disjunction of the negations of the Γ-formulas and the Θ-formulas:

(16) $$(\sim F'_1 \vee \ldots \vee \sim F'_f \vee F'_{f+1} \vee \ldots F'_g).$$

(16) is called the *origin formula*, abbreviated: *OF*.

§ 3.2. *The phases of a cycle of the procedure*

When starting a new cycle we call it the s:th cycle if the just preceding cycle was the (s-1):th cycle. (The first cycle has the number 1.) Let the cycle now to be described be the t:th cycle ($t=1,2,\ldots$).

Phase I. Transform an instance of the origin formula, OF, by substituting t for the third index in all the dummies and complex constants. The resulting formula is called OF_t.

Phase II. Form the t:th cycle formula, abbreviated: CF_t, defined as follows. The first cycle formula is OF_1 transformed to conjunctive normal form. If the s:th cycle formula is CF_s, then the $(s+1)$:th cycle formula is $(\mathrm{CF}_s \vee \mathrm{OF}_{s+1})$ transformed to conjunctive normal form. (CF_t is thus the result of transforming the formula $(\mathrm{OF}_1 \vee \mathrm{OF}_2 \vee \ldots \vee \mathrm{OF}_t)$ to conjunctive normal form.)

Phase III. (a) If there is a conjunction clause of CF_t in which no predicate symbol (with any number of arguments) occurs both negated and not negated, then the given sequent $\Gamma \rightarrow \Theta$ is not provable and the procedure is concluded. Otherwise we continue and apply (b) below.

Remark: If there is a conjunction clause of CF_t as described in (a) above, then clearly there is such a clause of every CF_s ($s=1, 2, \ldots$). Then no saf of the given sequent $\Gamma \rightarrow \Theta$ can be a tautology, and hence, according to Herbrand's (or Gentzen's) theorem, the sequent $\Gamma \rightarrow \Theta$ is not valid. (Cf. e.g. Church [5] p. 180, which states that a formula of the predicate calculus is not valid if its afp is not a tautology.)

(b) We now associate an *identity condition*, IC_i, with every conjunction clause, M_i, of CF_t ($i=1,2, \ldots, m$, where m is the number of conjunction clauses of CF_t), as follows.

(i) If M_i contains two formulas one of which is the negation of the other, then IC_i is $\mathrm{c}_1=\mathrm{c}_1$.

(ii) If M_i does not contain two formulas as in (i), then for every couple of formulas $Pa_1a_2 \ldots a_k$ and $\sim Pa'_1a'_2 \ldots a'_k$ in M_i (P is a predicate symbol with k arguments consisting of constants or dummies) we form an *identity list*, $\mathrm{IL}_{i,j}$ ($j=1,2,\ldots, n_i$; n_i is supposed to be the number of such couples occurring in M_i), of the form:

$$(a_1=a'_1 \;\&\; a_2=a'_2 \;\&\; \ldots\; a_k=a'_k). \tag{17}$$

IC_i is to be the disjunction of all the n_i identity lists:

$$(\mathrm{IL}_{i,1} \vee \mathrm{IL}_{i,2} \vee \ldots\; \mathrm{IL}_{i,n_i}) \tag{18}$$

Phase IV. The given sequent $\Gamma \rightarrow \Theta$ is provable in cycle t if and only if the conjunction of the identity conditions

$$(\mathrm{IL}_{1,1} \vee \mathrm{IL}_{1,2} \vee \ldots \vee \mathrm{IL}_{1,n_1}) \;\&\; (\mathrm{IL}_{2,1} \vee \mathrm{IL}_{2,2} \vee \ldots \vee \mathrm{IL}_{2,n_2}) \;\&\; \ldots \\ \&\; (\mathrm{IL}_{m,1} \vee \mathrm{IL}_{m,2} \vee \ldots \vee \mathrm{IL}_{m,n_m}) \tag{19}$$

(m is supposed to be the number of conjunction clauses of CF_t;

$n_1, n_2, \ldots, n_m$ are as in (18)) does not contradict the following three *restrictions*. If (19) contradicts the restrictions, we start a new cycle. – Below, k, p, q, s, and x are any positive integers.

Restriction 1. Two different constants are never equal if they are not complex and differ from each other only in respect to the third index. Thus, $c_k \neq c_{k'}$ if $k \neq k'$; $c_k \neq c(p,q,s)$; and $c(p,q,s) \neq$ $\neq c(p',q',s')$ if not $p=p'$ and $q=q'$.

Restriction 2. If $c(p,q,s)=c(p,q,s')$, then for all $x<q$: $d(p,x,s)=d(p,x,s')$.[15]

Restriction 3. If $q<q'$, then $d(p,q,s)<c(p,q',s)$.

Discussion. The identities can obviously be considered as formal expressions belonging to a formal system that comprises the ordinary rules for propositional calculus, equality and inequality (such as laws for the symmetry and transitivity of equality) and the additional rules and postulates stated in the restrictions 1–3. The question to be answered in phase IV is then whether we can infer a contradiction ($a=a'$ & $a \neq a'$) or ($a<a'$ & $a \not< a'$) from (19) in this formal system.

To determine whether (19) contradicts the restriction, we can rewrite (19) in disjunctive normal form. Every disjunction clause

$$(IL_{1,j_1} \,\&\, IL_{2,j_2} \,\&\, \ldots \,\&\, IL_{m,j_m})$$
$$(j_i=1, \ldots, n_i;\ i=1, \ldots, m;\ n_i \text{ as in (18)};\ m \text{ as in (19)})$$

of the so obtained expression that does not contradict the restrictions gives a usuable system of values of the dummies, i.e. a system that makes all conjunction clauses of CF_t tautological without contradicting the restrictions. In this way we get all such systems of the value dummies.

When determining in practice whether (19) contradicts the restrictions, several short cuts can be used. First, we can omit every identity list that in itself contradicts the restrictions. If all

[15] To simplify the soundness proof (§ 6.1), we give the following formulation of restriction 2:

Restriction 2'. Like restriction 2 except for the consequence to which we add: and $c(p,x,s)=c(p,x,s')$.

It is easily verified that the question whether (19) contradicts the restrictions is not effected if we use restriction 2' instead of restriction 2.

the identity lists of an identity condition are omitted in this way, then the given sequent is of course not provable in the cycle and we have to start another one (cf. also § 8 about not provable formulas).

Second, we can note that the same identity list occurs in a number of identity conditions at regular intervals (similar to the intervals defined in § 7.3). Using this fact when we rewrite (19) in disjunctive normal form, we can obtain a much shorter formula than the one we would get if we just applied the distributive laws of conjunction and disjunction in the ordinary way. As the problem of determining whether (19) contradicts the restrictions belongs to propositional calculus, we may also benefit from the many investigations in this field, e.g. the research in methods for simplifying truth functions to which Nelson [13] and Quine [17] among others have contributed.

Also worth noticing is the fact that it is sufficient to find *one* system of truth values of the truth constituent of (19) that makes (19) true (or, what is the same thing, to find *one* system of dummy values that makes CF_t tautological) but does not contradict the restrictions. We can therefore use the so-called feedback principle. Special logical machines have recently been built to solve just that kind of problem (see Bowden [4], ch. 15, especially pp. 188–192).

§ 4. *An example*

To demonstrate what we gain with the described proof method, I will consider a very simple example:

$$(20)\quad UxUyUzUuUvUw((Pxyzu \,\&\, Pxyzv \,\&\, Quvw) \supset Pxyzw) \rightarrow UxUyUzUuUvUw((Pxyzu \,\&\, {\sim}Pxyzv \,\&\, Pxyzw) \supset {\sim}Quwv)$$ [16]

[16] To make the example more interesting, we may interpret *Pxyzu* as saying that the point *u* is in the plane determined by the points *x*, *y*, and *z*, and *Qxyz* as saying that the point *z* is on the line determined by the points *x* and *y*. Then, (20) is the statement that from the premiss that any point (say *w*) on a line (determined by points *u* and *v*) is in any plane (say one determined by points *x*, *y*, and *z*) to which the line belongs, we can infer the consequence that a point not in a plane to which a line belongs, is not on the line.

The two formulas are equivalent as is easily seen. One is got from the other by a tautological transformation of the matrix and a change of the places of v and w only in the matrix.

If we apply the proof method to (20), we first substitute constants $c_1, \ldots, c_6$ for the variables of the second formula (preparatory step II) and substitute dummies $d(1,q,1)$, where $q=1,\ldots,6$, for the variables in the first formula (preparatory step III). To save space we omit all the indices of the dummies except the q-index and use $d_1, \ldots, d_6$ as dummies. The first cycle formula (phase II) is composed of four conjunction clauses, M_i, $i=1,\ldots,4$. The following four formulas occur in all the conjunction clauses:

$$Pc_1c_2c_3c_4,\ \sim Pc_1c_2c_3c_5,\ Pc_1c_2c_3c_6, \text{ and } Qc_4c_6c_5.$$

Besides these formulas M_1 contains the formula $\sim Pd_1d_2d_3d_4$, M_2, the formula $\sim Pd_1d_2d_3d_5$, M_3, the formula $\sim Qd_4d_5d_6$, and M_4, the formula $Pd_1d_2d_3d_6$.

When we form the identity lists (phase III), we can immediately omit the ones that contain an identity between constants (according to restriction 1 in phase IV). We then get the following identity conditions

$IC_1 : IL_{1,1} \vee IL_{1,2} : (d_1=c_1 \mathbin{\&} d_2=c_2 \mathbin{\&} d_3=c_3 \mathbin{\&} d_4=c_4) \vee (d_1=c_1 \mathbin{\&}$
$\mathbin{\&} d_2=c_2 \mathbin{\&} d_3=c_3 \mathbin{\&} d_4=c_6)$

$IC_2 : IL_{2,1} \vee IL_{2,2} : (d_1=c_1 \mathbin{\&} d_2=c_2 \mathbin{\&} d_3=c_3 \mathbin{\&} d_5=c_4) \vee (d_1=c_1 \mathbin{\&}$
$\mathbin{\&} d_2=c_2 \mathbin{\&} d_3=c_3 \mathbin{\&} d_5=c_6)$

$IC_3 : IL_{3,1} : (d_4=c_4 \mathbin{\&} d_5=c_6 \mathbin{\&} d_6=c_5)$

$IC_4 : IL_{4,1} : (d_1=c_1 \mathbin{\&} d_2=c_2 \mathbin{\&} d_3=c_3 \mathbin{\&} d_6=c_5)$

Every $IL_{i,j}$ above gives a system of dummy values that does not contradict the restrictions but makes M_i a tautology. If we form the conjunction of the identity conditions (like (19) in phase IV) and rewrite it in disjunctive normal form, we find that only one disjunction clause of the so obtained expression does not contradict the restrictions, namely $(IL_{1,1} \mathbin{\&} IL_{2,2} \mathbin{\&} IL_{3,1} \mathbin{\&} IL_{4,1})$; i.e. only the system $d_1=c_1$, $d_2=c_2$, $d_3=c_3$, $d_4=c_4$, $d_5=c_6$, $d_6=c_5$ makes all conjunction clauses of CF_1 tautologies without contradicting the restrictions.

If we had used any other proof method available in literature (cf. § 1), we could also have started with introducing six constants for the variables in the second formula. But when making substitutions in the first formula, we would have 6^6 different combinations of the constants to choose between. As we have seen, only one of these combinations of substitutions (namely the one of substituting $c_1, c_2, c_3, c_4, c_6, c_5$ in this order) gives a tautological saf of (20). It can also rather easily be shown that there is no valid saf of (20) that contains a number of formulas but not the formula obtained by the substitutions mentioned above. Thus, to find a valid saf of (20), we would have to form between 1 and 6^6 associated formulas of the first formula in (20). The exact number depends on the order in which we form the associated formulas; i.e. the order in which we make the substitutions in the first formula. Since we have no method to choose a favourable order, we reckon $\frac{6^6}{2} = 23{,}328$ as the mean number of the associated formulas that we have to form.[17] We also have to determine whether

[17] If we use the substitution order that is defined by Dreben [6] or Kanger [11], we have to form 39,842 associated formulas before finding the appropriate one (i.e. the one needed in a saf of (20)). (This number can be calculated by using the fact that the number of n-tuples of positive integers whose sum is less or equal to m, is $\binom{m}{n}$. The order defined by Dreben is such that of two combinations of constants the one comes before whose sum of the indices is less (the constants are supposed to be given in a list $c_1, c_2, \ldots$). Dreben also uses constants with indices greater than those of the constants introduced when replacing C-quantified variables; i.e. in this case indices greater than 6. Kanger uses the constant c_1 in place of such constants, but otherwise his order agrees with Dreben's. Thus, all the associated formulas of the first formula of (20) where the sums of the indices of the constants substituted are less than 21, come before the appropriate associated formula, and there are $\binom{20}{6} = 38{,}760$ such associated formulas. The combinations of constants where the sums of the indices agree, are ordered in lexicographic order, and there are 1,082 combinations where the sums are 21 and which come before the combination of constants substituted in the tautological associated formula. Thus, in total, there are $38{,}760 + 1{,}082 = 39{,}842$ associated formulas that come before the appropriate one.)

Beth [2] partially defines an order in which we would have to form be-the associated formulas formed are tautologies. Suppose that we use the first of the two methods discussed in § 1.2. We then get a saf of (20) that contains 23,329 formulas. When determining whether this saf is tautological, we get a formula with $4^{23,329}$ conjunction clauses if we use the method of rewriting the formula in conjunctive normal form. (As some conjunction clauses can be shown to be tautological at an early stage before the whole clause has been formed, the number of conjunction clauses can be somewhat diminished. It can however easily be shown that – using Beth's or Kanger's terminology – we get more than $2^{23,315}$ different semantical subtableaux or deduction branches, respectively, when proving (20). Even a fast electronic computer would need more time to carry out such a derivation than the earth is old.)

§ 5. *Logical machines*

The high speed of modern electronic computers makes it natural to attempt to use these machines for handling proof procedures. In the case of the propositional calculus, mechanical methods have been used for a long time.[18] The possibility to perform more extensive logical derivations mechanically, has been discussed several times, e.g. by Beth [2] and [3] (in Beth [3] references can also be found to discussions by Gelernter and Robinson). To realize this possibility for the predicate calculus, a mechanized method in the form of a pseudo-program was developed by D. Prawitz [15], building on Beth [2] and Kanger [11]. A

tween $5^6 = 15{,}625$ and $6^6 = 46{,}656$ associated formulas before finding the appropriate one. (In Beth's order all combinations of constants with indices less than n are substituted before any combination of constants containing a constant with the index n.)

[18] The first machine to solve problems in the propositional calculus was constructed by Jevons in 1869. Several logical machines have recently been built, which accomplish about the same thing as Jevon's machine but much faster (due to the development of electronical technique and sometimes due to the use of more subtle logical principles). More numerous than machines designed exclusively for logical problems, are all the existing programs designed to solve various problems in the propositional calculus by ordinary electronic digital computers. For further details see Bowden [5] ch. 15.

program for an electronic digital computer [19] was subsequently prepared by H. Prawitz in 1957 and modified and tested by N. Voghera in 1958.[20] A similar program has also been developed by P. C. Gilmore in 1959.[21] With some limitations pertaining to memory space, these programs are designed to prove any theorem and refute some non-theorems of the predicate calculus. As the programs rely upon the kind of proof procedures discussed in § 1 where one in some arbitrarily defined order has to try the different possibilities of applying the transformation rules to find the appropriate applications, the program can in practice be used to prove only very simple theorems, time being the restrictive factor.

The proof procedure developed in this paper is meant to be realizable in an electronic computer. Such a realization would considerably enlarge the range of theorems mechanically provable.[22] Still, many further improvements are necessary to obtain

---

[19] Facit EDB, AB Åtvidabergs Industrier, Stockholm.

[20] An account of the project will appear in Prawitz D., Prawitz H., and Voghera [16].

[21] Gilmore's work was presented at the International Conference on Information Processing arranged by Unesco in 1959. It was also mentioned at the conference that a similar work by H. Wang is forthcoming. (Added in proof: Since then, Gilmore and Wang have separately published papers on the subject in IBM Journal of Research and Development, vol. 4 (1960). Another program has been suggested but not yet programmed for machine by Davis and Putnam in "A computational proof procedure", Rensselaer Polytechnic Institute, AFOSR TR 59–124, October 1959. This program is superior to the other three in respect to the propositional part but is similar in the respect of being based on the same procedure of exhausting the different possibilities of applying the transformation rules.)

[22] Originating with Newell and Simon [14], there is some research in proof methods with a different approach favouring the use of anthropomorphic heuristics more than algorithms. A heuristic is then meant to be a method which sometimes leads astray but often gives short cuts as compared with the complete algorithms. After having remarked that a proof procedure building on an enumeration procedure of the kind described in n. 3 rapidly exhausts the capacity of the machine, Gelernter [8], concludes, "The remaining alternative is to have the machine rely upon heuristic methods".

This conclusion seems to be somewhat hasty (although the research in heuristics definitely seems to be very interesting), as we can not exclude the possibility of finding fast proof algorithms not relying upon enumerations of

a mechanical proof procedure able to solv more interesting problems. (An example of another improvement is a proof method developed by Kanger [12], which comprises rules for equality and signs for operations, such as addition and multiplication. Such rules and signs are advantageous when applying the proof method to most mathematical theories.)

PART II

§ 6.1. *The soundness of the proof method*

To see that the proof method is sound, i.e. that any sequent provable with the method is valid, we make use of Gentzen's theorem as stated in § 1.3 as follows.

Suppose that the given sequent $\Gamma \rightarrow \Theta$ is provable with the proof method in cycle t and suppose that the formulas are in prenex normal form. Then the conjunction of the identity conditions (19) obtained in phase IV of cycle t does not contradict the three restrictions listed in phase IV. Thus there is a conjunction of identity lists of the form

$$(21) \qquad (\mathrm{IL}_{1,j_1} \,\&\, \mathrm{IL}_{2,j_2} \,\&\, \ldots \,\&\, \mathrm{IL}_{m,j_m})$$

($j_i \leq n_i$; $i=1, \ldots, m$; n_i as in (18); m is the number of conjunction clauses of CF_t) that does not contradict the restrictions (cf. the discussion following phase IV in § 3.2). Make the following three transformations of CF_t (the cycle formula obtained in phase II of cycle t):

(a) If it follows from (21) and restriction 2′ (see n. 15) that

the hopeless kind. For example, the propositional part of the method of semantic tableaux (as developed by Beth [2]) could be considered as a systematization of heuristic ideas resulting in a very usuable algorithm. The machine programs mentioned above, relying upon this method, are superior also in their propositional parts to the heuristic methods developed and programmed by Newell and Simon; the former being both complete and faster. Beth's [3] assumption that the value of heuristic devices in the field of propositional calculus is insignificant seems justifiable. Another example of a proof algorithm is the proof method proposed in this paper. It seems reasonable to assume that it would be of no advantage to use heuristics in place of this algorithm.

two constants with different cycle numbers are equal, then substitute the constant with less cycle number for the one with higher cycle number in CF_t.

(b) If it follows from (21) and restriction 2 that a dummy is equal to a constant, then substitute the constant for the dummy in CF_t. If the dummy is equal to several constants, choose the one with the least cycle number. (Because of restriction 1, it can not follow from (21) that a dummy is equal to two constants which differ in other respects than the cycle number.)

(c) Substitute c_1 for the remaining dummies in CF_t transformed according to (a) and (b).

The result of so transforming CF_t is clearly a tautology and is also equivalent to the disjunction

$$(22) \qquad (OF'_1 \vee OF'_2 \vee \ldots \vee OF'_t)$$

(cf. phase II), where OF'_i $(i=1,\ldots,t)$ is like OF_i (see phase I) except for differences in the arguments of the predicate symbols (due to the transformations (a)–(c)). Thus, OF'_i $(i=1,\ldots,t)$ can be written

$$(23) \qquad (\sim F^i_1 \vee \ldots \vee \sim F^i_f \vee F^i_{f+1} \vee \ldots \vee F^i_g)$$

where F^i_j $(j=1,\ldots,g)$ is like F'_j in (15) (preparatory step IV) except for differences in the arguments of the predicate symbols. Let Γ_i $(i=1,\ldots,t)$ be the conjunction $(F^i_1 \,\&\, F^i_2 \,\&\ldots\&\, F^i_f)$ and let Θ_i be the disjunction $(F^i_{f+1} \vee F^i_{f+2} \vee \ldots \vee F^i_g)$. Then, as (22) is a tautology, it follows that so is

$$(25) \qquad \Gamma_1, \Gamma_2, \ldots, \Gamma_t \rightarrow \Theta_1, \Theta_2, \ldots, \Theta_t.$$

We now have to verify that the given sequent $\Gamma \rightarrow \Theta$ is obtainable from (25) by applying the six rules of proof of Gentzen's theorem (i.e. that (25) is a saf of $\Gamma \rightarrow \Theta$). (If there is a formula in Γ or Θ in which only C-quantifiers occur, it is necessary to extend the rules PR 3a and b to apply also to E- and U-quantifiers respectively. Below, I will suppose that the rules are so extended.)

Let us speak about a series of appropriate application of the proof rules to a formula in (25) meaning a series of such applications that would result in a formula in Γ or Θ; an *appropriate application* of the proof rules is then an application that could belong to a series of appropriate applications to a formula in (25).

Suppose now that no appropriate application of the proof rules can be made to (25) or that there are appropriate applications of the proof rules that applied to the sequent (25) give a sequent S which is different from $\Gamma \rightarrow \Theta$ but to which no appropriate application of the proof rules can be made. I will show that this assumption has the consequence that (21) contradicts the restrictions.

S contains a number of formulas, say $G_1, G_2, \ldots, G_n$, that are different from the formulas in Γ and Θ (otherwise applications of the extended PR 3 a or b would give $\Gamma \rightarrow \Theta$). The reason for there being no more appropriate applications of the proof rules to S must be that such a one would require an application of PR 1 a or b where the constant c, called the *critical constant*, has occurrences, called *critical occurrences*, besides the ones that should be quantified by the application of PR 1 a or b in question (otherwise there would be another appropriate application of the proof rules to S).

Call the critical constant in G_j $(1 \leq j \leq n)$ a_j. If there is a critical occurrence of a_j in G_j, then (21) contradicts the restrictions, which is seen as follows. The critical occurrence of a_j must have been substituted for a dummy, d, in G_j at the transformation (b) above, i.e. the identity $a_j = d$ follows from (21) (and restriction 2). This contradicts restriction 3 as $d < a_j$ according to that restriction.

Suppose there is no critical occurrence of a_j in G_j for any j. Then the critical occurrence of a_1 occurs in another formula than G_1, say G_2. Again, a_1 must have been substituted for a dummy d in G_2, and we have $a_1 = d$ (from (21)) and $d < a_2$ (restriction 3), which gives $a_1 < a_2$. If a_2 occurs in G_1 we also get $a_2 < a_1$. Hence, using the same argument n' times $(n' \leq n)$ we get $a_1 < a_2 < \ldots < a_n < a_j$ where $1 \leq j < n'$.

§ 6.2. *The completeness of the proof method*

To prove the completeness of the proof method, i.e. that every valid sequent is provable with the method, suppose that the sequent $\Gamma \rightarrow \Theta$ is valid and that the formulas are in prenex normal form. According to Gentzen's theorem (as stated in § 1.3), there is then a tautological saf of $\Gamma \rightarrow \Theta$. Let S be a tautological saf of $\Gamma \rightarrow \Theta$ and let n be the highest number of associated formulas of some formula in $\Gamma \rightarrow \Theta$ that occur in S. Choose S so that n is as small as possible. Add formulas to S so that the resulting sequent S' contains n associated formulas of every formula in $\Gamma \rightarrow \Theta$ in such a way that S' is still a saf of $\Gamma \rightarrow \Theta$. (In case $\Gamma \rightarrow \Theta$ contains a formula with only C-quantifiers it is necessary to extend PR 3 a and b to apply also to quantifier-free formulas.) Write the sequent S' as one formula in conjunctive normal form and call the result CF'_n. Clearly, every conjunction clause of CF'_n is a tautology.

Let CF_n be the n:th cycle formula obtained by applying the proof method to the sequent $\Gamma \rightarrow \Theta$ (phase II). It is easily verified that if CF_n is written in an appropriate conjunctive normal form CF_n and CF'_n differ from each other only in respect to the arguments of the atomics. Clearly, there is now a function, f, which takes the constants and dummies in CF_n as arguments and the constants in CF'_n as values in such a way that if a constant or dummy, a, has an occurrence o in CF_n, then $f(a)$ has such an occurrence o' in CF'_n that o and o' occupy corresponding places in CF_n and CF'_n.

As CF'_n is a tautology, there is a couple of disjunction clauses such that one is the negation of the other in every conjunction clause M'_i of CF'_n. Choose such a couple from every conjunction clause of CF'_n. Corresponding to such a couple there is an identity list, IL_{i,j_i}, that belongs to the identity clause M_i of CF_n (phase III). It can now be shown that the conjunction of these identity lists

(28) $$(IL_{1,j_1} \,\&\, IL_{2,j_2} \,\&\, \ldots \,\&\, IL_{m,j_m})$$

(m is the number of conjunction clauses of CF_n) does not contradict the restrictions listed in phase IV. Then the conjunction of the identity conditions (19) does not either contradict the restrictions, and hence the sequent $\Gamma \rightarrow \Theta$ is provable in cycle n with the proof method (phase IV).

That we can not infer any contradiction from (28) by applying restriction 1–3 in phase IV and the laws of the propositional calculus with equality and inequality, is seen as follows.

(a) If the identity $a=a'$ occurs in (28), then $f(a)=f(a')$ (because of the construction of (28)).

From the restriction on c in PR 1 a and b we get (b) and (c) below.

(b) If a and a' are two constants in CF_n such that $a \neq a'$ according to restriction 1, then $f(a) \neq f(a')$.

(c) If $f(a)=f(a')$ and a and a' are two constants in CF_n like the ones in the premis of restriction 2, then for all dummies a_1 and a'_1 in CF_n like the ones in the consequence of restriction 2, it holds that $f(a_1)=f(a'_1)$.

Let the constants in CF'_n be $c_1, c_2, \ldots$, and let us adopt the convention that the constant c in PR 1 a or b is the constant with the least index that does not occur in the sequent below the line. If $i<j$, then let us say that $c_i < c_j$. Then we get (d) below.

(d) If a is a dummy $d(p,q,s)$ and a' a constant $c(p,q',s)$ and $q<q'$, then $f(a)<f(a')$.

(e) Any application of the rules of propositional calculus with equality and inequality to equalities and inequalities between dummies and constants in CF_n is of course also possible to equations and inequalities between the corresponding values of the function f.

(a)–(e) give by induction that any inference about identities between constants and dummies $a_1, \ldots, a_k$ that can be drawn from (28) with the help of restrictions 1–3 and the propositional calculus with equality and inequality, also holds for $f(a_1), \ldots, f(a_k)$. Thus if a contradiction ($a=a'$ & $a \neq a'$) or ($a<a'$ & $a \not< a'$) were inferable from (28), then $f(a)=f(a')$ and $f(a) \neq f(a')$ or $f(a) < < f(a')$ and $f(a) \not< f(a')$. Hence no contradiction can be inferred from (28), which thus does not contradict the restrictions.

I will now in § 7 consider some modifications of the proof method.

§ 7.1. *Rules when transforming formulas to prenex normal form*

The proof procedure will be shortened if we follow certain rules when transforming the formulas to prenex normal form (preparatory step I). When using these rules, we have to know whether a quantifier will become a C- or D-quantifier when the formula is finally brought in prenex normal form. We call a U-quantifier that will become a C-quantifier a *CU-quantifier*, and in the same way we speak about *CE-*, *DU-*, and *DE-quantifier*. Before applying the rules, we contract the scoope of the quantifiers as much as possible according to the ordinary rules. We then transform to prenex normal form applying the following rules when possible:

Rule 1. When there is an alternative as regards the order in which to write a CU- and a DE- or a CE- and a DU-quantifier, place the CU- or the CE-quantifier to the left of the DE- or the DU-quantifier.[23]

Rule 2. Transform

(a) $(\mathrm{CU}xA \mathbin{\&} \mathrm{CU}yB)$ to $\mathrm{CU}z(A(z/x) \mathbin{\&} B(z/y))$
(b) $(\mathrm{DU}xA \mathbin{\&} \mathrm{DU}yB)$ to $\mathrm{DU}v \ldots \mathrm{DU}w(A(v/x) \mathbin{\&} B(w/x))$
(c) $(\mathrm{CE}xA \vee \mathrm{CE}yB)$ to $\mathrm{CE}z(A(z/x) \vee B(z/y))$
(d) $(\mathrm{DE}xA \vee \mathrm{DE}yB)$ to $\mathrm{DE}v \ldots \mathrm{DE}w(A(v/x) \vee B(w/y))$.

x, y, z, v, w denote variables, and A and B, formulas. z, v, and w are supposed not to appear free in A or B, and v and w are supposed to be distinct. In place of the dots in rule 2(b) and (d), there may be a number of other quantifiers. It is advantageous in 2(b) and (d) to place C-quantifiers between the D-quantifiers, thus applying rule 1. We may change the place of the two D-quantifiers for this end.

The advantage of rule 1 is easily seen by observing that restriction 3 is less often applicable if we follow this rule. Thus the number of cycles required for proving some sequents is lessened.

The effect of rule 2 is that different constant occurrences more often become occurrences of the same constant (rule 2(a) and 2(c)) and different dummy occurrences more often become occurrences of different dummies (rule 2(b) and 2(d)) (i.e. after the transformations in preparatory step II and III). The following situation will then be less frequent: a_1 and a_3 are constants such

[23] Rule 1 is also suggested in Quine [18].

that $a_1 \neq a_3$ according to restriction 1 in phase IV, a_2 is a dummy, and the identities $a_1 = a_2$ and $a_2 = a_3$ are members of the identity lists formed in phase III. As $a_1 = a_2$ and $a_2 = a_3$ give $a_1 = a_3$, the conjunction of the identity conditions will less often contradict restriction 1 if we apply rule 2.

Sometimes an application of rule 1 makes it impossible to apply rule 2(a) or (c). There seems to be no simple rule for choosing the most advantageous application in these cases.

§ 7.2. *Omitting the transformation to prenex normal form*

The procedure will sometimes be shorter if we do not transform the formulas to prenex normal form. Still, the best strategy is not always to contract the scope of the quantifiers as much as possible. Sometimes it will be more advantageous to apply one of the rules 2(a) and 2(c) in § 7.1. (The problem of choosing between these two alternatives, is the same as that of choosing between an application of rule 1 and 2(a) or 2(c) in § 7.1.) I will briefly outline how the proof method could be modified if preparatory step I was omitted.

A quantifier is then a C- or D-quantifier according as it would be a C- or D-quantifier if the formula was transformed to prenex normal form (cf. definition 1, § 3.1).

We introduce a fourth index of the dummies and the complex constants, called *scope index*. Let the third index, r, in $c(p,q,r,s)$ and $d(p,q,r,s)$ be the scope index and let the other indices be as before (preparatory step III). The scope index of a well-formed part of a formula F is defined as follows:

(a) F has the scope index 1.

(b) If G is a well-formed part of F and has the scope index n, then if G is $\sim H$, UxH, or ExH the scope index of H is also n, and if G is $(H \, o \, I)$ where o is any binary sentence connective the scope index of H is $10 \cdot n$ and the scope index of I is $10 \cdot n + 1$.

The scope index of a quantifier Q in a formula F is the same as the scope index of the scope of Q.

The q index of a dummy or a complex constant previously got by counting the quantifiers in a formula from the left to the right (preparatory step III), is now got by counting only the quantifiers in the formula that have the same scope index.

Example: In the formula $Ux(EyF \supset (G \vee Uy{\sim}EzH))$ the scope index of Ux is 1, of Ey, 10, and of Uy and Ez, 111. The q-index of all quantifiers except for Ez is 1. The q-index of Ez is 2.

The scope index is always a sequence of figures 0 and 1, which fact will now be employed when restriction 3 in phase IV is changed for.

Restriction 3'. If $r=r'$ and $q<q'$ or if $r\neq r'$ but the sequence r forms an initial part of the sequence r', then $d(p,q,r,s)< <c(p,q',r',s)$.

§ 7.3. *Interval index*

It is an advantage if two different dummy occurrences are occurrences of different dummies (as is the effect of rule 2(b) and (d) see § 7.1). A device having this effect will now be described.

We transform the origin formula (preparatory step IV) to conjunctive normal form in two steps as follows. First, we write every formula ${\sim}F'_i, i\leq f$, and $F'_j, f<j\leq g$, (index as in (15)), in this form. If indexed in an obvious way, we can then write the p:th formula

(29) $$(K_{p,1} \mathbin{\&} K_{p,2} \mathbin{\&} \ldots \mathbin{\&} K_{p,m_p})$$

where $K_{p,n}$ ($n=1, \ldots, m_p$; m_p is the number of conjunction clauses of the p:th formula or its negation transformed to conjunctive normal form) is a disjunction of atomics and negated atomics. A conjunction clause of the origin formula transformed to conjunctive normal form, called OF*, can be written as a disjunction of g different $K_{p,n}$, one from each of the g formulas. I.e. a conjunction clause of OF* has the form

$$(K_{1,n_1} \vee K_{2,n_2} \vee \ldots \vee K_{g,n_g})$$

where the sequence $n_1, n_2, \ldots, n_g$ is any sequence where $n_p\leq m_p$ ($p=1,\ldots,g$; m_p is as in (29)).

There are $\prod_{p=1}^{g} m_p$ such conjunction clauses of OF*, and we enumerate them so that their corresponding sequences $n_1, n_2, \ldots, n_g$

come in lexicographic order. Every conjunction clause of OF* gets thus a number, and with this enumeration in mind we define the *k:th interval of the p:th formula* as the sequence consisting of i conjunction clauses of OF* having consecutive numbers, such that the first number is j, $i = \prod_{x=p}^{g} m_x$ and $j = 1 + (k-1) \cdot i$.

The characteristic feature of an interval of the p:th formula is that its first member is a conjunction clause of OF* containing $K_{p,1}$ and that the different $K_{p,n}$ ($n=1, \ldots, m_p$; m_p is as in (29)) occur an equal number of times in the interval.

If a dummy d(p,q,1) in a formula F is such that there is no constants c(p,q',1) where $q' > q$, we now note that two occurrences of the dummy that are in different intervals of F, can be treated as occurrences of different dummies. We can add an *interval index* to such dummies. The value of the interval index of an occurrence of a dummy in a formula F is set when we form the conjunction clauses of OF*, so that it becomes equal to the number of the interval of F to which the occurrence of the dummy belongs. Dummies with different interval indices can then be regarded as different when determining whether the conjunction of the identity conditions contradicts the restrictions.

To see that the proof method remains sound if we use the device described in this section, we modify the soundness proof in § 6.1 as follows. OF'_i can now not be written as a disjunction (23). Instead we now make successive transformations of OF'_i starting in the following way. The conjunction of the members of the j:th interval of the g:th formula as they occur in OF'_i (i.e. as they have been transformed according to phase I and (a)–(c) in § 6.1) can be written

$$(I_g^{i,j} \vee K_{g,1}^{i,j}) \,\&\, (I_g^{i,j} \vee K_{g,2}^{i,j}) \,\&\ldots\&\, (I_g^{i,j} \vee K_{g,m_g}^{i,j})$$

where $K_{g,n}^{i,j}$ ($n=1, \ldots m_g$; m_g as in (29)) is like $K_{g,n}$ in (29) except for differences in the arguments of the predicate symbols. OF'_i is the conjunction of these conjunctions when $j=1, \ldots, h_g$, where h_g is the number of intervals of the g:th formula (i.e. $h_g = \prod_{x=1}^{g-1} m_x$; m_x is the number of conjunction clauses of the x:th formula as in (29)). If we apply the distributive law to the conjunction displayed above, we get

$$(I_g^{i,j} \vee (K_{g,1}^{i,j} \,\&\, K_{g,2}^{i,j} \,\&\ldots\&\, K_{g,m_g}^{i,j}))$$

or $(I_g^{i,j} \vee F_g^{i,j})$, where $F_g^{i,j}$ is like F'_g in (15) except for differences in the arguments of the predicate symbols. OF$'_i$ now becomes.

$$(30) \qquad (I_g^{i,1} \vee F_g^{i,1}) \,\&\, (I_g^{i,2} \vee F_g^{i,2}) \,\&\ldots\&\, (I_g^{i,h_g} \vee F_g^{i,h_g})$$

(30) is transformed by appropriately existentializing ("appropriately" is used in the same way as the notion of an appropriate application of the proof rules in § 6.1) all constants in every formula $F_g^{i,j}$ ($j=1, \ldots, h_g$) that have been substituted at the transformations (b) and (c) in § 6.1 for dummies with interval index. The quantified $F_g^{i,j}$ then become like F_g (as obtained after preparatory step I) in respect to the D-quantifiers that are not succeeded by any C-quantifier and the variables quantified by these quantifiers. All the quantified $F_g^{i,1}$, $F_g^{i,2}$, ..., F_g^{i,h_g} then become identical; call such a formula F'^i_g. (30) can then be written

$$((I_g^{i,1} \,\&\, I_g^{i,2} \,\&\ldots\&\, I_g^{i,h_g}) \vee F'^i_g).$$

We continue these transformations in a similar manner with the formulas $(g-1)$, $(g-2)$, ..., $(f+1)$, f, $(f-1)$, ..., 1 in place of the formula g, observing that the last f formulas are to be generalized instead of existentialized as they are negated. The result of so transforming OF$'_i$ can be written

$$(31) \qquad (\sim F'^i_1 \vee \ldots \vee \sim F'^i_f \vee F'^i_{f+1} \vee \ldots \vee F'^i_g).$$

The disjunction of all disjunctions (31) ($i=1, \ldots, t$) is of course valid. We then continue the argument as in § 6.1 with F'^i_p in place of F^i_p ($i=1, \ldots, t$; $p=1, \ldots, g$) except that the sequent (25) now obtained is not a saf of $\Gamma \rightarrow \Theta$ as the formulas of the sequent (25) now obtained contain all the D-quantifiers that are not succeeded by a C-quantifier in the sequent $\Gamma \rightarrow \Theta$. Clearly, the validity of the sequent (25) now obtained also implies the validity of $\Gamma \rightarrow \Theta$.

§ 7.4. *C-quantifiers not succeeded by any D-quantifier*

To simplify the problem in phase IV of determining whether the conjunction of the identity conditions contradicts the restrictions, we can strengthen restriction 2 by adding: If there is no dummy d$(p,q',1)$ where $q'>q$, then c$(p,q,s) \neq$ c(p,q,s').

By making this addition we spare ourselves from considering some unpromising values of the dummies without affecting the question whether the conjunction of the identity conditions contradicts the restrictions. That this is so, is evident from the fact that if there is a valid saf of $\Gamma \rightarrow \Theta$, then there is such a saf with the same number of formulas that is obtained by so applying the transformation rules that the same constant is never substituted at two applications of TR 2 a or b to the same formula. Since TR 3 a or b is applied only to formulas beginning with a D-quantifier, the following is then true: If there is no dummy d$(p,q',1)$ where $q'>q$ and if there is a valid saf of $\Gamma \rightarrow \Theta$, then there is such a saf with the same number of formulas where $f($c$(p,q,s)) \neq f($c$(p,q,s'))$ always holds (f is the same function as in § 6.2).

Hence, if no quantifier prefix of the formulas in Γ and Θ brought into prenex normal form is more complex than $Cx_1Cx_2 \ldots Cx_kDy_1Dy_2 \ldots Dy_mCz_1Cz_2 \ldots Cz_n$ (as is the case in Skolem normal form), then restriction 2 can be omitted and restriction 1 strengthen to exclude any identity between two constants.

§ 7.5. *A proof method without inequality*

We can modify restriction 3 so that we only have to deal with equalities and not with inequalities when determining the question in phase IV whether the conjunction of the identity conditions contradicts the restrictions. The effect of this is to fix the relative order between the applications of the transformation rules, which will often give longer derivations than before, i.e. we will often need more cycles than before to prove a sequent.

First we define the *degree* of a quantifier in the formulas resulting from step II. All D-quantifiers in a formula not succeeded by a C-quantifier have the degree 1, and all C-quantifiers in a formula not succeeded by a D-quantifier have the degree 2. All C-quantifiers succeeded by D-quantifiers of degree r but of no higher degree have the degree $r+1$, and all D-quantifiers succeeded by C-quantifiers of degree r but of no higher degree have the degree $r+1$.

To all dummies and complex constants we then add a fourth index, a degree index, denoting the degree of the corresponding quantifier. Let the third index in $c(p,q,r,s)$ and $d(p,q,r,s)$ be the degree index. We then change restriction 3 for

Restriction 3″. If $c(p,q,r,s)=d(p',q',r',s')$ and either $s=s'$ and $r<r'$ or $s>s'$, then

if $r<r'$ there is some $x<s'$ such that $c(p,q,r,s)=c(p,q,r,x)$, and if $r>r'$ there is some $x\leq s'$ such that $c(p,q,r,s)=c(p,q,r,x)$.

Thus, if $s=s'=1$ and $r<r'$, then $c(p,q,r,s)\neq c(p,q,r,s')$.

That the proof method remains sound after these modifications is seen as follows. Suppose the sequent $\Gamma\rightarrow\Theta$ is provable with the modified proof method. Then there is a conjunction (21) (see § 6.1) of identity lists that does not contradict restrictions 1, 2, 3″. Transform (21) in the following way:

If it follows from (21) and restrictions 2 and 3″ that there is an x, such that $x<s'$ and $c(p,q,r,s)=c(p,q,r,x)$, then choose an $s''<s'$ such that the conjunction of (21) and $c(p,q,r,s)=c(p,q,r,s'')$ does not contradict the restrictions (as (21) does not contradict the restrictions there exists such an s''). Transform (21) by forming the conjunction of (21) and the identity $c(p,q,r,s)=$ $=c(p,q,r,s'')$.

We can now order the constants and dummies as follows.

(i) If it follows from the transformed (21) and restriction 2 that some complex constants $a_1, a_2, \ldots, a_n$, are equal, then they are all to have the same order as the one with the lowest cycle index (i.e. the fourth index).

The constants not ordered by (i) are ordered by (ii)–(iv):

(ii) of two complex constants with different cycle numbers the one comes first whose cycle number is lowest

(iii) of two complex constants with the same cycle number the one comes first whose degree is highest

(iv) all simple constants come first in the enumeration.

(v) If it follows from the transformed (21) and restriction 2 that a dummy is equal to a constant, the dummy is to have the same order as the constant. The other dummies are to have the same order as c_1.

All the constants and dummies are now ordered in such a way that if it follows from (21) restriction 2 and 3 that $a > a'$, then a comes before a' in the order. Thus, (21) does not contradict the restrictions 1, 2, and 3.

To see that the proof method remains complete, we have to make use of another version of Herbrand's theorem than the one stated in § 1.1. The new version is called *Herbrand's theorem 2* (abbreviated: *HT* 2):

If A is in prenex normal form, then the sequent $\rightarrow A$ is valid if and only if there exists a valid saf of A

(32) $$\rightarrow A_1, A_2, \ldots, A_t$$

fulfilling the following conditions concerning the substitution of a constant c_k ($k = 1, 2, \ldots$) for a variable x quantified by a universal quantifier Q:

(i) c_k is different from all constants substituted for variables quantified by other universal quantifiers than Q, and also different from the constants that occur in A (if any);

(ii) if c_k is substituted for the variable x to obtain both A_i and A_j, then every variable quantified by a quantifier to the left of Q in A is replaced by the same constant in A_i and A_j;

(iii) if c_k is substituted for x to obtain A_i but not substituted for x to obtain any A_j where $j < i$, then for every c_m that occurs in the formula A_i as substituted for a variable quantified by a quantifier to the left of Q in A, or that occurs in a formula A_j where $j < i$, it holds that $k > m$.[24]

The completeness proof in § 6.2 can now be used with HT 2 instead of Gentzen's theorem and some other modifications. The

[24] For proof that the validity of $\rightarrow A$ implies the validity of a sequent (32), we can use Dreben [6], who states that if A is valid there exists a certain Herbrand Tautology. It is easily verified that a Herbrand Tautology satisfies the conditions of (32). (For proof of his statement Dereben refers to Gödel.)

construction of CF'_n is now a little different. Let $F_1, \ldots, F_f \rightarrow F_{f+1}, \ldots, F_g$ be the valid sequent $\Gamma \rightarrow \Theta$. The sequent

$$\rightarrow (\sim F_1 \vee \ldots \vee \sim F_f \vee F_{f+1} \vee \ldots \vee F_g)$$

consisting of only one formula, is then also valid. Rewrite this formula in prenex normal form in such a way that the quantifiers of greater degree come before the ones of less degree. Call the obtained formula A. According to HT 2 there is then a valid saf of A. Let the sequent $\rightarrow A_1, A_2, \ldots, A_n$ be one with the least possible n, and use the result of writing $(A_1 \vee A_2 \vee \ldots A_n)$ in conjunctive normal form as CF'_n.

(b) and (c) in § 6.2 now follow from the conditions (i) and (ii), respectively, in HT 2. Finally, (d) is changed for

(d') If $f(a) = f(a')$ and a is a constant in CF_n and a' is a dummy in CF_n both like the ones in the premiss of restriction 3", then it follows from condition (iii) in HT 2 that there exists a constant a_1 in CF_n such that $f(a) = f(a_1)$ and a_1 is like a except for the cycle number, which is less than that of a' if the degree of a is less than that of a' and which is equal or less than the cycle number of a' if the degree of a is greater than that of a'.

§ 8. *Not provable formulas*

In this section, I will show how the proof method can sometimes be used to prove that a given sequent $\Gamma \rightarrow \Theta$ does not hold, i.e. as a refutation method. (A very simple case is already included in (a) of phase III.) The case to be considered is essentially the special case of the decision problem solved by Herbrand for a formula in prenex normal form whose matrix is a disjunction of atomics and negated atomics (see e.g. Ackerman [1], p. 85, Church [5], p. 256, or Hilbert and Bernays [9], pp. 160 and 161). As will be shown below, if a formula is as just described, then it is either provable with our proof method in cycle 2 or it is not provable at all. The case to be treated below is an obvious extension obtained by the following observation: Let QM be a formula in prenex normal form where Q is the quantifier prefix and M is the quantifier-free matrix, and let M be in conjunctive normal form $(M_1 \& M_2 \& \ldots \& M_m)$ where M_i $(i=1, \ldots, m)$ is a disjunction of atomics and negated atomics. Then, QM is not provable if QM_i is not provable, and the latter formula is in the form required in Herbrand's case of the decision problem.

Let us first define an enumeration of the conjunction clauses of a cycle formula CF_s supposing that the origin formula, OF, is written in conjunctive normal form. If s is 1, the i:th conjunction clause, M_i, of CF_s is the same as the i:th conjunction clause of OF_1 (for an enumeration of OF we can use the one defined in § 7.3). If $s > 1$, M_i of CF_s is the disjunction of the j:th conjunction clause of CF_{s-1} and the k:th conjunction clause of OF_s, where $n \cdot (j-1) + k = i$ and n is the number of conjunction clauses of OF. (Thus, M_1 is the disjunction of the first conjunction clause of CF_{s-1} and the first conjunction clause of OF_s; M_2, the disjunction of the first and second conjunction clause of CF_{s-1} and of OF_s respectively; M_{n+1}, the disjunction of the second and first conjunction clause of CF_{s-1} and OF_s respectively, etc.)

Let us now introduce the notion of *simple conjunction clause of type i* as follows. A conjunction clause M_i of CF_1 is a simple conjunction clause of type i. If M_i of CF_s is a simple conjunction clause of type j, then M_k of CF_{s+1} is a simple conjunction clause of type j if $k = (i-1) \cdot n + j$ where n is the number of conjunction clauses of OF transformed to conjunctive normal form. In other words, a simple conjunction clause of type i of CF_s, $s > 1$, is a disjunction

$$(M_i \vee M^2 \vee M^3 \vee \ldots \vee M^s)$$

where M_i is a conjunction clause of CF_1 and M^j $(j=2, \ldots, s)$ can be obtained from M_i by just substituting j for the cycle index (i.e. the last index) of the dummies and complex constants. (A conjunction clause that is not simple can not be obtained from just one conjunction clause of CF_1 in this way.)

The following now holds:

If there is a simple conjunction clause M_i of CF_2, whose associated identity condition IC_i contradicts the restrictions in phase IV, then the given sequent $\Gamma \rightarrow \Theta$ does not hold.

Proof. Suppose that there is a simple conjunction clause M_i of type t of CF_2 such that IC_i contradicts the restrictions. Then, every identity list $IL_{i,j}$ ($j=1, \ldots, n_i$; n_i as in (18)) contradicts the restrictions. $IL_{i,j}$ is a conjunction

$$(32) \qquad (a_1=a'_1 \ \& \ a_2=a'_2 \ \& \ \ldots \ \& \ a_k=a'_k)$$

(cf. (17)), where the cycle indices of all the a_h ($h=1, \ldots, k$) are identical and similarly for all the a'_h.

We now consider another simple conjunction clause, $M_{i'}$, of type t of some arbitrary cycle formula CF_s. Then we notice that every identity list $IL_{i',j}$ of $IC_{i'}$ in cycle s can be obtained from some identity list (like (32)) of IC_i in cycle 2 where the cycle numbers of a_h and a'_h ($h=1, \ldots, k$) are different (i.e. one is 1 and the other is 2) by substituting and appropriate number s' for the cycle index of every a_h in (32) and an appropriate number s'' for the cycle index of every a'_h in (32) ($h=1, \ldots, k$).

But from the identity list (32) of IC_i in cycle 2, we can infer a contradiction by applying the restrictions 1–3 and rules for the propositional calculus, equality and inequality. It is then easily verified that we can infer a similar contradiction from an identity list that is obtained from (32) by the substitutions mentioned. Indeed, suppose $IL_{i',j'}$ (in cycle s) is obtainable from $IL_{i,j}$ (in cycle 2) by the substitution of s' for the cycle index 1 and s'' for the cycle index 2 in the dummies and complex constants in $IL_{i,j}$. Make the same substitutions in all the dummies and complex constants in the inference (of the kind described above) from $IL_{i,j}$ which ends with a contradiction. We then obtain an inference of the same kind from $IL_{i',j'}$ that ends with the contradiction. Hence all the identity lists of $IC_{i'}$ in cycle s contradict the restrictions, and according to phase IV, the given sequent $\Gamma \rightarrow \Theta$ is then not provable in cycle s.

BIBLIOGRAPHICAL REFERENCES

[1] Ackerman, W., Solvable cases of the decision problem, Amsterdam 1954.
[2] Beth, E. W., Semantic entailment and formal derivability, Medel. der Kon. Nederl. Akad. van Wetensch., deel 18, no. 13, Amsterdam 1955.
[3] Beth, E. W., On machines which prove theorems, Simon Stevin (Wis-en Natuurkundig Tijdschrift), vol. 32 (1958), pp. 49–60.

[4] Bowden, B. V., (ed.), Faster than thought, London 1953.
[5] Church, A., Introduction to mathematical logic, vol. I, Princeton 1956.
[6] Dreben, B., On the completeness of quantification theory, Proc. National Academy of Sciences, vol. 38 (1952), pp. 1047–1052.
[7] Gentzen, G., Untersuchungen über das logische Schliessen, Matematische Zeitschrift, vol. 39 (1934), pp. 176–210 and 405–431.
[8] Gelernter, H., Intelligent behavior in Problem-Solving Machine, IBM Journal of research and development, vol. 2 (1958).
[9] Hilbert, D. and Bernays, P., Grundlagen der Mathematik, vol. 2, Berlin 1939.
[10] Hintikka, K. J. K., Two papers on Symbolic logic (Form and content in quantification theory), Acta Philosophica Fennica, Fasc. VIII, Helsinki 1955.
[11] Kanger, S., Provability in logic, Stockholm 1957.
[12] – Handbok i logik, Filosofiska studier utgivna av seminariet för filosofi vid Stockholms universitet (to be mimeographed in 1959).
[13] Nelson, R., Simplest normal truth function, The journal of symbolic logic, vol. 20 (1955), pp. 105–108.
[14] Newell, A. and Simon, H., The logic theory machine, IRE Transactions on information theory, vol. IT-2, no. 3 (1956), pp. 61–79.
[15] Prawitz, D., Mekanisk bevisföring i predikatkalkylen, Uppsats för seminariet i teoretisk filosofi (mimeographed), Stockholm 1957.
[16] Prawitz, D., Prawitz, H. and Voghera, N., A mechanical proof procedure and its realization in an electronic computer, The journal of the association for computing machinery vol. 7 (1960), pp. 102–128.
[17] Quine, W. V., The problem of simplifying truth functions, American mathematical monthly, vol. 59 (1952), pp. 521–531.
[18] – A proof procedure for quantification theory, The journal of symbolic logic, vol. 20 (1955), pp. 141–149.
[19] Schütte, K., Ein System des verknüpfenden Schliessens, Archiv für matematische Logik und Grundlagenforschung, vol. 2 (1956), pp. 55–67.
[20] Skolem, T., Über die mathematische Logik, Norsk matematisk tidskrift, vol. 10 (1928), pp. 125–142.
[21] – Über einige Grundlagenfragen der Mathematik, Skrifter utgitt av det Norske Videnskaps-Akademie i Oslo, I. Mat.-naturv. klasse 1929, no. 4, Oslo.

Commentary by the Author

Preface to 'A Mechanical Proof Procedure ...'

In 1955 - 57 four new proofs of the completeness of predicate logic appeared, which were all based on essentially the same idea of establishing that a formula is valid if and only if a search procedure terminates in a certain way. In two of them, given by Kanger and Schütte, one searched for a proof of a specific kind in a Gentzen calculus of sequents or a similar system. The search was by applying the rules of the calculus backwards in such a way that if the search was successful, a so called cut-free proof was found, while, if it was unsuccessful, this fact was in itself a datum for a counter-example to the given formula. In the other two proofs, given by Beth and Hintikka, one searched instead for a counter-example to a given formula in such a way that the process would terminate unsuccessfully if and only if no counter-example to the formula existed. Hence, the termination constituted in itself a proof of the validity of the formula; futhermore, the calculations made in such a search were easily converted (essentially by it upside down) to a cut-free proof in a Gentzen-type calculus. The four proofs thus established the completeness of one and the same specific proof procedure, which seemed reasonably efficient, and in this respect they differed from Henkin's wellknown completeness proof.

This theoretical development came not long after the technical development of computers had started to impress people, and these two developments made it seem reasonable to several people to investigate the possibility of proving theorems in predicate logic on a machine. When Kanger submitted his proof as a doctoral thesis in 1957, the teacher of the first year logic course that I was following at the same university, Aridus Wedberg, mentoned this possibility, which I set myself to realize.

Since Beth's system of semantic tableaux seemed to offer the least redundant notation, I relied essentially on it with some modifications such as instead of constructing parallel subtableaux, I assigned truth and falsity to the formulas in a tree formed order (a device later taken over by Smullyan in his textbook, First order Logic) as described in section I.3 of the paper. No programming Language suitable for my purpose existed at that time (at least to my knowledge), and I therefore developed my own language in which I described the procedure in form of what I called a pseudo-program. It was presented in a seminar essay in 1957 and is included in the present paper in sections I.6 and I.7 after some explanations in section I.4 and I.5. The pseudo-program was translated to a machine program by Håkan Prawitz in the same year, and it was in turn worked over and run on the machine by Neri Voghera in the year after, 1958, as described in parts II and III.

At the First International Conference on Information Processing, organized by UNESCO in Paris in 1959, we became aware of parallel developments, which are recordered in the note of the introduction of the paper.

It must be said that neither our nor the other programs that had been worked out at about the same time yield very much. It did not matter much that we had used the elegant and recent procedures provided by Beth, Hintikka, Kanger and Schütte, whereas the other programs essentially relied upon procedures that were available already in Herbrand's theorem or in Skolem's papers from the twenties (utilized, in fact, in Gödel's original completeness proof), because the all-absorbing difficulty, which we became aware of, was the random generation of substitution instances of quantified formulas common to all the programs and discussed in section III.1 of our paper.

In retrospect, the value of the enterprise - except for having been perhaps the first automated theorem proving in predicate logic - seems to have been the negative one of learning the limitations of the existing procedures, some of which turned out to be possible to overcome; however, I think that the need of improving the way of generating substitution instances could have been learned in a less costly way.

Dag Prawitz

A Mechanical Proof Procedure and its Realization in an Electronic Computer

D. Prawitz, H. Prawitz, N. Voghera

Introduction

In this paper we will report on the *realization* in an electronic *digital computer* of an effective proof procedure for the first order *predicate calculus.*

The only logic to which electronic computers have so far been applied is the one contained in the propositional calculus (also known as sentential calculus or Boolean algebra).[1] However, propositional calculus is insufficient for the deductions needed in mathematics and other sciences. There has recently been some interest in using computers for such deductions (see e.g. discussions by Gelernter 1957 and Robinson 1957).[2] It is possible to use the predicate calculus (also known as functional calculus) for this purpose. A discussion of this may be found in Beth 1958.[3] The realization of this possibility, reached in 1957–58, is the topic of the present paper.

Unlike propositional calculus, predicate calculus (which contains propositional calculus as a part) is sufficient to express most mathematical theories in such a way that a proof procedure for the predicate calculus also becomes a proof procedure for these theories. (The question whether a sentence S is a theorem in such a theory with the axioms A is equivalent to the question whether the sentence *if* A *then* S is a theorem in the predicate calculus.) Several effective proof procedures for the predicate calculus can be found in the logical literature. Recently published ones are Beth 1955, Hintikka 1955, Kanger 1957, Quine 1955, and Schütte 1954–56.

The work reported below embraces three major developments. A mechanical proof procedure for the predicate calculus in the form of a pseudo-program was developed by Dag Prawitz 1957, building on Beth 1955 and Kanger 1957. A

Received June, 1959. The later stages of this research were supported by Statens tekniska forskningsråd (Sweden). We are also indebted to AB Åtvidabergs Industrier, Stockholm, for allowing us to use their electronic computer Facit EDB.

Neri Voghera is at present with Swedish Board for Computing Machinery, Stockholm.

[1] The first machine for solving problems in the propositional calculus was constructed by Jevons in 1869. Several electronic machines for the same purpose have recently been constructed as well as programs for solving such problems in digital computers; see e.g. Bowden 1953, ch. 15. (A date after a name is a reference to the bibliography at the end of the paper.)

[2] This interest has often had connection with an interest in complex information processing systems. The first to treat proof procedures with such an approach seem to have been Newell and Simon 1956, who studied a proof procedure for some parts of the propositional calculus. A project with a similar approach for applying an electronic computer to a proof procedure for geometry was reported by Gelernter and Rochester 1958.

[3] A brief discussion is already included in Beth 1955.

Article from: Journal of the Association for Computing Machinery, Volume 7, Number 2, April 1960.

program for an electronic digital computer[4] was subsequently prepared by Håkan Prawitz in 1957. This program was modified and tested by Neri Voghera in 1958.

The present paper is divided into three parts. Part I outlines the logical theory behind the proof method (§1–§3), gives a suitable language in which to write the logical formulas in order to simplify the processing of them (§5), lists the operations used in the procedure (§6), and, employing these operations, finally gives the proof method in the form of a pseudo-program (§7). Part II deals with the problem of realizing the pseudo-program in an electronic digital computer. Part III discusses the applications of the method and gives some examples of theorems actually proved in a machine using the program.

NOTE ADDED IN PROOF: Since this was written, a number of reports on mechanical proof procedures in predicate logic have been published. (Concerning the dates of our work we refer to the ones above.) P. C. Gilmore has published two articles about a program of his: (1) "A program for the production of proofs for theorems derivable within the first order predicate calculus from axioms," *Proceedings of the First International Conference on Information Processing*, Paris, 1959; and (2) "A proof method for quantification theory: Its justification and realization," *IBM J. Res. Devel.* **4** (1960), 28–35. The main differences between Gilmore's program and ours were discussed by D. Prawitz at the session on theorem proving at the First International Conference on Information Processing, Paris, 1959 (the discussion appears in the Proceedings of the conference). The present paper was available in mimeographed form at the conference.

Hao Wang in "Towards mechanical mathematics," *IBM J. Res. Devel.* **4** (1960), 2–22, discusses a program (planned 1958 and realized 1959) which agrees with ours in the respect that the formulas do not have to be written in prenex normal form, but which otherwise differs from ours in about the same respects as Gilmore's. Finally, Martin Davies and Hillary Putnam in "A computational proof procedure," Rensselaer Polytechnic Institute (AFOSR TR 59–124), 1959, present a proof procedure (not yet programmed) which is similar to Gilmore's and Wang's as far as quantification theory is concerned but which is superior to these and our proof procedure in the truth-functional part (due to an idea previously used by B. Dunham, R. Fridshal, and G. L. Sward: "A nonheuristic program for proving elementary logical theorems," *Proceedings of the First International Conference on Information Processing*, Paris, 1959).

Part I

1. *Logical Calculi*

Logic is generally studied by employing a formalized language or a *calculus*. A calculus is usually set up in the following way.

(i) We give a list of the *signs* that are to be used in the calculus.

(ii) In *formation rules* we specify how a *formula* (sometimes called well-formed formula) is to be constructed as a finite sequence or *string* of signs. The rules are to be such that we can decide effectively whether a given string of signs is or is not a formula.

(iii) In *derivation rules* we specify how a *proof of a formula* is to be constructed as a finite sequence of formulas. The rules are to be such that we can decide effectively whether a given sequence of formulas is or is not a proof.

[4] Facit EDB, AB Åtvidabergs Industrier, Stockholm.

(iv) A *theorem* is defined as a formula F such that there exists a proof of F.

To say that a question is effectively decidable is to say that it is possible to construct an algorithm or—equivalently—a machine for deciding the question.[5] (The application of mechanical devices to logical calculi is thus very natural.)

Although the question whether a given sequence of formulas is a proof is effectively decidable, it does not follow that the question whether a given formula is a theorem (i.e. whether there *exists* a proof of the formula) is effectively decidable. In the case of the predicate calculus the last question is actually not effectively decidable (Church's theorem, see e.g. Kleene 1952).

If a formula F is a theorem, however, it is always possible to find a proof of F by searching through the sequences of formulas of the calculus. (The class of sequences of formulas is of course denumerable; hence by scanning these sequences in some appropriate order we will sooner or later find a sequence which is a proof of the formula F if this formula, as supposed, is a theorem; it is of course essential here that we can decide effectively for every sequence of formulas whether it is or is not a proof.) Hence for every calculus there exists an effective proof procedure, i.e. a method for finding a proof of any theorem of the calculus. The method just described is of course not possible to use in practice[6] but, thanks to recent developments in logic, proof procedures now exist that are both simple and elegant; one will be described in §3.

2. *Predicate Calculus*

A version of the predicate calculus adapted for mechanical use will be given by the formation rules stated in §4 and the derivation rules stated in the form of a pseudo-program in §7. Let us first, however, outline the predicate logic somewhat more informally.

The *signs* needed are of different kinds.

(a) To denote *individuals* let us use the signs $c_1, c_2, \cdots$ and $x_1, x_2, \cdots$. The former are called *individual constants* and the latter *individual variables*. As examples of individuals we may consider the integers 1, 2, $\cdots$, or geometrical points.

(b) To denote arbitrary *classes* of individuals or *relations* between individuals we use the signs $P_1, P_2, \cdots$, called *predicate symbols*. An example of a class is the one of natural numbers and of a relation, the relation "greater than".

(c) As *sentence connectives* we use the *singulary* $\sim$ (read: not) and the *binary* $\wedge$ (and), $\vee$ (or), $\supset$ (only if), and $\equiv$ (if and only if).

(d) To *quantify* expressions we use the *quantifiers* A (for all), and E (there exists).

(e) Finally, we use two *technical signs*, namely the left and right parentheses.

Atomic formulas are formed by suffixing any number of individual constants

[5] A rigorous definition of the notion "effectively decidable" in terms of machine processes was developed by Turing in 1936 (see e.g. Kleene 1952).

[6] That kind of procedures is known as the "British museum method" or the "fifty million monkeys technique".

to a predicate symbol; the constants are said to be *arguments of the predicate symbol*. Examples of atomic formulas: P_1c_2 (read: the property P_1 holds for c_2), $P_1c_2c_1$ (read: the relation P_1 holds between the individuals c_2 and c_1). From simpler formulas we can form more complex ones, *composite formulas*, by using the sentence connectives and the quantifiers. From a formula F we can form the formula $\sim F$, and from two formulas F and G we can form the formulas $(F \wedge G)$, $(F \vee G)$, $(F \supset G)$, and $(F \equiv G)$; in these formulas F and G are said to be *arguments of the binary sentence connective*. We will use the notation $F(a/b)$, where F is a formula and a and b are individual constants or variables, to denote the result of substituting an occurrence of a for each occurrence of b in F. Then from a formula F in which the variable x_i does not occur we can form the formulas $Ax_iF(x_i/c_j)$ and $Ex_iF(x_i/c_j)$ (i and j are integers). A formula F is thus of one and only one of the following eight forms: $P_ic_{i_1} \cdots c_{i_n}$, $\sim G$, $(G \wedge H)$, $(G \vee H)$, $(G \supset H)$, $(G \equiv H)$, Ax_iG, and Ex_iG (i, n, and indexed i are integers, G and H are formulas); and in each case F is of one of these forms in only one way. The predicate symbol, sentence connective, or quantifier displayed in the form in question is called the *principal sign of F*. The formulas of which a composite formula, F, is built up, are called the *subformulas of F*. If F contains a subformula Qx_iG (Q is a quantifier, and i, an integer), then all formulas $G(c_j/x_i)$ ($j = 1, 2, \cdots$) are considered subformulas of F.

Formulas in which no constants occur we call *sentences*. An example of a sentence is $(Ex_1Ax_2P_1x_1x_2 \supset Ax_2Ex_1P_1x_1x_2)$, which can be read: if there is an individual x_1 so that for every individual x_2 the relation P_1 holds between x_1 and x_2, then for every individual x_2 there is an individual x_1, so that P_1 holds between x_1 and x_2. As is seen, this sentence happens to be true for all relations—i.e. whatever relation P_1 may be—and is therefore called a *valid* sentence. It is possible to construct the derivation rules of the predicate calculus in such a way that all valid sentences become theorems. Indeed, by the proof procedure in §3 all valid sentences can be proved to be valid.

3. *A Proof Procedure for the Predicate Calculus*

To prove that a sentence S is valid we first assume that it is not, i.e. we assume that it is false (or, more correctly, we assume that S is possible to falsify). From this assumption we can draw certain conclusions about the truth and falsehood of the subformulas of S. For example, if S is of the form $(F \vee G)$ and is false, then both F and G are false. If we, continuing in this way, obtain the contradiction that a subformula of S is both true and false, then we have to reject our initial assumption that S is false, and have hence proved the validity of S.

The proof method to be considered here is a systematization of the idea just described. The procedure is carried out in the following way:

First, we assign the value falsehood to the sentence S that we want to prove. Then, we assign truth values (truth or falsehood) to the subformulas of S in some order (which order will be specified in §5) by applying the following rules, called *truth rules.*

NOTATIONS: F and G are formulas, t and f are abbreviates of truth and falsehood, i and j are integers.

T1. If $\sim F$ has the value t, then assign f to F.

T2. If $(F \wedge G)$ has the value t, then assign t to both F and G.

T3. If $(F \vee G)$ has the value t, then consider two cases: (a) assign t to F, (b) assign t to G.

T4. If $(F \supset G)$ has the value t, then consider two cases: (a) assign f to F, (b) assign t to G.

T5. If $(F \equiv G)$ has the value t, then consider two cases: (a) assign t to both F and G, (b) assign f to both F and G.

T6. If $\mathrm{Ax}_i F$ has the value t, then assign the value t to $F(\mathrm{c}_j/\mathrm{x}_i)$ where c_j is any constant that has been introduced by earlier application of T7 or F6 (or if there is no constant introduced, c_j is c_1). (Using the notion of *thread* introduced below, we can restrict the constants c_j to those constants introduced in the same thread as $\mathrm{Ax}_i F$ belongs to.)

T7. If $\mathrm{Ex}_i F$ has the value t and for no constant c_j has $F(\mathrm{c}_j/\mathrm{x}_i)$ obtained the value t, then introduce the constant c_k that has the least index of the constants that has not occurred in earlier applications of the truth rules, and assign t to $F(\mathrm{c}_k/\mathrm{x}_i)$.

F1. If $\sim F$ has the value f, then assign t to F.

F2. If $(F \wedge G)$ has the value f, then consider two cases: (a) assign f to F, (b) assign f to G.

F3. If $(F \vee G)$ has the value f, then assign f to both F and G.

F4. If $(F \supset G)$ has the value f, then assign t to F and f to G.

F5. If $(F \equiv G)$ has the value f, then consider two cases: (a) assign t to F and f to G, (b) assign f to F and t to G.

F6. If $\mathrm{Ax}_i F$ has the value f and for no constant c_j has $F(\mathrm{c}_j/\mathrm{x}_i)$ obtained the value f, then introduce the constant c_k that has the least index of the constants that has not occurred in earlier applications of the truth rules, and assign f to $F(\mathrm{c}_k/\mathrm{x}_i)$.

F7. If $\mathrm{Ex}_i F$ has the value f, then assign f to $F(\mathrm{c}_j/\mathrm{x}_i)$ where c_j is any constant that has been introduced in earlier applications of T7 or F6 (or if there is no constant introduced, c_j is c_1). (Using the notion of *thread* introduced below, we can restrict the constant c_j to those constants introduced in the same thread as $\mathrm{Ex}_i F$ belongs to.)

We have to consider two cases when we apply one of the rules T3, T4, T5, F2, and F5. Let us call these rules *branching rules.* We can visualize the proof procedure as the construction of a *tree.* In the *root* of the tree we assign the value falsehood to S. We then proceed upwards assigning truth values to the subformulas of S. Every time when we apply one of the branching rules, the tree branches.

A *thread* of the tree is obtained by starting at the root and proceeding upwards as far as possible, choosing one of the branches at every branching point. A thread is said to be *contradictory* if it contains two identical subformulas of S where one has obtained the value truth and the other the value falsehood. In order that our initial assumption that S is false shall hold, it is necessary and sufficient that some thread of the tree is not contradictory.

Thus, when, at the top of a thread, a truth value is assigned to a formula that has obtained the opposite truth value below in the same thread, we terminate the construction of that thread. If every thread is terminated in this way, then we have to reject our assumption that S is false, and have then demonstrated the validity of S; hence S is a theorem.

If the construction of a thread has to be terminated because there is no more way in which we can apply the rules to the formulas in the thread, then we can not reject our assumption that S is possible to falsify; then S is not valid and is no theorem.[7]

By the same procedure we can prove that a sentence S is a logical consequence of the sentences $S_1, S_2, \cdots, S_n$. We first assume the contrary, i.e. we assign the value truth to the sentences $S_1, S_2, \cdots, S_n$ and the value falsehood to S. Then we use the same procedure as before to assign truth values to the subformulas of $S_1, S_2, \cdots, S_n$, and S; the only difference is that we now have a number of initial formulas instead of only one. (As is easily seen, we get the same result if we apply the described method of proving valid sentences to the formula $((S_1 \wedge S_2 \wedge \cdots \wedge S_n) \supset S)$.)

4. *Adapting the Language to Mechanization Requirements*

The proof procedure in §3 can obviously be mechanized. We will now show how to write the formulas in a notation designed to facilitate the mechanical processing of them; thus, we will modify the language of the predicate calculus given in §2.

Before we list the signs used in the formulas and state the formation rules, we will give some preliminary explanations.

A. *Numerals.* For various purposes we will make use of numerals when writing the formulas. The number system we use here is the hexadecimal system (in order to facilitate the programming of the machine in which the pseudo-program is realized). Thus, among the signs that the formulas are made up of, there will be sixteen *figures* 0, 1, 2, $\cdots$, 9, a, b, c, d, e, and f, denoting the first sixteen non-negative integers. A string of n figures ($n = 1, 2, \cdots$) is called an n-digit numeral.

B. *Truth signs.* To know whether a T- or F-rule (see the truth rules in §3) is applicable to a given formula, we have to know its truth value. When we assign a truth value to a formula, we therefore write a figure 1 or 2, before the formula to denote that we have assigned the value truth or falsehood, respec-

[7] If S is a sentence the first signs of which are, e.g., $\mathrm{Ex}_i\mathrm{Ax}_j$ (i, j are integers), then the tree can never terminate in this way as is easily seen from the rules F7 and F6. Hence, if S is such a sentence and is not valid, then the procedure will never terminate. This is in agreement with the remark in §1 that there is no procedure for deciding in the general case whether a formula is or is not a theorem of the predicate calculus.

tively, to the formula. A figure 1 or 2 having this function is called a *truth sign*. The result of prefixing a truth sign to a formula is called an *evaluated formula* (abbreviated: *e-formula*).

C. *Principal signs*. To know which of the seven rules in either the T- or F-group of rules is applicable to a given formula, we have to know the principal sign of the formula. If we use the notation of §2, this sign is easily found when it is the singulary sentence connective $\sim$ or a quantifier as it then heads the formula, but when it is a binary sentence connective, it can only be found by mating parentheses. In a machine such a procedure would be time-consuming, and therefore we also write the binary sentence connectives before their arguments instead of between them.[8] For example, instead of writing $(P_1c_2c_1 \wedge P_2c_1)$, we now write $\wedge P_1c_2c_1P_2c_1$.

D. *Connective Index*. When applying the proper truth rule to a formula whose principal sign is a binary sentence connective, we split the formula into the two arguments of the connective. In the notation of §2, parentheses marked the scope of the arguments of a binary sentence connective. These parentheses become superfluous when using the device introduced in C above, because the scopes of the arguments can then be determined by analyzing the subformulas succeeding the connective. However, as such a procedure would be time-consuming, we insert a numeral, called a *connective index*, after a binary sentence connective to denote the number of signs of the first argument of the connective. This of course facilitates the splitting of these formulas. It will be advantageous to fix the length of the connective index (cf. H below). We will here use 2-digit numerals as connective indices.[9] The example above in C would thus be written $\wedge 03P_1c_2c_1P_2c_1$.

E. *Quantification*. When applying the proper truth rule to a formula QxF, whose principal sign, Q, is a quantifier, we substitute a constant for all occurrences of x in F. To carry out such a substitution in a machine we would have to scan all the signs of F and determine for every sign whether it is an occurrence of x. Such a procedure would be time-consuming. After the quantifier we will therefore put a *quantifier index*, which, for every occurrence of x in F, contains a numeral denoting the *position number* of x in F. The position number of an occurrence of a sign in a formula F is the number of signs in F standing before the occurrence in question. We will here use 2-digit numerals to denote position numbers.[10] The quantifier index will further contain a numeral which heads the index and denotes the number of occurrences of x in F. This numeral will here be a 1-digit numeral.[11]

Special signs for variables are not necessary now. The first occurrence of x in QxF is replaced by the quantifier index, and for the other occurrences of x in F we may now put an arbitrary sign just as a place holder until we substitute a

[8] This device was first used by Łukasiewicz (see, e.g., Church 1956, n. 91).

[9] This means that we put a limit on the length of the formulas. Using the hexadecimal system, we find that no first argument of a binary sentence connective may contain more than 255 signs.

[10] We can thus not quantify a constant, c, in a formula, F, if the position number of an occurrence of c in F is greater than 255 (cf. fn. 9).

[11] A quantified variable may thus have at most 15 occurrences in the formula.

constant; we will here use the figure 0. For example, instead of quantifying c_1 in $\wedge 03P_1c_2c_1P_2c_1$ by writing $Ax_1\wedge 03P_1c_2x_1P_2x_1$, we now write $A20507\wedge 03P_1c_20P_20$.

F. *Individual Constants.* When we apply the rule T7 or F6 we introduce a new individual constant. As constants we will use all the 1-digit numerals except 0, and we introduce them in the natural order 1, 2, $\cdots$. (By limiting the number of constants the proof procedure loses its universality as the proofs of some formulas require more than 15 applications of T7 and F6. This is, however, of little practical importance because in practice the proof procedure here considered would usually be unmanageable if we had to introduce more than 15 constants.)

G. *Counters.* The rules T6 and F7 can be applied to the same formula many times—as many times as there are constants introduced, i.e. the total number of applications of T7 and F6. Therefore we cannot erase a formula QiF (Q is a quantifier; i, a quantifier index; and F, a formula) to which T6 or F7 is applicable before we have assigned the proper truth value to $F(c/i)$ (i.e. the result of putting c in those positions in F that have the position numbers stated by i) for every constant c that has or will be introduced. To keep an account of the constants c for which we have assigned the proper truth value to $F(c/i)$, we insert a *counter*, a 1-digit numeral initially 0, between all quantifiers and quantifier indices. We then assign the proper truth value to $F(1/i)$, $F(2/i)$, $\cdots$, in the order the constants are introduced, and every time we do this we change the counter so that the number it denotes is increased by one. The formula considered above in E would thus be written $A020507\wedge 03P_1c_20P_20$.

H. *The Signs Used.* We need signs that have the following five different functions:

(i) designating the two truth values, truth and falsehood (cf. B);

(ii) functioning as predicate symbols, sentence connectives, and quantifiers (cf. §2 (b), (c), and (d));

(iii) functioning as individual constants (cf. F);

(iv) functioning as place holders instead of variables (cf. E);

(v) functioning as technical signs (as connective and quantifier indices and counters; cf. D, E, and G).

We will write the formulas in such a way that we can determine which of the five kinds of functions an occurrence of a sign has in a formula by just knowing the relative position of the occurrence in the formula; i.e. without knowing the shape of the sign.

If F is an e-formula, we can analyze F in the following way:

(a) the first sign of F is a truth sign (function (i));

(b) the second sign of F is the principal sign of F (function (ii));

(c) if the second sign of F is the singulary sentence connective $\sim$, then the string of signs following this sign is a formula;

(d) if the second sign of F is a binary sentence connective, then the third and fourth signs constitute a connective index n (function (v)), and both the sequence of the n following signs and the remaining string of signs of F are formulas;

(e) if the second sign of F is a quantifier, then the third sign is a counter, and the signs from the fourth to the $(n + 3)$-th (incl.), where n is the fourth sign of F, constitute a quantifier index (function (v)) and the sign string following the index is a formula;

(f) if the second sign of F is a predicate symbol, then F is atomic and all the signs following the predicate symbol are constants (function (iii)).

Hence, the same signs can be used for all the five different kinds of functions listed above, and the signs we will use are the 16 figures in the hexadecimal system (see A). To facilitate the reading, however, we will in this paper keep the notations for sentence connectives and quantifiers introduced in §2 (c) and (d), and use the eight letters P, Q, R, S, T, U, V, and W as predicate symbols. (If we did not limit the number of predicate letters we would have to let sign strings of variable length function as predicate symbols.) In the machine program these 15 sentence connectives, quantifiers, and predicate symbols are replaced each by its own hexadecimal figure.

After these explanations we now state the *formation rules*:

(1) A predicate symbol followed by a number of figures is a (atomic) formula.

(2) If F is a formula, then $\sim F$ is a formula.

(3) If F and G are formulas and p is a 2-digit numeral denoting the number of signs in F, then $\wedge pFG$, $\vee pFG$, $\supset pFG$, and $\equiv pFG$ are formulas.

(4) If F is a formula, then $\mathrm{A}0np_1 \cdots p_nF$ and $\mathrm{E}0np_1 \cdots p_nF$ are formulas if n is a figure and p_i $(i = 1, \cdots, n)$ is a 2-digit numeral denoting the position number of a *free* (see definition 1 below) occurrence of 0 in F.

(5) No other strings of signs than the ones constructed according to (1)–(4) above, are formulas.

Definition 1: An occurrence of 0 is *bound* in F if it stands in a part of F of the form $Q0np_1 \cdots p_nG$ (Q is A or E, n is a 1-digit numeral, and $p_1, \cdots, p_n$ are n 2-digit numerals) where some p_i $(i = 1, \cdots, n)$ denotes the position number of the occurrence of 0 in question in G. An occurrence of 0 is free in F if it stands in an atomic part of F and is not bound.

Definition 2: A formula in which all atomic parts contain only bound occurrences of 0 besides the predicate symbol is called a *sentence*.

5. *Some Other Problems when Mechanizing the Proof Procedure*

As described in §3, the proof procedure is a series of assignments of truth values to the subformulas of the initial formula(s). We will now prescribe a certain order in which we are to carry out these assignments.[12]

(i) When there is a formula to which one of the rules T1, T2, T6, F1, F3, F4, and F7 is applicable, we apply that rule to the formula.

(ii) When there is no formula as in (i) but there is a formula to which a

[12] If we had not limited the number of constants available for substitutions (§4, F), the number of subformulas to which we could have assigned truth values would have been infinite in the general case (cf. fn. 7). The order in which we assign truth values to the subformulas would then have been essential, because, then, if we had not chosen an appropriate order, we might have assigned truth values only to some of the subformulas leaving the others out.

branching rule (i.e. T3, T4, T5, F2, or F5) is applicable, we apply that rule to the formula.

(iii) When there is no formula as in (i) or (ii) but there is a formula to which the rule T7 or F6 is applicable, we apply that rule to the formula and hence introduce a new individual constant.

When the tree branches, we continue by first constructing one branch and one thread of the tree until its termination (if any) before constructing another. To keep track of the different branches we assign numbers to them. The root has the number 1; its branches have the numbers 11 and 12; their branches have the numbers 111, 112 and 121, 122, respectively, and so on.

To determine whether a thread is contradictory we have to compare the e-formulas in the thread, examining whether there are two e-formulas that differ only in respect to the truth signs. We will compare only atomic e-formulas in this way. This has the consequence that we will fail to notice two composite e-formulas that differ only in respect to the truth signs, and hence fail to terminate the corresponding thread. In that case we can of course always terminate the thread at a later stage as we, from the two e-formulas in question, sooner or later will obtain two atomic e-formulas that differ only in respect to the truth sign. The disadvantage of the prolongation of the procedure in the respect just described is balanced, however, by two advantages: we usually decrease the number of comparisons we have to make and we can save space by erasing all e-formulas, except those to which the rule T6 or F7 is applicable (cf. §4, G), as soon as we have applied the proper truth rule to them.

To separate different kinds of e-formulas during the procedure, we will store them in different *memories*. For instance, to each of the three kinds of formulas classified by (i), (ii), and (iii) above there will be one memory.

Besides e-formulas we will have to store certain auxiliary information in special memories, e.g. registrations telling which branches the e-formulas belong to. In total we will use 21 different memories whose names and functions are given in the table below.

Name	*Storing*
M1a	initial e-formulas and e-formulas not stored in other memories
M2a	e-formulas to which T7 or F6 is applicable (requiring the introduction of a new constant)
M3a	e-formulas to which T6 or F7 is applicable (which e-formulas must not be erased; cf. §4, G)
M4a	atomic e-formulas
M1b, M1c, M1d M2b, M2c, M2d M3b, M4b	auxiliary information showing to which branches the e-formulas in respective memory Mia, $i = 1, 2, 3, 4$, belong
M3c	auxiliary information showing to which formulas in M3a the rule T6 or F7 has so been applied that the last introduced constant was substituted
M5	auxiliary information showing which individual constants have been introduced in different branches
M6	auxiliary information showing which branch is under construction

M7 e-formulas which belong to the branch under construction and to which a branching rule is applicable

M8 e-formulas not belonging to the thread under construction

OP1, OP2 (= "operation place") formulas to which some truth rule is being applied

OP3 auxiliary information

Out (= "output") the results of the truth assignments constituting the tree (including the numbers of the branches)

6. *Operations Needed for the Procedure and Notations Used in the Pseudo-program*

We will in §7 describe the proof procedure as a mechanical manipulation of signs (listed in §4, H) and strings of signs (the sign strings are either e-formulas formed according to the formation rules in §4 or auxiliary information represented by the numerals introduced in §4, A). The description, called the *pseudo-program*, consists of a number of *instructions*. To formulate the instructions we need names of the instructions, names of the signs (listed in §4, H) and strings of these signs (i.e. e-formulas and numerals), names of *locations*, and names of *conditions* and *operations* listed below.

A. *Names of instructions.* The instructions are numbered in order 1, 2, $\cdots$, and named by corresponding numerals.

B. *Names of locations.* As explained in §5, we write the e-formulas and the auxiliary information (i.e. certain numerals) in 21 different memories, named as in the table in §5. The memories are divided into lines except for M1c, M2c, M6, 0P1, and 0P2, that have only one line each. The lines are numbered in order 1, 2, $\cdots$ and named by corresponding numerals. A location name has one of the following three forms, where M is a memory: (a) M, denoting the memory M; (b) M ln, $n = 1, 2, \cdots$, denoting the line n in the memory M; (c) M fel, denoting the first empty line in the memory M, i.e. the line with the least number on which no signs are written.

C. *Names of signs and strings of signs.* To name the occurrences[12a] of the signs and the strings of signs which we are manipulating we use

(a) the very same signs (only used to refer to a single sign);

(b) the notation num(M), where M is a memory, to name the numeral that denotes the last occupied line in the memory M, i.e. the line in M with the highest number on which signs are written, or, if the memory M is empty, to name the numeral 0.

(c) *location references* (abbreviated: *loc. ref.*) which name an occurrence of a sign or a string of signs by referring to the location where the occurrence is stored. If the location to which a loc. ref. refers is empty, the loc. ref. is to name the numeral 0.

The location references will be of different kinds:

(1) A loc. ref. consisting of the name of a memory and the name of a line, denotes the sign or string of signs standing in the memory on the line in question. As names of lines we use

[12a] In cases (a) and (b), a sign or sign string rather than one of its occurrences is named, as it is immaterial which occurrence of the sign or sign string we consider.

(1.1) ln, where n is a number, to denote the line with the number n;

(1.2) lol, to denote the last occupied line (note: if M is a memory, num(M) denotes the numeral that denotes the last occupied line in M, while M lol denotes the signs written on the last occupied line in M);

(1.3) l(x), where x is a loc. ref. naming a numeral, to denote the line with the number denoted by x;

(1.4) if the memory has only one line, we leave out the name of the line.

(2) A loc. ref. consisting of the name of a memory M, the name of a line l, and the notation sn, where n is a number, denotes the nth sign (counting from the left to the right) in the sign string named by Ml. Instead of sn we will sometimes use the notation *last sign*.

D. *Conditions*. The conditions are of one of the following five forms. (Below s and t are occurrences of signs or strings of signs (i.e. e-formulas or numerals named according to C above), n and m are occurrences of numerals (named according to C), and M is a memory.)

1. $=(s, t)$, meaning: s and t are identical.
2. $\neq(s, t)$, meaning: s and t are not identical.
3. $<(n, m)$, meaning: the number denoted by n is less, than the number denoted by m.
4. Empty(M), meaning: M is empty.
5. Exist(s, M), meaning: there is an occurrence of s in M.

E. *Operations*. An instruction will consist of an operation having one of the following 15 forms. (Below, s, t, n, m, and M are used as in paragraph D and N is a location (named according to B).)

1. Wr(s, N), meaning: write a copy of s in N.
2. Er(s), meaning: erase s.
3. Er> (M, n), meaning: erase the content on the lines in M that have numbers greater than the one denoted by n.
4. Rp(s, t), meaning: replace s with a copy of t.
5. Mo(s, N), meaning: move s to N, i.e. Wr(s, N) and Er(s).
6. Mo(M7, M1a), meaning: move the content of each line in M7 to the consecutive last empty lines in M1a.
7. Pr(s, t), meaning: prefix the sign t to s.
8. Suf(s, t), meaning: suffix the sign t to s, or if s is 0, Rp(s, t).
9. Add(n, m), meaning: change n so that it will denote the sum of the two numbers denoted by n and m.
10. Sub(n, m), meaning: change n so that it will denote the difference between the numbers denoted by n and m.
11. Split, meaning: let n be the number denoted by the numeral that consists of the third and fourth sign of the sign string in 0P1 and let s be the part of the sign string in 0P1 that consists of the signs from the fifth to the $(n + 4)$-th (incl.) of the sign string in 0P1, and write s in 0P2 and erase the first $n + 4$ signs in 0P1.
12. Substitute(m), meaning: let n be the number denoted by the numeral standing as fourth sign in 0P1 and let p_1, p_2, $\cdots$, and p_n be the numbers denoted by the numerals that consists of the 5th and 6th signs, the 7th and 8th

signs, $\cdots$, and the $(n + 4)$th and $(n + 5)$th signs, respectively, in OP1. Then, for every i $(i=1, \cdots, n)$, add a copy of m to the $(4 + 2n + p_i)$-th sign of the e-formula standing in OP1 and then erase the first $4 + 2n$ signs of the formula.

13. Stop.

14. Con(i), where i is a number of an instruction (see A), meaning: continue with instruction i.

15. Con($C_1 \rightarrow i_1$, $C_2 \rightarrow i_2$, $\cdots$ $C_j \rightarrow i_j$), where C_1, C_2, $\cdots$, C_j are conditions (see D above) that are mutually exclusive (i.e. two of them can not hold simultaneously) and i_1, i_2, $\cdots$, i_j are names of instructions, meaning: if C_k $(k = 1, \cdots, j)$ holds then continue with instruction i_k (if no condition C_k holds, then no operation is to be performed).

7. *The Pseudo-program*

(This section is not essential for the understanding of part II and part III.)

We now suppose that we will prove that the sentence S is a logical consequence of the sentences S_1, S_2, $\cdots$, S_n $(n = 1, 2, \cdots)$, all sentences formed according to the formation rules in §4. Then we first prefix the truth sign 1 to all the sentences S_1, S_2, $\cdots$, S_n, and the truth sign 2 to the sentence S. The so-obtained e-formulas are stored in the first lines in M1a; every e-formula gets its own line. All other lines of the memories are supposed to be empty. We then carry out the instructions in the order they are given in the pseudo-program (except when the instructions instruct otherwise).

(To facilitate the reading we will give some explanations in a column to the right of the instructions. The pseudo-program could be shortened (cf. part II) though that would make it less lucid.)

I. Processing Formula from M1a

1. Con(= (M1b lol, num(M1a)) → 64) — Succeeds instructions 12, 23, 30, 38, 46, 51, 63, and 134. Only branching rules are applicable to the e-formulas that are stored in M1a on lines with numbers less or equal to the number denoted by the number in M1b lol.

2. Mo(M1a lol, OP1)

II. Application of Truth Rules

3. Wr(OP1, Out) — Succeeds also instructions 16, 21, 36, 49, 56, 61, 68, 90, 95, 100, 114, 128, 158, and 161.

4. Con(= (OP1 s1, 2) → 39) — OP1 s1 is a truth sign.

IIa. Application of a T-rule

5. Con(= (OP1 s2, ∼) → 13,
= (OP1 s2, ∧) → 17,
= (OP1 s2, ∨) → 22,
= (OP1 s2, ⊃) → 22,
= (OP1 s2, ≡) → 22,
= (OP1 s2, A) → 24,
= (OP1 s2, E) → 37) — OP1 s2 is the principal sign of the e-formula in OP1.

6. Con(Exist(OP1, M4a) → 11)	Instruction 6 is carried out when the principal sign of the e-formula stored in OP1 is a predicate symbol.
7. Rp(OP1 s2, 2)	
8. Con(Exist(OP1, M4a) → 135)	
9. Rp(OP1 s1, 1)	
10. Wr(OP1, M4a fel)	
11. Er(OP1)	Succeeds also instruction 6.
12. Con(1)	
13. Er(OP1 s1)	Application of T1; succeeds instruction 5.
14. Er(OP1 s1)	
15. Pr(OP1, 2)	
16. Con(3)	
17. Split.	Application of T2; succeeds instruction 5.
18. Pr(OP2, 1)	
19. Mo(OP2, M1a fel)	
20. Pr(OP1, 1)	
21. Con(3)	
22. Mo(OP1, M7 fel)	Postponement of applications of T3, T4, and T5; succeeds instruction 5.
23. Con(1)	
24. Wr(OP1 s1, OP3 fel)	Application of T6 and F7; succeeds instruction 5 and 39.
25. Con($\neq$(OP1 s3, 0) → 28)	OP1 s3 is a counter.
26. Wr(OP1, M3a fel)	
27. Add(OP1 s3, 1)	
28. Con($<$(OP1 s3, M5 lol) → 31, $=$(OP1 s3, M5 lol) → 33)	Succeeds also instruction 25.
29. Er(OP1)	
30. Con(1)	
31. Wr(OP1, M1a fel)	Succeeds instruction 28.
32. Add(M1a lol s3, 1)	
33. Substitute(OP1 s3)	Succeeds also instruction 28.
34. Pr(OP1, OP3 lol)	
35. Er(OP3 lol)	
36. Con(3)	
37. Mo(OP1, M2a fel)	Postponement of application of T7; succeeds instruction 5.
38. Con(1)	

IIb. Application of an F-rule

39. Con($=$(OP1 s2, $\sim$) → 47, $=$(OP1 s2, $\wedge$) → 50, $=$(OP1 s2, $\vee$) → 52, $=$(OP1 s2, $\supset$) → 57, $=$(OP1 s2, $\equiv$) → 50, $=$(OP1 s2, A) → 62, $=$(OP1 s2, E) → 24)	Succeeds instruction 4. OP1 s2 is the principal sign of the e-formula stored in OP1.
40. Con(Exist(OP1, M4a) → 45)	Instruction 40 is carried out when the principal sign of the e-formula stored in OP1 is a predicate symbol.
41. Rp(OP1 s1, 1)	
42. Con(Exist(OP1, M4a) → 135)	
43. Rp(OP1 s1, 2)	

44. Wr(OP1, M4a fel)	
45. Er(OP1)	Succeeds also instruction 40.
46. Con(1)	
47. Er(OP1 s1)	Application of F1; succeeds instruction 39.
48. Rp(OP1 s1, 1)	
49. Con(3)	
50. Mo(OP1, M7 fel)	Postponement of application of F2 and F5; succeeds instruction 39.
51. Con(1)	
52. Split.	Application of F3; succeeds instruction 39.
53. Pr(OP2, 2)	
54. Mo(OP2, M1a fel)	
55. Pr(OP1, 2)	
56. Con(3)	
57. Split	Application of F4; succeeds instruction 39.
58. Pr(OP2, 1)	
59. Mo(OP2, M1a fel)	
60. Pr(OP1, 2)	
61. Con(3)	
62. Mo(OP1, M2a fel)	Postponement of application of F6; succeeds instruction 39.
63. Con(1)	

III. Processing Formula from M3a

64. Con(= (M3c l1, M3c l2) → 69)	Succeeds instruction 1. The numeral ln M3c l2 denotes the last line in M3a on which an e-formula was stored before the last constant was introduced. Of the lines in M3a that have numbers less or equal to the number denoted by the numeral in M3c l2, the numeral in M3c l1 denotes the last line that stores a formula to which we have applied the rules T6 or F7 substituting the last introduced constant.
65. Add(M3c l1, 1)	
66. Wr(M3a l(M3c l1), OP1)	
67. Add(OP1 s3, M5 lol)	
68. Con(3)	

IV. Application of a Branching Rule

69. Con(Empty(M7) → 73)	Succeeds instruction 64.
70. Mo(M7 l1, OP1)	
71. Wr(M7, M1a)	
72. Con(76)	
73. Con(= (M1c, M1b lol) → 115)	Succeeds instruction 69. Of the lines in M1a that have numbers less than or equal to the number denoted by the numeral in M1b lol, the numeral in M1c denotes the last line that stores an e-formula to which the proper branching rule (cf. comment to instruction 1) has been applied in the thread under construction.
74. Add(M1c, 1)	
75. Wr(M1a l(M1c),OP1)	
76. Wr(OP1, Out)	Succeeds also instruction 72.
76a. Suf(M6, 1)	

77. Wr(M6, Out)	
78. Wr(num(M1a), M1b)	The numeral in M*i*b lol, *i* = 1, 2, 3, or 4, denotes the last line in M*i*a that stores an e-formula which belongs to the part of the tree below the last branching point, i.e. which belongs to the common part of the two threads separated by the last branching point.
79. Wr(num(M2a), M2b)	
80. Wr(num(M3a), M3b)	
81. Wr(num(M4a), M4b)	
82. Wr(M1c, M1d fel)	
83. Wr(M2c, M2d fel)	
84. Wr(M5 lol, M5 fel)	
85. Con(= (OP1 s2, ⊃) → 91, = (OP1 s2, ∧) → 96, = (OP1 s2, ≡) → 101)	OP1 s2 is the principal sign of the e-formula in OP1.
86. Split.	Application of T3.
87. Pr(OP2, 1)	
88. Mo(OP2, M8 fel)	
89. Pr(OP1, 1)	
90. Con(3)	
91. Split.	Application of T4; succeeds instruction 85.
92. Pr(OP2, 2)	
93. Mo(OP2, M8 fel)	
94. Pr(OP1, 1)	
95. Con(3)	
96. Split.	Application of F2; succeeds instruction 85.
97. Pr(OP2, 2)	
98. Mo(OP2, M8 fel)	
99. Pr(OP1, 2)	
100. Con(3)	
101. Wr(OP1 s1, OP3 fel)	Application of T5 and F5; succeeds instruction 95.
102. Wr(3, OP3 fel)	
103. Sub(OP3 lol, OP1 s1)	
104. Split.	
105. Wr(OP2, M1a fel)	
106. Pr(M1a lol, 1)	
107. Wr(OP1, M8 fel)	
108. Pr(M8 lol, OP3 lol)	
109. Er(OP3 lol)	
110. Pr(OP2, 3)	
111. Mo(OP2, M8 fel)	The first sign of the sign string in M8 lol is now 3, hence the sign string is not a proper e-formula. The sign 3 denotes, however, that we have stored two formulas in M8 at this branching point. 3 is later replaced with the sign 2; cf. instructions 158, 159, and 160.
112. Pr(OP1, OP3 lol)	
113. Er(OP3 lol)	
114. Con(3)	

V. The Introduction of a New Constant

115. Con(= (M2c, num(M2a)) → 129)	Succeeds instruction 73. The numeral in M2c denotes the last line in M2a that stores an e-formula to which we have applied the proper truth rule, T7 or F6, in the thread under construction.

116. Con(≠(M5 lol, f) → 119)	
117. Wr(f, Out)	
118. Stop.	
119. Add(M5 lol, 1)	Succeeds instruction 116.
120. Add(M2c, 1)	
121. Wr(M2a l(M2c), OP1)	
122. Rp(M3c l1, 0)	
123. Rp(M3c l2, num(M3a))	
124. Wr(OP1 s1, OP3 lol)	
125. Substitute(M5 lol)	Application of T7 and F6.
126. Pr(OP1, OP3)	
127. Er(OP3 lol)	
128. Con(3)	
129. Con(=(M5 lol, 0) → 132)	Succeeds instruction 115.
130. Wr(0, Out)	
131. Stop.	
132. Add(M5 lol, 1)	Succeeds instruction 129.
133. Rp(M3c l2, num(M3a))	
134. Con(1)	

VI. Terminating a Contradictory Thread

135. Con(Empty(M8) → 162, =(M6 last sign, 1) → 138)	Succeeds instruction 8, 42 and 137.
136. Er(M6 last sign)	
137. Con(135)	
138. Add(M6 last sign, 1)	Succeeds instruction 135.
139. Wr(M6, Out)	
140. Er>(M1a, M1b lol)	Cf. comments to instruction 78–81.
141. Er>(M2a, M2b lol)	
142. Er>(M3a, M3b lol)	
143. Er>(M4a, M4b lol)	
144. Er(M1b lol)	
145. Er(M2b lol)	
146. Er(M3b lol)	
147. Er(M4b lol)	
148. Er(M5 lol)	
149. Er(M7)	
150. Er(OP1)	
151. Rp(M1c, M1d lol)	
152. Rp(M2c, M2d lol)	
153. Er(M1d lol)	
154. Er(M2d lol)	
155. Rp(M3c l1, 0)	
156. Rp(M3c l2, 0)	
157. Mo(M8 lol, OP1)	
158. Con(≠(OP1 s1, 3) → 3)	Cf. comment to instruction 111.
159. Rp(OP1 s1, 2)	
160. Mo(M8 lol, M1a fel)	
161. Con(3)	
162. Wr(1, Out)	Succeeds instruction 135.
163. Stop.	

As is seen from the above, there are three instructions that stop the procedure. When the last sign in the output is 0, then S is not a logical consequence of the sentences $S_1, S_2, \cdots, S_n$, and when the last sign is 1, then S is a logical consequence of the sentences. When the last sign in the output is f, then a continuation of the proof procedure would require the introduction of more than 15 individual constants.

Part II

To realize the pseudo-program given in part I, §7 a program was written for a computer, named Facit EDB, manufactured by AB Åtvidabergs Industrier. This machine[13] has a fast access storage containing 2048 cells, each one holding a word consisting of 40 binary or 10 hexadecimal signs. A slow access storage (magnetic drums) contains 8192 cells, grouped into 256 separately accessible channels, each one containing 32 cells (a *block* of words). Below, we will use the denotations *K-storage* (core-) and *D-storage* (drum-) for these two storages.

Below, the words "sign", "figure", "position" are to be understood in the hexadecimal sense, if the opposite is not stated.

1. *m-Formulas*

In part I it was imagined that the formulas were written on separate lines of the memories. In the machine, such a representation would not be advisable because it would involve a waste of space. Therefore, the formulas will be stored one immediately after the other in the memories corresponding to those specified in part I, §5. To facilitate the separation of the individual formulas we will use a modified type of formulas, called *m-formulas* (machine-). An m-formula consists of:

(1) a 2-digit numeral called the *length-index*,
(2) an e-formula,
(3) possibly a number of empty positions (zeros),
(4) the length-index anew.

The length-index states the number of positions occupied by (1) and (2). The fact that a 2-digit numeral is used as length-index implies that an e-formula must not contain more than 253 signs. The number of empty positions according to (3) is at most nine and is chosen so that the total number of positions in (1)–(4) is an integral multiple of 10. Thus, an m-formula consists of an integral number of words.

In the machine, an m-formula is always placed so that it starts at the beginning of a cell. Then, the terminal length-index (4) will always occupy the two last positions of the last cell occupied by the formula. With the aid of the length-indices, a formula can be extracted from a memory if the address of its first or last word is known.

2. *The Memories*

It appears from part I, §5, that there are two different kinds of memories: such for storing e-formulas (*F-memories*), and such for storing numerals (*N-*

[13] The description refers to the machine version available at the time when the program was written.

memories). The memory "Out" is intended to take formulas as well as numerals. The output of the machine is used as this memory, because no access to it is required during the machine process. F-memories are M1a, M2a, M3a, M4a, M7, M8, 0P1, and 0P2. The others are N-memories.

The N-memories require only little space, and it is therefore possible to place them in the K-storage. 0P1 and 0P2 contain at most one formula, and they are therefore placed in the K-storage also. For the other F-memories, to which everything said below refers, the D-storage is used. In order to save processing time, however, a domain in the K-storage is reserved for each F-memory as an easily accessible part of it. Thus, we will speak of the *K-part* and the *D-part* of an F-memory. The formulas are first stored in the K-part, and nothing is transferred from the K-part to the D-part until a block of words, filling a channel, can be moved.

As to some of the F-memories, the machine has to deal with the last formula in the memory. For such a memory a K-part holding at least 57 words has to be reserved. Nothing is to be transferred to the D-part until the moment when the storing of a formula will cause the capacity of the K-part to be exceeded. Then, the first 32 words in the K-part are transferred to the D-part and the remainder, at least 26 words, is shifted to the beginning of the K-part. Erasing of formulas will concern only the last formulas of a memory. If, after some formulas at the end of a memory of the kind in question having been erased, the K-part should contain less than 26 words, a block will be moved from the D-part to the K-part. Thus, if 26 words or more have been stored in such a memory, the K-part will always contain at least 26 words and, consequently, the last formula.

The cells of the F-memories are numbered sequentially, for each memory starting from 0. The numbers, thus introduced, will be called *C-numbers*. A given C-number can denote a cell in the D-part or the K-part, depending on the number of words stored in the memory. The C-number of the first word of an m-formula is used in the location references (see part I, §6 C (c)) instead of the line number.

For N-memories intended to take only one or two numerals, one or two cells, respectively, are used. In N-memories intended to take more than two numerals, which memories are referred to in the sequel, if the contrary is not stated the numerals are stored in such a way that no numeral is divided into parts falling within different cells. Therefore, some positions at the beginning of each cell of the memory may be left unused. Concerning an N-memory of the kind now regarded, a place intended for taking a numeral is called a *line*. The lines in an N-memory of the kind in view correspond to the lines considered in part I, §6 B. They are numbered by *L-numbers*, for each memory starting with the number 0.

For each memory some important data are brought together in a *register*. The registers of the F-memories embrace three words and contain the following data:

(1) The address of the first cell of the K-part,
(2) the address of the last cell of the K-part,

(3) the number of the first channel of the D-part,

(4) the total number of cells (i.e. words which can be stored) in the memory,

(5) the current C-number of the first word of the last formula,

(6) the current C-number of the first empty cell in the memory.

The data (1)–(4) are given in advance, while the data (5) and (6) are changed when a formula is stored or erased.

The register of an N-memory embraces two words and contains the following data:

(1) The address of the first cell of the memory,

(2) the address of the last cell of the memory,

(3) the number of lines in a cell,

(4) the number of digits in each stored numeral,

(5) the current L-number of the first empty line.

Here, the data (1)–(4) are given in advance while the datum (5) is variable.

The numbers introduced in part I, §5, in order to keep account of the different branches have been modified in the program. Instead of the figures 1 and 2, the figures 0 and 1 have been used. It is then possible to represent such a number by a binary numeral. When such a number is stored in M6, it is, however, necessary to make it possible to distinguish zeros at the beginning of the numeral from empty positions. To this end the number of digits in the numeral is stored in a register of M6.

3. *Subroutines*

In the program, thirteen subroutines are used. In the cases when the routines have to deal with some memory, the transition to the routine is made with the aid of a composite jump-instruction, where the address of the first word of the register of the memory in view is noted.

The subroutines have the following purposes.

1. *For an F-memory, searching (a) the address of the first empty cell, (b) the address of the first word of the last formula stored, and (c) the number of the first empty channel of the D-part.*

2. *For an N-memory, searching the first empty line and copying the last numeral stored.* The address of the cell containing the first empty line and the position of this line in the cell are to be found.

The data searched in 1 and 2 can be found by treating merely the register of the memory in question. The determination of the data mentioned is a necessary preparation for other routines.

3. *Storing a formula in an F-memory.* The address of the first word of the formula to be stored is given in the composite jump-instruction. The formula will be stored immediately after the last formula in the memory. If necessary a block of words will be moved to the D-part before the storing of the formula.

4. *Storing a numeral in an N-memory.* A numeral placed in a given cell will be stored in the first empty line.

The routines 3 and 4 correspond to the instruction Wr(s, M fel) (see I, §6E1).

5. *Erasing the last formula or formulas in an F-memory.* If the C-number of the first cell to be emptied is placed in a given cell, the routine will perform the desired erasure and transfer, if necessary, a block of words from the D-part to the K-part.

6. *Erasing numbers in an N-memory.* If an L-number is placed in a given cell, the routine will cause erasure of the numerals stored in lines with this or higher L-number.

The routines 5 and 6 correspond to the instruction Er $> (M, n)$ (see I, §6E3).

7. *Erasing signs in the beginning of a formula and shifting the remainder leftwards.* According to certain instructions in the pseudo-code (I, §7) some signs at the beginning of a formula are to be erased. The sense of a sign depends on its position (cf. I, §5). Therefore, if some signs are to be erased, the following signs must be shifted leftwards. In the third position of the first cell a zero is entered in order to simplify the insertion of a new truth sign. The number of shift-steps is given in the jump-instruction.

8. *Splitting a formula.* As is seen from the pseudo-code, only formulas in 0P1 are to be split. The data necessary for the splitting are contained in the formula itself (cf. I, §5C).

9. *Substitution* (see I, §6E12) is carried out only for a formula placed in 0P1. The figure that shall be substituted has to be placed in a given cell.

10. *Comparison.* A formula placed in 0P1 is compared with each formula stored in M4a. The subroutine corresponds to the sequence of instructions appearing in the pseudo-code:

$$\begin{array}{l}\text{Con(Exist(0P1, M4a))} \rightarrow i_1)\\ \text{Rp(0P1 s1}, t_1)\\ \text{Con(Exist(0P1, M4a))} \rightarrow i_2)\\ \text{Rp(0P1 s1}, t_2)\end{array}$$

where t_1, t_2 are truth signs and i_1, i_2 are names of instructions. The routine is, however, carried out in another way in so far that the first and third instructions are performed simultaneously.

11. *Copying a formula from a memory.* The C-number of the first word of the formula to be copied is placed in a given cell. If the formula wholly or partly is stored in the D-part of the memory, the subroutine will copy the relevant blocks of the D-part in a given part of K-storage and will place the address of the first word of the formula in a given cell. Furthermore, the address and the C-number of the first word of the next formula will be placed in another cell.

12. *Output of M6.* The output is marked by two asterisks before and two afterwards.

13. *Output of 0P1.* The output is marked by one asterisk before and one asterisk afterwards.

The output of 0, 1, or f as the final sign according to the pseudo-code is marked by three asterisks before and three asterisks afterwards.

The main program follows the pseudo-code relatively closely. A simplification is made so far as some parts of the program are used for separate but sufficiently similar sequences of operations in cases when only a slight modification is needed.

The procedure will stop in the three cases mentioned in I, §7. Further stops occur whenever a memory becomes full. In this case an output shows which memory this has occurred for and the extent to which the other memories have been made use of. Especially, M6 is limited to 40 binary figures and therefore the procedure will stop if any thread of the tree contains more than 40 branching points.

A limitation, besides those accounted for above, has been made concerning the atomic m-formulas, which are assumed not to embrace more than 10 signs.

Part III

We will here consider the application of the machine program (described in part II) to proof problems. To test the program, we used it to derive the proofs of a number of well-known theorems of the predicate calculus, the theory of classes, and the theory of relations. In §4 below, we cite some of them and in an appendix we reproduce the complete proof of one of them as obtained with the machine Facit EDB. The program easily proves theorems of the kind found in elementary textbooks on symbolic logic. It is hardly possible, however, to use it for treating proof problems of much greater complexity. The factors causing this situation will be examined in §§1–3.

1. *Methodological Factors Limiting the Applicability of the Program*

In principle, all solvable proof problems, whose formulations and solutions (proofs) lie within the limits stated in parts I and II, could be solved by the machine using the program described. Of these problems, however, only relatively few are solvable within reasonable time, even if the fastest machines available were used. The main cause of this situation resides in the proof procedure itself as may be seen from the following argument.

Suppose that the solution of a given proof problem requires the introduction of m individual constants and that, during the proof procedure, we obtain a formula F of the form $Q\mathrm{x}_1 Q\mathrm{x}_2 \cdots Q\mathrm{x}_n G$ where either F has the value truth and Q is A, or F has the value falsehood and Q is E (such a formula will occur in all nontrivial proof problems). Let $G!_m$ be the result of, for every x_i $(i = 1, \cdots, n)$, substituting one of the first m constants in G, i.e. $G!_m$ is an abbreviate of the formula[14] $G(\mathrm{c}_{i_1}/\mathrm{x}_1)(\mathrm{c}_{i_2}/\mathrm{x}_2) \cdots (\mathrm{c}_{i_n}/\mathrm{x}_n)$, and is obtainable from F by n successive applications of either T6 or F7. Further, suppose that we, in order to be able to terminate the proof process, have to generate some specific $G!_m$ (called the appropriate $G!_m$) that contain at least one occurrence of c_m. Then we observe that, before we introduce the constant c_m, we have to introduce the constants $\mathrm{c}_1, \mathrm{c}_2, \cdots, \mathrm{c}_{m-1}$ and generate all $G!_{m-1}$ (cf. the order in which we assign truth values defined in part I, §5). There are $(m-1)^n$ different $G!_{m-1}$.

[14] An i with index is one of the numbers 1, 2, $\cdots$, or m,

When we have introduced the constant c_m, there are $m^n - (m - 1)^n$ additional $G!_m$ that can be generated and the appropriate $G!_m$ are to be found among them. This can be done only by generating the various $G!_m$ in the order defined by the program until the appropriate ones are found. The order defined is quite arbitrary; there is no known method for choosing a particularly favorable order for finding the appropriate $G!_m$.

This argument shows the extreme unwieldiness of the adopted technique of choosing the constant when applying T6 or F7. To date, no published proof procedures contain any essentially more efficient devices for making this choice.

2. *Methodological Developments for Increased Applicability*

Obviously, the elementary exhaustion technique discussed above must be replaced by more efficient devices before any significant advances towards practical machine proof methods can be made. Such devices will, however, be reported in Kanger 1959[15] and Prawitz 1959–60, and their adoption will considerably widen the range of application of proof procedures as performed even by existing machines.

There are, however, many other directions in which important advances can be made. In general, much information available or latent at each stage of a procedure of this type is discarded (to save memory space only absolutely necessary information is retained) although this information might be helpful in speeding up the processing. Thus, one can include a comparison process, making it possible to avoid the construction of subtrees identical with such as are already obtained.

Another obvious characteristic of such algorithms as are treated here is their complete lack of all "learning" facilities; such as, e.g., simple accumulation and use of results obtained in applications. It may be observed, however, that the machines available today are in certain respects insufficient for large systems containing learning processes.

3. *Machine Properties Influencing Applicability*

Considering the influence of machine "constitution" upon the performance of the present proof program, it must be pointed out that most "general purpose" computers available today are in fact specifically designed to perform arithmetical processing of numerical information in a simple and efficient manner. In most details, however, the requirements of such processing are in rather sharp contrast to the logical processing of general (alphabetical) information. E.g., in

[15] Kanger 1959 gives a proof procedure which also includes the theory of identity and allows the use of symbols for operations, e.g. addition and multiplication. This results in heavy reduction of the work and time needed in applications to problems within most mathematical theories.

numerical information processing it is technically and operationally convenient to specify the information units ("words") as numerals with a fixed number of figures, while in logical information processing the natural information units are formulas of variable (eventually large) length. Programming today's fixed-word-length computers to handle variable-length information units naturally results in low processing speed relative to that attainable with machines of a truly general design.

Another important point concerns the output requirements. For many reasons it is generally desired to have, on an output medium, a complete recording of the proof produced in an application of the procedure. It is advantageous to make this recording directly when the relevant information is obtained in the procedure, so that no memory space is required for this purpose. Such a policy, however, ties the processing speed directly to the speed of the output recording equipment, which, in the Facit EDB machine we have been using, consisted of a high-speed paper tape punch working at the rate of 150 symbols a second. This rate is low by a factor of about 12 to the actual rate of production of symbols for output. To show the effect of this mismatch, we made double runs of each problem in the test series. The first run of each problem was made with complete proof output; in the second run, no proof output was taken.

4. *Examples*

We now list some illustrative examples among the theorems proved with the present program. To ease the reading and to provide information about the proofs, these being too long to be published in extenso here, clarifying comments are adjoined. Two processing times (in seconds) are stated for each example: the first holding for full proof output on punched paper tape; the second, within parentheses, holding for no proof output.

Ex. 1: The formula $(Ex_1Ax_2Px_1x_2 \supset Ax_2Ex_1Px_1x_2)$ considered in part I, §2 is, as explained there, a valid sentence and hence a theorem of the predicate calculus. Machine proof times: 3 (0, 5) sec.

Ex. 2: $(Ax_1(G \vee Px_1) \supset (G \vee Ax_1Px_1))$, where G is an arbitrary sentence not containing x_1 as a free variable, is a theorem of the predicate calculus. In the machine, G is represented simply as a predicate symbol with no argument places. Machine proof times: 3 (0, 7) sec.

Ex. 3: $(Ax_1(Px_1 \wedge Qx_1) \equiv (Ax_1Px_1 \wedge Ax_1Qx_1))$ is a theorem of the predicate calculus. The proof has 6 branches and 4 threads and the machine needs 8 (2) seconds to complete it.

Ex. 4: The formula $Ex_1Ax_2(Px_2x_1 \equiv \sim Ex_3(Px_2x_3 \wedge Px_3x_2))$ is contradictory, i.e. its negation is valid. If P is the relation "to be a member of", the formula can be regarded as stating a version of Russel's paradox concerning the theory of classes. The formula then states the existence of a class x_1 embracing all classes x_2 that are not members of members of themselves (i.e. that are not members of classes that are members of x_2). To show that the formula is contradictory we can prove the validity of the negation of the formula, or, what amounts to the same thing, we can assign the value truth to the formula and then apply the truth rules to the so obtained e-formula. The proof has 8 branches and 5 threads and takes 9,8 (2,3) seconds to complete in the machine.

Ex. 5: A well-known theorem of relation theory says that a transitive and irreflexive relation must also be asymmetric. In this theorem the two premises get the formulations

$$Ax_1Ax_2((Rx_1x_2 \wedge Rx_2x_3) \supset Rx_1x_3),$$
$$Ax_1 \sim Rx_1x_1\ ,$$

and the conclusion

$$Ax_1Ax_2(Rx_1x_2 \supset \sim Rx_2x_1).$$

The proof of this theorem has 60 branches and the machine needs 48 (12) seconds to complete it.

Appendix

The left column below reproduces the complete machine proof of the predicate calculus theorem $(Ax_1(Px_1 \supset Qx_1) \supset (Ax_1Px_1 \supset Ax_1Qx_1))$. To the right, the m-formulas are translated to the notation of part I, §2, and some clarifying comments are added. It took 4 seconds to prove the theorem (with tape output).

25270E2020 406702C0D0 70720101C0 20101D0025	
	$(Ax_1(Px_1 \supset Qx_1) \supset (Ax_1Px_1 \supset Ax_1Qx_1))$ with the value f (= 2, the 3rd m-formula sign) and principal sign $\supset$ (= 7, the 4th m-formula sign) is the initial e-formula. Applying the procedure, this e-formula is split according to rule F4 and its second subformula
1427072010 1C020101D0 0000000014	$(Ax_1Px_1 \supset Ax_1Qx_1)$ gets the value f. This e-formula is split in turn, and its second subformula
0A220101D0 000000000A	Ax_1Qx_1, getting the value f and having the principal sign A (= 2), is then transferred to M2a. Its first subformula
0A120101C0 000000000A	Ax_1Px_1, having got the value t (= 1) and having the principal sign A is then transferred to M3a. Now the first subformula
1112020406 702C0D0011	$Ax_1(Px_1 \supset Qx_1)$ of the initial formula, having got the value t, is examined and is also transferred to M3a. Now we must introduce a constant, so the e-formula
0A220101D0 000000000A	Ax_1Qx_1 with the value f is taken from M2a and the constant introduced, c_1 (= 1), is substituted into it according to rule F6, thus obtaining

052D100005	Q1 with the value f and principal sign Q (= D). This is our first atomic e-formula and is transferred to M4a. Now we can apply rule T6 to the first e-formula in M3a.
0A121101C0 000000000A	Ax_1Px_1 with the value t, obtaining our second atomic e-formula,
051C100005	P1 with the value t and principal sign P (= C), which, after comparison with our first atomic e-formula, is also transferred to M4a. Now we must use rule T6 on the second e-formula in M3a,
1112120406 702C0D0011	$Ax_1(Px_1 \supset Qx_1)$ with the value t, obtaining
0A1702C1D1 000000000A	(P1 $\supset$ Q1) with the value t and principal sign $\supset$. This e-formula requires the application of branching rule T4 and is thus transferred to M7. However, no possibility of applying a simpler rule remains, so
0A1702C1D1 000000000A	(P1 $\supset$ Q1) with value t is taken from M7, and rule T4 is applied. The first, left branch, named
0	0, then comes to contain the atomic e-formula
051D100005	Q1 with the value t, which upon comparison is found to contradict the first e-formula in M4a. Thus this branch is terminated and now the other branch named
1	1, is treated. It contains the atomic e-formula
052C100005	P1 with the value f, which upon comparison is found to contradict our second e-formula in M4a, and so
1	*the theorem is proved*, since no more branches remain to be treated.

REFERENCES

E. W. Beth 1955, Semantic entailment and formal derivability, Med. der Kon. Nederl. Akad. van Wetensch., deel 18, no. 13, Amsterdam.

——— 1958, On machines which prove theorems, Simon Stevin (Wis- en Natuurkundig Tijdschrift), vol. 32, pp. 49–60.

B. V. Bowden (e.d.) 1953, *Faster Than Thought*, London.

A. Church 1956, *Introduction to Mathematical Logic*, vol. I, Princeton.

H. Gelernter 1957, Theorem proving by machine, Summaries of talks presented at the Summer Inst. of Symbolic Logic in 1957 at Cornell University (mimeographed). (This reference is taken from Beth 1958.)

H. Gelernter and N. Rochester 1958, Intelligent behavior in Problem-Solving Machine, *IBM J. Research Develop.*, vol. 2

K. J. K. Hintikka 1955, Form and content in quantification theory, Two papers on symbolic logic, Acta Philosophica Fennica, fasc. VIII.

S. KANGER 1957, *Provability in Logic*, Stockholm.

——— 1959, Handbok i logik, Filosofiska studier utgivna av seminariet för teoretisk filosofi vid Stockholms universitet (to be mimeographed 1959).

S. C. KLEENE 1952, *Introduction to Metamathematics*, Amsterdam.

A. NEWELL and H. SIMON 1956, The logic theory machine, *IRE Trans. Infor. Theory*, vol. IT-2, no. 3, pp. 61–79.

D. PRAWITZ 1957, Mekanisk bevisföring i predikatkalkylen, Uppsats för seminariet i teoretisk filosofi (mimeographed), Stockholm.

——— 1959–60, An improved proof procedure, to appear in Theoria (Lund).

A. ROBINSON 1957, Proving a theorem (as done by man, logician, or machine), Transcription of the Proc. of the 1957 Cornell Summer Inst. of Logic, Ithaca. (This reference is taken from Gelernter 1958.)

K. SCHÜTTE 1954–56, Ein System des verknüpfenden Schliessens, *Arch. matem. Logik Grundl.*, vol. 2, pp. 55–67.

Proving Theorems by Pattern Recognition – I

H. Wang

1. Introduction

Certain preliminary results on doing mathematics by machines ("mechanical mathematics") were reported in an earlier paper [20]. The writer suggested developing inferential analysis as a branch of applied logic and as a sister discipline of numerical analysis. This analogy rests on the basic distinction of pure existence proofs, elegant procedures which in theory always terminate, and efficient procedures which are more complex to describe but can more feasibly be carried out in practice. In contrast with pure logic, the chief emphasis of inferential analysis is on the efficiency of algorithms, which is usually attained by paying a great deal more attention to the detailed structures of the problems and their solutions, to take advantage of possible systematic short cuts. The possibilities of much more elaborate calculations by machines provide an incentive to studying a group of rather minute questions which were formerly regarded as of small theoretical interest. When the range of actual human computation was narrow, there seemed little point in obtaining faster procedures which were still far beyond what was feasible. Furthermore, on account of the versatility of machines, it now appears that as more progress is made, strategies in the search for proofs, or what are often called heuristic methods, will also gradually become part of the subject matter of inferential analysis. An analogous situation in numerical analysis would be, for example, to make the machine choose to apply different tricks such as taking the Fourier transform to obtain a solution of some differential equation.

The present paper is devoted to a report on further results by machines and an outline of a fairly concrete plan for carrying the work to more difficult regions. A fundamentally new feature beyond the previous paper is a suggestion to replace essentially exhaustive methods by a study of the patterns according to which extensions involved in the search for a proof (or disproof) are continued. The writer feels that the use of pattern recognition, which is in the cases relevant here quite directly mechanizable, will greatly extend the range of theorems provable by machines.

As is to be expected, the actual realization of the plan requires a large amount of detailed work in coding and its more immediate preparations. The machine program P completed so far on an IBM 704 contains only a groundwork for developing the method of pattern recognition. It already is rather impressive insofar as ordinary logic is concerned but has yet a long way to go before truly significant mathematical theorems can be proved. For example, the program P has to be extended in several basic directions before a proof can be obtained for the theorem that the square root of 2 is not a rational number. On the other hand, theorems in the logical calculus can be proved very quickly by P. There are in *Principia Mathematica* altogether over 350 theorems strictly in the domain of logic, viz., the predicate calculus with equality, falling in 9 chapters (1 to 13, since there are no 6, 7, 8, and since 12 contains no theorems). The totality of these is proved with detailed proofs printed out by the program P in about 8.4 minutes. To prove these theorems, only about half—and the easier half—of P is needed. The other half of P can prove and disprove considerably harder statements and provides at the same time groundworks for handling all inferential statements. This program P will be described in section 2.

Since the central method to be discussed is primarily concerned with the predicate calculus, its wider signifi-

* On leave of absence from University of Oxford, Oxford, England, for 1959–60.

Article from: Communications of the Association for Computing Machinery, Vol. 3, No. 4, April 1960.

cance may be appreciated better, if we review briefly certain familiar facts about the relation of the predicate calculus to mathematics in general.

Thus, it is well known among logicians that if we add equality and the quantifiers "for all x," "for some y" to the propositional connectives "and," "if," "or," "not," etc., we obtain the predicate calculus in which every usual mathematical discipline can be formulated so that each theorem T in the latter becomes one in the former when the mathematical axioms A applied are added as premises. That is to say, if T is the theorem in the mathematical discipline, "if A, then T" is one of logic. This, rather than the constructions of Frege and Dedekind, is the significant sense in which mathematics is reducible to logic. From this fact it is clear that in order to prove mathematical theorems by machines a major step is to deal with theorems of the predicate calculus.

There is a natural uneasy feeling that this cannot be a feasible way of handling mathematics since we expect the methods to be largely dictated by the peculiar mathematical content of each individual branch, which presumably gets partly lost when the disciplines are thus uniformly incorporated into the predicate calculus by formalization and abstraction. This is quite true, and indeed we have to add special methods for each special mathematical discipline. But the point often neglected is that an adequate treatment of the predicate calculus is of dominating importance and that for each discipline the basic additional special methods required are relatively homogeneous. For number theory, the essential new feature is largely concentrated in mathematical induction as a method of proof and as one of definition; for set theory, in the axiom of comprehension, i.e., the axiom specifying all the conditions which define sets. Hence, there is the problem of choosing the formula to make induction on, or the condition for defining a set. While it seems doubtful that there is any uniform efficient mechanical method for making such selections, there are often quite feasible partial methods. For example, for making such selections in number theory the obvious uninspired method of trying a conclusion and its subformulae as the induction formula should suffice in many cases.

Thus, it would seem that, once a feasible way of doing logic is given, fairly simple additional methods could carry us quite some way into special mathematical disciplines. Moreover, the method of pattern recognition is basically number-theoretic, and as such recovers a considerable amount of the mathematical content of each branch of mathematics. This is so because, in order to establish, e.g., a conclusion (x)(Ey)Rxy, it aims at choosing a simple correct function f such that (x)Rxfx. And it seems not unreasonable to contend that a good deal of originality in mathematics consists precisely in the ability to find such functions.

The proposed method for doing logic always begins from scratch for each theorem. This is quite different from the type of proof we encounter in Euclid, where it is essential that later theorems are proved with the help of earlier ones. While this problem of selecting relevant earlier theorems to apply appears unimportant in the domain of logic as dealt with by the method to be described, it has to be faced at some stage, and the writer does not have a ready general solution of it. Two remarks, however, seem to be relevant. In the first place, because of the greater speed of calculations by machines, it is natural to expect that it is often faster to prove an easy old theorem anew rather than look it up. Hence, we may neglect easy theorems and record only hard ones, perhaps as new axioms. In this way we arrive at a conception of expanding axiom systems which include difficult new theorems as additional axioms. Here it is irrelevant that the new axioms are not independent, since the goal is to prove other new theorems as quickly as possible. When we use such expanding axiom systems, we arrive at a compromise between pedantry and ignorance. This is not much different from the practice of a good mathematician who remembers only a number of important theorems and works out simple consequences as he is in need of them. In the second place, although it is of interest to extend the range of problems which the machine can do without human intervention, it is fair to expect that when we arrive at the stage of having machines try to prove theorems which we cannot prove, we shall not hesitate to feed the machine all the useful suggestions we can think of. Eventually machines are to be an aid to mathematical research and not a substitute for it; there is no point of running a handicap race by refusing to lend the machine a hand to complement its shortcomings. In fact, once the general framework is available, one would expect that, compared with a mere expert of the general techniques, a mathematician working on a particular problem will more likely succeed in using the framework with additional hunches appended to get a proof of the desired theorem by machine.

Another question is that the axiom of induction and the axiom of comprehension both have infinitely many instances. Or, in the usual formulation of number theory and set theory, each contains infinitely many axioms beyond the predicate calculus. Hence, if we ask whether, e.g., a statement T is a theorem of number theory, we are actually asking whether it is a logical consequence of the infinitely many axioms. If there are only finitely many axioms, we can write their conjunction A, and ask simply whether $A \rightarrow T$ is a theorem of the predicate calculus. This trick is denied us when the axioms are infinite in number. It, therefore, seems desirable to use only finitely many axioms when possible, and indeed there are standard methods for reducing usual sets of axioms to finite sets (see, e.g., [18] for one such formulation of number theory). The matter is, however, not very clear since we have to make selections from the axioms anyhow and the finite set only gives an enumeration of the infinite set, introducing meanwhile complexities through another avenue. Finite sets of axioms are, however, undoubtedly useful for many purposes of mechanization, e.g., when one comes to classifying theorems according to their logical forms.

Since most of us learned Euclid and number theory without worrying about the predicate calculus, it might seem that the natural course is to bypass logic and go directly to mathematics. The writer is opposed to such an approach if the aim is to prove more and harder theorems rather than to study the psychology and history of mathematical thinking. Obviously what is natural for man need not be natural for the machine. More specifically, if logic is not treated in an explicit and systematic manner, constant additions of ad hoc new devices make the progress toward less trivial theorems slower and slower, as well as more and more confusing. As a result, one may, e.g., even mistake the introduction of familiar logical principles for genuinely giant steps. Devising a vast machinery specifically designed to obtain a few easy theorems is wasteful. The writer feels that results obtained from different approaches ought to be measured against the generality and economy of the machinery behind them, and that preliminary steps should be capable of supporting large superstructures yet to be erected. It is the writer's conviction that the alternative approach of treating logic only by the way would score very poorly by both criteria.

This is, however, not to deny that some of the problems encountered in dealing directly with mathematics will still have to be faced by the present approach. It is merely contended that the alternative approach does not take advantage of the possibility of "divide and conquer." As a result, what could be handled simply with the help of known techniques is mixed up with the less easily manageable further details, so that an intrinsically complex problem is made even more complex than necessary. The present attempt is concerned less with obtaining partial results which immediately excite man's undisciplined imagination, but rather more with setting up a framework capable of yielding rich results in the long run. There is a third approach which concentrates on coding known decision procedures for isolated areas such as elementary geometry, or arithmetic with only multiplication. Since these areas do not include very many interesting theorems and do not form organic parts of proof procedures for more interesting areas, the writer feels they are not of central importance to the program of proving theorems by machines. At a later stage they may serve as useful auxiliary devices to assist th more basic techniques. It cannot be denied, however, that this type of problem has the advantage that only more restricted theoretical considerations are needed for their mechanical implementation.

In the previous paper [20], the writer has suggested an Herbrand-Gentzen type proof procedure which is also an efficient decision procedure in the realms of the propositional calculus and the AE predicate calculus (i.e., those formulae which can be transformed to ones with prefix $(x_1) \cdots (x_m)(Ey_1) \cdots (Ey_n)$). In the more general case, there is the well-known unbounded search procedure illustrated in the following simple example.

EXAMPLE (1).

$$(x)(Ey)(Gyy \ \&\ Gxx) \supset (Ex)(z)(Gzx \ \&\ Gzz);$$

or, alternatively,

$$(Ex)(y)(z)[(Gyy \ \&\ Gxx) \supset (Gzx \ \&\ Gzz)].$$

According to Herbrand's theorem to be described in Part II, (1) is a theorem if and only if there exsts some N such that $S_1 v \cdots v S_N$ is a truth-functional tautology, where the S_i's are:

$$S_1 : (x, y, z) = (1, 2, 3): (G22 \ \&\ G11) \supset (G31 \ \&\ G33)$$

$$S_2 : (x, y, z) = (2, 4, 5): (G44 \ \&\ G22) \supset (G52 \ \&\ G55)$$

$$S_3 : (x, y, z) = (3, 6, 7): (G66 \ \&\ G33) \supset (G73 \ \&\ G77)$$

$$S_4 : (x, y, z) = (4, 8, 9): (G88 \ \&\ G44) \supset (G94 \ \&\ G99)$$

...

Since (1) is not a theorem, there can exist no tautologous disjunction $S_1 v \cdots vS_n$, or briefly, D_n. If we are to test successively the disjunctions D_1, D_2, etc., we can never reach an answer. One can undoubtedly use some special argument to show that (1) is not a theorem, but then there is the question of formalizing the argument and generalizing it to apply to some wide range of cases. As it happens, (1) falls under a simple decidable class, viz., the E_1A case of all formulae beginning with a prefix $(Ex)(y_1) \cdots (y_n)$, and it has been shown that for each formula A in the class, one can find some N, such that either D_N is tautologous or A is not a theorem. Hence, it may seem that simply adding the method of calculating N to the unbounded search procedure would already provide a decision procedure for the class in question. This is, however, only a theoretical possibility and hardly feasible even on machines. For example, according to Ackermann's evaluation for N (see [3], p. 265) the value of N for the simple example (1) is no less than $2^{48} - 1$. Of course, the bounds for more complex formulae in the class, and formulae in more complex decidable classes, are much higher according to the traditional decision procedures. This situation led the writer to envisage in [20] the prospect of not using more decision procedures but trying to simplify directly the brute force search procedure of proof as soon as we get beyond the AE predicate calculus.

More recently, steps along such a direction have been taken by Gilmore, Davis and Putnam. Gilmore has written a program using essentially the brute force method and tested a small sample of examples [8]. Davis and Putnam have in [4] devised efficient techniques for testing whether a given disjunction $S_1 v \cdots vS_N$ is tautologous. An efficient test for truth-functional tautologies in general, proposed earlier by Dunham-Fridshal-Sward, has also been coded and run on a machine [7].

The writer feels that Gilmore's result is basically negative, i.e., it shows that without fundamental improvements the brute force method will not do. Perhaps the two most interesting examples, a nontheorem (2) and a theorem (3), which his program fails to decide, are fairly simple and can indeed be decided by the method of pattern recognition quite easily, as will be shown in Part II. His examples are (drawn from [3], p. 262):

EXAMPLE (2).

$$(Ex)(Ey)(z)\{[Gxz \equiv Gzy) \;\&\; (Gzy \equiv Gzz) \;\&\; (Gxy \equiv Gyx)] \supset (Gxy \equiv Gxz)\}.$$

EXAMPLE (3).

$$(Ex)(Ey)(z)\{[Gxy \supset (Gyz \;\&\; Gzz)] \;\&\; [(Gxy \;\&\; Hxy) \supset (Hxz \;\&\; Hzz)]\}.$$

Davis and Putnam have indicated that by their improved method of testing for tautologies, a treatment of (3) becomes feasible. Since their method is concerned only with the last stage, viz., that of testing each disjunction, it can of course do nothing with nontheorems such as (2). Moreover, since it provides no device for deleting useless terms among S_1, S_2, etc., it is not likely to be of use even when a formula is indeed a theorem but the smallest n for which $S_1 v \cdots vS_n$ is tautologous is large. For example, with regard to (3), $S_1 v \cdots vS_{25}$ is the earliest tautology; in 21 minutes on an IBM 704, only $S_1 v \cdots vS_7$ has been handled by Gilmore's program. Although the particular example appears to be mechanically manageable by the method of Davis and Putnam, one would expect that expressions can easily become too long to handle by this method.

The writer now feels that a more basic step is to eliminate in advance useless terms among S_1, S_2, etc., or, alternatively, instead of actually constructing and testing the disjunctions, examine in advance, for each given problem, all the possible courses along which counterexamples to S_1, $S_1 vS_2$, etc. may be continued. Using the second alternative, we obtain at the same time a disproving procedure for most cases. The detailed techniques for achieving these goals are here called the method of proving theorems (and disproving nontheorems) by pattern recognition, or, more specifically, the method of sequential tables. When applied to Example (1), the method gives the desired answer in the following manner. When we substitute numbers for the variables, each elementary part Gyy, Gxx, Gzx, Gzz gives way to infinitely many new elementary parts which occur in S_1, S_2, etc. We now ask whether we can so assign truth values (true or false) to the infinitely many elementary parts that S_1, S_2, etc. all become false. If that is impossible, (1) is a theorem, otherwise we get a counterexample and (1) is not a theorem. If we look at the matrix of (1):

$$(Gyy \;\&\; Gxx) \supset (Gzx \;\&\; Gzz),$$

we see that it is false only when:

Gxx	Gyy	Gzx	Gzz
t	t	f	t
t	t	t	f
t	t	f	f

Since, as happens in this case, different variables are always replaced by different numbers, each of the above rows can make each of S_1, S_2, etc. false if we imagine that the variables are replaced by their corresponding numbers. The problem is whether we can select simultaneously one row for each S_i which, taken together, will not conflict with one another. For example, although each row can falsify S_1, and even both S_1 and S_2, S_3 becomes true when G33 gets the value f. Hence, to falsify S_1, S_2, and S_3, we must take the first row for S_1:

G11	G22	G31	G33
t	t	f	t

Then any row can falsify S_2 and S_3. But in order to falsify also S_5, G55 has to be t, so that again we can use only the first row for S_2:

G22	G44	G52	G55
t	t	f	t

Similarly, in order to falsify S_7 and S_3 one has to use the first row for S_3. It is clear that by always using the first row we can simultaneously falsify all S_i's since the constraints imposed on later S_i's by earlier S_i's are uniform. Hence, we conclude that (1) is not a theorem. In fact, as will be discussed in section 3, all we have to do is to cross out every falsifying row in which Gyy or Gzz gets a value which Gxx does not get in any row. After repeated application of this operation, either no row is left and then the original statement is a theorem, or else some row is left and then a countermodel is possible. This last method, called the method of sequential tables, seems to be a new feature that goes beyond the general method of pattern cognition.

As will be shown in Part II, the method can be generalized and rigorously justified for a number of broad classes. The type of considerations involved in such a method should be clear, however, from the above example.

The basic ideas of the general method of pattern recognition, though not the special addition of the method of sequential tables directed at efficiency, go back to Herbrand [10] and, in a less general form, also to Skolem [16]. By this method, Herbrand was able to give in a uniform way a treatment of most solvable cases of the decision problem (for logic) known at his time, and to discover two interesting new cases, viz., a generalization of the E_1A case and the disjunctive predicate calculus dealing with formulae with a matrix that is a disjunction of elementary parts and their negations. Church gave along a similar line a more exact treatment of these same cases plus two cases by Skolem but minus Herbrand's generalization of the E_1A case [2, 3]. The chief additional case at first obtained by the more usual sort of technique, shortly after Herbrand's treatment, has recently been handled by Klaua [12] with this general method. Dreben has pursued the matter further and announced in general terms a number of results [6]. We understand Dreben is writing a monograph on the subject.

The writer believes that in several directions the important implications of the method has not yet been fully exploited in the works just cited. First, the method can be used to give decision procedures for well-known unsolved cases. The writer has found a partial solution of the decision problem for the class of formulae with the prefix (Ex)(y)(Ez) (this open problem is mentioned, e.g., by Church, [2, p. 271], and by Ackermann, [1, p. 85]; it seems to go back to the early thirties). The solution will be given in Part II. This case is of special interest since it is a natural class and includes simple examples which are nontheorems but have no finite countermodels. A well-known example due to Schütte [1, p. 83] is:

EXAMPLE (4).

$$(Ex)(y)(Ez)\{Gxy \supset [Gxx \vee (Gyz \;\&\; \sim Gxz)]\}.$$

The negation of this is an axiom of infinity, i.e. a statement satisfiable only in an infinite domain. A related but more familiar form of axioms of infinity is the conjunction of:

(i) $(x) \sim Gxx$; (ii) $(x)(Ey)Gxy$;

(iii) $(Gxy \;\&\; Gyz) \supset Gxz$.

Secondly, the method and ideas of pattern recognition can be extended to give some quasi-decision procedure for the whole predicate calculus. By this is meant a procedure which in theory always gives a proof if the given formula is indeed a theorem, and which in "most" cases gives also a counterexample if the given formula is not a theorem, so that the undecidable formulae become, one might say, points of singularity. In general, it is no longer a question whether a nontheorem has finite countermodels but whether it has either finite or simple infinite countermodels. For example, if a nontheorem has no recursive countermodels, it is to be expected that a natural quasi-decision procedure will not be able to refute it. It is possible to design different quasi-decision procedures which have different ranges of application. The way to get such procedures is roughly to apply the consideration of patterns to all formulae or to all in a reduction class, i.e., a class such as all formulae with the prefix (Ex)(Ey)(Ez)(w), such that there is an effective method by which each formula can be transformed into an equivalent one in the class. There are many ramifications in carrying out the matter in detail.

Thirdly, as is only natural, not sufficient attention has been paid to the question of efficiency, or the difficulties in actually applying the procedures by man or by machines. In particular, the method of sequential tables is an example of the possible ways to improve efficiency. A related minor point is that the more difficult decision procedures are usually not illustrated by examples.

In view of these three explorable areas, the writer feels that there is a good deal of interesting theoretical work which is yet to be done. This is one of the reasons why it seems to the writer difficult to make definite estimates and predictions as to how fast and how far theorem proving can be mechanized. At present, it appears that there are a succession of rather difficult but by no means humanly impossible steps yet to be taken which do not embody any known limitations. Man will have to devise methods or methods for devising methods, but the machine will use the methods to do things which man cannot feasibly do. There is nothing paradoxical in this. Even today long multiplications and other calculations provide ample examples. Since we should, as a fundamental methodological principle, expect no miracles, the fact that there is much yet to be done and that we have a fairly good idea of the sort of thing to be done seems a very good indication that we are not after a will-o'-the-wisp.

After a longer paper had been nearly completed, the writer learned of the restriction on the length of the paper. As a result, the paper is divided into two parts. Nearly all detailed theoretical considerations are given in Part II, which will be made a memorandum at the Bell Laboratories and presumably published eventually.

In this first part, section 2 gives a general description of the completed machine program mentioned above. Sketches are given of results on elementary domains such as the restricted AE predicate calculus (similar to Qp in [20, p. 10], except for a method of eliminating functors) and the AE calculus, as well as general preliminary steps useful for extension of the program to the whole predicate calculus. In particular, devices will be stated which are useful for the systematic simplification of formulae so that many additional formulae are reduced to members in classes known to be decidable. It should be emphasized that in the program actually completed procedures using the method of pattern recognition in the specific sense explained above have not been included, although the program is oriented toward a systematic preparation for the treatment of such procedures.

Section 3 gives a solution of the simple E_1A case as an illustration of the general method of pattern recognition, and, more specifically, for the method of sequential tables. In Part II, all the main decidable cases, the new case (Ex)(y)(Ez), as well as quasi-decision procedures, will be considered, all along a line similar to that followed by the method in section 3.

Finally, section 4 contains a number of general remarks.

2. A Program That Does 9 Chapters of Principia in 9 Minutes

The running program P on an IBM 704 accepts any sequent S in the predicate calculus (with equality), in particular, any sequent expressing that a theorem follows from certain axioms in some special mathematical discipline; reduces it to a finite set of atomic sequents (in other words, gives its quantifier-free matrix in a conjunctive normal form); and compiles an economic quantifier tree (see below) for S which can be used directly or as a basis for selecting a most favorable prefix (i.e., a quantifier string that does not violate the relations of dominance in the tree). When no negative variables [20, p. 9] occur in the set of atomic sequents, or briefly, the matrix, of S, the program P can decide always whether S is a theorem, and give a proof or a counterexample. The program P can often, though not always, do the same for S, when no positive variables (U-variables) are governed by negative variables (E-variables) in the matrix of S, i.e., when S is reducible to the AE-form. It is fairly easy to extend the program P to include a procedure for dealing with all AE cases. We have not done this so far because of a rather paradoxical situation. On the one hand, the restricted AE method which is included in P already suffices to decide a clear majority of the examples encountered in books on logic, and there are few actual examples which are undecidable by the restricted AE method but decidable by the full AE method. On the other hand, considered in the abstract, even the full AE method can deal with only a very restricted class of sequents which are of interest to us.

Hence, on the one hand, for the simple purpose of illustrating the surprising ease with which machines can be employed to prove and disprove common examples in logic, the theoretically narrow range of the restricted AE method is more remarkable than some more extended method. On the other hand, when we wish to use examples obtained by formalizing the statement of mathematical theorems as consequences of certain axioms, our needs will quickly go beyond even the full AE method. That is why considerable thought has been given to the question of reducing a given sequent S to the simplest possible form as a uniform basis for the treatment of many diverse cases. Most pieces in the program P are designed in such a way that they can be efficiently useful in handling all more complicated cases. Thus, the elimination of logic connectives, the reduction of each problem to as many simpler subproblems as possible, the construction of the simplest quantifier trees, and the relatively fast comparison routine for deciding atomic sequents: all these are designed as a common part in further extensions.

Familiarity with [20] should be helpful, though not necessary, for understanding the following more detailed description of P. Even though P is essentially an extension of the program III as described in [20], and we shall try to avoid repetitions, there are a number of differences in those parts which are dealt with in both programs. Some of these differences should be mentioned in advance to prevent misunderstandings. While negative variables (E-variables) are replaced by numbers in [20], it has been decided to use in P the more natural course of replacing initial positive variables (U-variables) by numbers. The reduction to the miniscope form envisaged in [20] has in part been abandoned; in P, what is taken as a better procedure is used instead. Functors with explicit arguments attached to them are not used in P; rather the governing relations among the letters which replace the variables are tabulated separately.

We proceed to give a more detailed description of the program P. The program is written entirely in the language of SAP except that the subroutines of reading and writing tapes are from the Bell Monitoring System. This impurity could, if one wishes, be gotten rid of by changing just a few instructions. The whole symbolic deck of the program contains about 3200 cards. About 13,000 words of the core storage are assigned for use by the program, although a lot of these words are only reserved spaces for handling more complex problems. For the problems actually run so far, it should be easy to fit everything into a machine with 8000 words. Auxiliary storages are not needed except that, as a convenience, tapes are used to avoid going through on-line input-output equipments.

At present, there are two somewhat irksome restrictions on the program. It can only deal with a problem expressible with no more than 72 characters (i.e., one card long). This is quite adequate for handling common theorems of logic, but insufficient when we wish to apply the program to, e.g., number theory or elementary geometry. It is highly desirable to remove this restriction, which is the sort of thing that has been taken care of in several systems for symbol manipulation. Since we have no immediate plan for using such systems, we only envisage a modification that will accept, say, a problem 10 cards long.

A less fundamental restriction is the use of a sort of Polish notation, partly to speed up operations, partly to reduce the length of the sequent stating a given problem, and partly necessitated by the fact that machine printers do not have the familiar logic symbols. This has the consequence that it is not so easy to read the outputs. A translation routine could be added to bring the outputs (and, if one wishes, also the inputs) into a form which resembles some ordinary notation more closely. To assist exposition, we shall, in what follows, neglect this notation feature, and speak always as if everything had been done in a more familiar notation. (For a "dictionary," see [20, p. 6].)

A readily quotable, albeit misleading, indication of the power of the program P is the fact that it disposed of nine chapters of *Principia* in about 8.4 minutes, with an output of about 110 pages of 60 lines each, containing full proofs of all the theorems (over 350). This is misleading for two reasons. On the one hand, proving these theorems does not require the full strength of the program which can do considerably more things that are basically different from this particular task. Hence, this does not give a fair summary of what the program is capable of doing. On the other hand, while many college and graduate students, especially in philosophy, find it not so easy to prove these theorems in their homeworks and examinations, the methods we use are a bit easier than the usual methods and of the type that is specially suitable for machines. As a result, the theorems in *Principia* are far easier to prove than expected, and it is not very remarkable that they can be proved in a reasonably short time. The actual time required came, however, as a bit of a surprise. At the very outset, the writer guessed 20 hours on an IBM 704 as the probable time required to prove these (over 350) theorems. In [20], the theorems of the propositional calculus (over 200) were proved in about 37 minutes with the on-line printer, and it was estimated that the computing time was only about 3 minutes; the majority of the theorems with quantifiers (over 150) were also proved then, and it was conjectured that about 80 minutes would be needed to prove the lot. The final result with the program P is that the 200 strong theorems in the propositional calculus took about 5 minutes, while the 150 strong theorems with quantifiers took less than 4 minutes. (The writer has not been able to determine how much of the time was spent on input and output operations.) In every case, it seems that the machine did better than expected. While this fact presumably means very little, it was natural that one felt encouraged by it.

On second thought, there is an uneasy feeling that the efforts to improve the restricted AE method were a bit wasted. At the time when [20] was written, it was already quite clear that even with rather slight changes, the program available then would yield proofs of the desired theorems in a few hours. What is the point of spending many weeks' efforts to bring the time down to a few minutes? Would it not have been better if the efforts had been spent on studying more difficult cases? As a matter of fact, however, in the process of trying to handle the restricted case more efficiently, one also got a clearer view of more general questions. Moreover, while the difference between a few minutes and a few hours is not very important, the difference between a few hours and several hundred hours may prove to be decisive; and it is of interest to have a definite example of the degree of increase in speed which one can expect from improved methods.

A possible objection to the present approach to the problems of coding is that a good deal of time has been spent in those parts of the procedures which are relatively easy to carry out even by hand. The more sensible alternative approach would seem to be a concentration of efforts on testing those parts where there is a serious doubt whether machines can feasibly succeed at all. To this objection, the answer can only be that the gradual approach adopted here is meant as the beginning of a long range scientific project rather than a quick test whether crude standard methods can already produce amazing results. In fact, it is fair to say that we have learned enough to see a healthy situation with regard to the question of proving theorems by machine: the prospects are encouraging, but one has no right to expect fast miracles. This being so, the gradual approach has the advantage that we can more easily test a large number of sample problems because little work is needed in preparing them by hand.

The master control of the program P has 45 instructions. Subroutines are heavily employed. When a problem on a single card is accepted, the program first takes the following preliminary steps:

S1. If the first 6 characters of the card all are blanks, the program P interprets this as an indication that no more data cards are to be accepted. The machine stops or goes to some other job (e.g., run or compile the next program in waiting). Otherwise, P searches for the arrow sign → (actually the sign / could be used to serve the same purpose).

S2. If → does not occur, the input is treated as ordinary prose. It is printed out without comment and P proceeds to receive the next data card. If the arrow sign does occur, P searches for quantifiers. By the way, standard BCD representation of characters is used in the cores, except that the BCD representations of zero and blank are interchanged.

S3. If quantifiers do not occur (the equal sign may occur), proceed more or less in the same way as in the treatment of the quantifier-free case in [20]. Otherwise, we have the principal case.

S4. When the input problem contains quantifiers, the following preliminary simplifications are made. (i) All free variables are replaced by numbers, distinct numbers for distinct variables. (ii) Vacuous quantifiers, i.e., quantifiers whose variables do not occur in their scopes, are deleted. (iii) Different quantifiers are to get distinct variables; for example, if (x) occurs twice, one of its occurrences is replaced by (z), z being a new variable. This last step of modification is specially useful when occurrences of a same quantifier are eliminated more than once at different stages.

S5. After the above preliminary simplifications, each problem is reduced to as many subproblems as possible in the following manner. (i) Eliminate in the usual manner every truth-functional connective which is not governed by any quantifiers. (ii) Drop every initial positive quantifier (i.e., universal in the consequent or existential in the antecedent that is not in the scope of any other quantifier) and treat its variable as free, i.e., replace all its occurrences by those of a new number. (i) and (ii) are repeated for as long as possible. As a final result of this step, each problem is reduced to a finite set of subproblems such that the problem is a theorem if and only if all the subproblems are.

To illustrate the steps S1 to S5, we give in an ordinary notation a proof obtained by the program P:

11*53/→ (x)(y)(Gx ⊃ Hy) ≡ ((Ex)Gx ⊃ (y)Hy)		
/→ (x)(y)(Gx ⊃ Hy) ≡ ((Ez)Gz ⊃ (w)Hw)		(1)
1/(x)(y)(Gx ⊃ Hy) → (Ez)Gz ⊃ (w)Hw		(2)
2/G1, (x)(y)(Gx ⊃ Hy) → H2		(3)
1/(Ez)Gz ⊃ (w)Hw → (x)(y)(Gx ⊃ Hy)		(4)
4/(w)Hw → G3 ⊃ H4		(5)
5/G3, (w)Hw → H4		(6)
4/ → G3 ⊃ H4, (Ez)Gz		(7)
7/G3 → H4, (Ez)Gz		(8)
8/G1 → H2, (Ez)Gz		(1)
1/G1 → H2, G1	SVA	(2)
	PQED	
6/G1, (w)Hw → H2		(1)
1/G1, H2 → H2	SVA	(2)
	PQED	
3/G1, (x)(y)(Gx ⊃ Hy) → H2		(1)
1/G1, (y)(Gx ⊃ Hy) → H2		(2)
2/G1, Gx ⊃ Hy → H2		(3)
3/G1 → H2, G1	SVA	(5)
3/G1, H2 → H2	SVA	(4)
	QED	

In the above example, all quantifiers are made distinct in (1). By (i) of S5, (1) is reduced to (2) and (4). By (i) and (ii) of S5, (2) is reduced to (3), which can be reduced no further by (i) or (ii) of S5. By (i) of S5 and then (ii) of S5, (4) is reduced to (5) and (7). Finally, by (i) of S5, (5) and (7) are respectively reduced to (6) and (8). Hence, the original problem 11*53 is reduced to the 3 subproblems (3), (6), (8). It is possible to show the following:

T2.1. *The original problem is a theorem if and only if all its subproblems (in the above sense) are.*

Hence, if any subproblem is refuted, then the original problem is also. If no subproblem is refuted, but some subproblem is undecidable by some restricted method, then the original problem is undecidable by the same restricted method.

Hence, the remaining problem is to study each subproblem (in the above sense). In theory, this reduction to subproblems is rather wasteful, since it could be automatically taken care of by tackling the whole problem directly in a uniform manner similar to the way in which each subproblem is tackled. In practice, however, it is clearly desirable to isolate separate problems whenever possible. It is to be noted that further reductions are of a different nature because the subproblems would be interconnected through variables attached to some common quantifiers. This will soon become clear.

Now, e.g., we may study (3), (6), (8) in the above example each as a separate problem in itself. Each is stored away temporarily until we have obtained the last, (8) in the example. Then (8) is taken as a new first line and treated. Afterwards, (6) and (3) are called back and handled similarly. Note also that the numbers in each subproblem begin from 1 both at the end of each line and inside the body of the proofs. Each of (3) (6), (8) happens to be a theorem, so we conclude at the end that 11*53 is itself a theorem.

We now explain how each subproblem is to be treated. In order to do this, we have to describe first how quantifiers in general are handled, as well as how different comparison procedures are performed on atomic sequents.

Variables can be replaced by 3 kinds of symbol according to the status of their quantifiers.

(i) Free variables and initial U-variables: by 1, 2, 3, $\cdots$, 9 (numbers).

(ii) All E-variables: by s, t, u, v, w, x, y, z. (In fact, unchanged.)

(iii) U-variables governed by E-variables: k, l, m, n, o (functors).

With U-variables governed by E-variables and E-variables governed by such U-variables, a record is kept separately of the letters which govern them.

The present method avoids the necessity of reducing a formula first to a prenex normal form, as well as an unpleasant feature about $\equiv$. For example, if we have a formula (Ex)Gx $\equiv$ (y)Hy, then we should get 4 quantifiers when $\equiv$ is eliminated. The situation may be seen from the reduction of (1) to (2) and (4) in 11*53. Now, since it is necessary, for certain purposes, to have distinct variables for distinct quantifiers, we may feel we have to double the number of variables in such cases. Since, however, the two quantifiers resulting from one quantifier always have different signs, the above convention about the replacement of variables automatically assures us that the two new quantifiers get different symbols. Thus, in 11*53, the variable x in (4), being an initial U-variable, is replaced by the number 3 in (5), while the variable x in (2), being an E-variable, will remain unchanged after the quantifier (x) is dropped, as is seen in (2) in the last part of the whole proof.

In determining what variables or functors govern a given quantifier Q, we use a somewhat more economic criterion. Instead of recording all quantifiers which contain Q in their scopes, we use all the variables (and functors) which are free in the scope of Q and distinct from the variable of Q. This requires a theoretical justification that can be stated (true only under restrictions):

T2.2. *We can separate out Q and its scope from those quantifiers whose variables do not occur in the scope of Q.*

Another device is employed to simplify the governing relations among variables and functors when one subproblem is reduced to a finite set of atomic sequents (a matrix). Two symbols, each a variable or a functor, are connected if there is an elementary part in the matrix which contains both symbols or contains one of the two symbols as well as a variable or functor connected to the other. Then a variable or functor is really governed by another if its quantifier was originally governed by the latter and they are connected. A subroutine EFCTR serves to reduce the governing relations in this way. This will be justified in Part II by:

T2.3. *If two symbols, each a functor or a variable, are not connected in the final matrix, we can always so transform the original sequent as to separate the two quantifiers which give way to them.*

We make use of several kinds of comparison procedure in deciding an atomic sequent. Given an atomic sequent, we first compare the antecedent with the consequent as if no quantifiers occur, i.e., whether a same atomic formula occurs on both sides, or, if = occurs, whether a self-identity occurs in the consequent or substituting equals for equals would yield an atomic formula on both sides. This is COMP, a procedure described in [20]. If the answer is yes, then the atomic sequent is a theorem, and we put a VA on it and print it out. We do not have to worry about it anymore.

If the answer is no, we generally go to a different comparison routine COMQ, which permits us to make substitutions on the variables: each variable can be replaced by any number, as well as by any functor not governed by the variable, or by another variable. If in this way, we can obtain a result valid by the previous criterion, we put tentatively the label SVA on and store the sequent away. This comparison routine is quite complex because we require that the same variable get the same substituent not only in each atomic sequent but in all the atomic sequents which come from one given subproblem. The presence of = makes this part doubly complex.

If every atomic sequent from a subproblem gets SVA by compatible substitutions, the subproblem is proved. If at one stage, an atomic sequent fails to get SVA with any substitution compatible with earlier substitutions, we shall test no more atomic sequents by substitution until we have, if possible, simplified the governing relations of variables and functors with regard to the whole set of atomic sequents obtained from the original subproblem.

The substitutions are made in a sensible way in the sense that we do not try all possible substitutions but try only the most likely ones (compare [20, p. 11]). This is an important factor in making P more efficient than the earlier program.

At a later stage there is also a negative comparison test called NTEST in which each atomic sequent is tested separately by substitution, possible conflicts with substitutions for other atomic sequents being neglected. When this is not possible and the atomic sequent contains no functors, the atomic sequent, and therewith the original subproblem, is refuted. This step again requires a theoretical justification.

Let us now give some examples (at the right) from the outputs of P and then summarize the main steps.

These should be a fair sample of the shorter results among the problems beyond *Principia* which have been handled by the program P. We now give a summary of the steps needed in solving each subproblem and illustrate them by the above examples. When a problem has only one subproblem the subproblem is of course the problem itself.

S6. Eliminate quantifiers and truth-functional connectives whenever possible, i.e., whenever a sequent under consideration is not an atomic sequent. By the way, before all subproblems were obtained, atomic sequents were not dealt with, e.g., (5) in 19*10.

S7. If the sequent is atomic, try to decide it by COMP. Put on VA if it is valid and continue with next sequent. If it is not valid, put on NO and finish a subproblem (hence, also the problem) which contains no quantifiers. This is the case with the subproblem (8) in 19*10.

S8. If the subproblem contains quantifiers, go to COMQ. If this makes it valid, i.e., there are acceptable substitutions to make the sequent valid, put on SVA and store it away. If this is the last atomic sequent of a subproblem, then we have proved it. We put the line out together with all earlier SVA sequents which have been stored away. For example, this is the case with 14*4 and 15*16.

S9. If this cannot make the atomic sequent valid, we store it away and record the fact. We then continue with the problem but test no more atomic sequents beyond COMP. When all the atomic sequents are obtained, we use EFCTR to simplify the governing relations between the functions and variables. There are four possibilities given under S10, S11, S12, S13.

S10. If there is no governing relations in the result, i.e., the result is in the AE form; then either this was so all along, or this is so only because certain functors could be eliminated (i.e., not really governed and can therefore be replaced by new numbers). In the first case, test whether there is only one undecided atomic sequent, or only one number occurs in the undecided atomic sequents. In either case, the restricted AE method is sufficient, and

14*4/(x)((Hx & Hy) ⊃ Gx), p, (x)Hx → Gy		
/(x)((Hx & H1) ⊃ Gx), p, (z)Hz → G1		(1)
1/(Hx & H1) ⊃ Gx, p, (z)Hz → G1		(2)
2/Gx, p, (z)Hz → G1		(3)
2/p, (z)Hz → G1, Hx & H1		(5)
5/p, Hz → G1, Hx & H1		(6)
6/p, H1 → G1, H1	SVA	(8)
6/p, H1 → G1, H1	SVA	(7)
3/G1, p, H1 → G1	SVA	(4)
	QED	

14*15/(Ex)(y)Gxy → (x)(Ey)Gxy		
/(y)G1y → (Ew)G2w		(1)
1/G1y → (Ew)G2w		(2)
2/G1y → G2w	SNO	(3)
	NOT VALID	

14*6/ → (Ex)(y)(z)((∼Gxw ⊃ Gwy) ⊃ (Gzw ⊃ Gwz))		
/ → (y)(z)((∼Gx1 ⊃ G1y) ⊃ (Gz1 ⊃ G1z))		(1)
1/ → (y)(z)((∼Gx1 ⊃ G1y) ⊃ (Gz1 ⊃ G1z))		(2)
2/ → (z)((∼Gx1 ⊃ G1k) ⊃ (Gz1 ⊃ G1z))		(3)
3/ → (∼Gx1 ⊃ G1k) ⊃ (Gm1 ⊃ G1m)		(4)
4/∼Gx1 ⊃ G1k → Gm1 ⊃ G1m		(5)
5/G1k → Gm1 ⊃ G1m		(6)
5/ → Gm1 ⊃ G1m, ∼Gx1		(8)
8/Gm1 → G1m, ∼Gx1		(9)
6/G31, G12 → G13	NO	(7)
	NOT VALID	

14*16/(x)Gxu → (Ew)(Gyw & Gzw)		
/(x)Gx1 → (Ew)(G2w & G3w)		(1)
1/Gx1 → (Ew)(G2w & G3w)		(2)
2/Gx1 → G2w & G3w		(3)
3/Gx1 → G2w	F	(4)
4/Gx1 → G3w	F	(5)
	NONE	

15*16/(x)x = x, (Ey)(Ez)y ≠ z → (Ex)(Ey)(Ez)(x ≠ y & y ≠ z)		
/(x)x = x, 1 ≠ 2 → (Ew)(Ev)(Eu)(w ≠ v & v ≠ u)		(1)
1/(x)x = x → (Ew)(Ev)(Eu)(w ≠ v & v ≠ u), 1 = 2		(2)
2/x = x → (Ew)(Ev)(Eu)(w ≠ v & v ≠ u), 1 = 2		(3)
3/x = x → w ≠ v & v ≠ u, 1 = 2		(4)
4/x = x → w ≠ v, 1 = 2		(5)
4/x = x → v ≠ u, 1 = 2		(7)
7/2 = 1, x = x → 1 = 2	SVA	(8)
5/1 = 2, x = x → 1 = 2	SVA	(6)
	QED	

19*10/ → ((Ey)Guy & ∼Guu) & (Guv ⊃ (Gvw ⊃ Guw))		
/ → ((Ey)G1y & ∼G11) & (G12 ⊃ (G23 ⊃ G13))		(1)
1/ → (Ey)G1y & ∼G11		(2)
2/ → (Ey)G1y		(3)
2/ → ∼G11		(4)
4/G11 →		(5)
1/ → G12 ⊃ (G23 ⊃ G13)		(6)
6/G12 → G23 ⊃ G13		(7)
7/G23, G12 → G13		(8)
8/G23, G12 → G13	NO	(1)
	NOT VALID	

19*13/ → (Ex)(y)(Ez)((∼GxyvGxx)v(Gzx & ∼Gzy))		
/ → (Ex)(y)(Ez)((∼GxyvGxx)v(Gzx & ∼Gzy))		(1)
1/ → (y)(Ez)((∼GxyvGxx)v(Gzx & ∼Gzy))		(2)
2/ → (Ez)((∼GxkvGxx)v(Gzx & ∼Gzk))		(3)
3/ → (∼GxkvGxx)v(Gzx & ∼Gzk)		(4)
4/ → ∼GxkvGxx, Gzx & ∼Gzk		(5)
5/ → ∼Gxk, Gxx, Gzx & ∼Gzk		(6)
6/Gxk → Gxx, Gzx & ∼Gzk		(7)
7/Gxk → Gxx, ∼Gzk		(9)
7/Gxk → Gxx, Gzx	FNO	(8)
9/Gzk, Gxk → Gxx	FNO	(10)
	k by x	
	z by k	

we have sufficient data to conclude that the subproblem is not a theorem. This is the case with 14*15. Otherwise, we go to S14. This is the case with 14*16.

S11. If the result contains no more functors after elimination, we know that the subproblem can be treated by the full AE method. In general, it is, however, necessary to first transform the matrix by a procedure similar to the reduction to a miniscope form (see [20]). Such a procedure is not included in the program P. Instead, we use the original matrix with functors replaced by numbers and go directly to S14 for a negative test only. 14*6 is an example.

S12. If the result is not in AE form and no eliminations have been made, go to S14 directly. This is the case with 19*13.

S13. If the result is not in the AE form but some eliminations have been done, repeat S8 and S9 except that upon failure the program goes to S14.

S14. Make a negative test. If some atomic formula with no functors cannot be made valid even by NTEST, append NO to it, and refute the whole problem. This is the case with 14*6.

S15. Otherwise, the question is undecided. In this case, restore functors if eliminations have been made. And then, put F after each atomic sequent. If an atomic sequent could not be made valid by NTEST but contains functors, add also NO after F. The two cases are seen in the last lines of 14*16 and 19*13.

S16. Finally, print out the best possible governing relations among the functors and variables. In the case of 14*16, there is none. This means the problem can be settled by the full AE method. In the case of 19*13, we have an irreducible string (Ex)(y)(Ez) of quantifiers.

This completes the summary of the program P. It is clear that at the end we are ready to add more powerful methods which are usually classified according to the string of quantifiers. The governing relations we list in general give quantifier trees ("trees" in an intuitive sense). Thus, if x governs k, y governs m, m governs z, and we treat k, m as variables for the moment, then we are free to use several different strings as long as the governing relations are preserved, e.g., (Ex)(k)(Ey)(m)(z), (Ex)(Ey)(k)(m)(z), (Ey)(Ex)(k)(m)(z), etc. That is why quantifier trees give us in general a better reduction of a given problem.

It should be emphasized that although the methods of the program P are essentially confined to a subdomain of the AE method, it can solve problems which are not solvable by the ordinary full AE method. Quite a number of problems can be decided by the auxiliary procedures introduced along the way. So far we have only been able to give a small sample of shorter problems. Now we list below a number of further examples which have been definitely proved or disproved by the program P. We shall not list problems for which the program P has given no complete solutions. (All the examples are drawn mainly from [3] and [1].)

It seems fair to say that the program P can decide some rather complex propositions. With a more advanced program, one can naturally expect mechanical proofs of more elaborate theorems of logic. Since only quite simple logical principles are employed in actual mathematics, it seems reasonable to expect that machines will often turn out proofs rather different from those obtainable by man. In fact some fairly complex but quite useful logical principles might be suggested by mechanical proofs of even familiar mathematical theorems.

We discuss now briefly the possible full AE methods. Among the examples given earlier, 14*16 is not solvable by the restricted AE method but solvable by a full AE method. One method is this. In general, whenever all functors can be eliminated, we simply take the conjunction of all the undecided atomic sequents and make all possible substitutions of numbers for variables, and test the disjunction of all the instances. In the case of 14*16, we have:

x = 1, w = 1 : G11 → G21 ; G11 → G 31

and

(x, w) = (1, 2), (2, 1), (2, 2), (1, 3), (3, 1), (2, 3), (3, 2), (3, 3)

This is like Qr [20, p. 12], except that the elimination of functors extends the range beyond Qr.

It is, however, clear that this is not efficient and we can improve the method by using pattern recognition. In cases, however, like 14*16, which can be seen to be of the AE form at the beginning, we may proceed simply as follows. From the original sequent, we see that there are 3 initial positive quantifiers and 2 negative quantifiers. Hence, we need to consider just:

G11, G21, G31 → G21 & G31, G22 & G32, G23 & G33

which is easily seen to be a tautology. (Compare Qq, [20, p. 12].)

In the original version of the program P, step S11 contained also an erroneous method of proving a subproblem directly after the elimination of all functors. As a

LIST I. *Theorems Proved by* P

14*7/(Ez)Hxz ⊃ (z)Gxz, (z)(Gzz ⊃ Hzy) → Hxy ≡ (z)Gxz
15*3/ → (Ex)(Ey)[(x = u & y = v) ⊃ (Gu ⊃ Gv)]
15*4/ → (Ew)(Ex)(y)(z){(Gvy & Gwz) ⊃ [(Gwy & Gxy)v(Gvz & Gxz)]}
15*6/ → (Eu)(v)(Ew)(x){[((Gux ≡ Gxw) ≡ Gwx) ≡ Gxu] & [((Gvx ≡ Gxw) ≡ Gwx) ≡ Gxv]}
15*7/ → (Eu)(v)(Ew)(x){(Gux ≡ Gvx) ⊃ [((Gux ≡ Gxw) ≡ Gwx) ≡ Gxv]}
15*9/ → (Eu)(Ev)(z){[(Gxu ⊃ Gzx) ⊃ Gxx] ⊃ (Gxx & Guv)}
15*18/ → (Ex)(y)(Ez){Gyy ⊃ [Gxxv((GxzvGyz) & GzxvGzy))]}
19*2/ → ~(Ex)(y)(Gyx ≡ ~Gyy)

LIST II. *Nontheorems Disproved by the Program* P

14*12/(y){[Jx ≡ (Jy ⊃ Gy)] & [Gx ≡ (Jy ⊃ Hy)] & [Hx ≡ ((Jy ⊃ Gy) ⊃ Hy)]} → Hz & Gz & Jz
15*5/ → (Ex)(y)(Ez)[(GxyvGxz) & (~Gxyv ~ Gxz)]
15*10/ → (Ey)(z)(x ≠ zvy ≠ z)
15*13/ → (Ex)(Ey)(Gxy ⊃ p) ≡ (x)(y)(Gxy ⊃ p)
19*5/ → (Ey)(x)[Gxy ≡ (z) ~ (Gxz & Gzx)]
19*12/ → (Ey)(z)[(Gxy & ~Gxx) & (Gzx ⊃ Gzy)]
19*14/ → (Ey)Gxy & [Gxy ⊃ (z)(Gxz ⊃ y = z)] & [Gyx ⊃ (Gzx ⊃ y = z)] & (Ex)(y) ~ Gyx
19*17/ → (Ey)Gxy & (Ex)(y) ~ Gyx & [Gyx ⊃ (Gzx ⊃ y = z)]

result, some nontheorems were asserted to be theorems in the print-out. This fact was noticed by John McCarthy, and has led to the revised S11.

3. The E_1A Case Solved With Sequential Tables

The E_1A case is simple because, as can be seen from example (1) in the introduction, we need only worry about those elementary parts each of which contains only occurrences of a same variable, and then only the relations between the U-variables and the single E-variable require considerations. A simple subcase is explained quite thoroughly in [3, p. 259].

In general, let us consider:

(3.1) $(Ex)(y_1) \cdots (y_n)Mx \cdots y_n$, M *containing* N *predicates* $G_1, \cdots, G_N$.

To form S_1, S_2, etc., we need only replace $(x, y_1, \cdots, y_n)$ by $(1, 2, \cdots, n + 1)$, $(2, n + 2, \cdots, 2n + 1)$, etc. The number of possible elementary parts in M depends of course on the number of places of the predicates $G_1, \cdots, G_N$. If, e.g., they are all dyadic, then there are 3^2N possible elementary parts G_ixx, G_ixy, G_iyx, G_iyy, G_ixz, G_izx, G_iyz, G_izy, G_izz $(i = 1, \cdots, N)$. In the present case, the number of places of each predicate is immaterial since we have to consider only $G_ix \cdots x$, $G_iy \cdots y$, $G_iz \cdots z$, and each predicate behaves like a monadic one.

Thus, for example (1) in the introduction, we need consider only the following table T of all possible assignments of t and f to Gxx, Gyy, Gzz which would make the matrix of (1) false:

Gxx	Gyy	Gzz
t	t	t
t	t	f

Since S_1, S_2, S_3, etc., are obtained by substituting (1, 2, 3), (2, 4, 5), (3, 6, 7), etc., for (x, y, z), we get a tree structure:

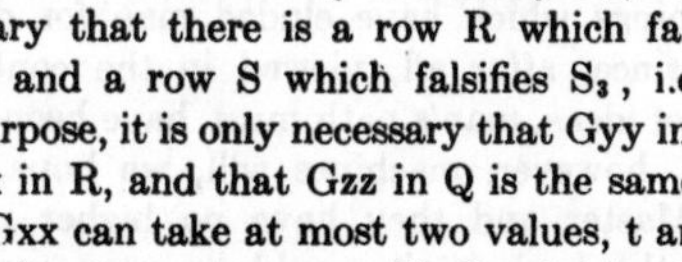

In order that a row Q can falsify S_1, i.e., M123, it is necessary that there is a row R which falsifies S_2, i.e., M245, and a row S which falsifies S_3, i.e., M367. For this purpose, it is only necessary that Gyy in Q is the same as Gxx in R, and that Gzz in Q is the same as Gxx in S. Since Gxx can take at most two values, t and f, if a table T contains two rows, one with Gxx taking t as value, one with Gxx taking f as value, we can always find a countermodel because, for each row falsifying S_1, we can always find two more rows which together with it falsify simultaneously S_1, S_2, S_3; and the same is true for any S_i and, therefore, for any D_i. If in every row of the table T, Gxx always takes one value, say t (or f), then it is necessary and sufficient to have one row in which both Gyy and Gzz take the same value, viz., t (or f). Hence, it is very easy to decide whether a statement (Ex)(y)(z)M, with a single predicate is a theorem, since, by the fundamental theorem of logic, it is a theorem if and only if there is some k such that D_k is a tautology.

In general, if a statement $(Ex)(y_1) \cdots (y_n)M$ contains a single predicate, the criterion is the same, viz.,

(3.2) *It is not a theorem if and only if either (i) its table* T *contains two rows in which* $Gx \cdots x$ *get different values, or (ii) it contains one row in which* $Gx \cdots x$, $Gy_1 \cdots y_1, \cdots, Gy_n \cdots y_n$ *all get the same value.*

If now there are N predicates $G_1, \cdots, G_N$ in the matrix of $(Ex)(y_1) \cdots (y_n)Mx \cdots y_n$, then the matter is a little more complex, because we have to consider a table T with $(n + 1)N$ columns:

$$G_1x \cdots x \cdots G_Nx \cdots xG_1y_1 \cdots y_1 \cdots G_Ny_1 \cdots y_1 \cdots G_1y_n \cdots y_n \cdots G_Ny_n \cdots y_n$$

In this case, $G_1x \cdots x, \cdots, G_Nx \cdots x$ together have 2^N possible sets of values, if every set occurs in some row of T, then the original formula is of course not a theorem. In general, use the following "sequential method."

Examine each row R of T and determine for each i $(i = 1, \cdots, n)$, whether there is a row R_i such that the values which $G_1x \cdots x, \cdots, G_Nx \cdots x$ take in R_i are respectively the same as the values which $G_1y_i \cdots y_i, \cdots, G_Ny_i \cdots y_i$ take in R. If there is one such R_i for each i, retain R, otherwise, cross out R. Each time a row is crossed out, the same process is repeated with the reduced table until either the table is empty or the table is not empty but no further reduction is possible. Using this procedure, it is easy to prove the following theorem:

(3.3) *The formula* $(Ex)(y_1) \cdots (y_n)M$ *is a theorem if and only if its reduced truth table is empty.*

This method seems considerably more efficient than existing alternatives in the literature which are usually based on a determination of some constant K such that the given formula is a theorem if and only if D_K is tautologous.

When all the predicates are dyadic, Ackermann gives the bound J in terms of validity in a domain of J members as a sufficient condition of general validity. If we recall that $nK + 1$ numbers occur in D_K, we can calculate that his bound for K is no better than:

$$n > 1, \quad K = \frac{n^{3k} - 1}{(n - 1)}, \quad \text{where} \quad k = 2^{Nn^2},$$

when

$$n = 1, \quad K = 3(2^N) - 1.$$

Church does not give the general bound, but calculates that [2, p. 213]; [3, pp. 260, 261]:

$$n = 1, \quad K = 2^N,$$

when

$$n = 2, \quad K = 2^k - 1, \quad \text{where } k = 2^N.$$

It appears that if one extends Church's argument, by using a tree with n branches at each node and of height $2^N - 1$, the general bound would be (compare Herbrand [10, p. 46]):

$$n > 1, \quad K = \frac{n^k - 1}{n - 1}, \quad k = 2^N$$

In particular, when $N = 1$, $K = n + 1$.

It seems quite clear that the sequential method is faster than testing the Herbrand disjunctions D_1, D_2, etc. Take a simple example with $n = 2$, $N = 4$ and a table T with 16 rows such that in R_1, Gx, Hx, Px, Qx gets tttt or 0000 = 0, but Gy, Hy, Py, Qy, as well as Gz, Hz, Pz, Qz all get tttf or 0001 = 1, and similarly in every R_i, Gx, Hx, Px, Qx, get the truth values corresponding to the binary notation $i - 1$, where the y, z parts both get the values corresponding to i. In such a case, it is easily seen by the sequential method that the formula is not a theorem since the reduced table is the same as the original table, yet by the alternative methods, we have to test D_K with $K = 8^{65536} - 1$ by one method and $K = 65535$ by the other. This example incidentally illustrates that just speeding up the method of testing each D_j is not sufficient to handle many interesting formulae.

Incidentally, there is a striking similarity between the type of argument involved in these decision procedures and the method of "sequential tables" developed in [19]. The similarity suggests the question of a more abstract mathematical treatment of more basic underlying principles which govern such sequential methods. The writer, however, has no inkling as to whether results will be obtained on this question and, if so, how interesting they will be.

4. General Remarks

One is naturally curious to know how far we are from machine proofs of truly significant mathematical theorems in different domains. The writer cannot see sufficiently far and clearly into the future to make any responsible predictions, except that the simplicity of all theorems of *Principia* in the predicate calculus came as a great surprise, suggesting the opinion that one could be too conservative in estimating the potentialities of machines in theorem proving. It seems that several types of objectives are likely to be achieved with just a few more months of programming efforts along the present approach which precludes ad hoc measures designed specially for a few immediate specific problems. Among these are proving a large portion of theorems in Landau's booklet [14] on the number systems, proving a fair number of theorems in high school algebra and geometry, formalizing fairly interesting theorems in set theory. In the last category, it seems likely that the machine will soon be able, e.g., to do the tedious but less inspired part of the work needed to establish the main conclusion of [11], viz., to derive the contradiction from the few axioms chosen in advance by man. Those who have worked on this type of problem would appreciate that such assistance is not to be despised. A considerably more difficult and remote task would be to formalize Specker's derivation of a contradiction [17] in Quine's "New Foundations" plus the axiom of choice, which is a more complex system. However, in view of the apparent artificiality of the formal system concerned, the advantage of man over the machine is greatly reduced when results on such systems are to be established.

On the whole, it seems reasonable to think that machines will more quickly excel in areas where man's intuition is not so strong. Hence, the author is now inclined to feel that difficult theorems in analysis and set theory will more easily be proved by machines than those in number theory. For example, the writer feels that among possible targets for the next year or two the irrationality of $\sqrt{2}$ and the unique factorization theorem may tax the ingenuities of machines and their programs equally heavily as theorems in set theory and analysis, such as the Heine-Borel theorem and the Bernstein theorem, which man finds considerably harder to understand. This may appear rather nonsensical in view of the great conceptual difficulties we have with the continuum and higher infinities. In the writer's opinion, however, the decisive factor is rather the fact that we are capable of making much more varied and extended moves when general theorems about natural numbers are being considered; this is likely to make it harder for machines to catch up with us. On the other hand, it is well known that machines are good at dealing with essentially combinatory problems, which are, however, not the chief concern of number theory.

When imagination is given free rein, Fermat's and Goldbach's conjectures, the Riemann hypothesis, the four-color problem, the consistency of impredicative analysis, the continuum hypothesis, and other famous overwhelmers all come to mind. Of these celebrated problems, it seems fair to concede that at present we have no idea how machines might assist in arriving at a settlement of any. It is of course possible that machines may get hold of things which have eluded man for decades or centuries, since, after all, viewed in the context of Plato's realm of ideas, man's path must have been pretty narrow. Since, however, machines will, we hope, never become the Master and they have no higher masters than man in the horizon, it would be very surprising indeed if they should quickly surpass man in areas which demand the highest human creative genius and prove a whole lot of theorems which the best mathematicians have strived and failed to establish for very long. At any rate, it is a happier thought that machines will increase the power of mathematicians rather than eliminate them, and there is no evidence at present that the latter alternative will ever materialize.

The writer, as an amateur programmer and as one who has given little thought to designing monitoring systems for symbol manipulation, has little to say on the theory of programming. One obvious suggestion is that the writer's program P as it exists now should be used as a guinea pig for testing symbol manipulation systems: see how much easier it is to rewrite the program in each system and determine how much slower the new program runs. A careful study of the program P may also suggest to the experts to add or modify certain devices in their systems in order to meet the natural demands by neutral programs which involve a good deal of symbol manipulation.

(After the above paragraph had been written, the writer saw John McCarthy's "The Wang algorithm for the propositional calculus programmed in LISP," Artificial Intelligence Project, MIT, Symbol Manipulating Language, Memo 14. This deals roughly with program I of [20]. "It took about two hours to write the program and it ran on the fourth try." Even making allowance for the fact that this part of the program P is relatively simple and specially suited to LISP, the coding time required is still amazingly short. The running time is also much better than the writer had expected. Apparently, for the same problems, the LISP program takes no more than 10 times longer than the original SAP program; and presumably this can be further improved. Moreover, it is stated in the memo that the algorithm has also been useful in suggesting a general concept of ambiguous functions to be used in computations.)

Like many people embarked on ambitious computing projects, the writer used to think how nice it would be to have at disposal a STRETCH computer or something beyond. Recently, however, the writer has come to feel a bit differently. Of course a larger machine can handle more difficult problems and one can afford to use more crude methods which are hopeless on some slower machine. But so often the difference in speed between different methods is so great that even an increase by 100 is quite inadequate to compensate for the deficiencies in a more crude method. How far can 100 go when people indulge in exponentiations? For example, if one were to use the so-called British Museum algorithm or even the less wasteful brute force search method mentioned above, a simple formula, say with a conjunction of two disjunctions of elementary parts as its quantifier-free matrix, would have for $S_1 v \cdots v S_{100}$ a formula with 2^{100} long disjunctions to test.

If, however, one is to seek more efficient methods, it often happens that we can only begin with simpler cases and proceed to more complex cases. Then it is likely that a good deal of effort is needed before one can do full justice to a large machine. For example, the writer has made no special efforts to economize storage and feels sorry to have been able to use only less than half of the 32,000 words on the particular IBM 704. Similarly, the writer is sorry to have found no natural problems for the program which actually require a long running time that is justified by the intrinsic interest of the problems. The building of more powerful machines and the designing of more efficient methods to use them both require time and human efforts. A big lagging behind in either direction causes waste. At least financially, a long lag in the using direction is the worse of two evils. When the best existing method does demand the full capacity of the largest existing machines, we have the happy situation of a harmonious coordination. This is truly nice only when there is no immediate prospect for improving the method. Under those circumstances, the problems solved must be quite interesting, because otherwise we should say that the method is not yet good enough for use on machines, or that the problems are not yet suitable for machine treatment. It seems that the machines, when cursed for being too slow or too small, may often with justification demand in turn that the user do some more thinking.

We have a feeling that there are things which machines can do and things they cannot do, things for which they are especially adapted and things for which they are not quite suitable. The familiar ambiguities of the word "can" seem to give at least two different conclusions. In one sense, it is silly to make machines do what they cannot do. In another sense, this is precisely the exciting thing. Compared with numerical calculations, proving theorems seems like forcing machines to do what they cannot do. Compared with proving theorems, mechanical translation, for example, seems even more remote from the natural aptitudes of machines.

Quite justifiably, mechanical translation has a wider appeal than mechanical mathematics. For one thing, more people use languages than mathematics. However, so far as the near future of the two projects is concerned, it would seem fair to say that mechanical mathematics has a brighter prospect. While inferential analysis can draw from a body of profound exact results in theoretical logic, a mathematical treatment of language is very much something yet to come and seems to involve fundamental intrinsic difficulties, particularly in dealing with meaning and ambiguities. Any substantial progress in mechanical translation would be impressive in so far as machines would be doing something for which they are apparently not suitable. It is, however, highly unlikely that machines can give better translations than expert human translators. On the other hand, machine proof of theorems is quite likely to yield impressive results in the more absolute sense: viz., the doing of things which lie beyond unassisted human capabilities.

It seems undeniable that computers have changed somewhat the face of applied mathematics. The writer feels that the cross-fertilization of logic and computers ought to produce in the long run some fundamental change in the nature of all mathematical activity. Such a development will not only make pure mathematicians take computers more seriously but provide proud applications for the parts of logic which are logicians' primary concern. Modern logic is intrinsically interesting, yet it is customary among mathematicians to think of it as a bit irrelevant. As logic matures, the relevance is being felt, e.g., in the general study of effective procedures and the analysis of existence proofs. Advances in mechanical mathematics will spread the influence of logic further and bring to the forefront detailed works on logic in the narrower sense of dealing with inferences in the first place.

The relevance of logic is perhaps rather gratifying to experts in advanced programming which seems to attract capable people who enjoy thinking but dislike extensive implicit presuppositions. The energy, created by their unfulfilled desire for solid theories unhampered by such presuppositions, finds an outlet in logic, and with the increasing relevance of logic they can now spend time to learn the new trade with a clear conscience. On the other hand, thinking on many logic problems can be assisted when the more exacting demands of machines and programming are kept in mind.

As far as we know, machines can only do significant things by means of algorithms. But in the human mathematical activity one also speaks of intuition, insight, and heuristic methods, which do not seem to possess easy exact definitions. When these terms are applied to machines, the unity of contraries sounds distinctly paradoxical. In particular, the term "heuristic method" has gained some currency among sophisticated users of machines, since Polya revived the term in a rather orthodox context.

In ancient times Archimedes distinguished a method of proof (the method of exhaustion in this case) from a method of discovery (the heuristic method) in finding out the area or volume of a configuration. His heuristic method includes considerations of the law of the lever and the center of gravity, and also of intuition such as that cylinders, spheres, and cones are made up of parallel circular discs. (An extensive discussion of the matter is given in T. L. Heath, *The Methods of Archimedes*, 1912.) In modern times, Euler was perhaps the best-known mathematician who told of how his theorems were often first discovered by empirical and formalistic experimentations.

In these examples, there appears to be an element of inexactness in the methods so that they cannot be rigorously formulated, cannot be taught in words alone but only by awakening certain latent understanding implicit in the pupil's mind. For the same reason heuristic methods are usually not taught explicitly, and while, e.g., Polya's attempt to teach them is very likely a good pedagogic idea, one would expect that pupils in the same class would achieve very different results which are determined largely by the ability of each pupil. There is nothing important in this trite prediction except the equally obvious conclusion that a machine in such a class can hardly be expected to do as well as the average pupil. This would seem to indicate that there are rather serious difficulties in teaching machines heuristic methods. It is very doubtful that any completely mechanized method, available so far, can properly be called a heuristic one. Much more elaborate instructions are necessary in order to produce the impression that the machine is actively participating as a creative agent in the search for a mathematical proof.

Sometimes it is suggested the heuristic methods are just the methods which a man would find natural to use. This is not very helpful. For example, it is not hard to contend that the method of doing propositional calculus described in [20] is quite natural, but by no stretch of imagination can it be called a heuristic method.

It can be a useful thing to fill an old bottle with new wine. But if this is being done when specific mechanical methods are said to be heuristic, it would perhaps be less misleading if the constituents of the new wine are explained a little more exactly. One might think it silly to make heavy weather on what is just a trivial terminological matter. The truth is, however, that such expansive obscurity can arouse useless enthusiasm in some lighthearted quarters and generate harmful suspicion among the more seriously-minded scientists.

If we leave aside at the present rather primitive stage the emotion-laden term "heuristic method," there are a few more prosaic distinctions which can be made among the algorithms for proving theorems and disproving nontheorems. A partial algorithm may be able to give yes and no answers only in some part of a domain, e.g., a partial decision procedure such as the monadic within the predicate calculus, or only yes as answer, e.g., the proof procedure for the predicate calculus. Among partial decision procedures, there are those for which we know in advance the ranges of application, i.e., there are some simpler effective methods by which we can test whether any given problem is decidable by the method. There are also those with ranges of application which are undecidable in advance or simply undecidable. All the procedures which are commonly studied have, however, one thing in common, viz., when the method does give a decisive answer, the answer should be correct. In actual research, however, we also use alternative methods which tend to give us an answer more quickly, and often although not always correctly. Then we can use more elaborate methods to verify whether the tentative answers are correct. Such methods, when mechanizable, would resemble the methods of approximation in numerical analysis and have, indeed, some flavor of the heuristic methods.

On the whole, we need interlocked hierarchies of methods, and the mysterious elements in the creative activity seem likely to be replaced by a complex web of clearly understood, definite and deterministic algorithms, rather than random elements or obscure machine programs. If man gets results by intuitive methods we cannot easily formalize, it does not mean that by introducing uncontrollable elements, a machine will more likely behave like man. If we do not understand how certain turns are made at crucial junctures, casting a die each time to guide the machine will scarcely ever produce the desired final effect. With machines, large masses of well-organized minute details seem to be the only sure way to make the correct surprises emerge.

REFERENCES

1. W. Ackermann. *Solvable Cases of the Decision Problem.* 114 pp., 1954, Amsterdam.
2. Alonzo Church. Special cases of the decision problem, *Revue philosophique de Louvain 49* (1951), 203–221; a correction, ibid., *50* (1952), 270–272.
3. Alonzo Church. *Introduction to Mathematical Logic, I.* 376 pp., 1956, Princeton.
4. M. Davis and H. Putnam. A computational proof procedure. AFOSR, 1959 (submitted to *Journal of Association for Computing Machinery*).
5. B. Dreben. On the completeness of quantification theory. *Proc. Nat. Acad. Sci, USA, 38* (1952), 1047–1052.
6. B. Dreben. Systematic treatment of the decision problem. Summaries of talks at the Summer Institute of Symbolic Logic, p. 363, 1957, Cornell.
7. B. Dunham, R. Fridshal, and G. L. Sward. A nonheuristic program for proving elementary logical theorems (abstract). *Comm. ACM, 2* (1959), 19–20.
8. P. C. Gilmore. A proof method for quantification theory: its justification and realization. *IBM J. Res. Develop. 4* (1960), 28–35.
9. J. Herbrand. *Recherches sur la Théorie de la Démonstration.* 128 pp., 1930, Warsaw.
10. J. Herbrand. Sur le problème fondamental de la logique mathématique, *Compt. rend. Soc. Sci. Lettres Varsovie,* Classe III, *24* (1931), 12–56.
11. K. J. J. Hintikka. Vicious circle principle and the paradoxes. *J. Symbol. Logic, 22* (1957), 245–249.
12. L. Kalmar. Über die Erfüllbarkeit derjenigen Zählausdrücke, welche in der Normalform zwei benachbarte Allzeichen enthalten. *Math. Ann. 108* (1933), 466–484.
13. Dieter Klaua. Systematische Behandlung der lösbaren Fälle des Entscheidungsproblems für den Prädikatenkalkül der ersten Stufe. *Zeit. math. Logik Grundl. Math. 1* (1955), 264–270.
14. E. Landau. *Grundlagen der Analysis,* 1930, Leipzig.
15. W. V. Quine. *Methods of Logic.* 1950 and 1958, New York.
16. T. Skolem. Über die mathematische Logik. *Norsk Mat. Tidsskrift, 10* (1928), 125–142.
17. E. Specker. The axiom of choice in Quine's new foundations for mathematical logic. *Proc. Nat. Acad. Sci. USA, 39* (1953), 972–975.
18. Hao Wang. A theory of constructive types. *Methodos 1* (1949), 374–384.
19. Hao Wang. Circuit synthesis by solving sequential Boolean equations. *Zeit. math. Logik Grundl. Mathe. 5* (1959), 291–322.
20. Hao Wang. Toward mechanical mathematics. *IBM J. Res. Develop. 4* (1960), 2–22.

Toward Mechanical Màthematics

H. Wang

Abstract: Results are reported here of a rather successful attempt at proving all theorems, totalling near 400, of *Principia Mathematica* which are strictly in the realm of logic, viz., the restricted predicate calculus with equality. A number of other problems of the same type are discussed. It is suggested that the time is ripe for a new branch of applied logic which may be called "inferential" analysis, which treats proofs as numerical analysis does calculations. This discipline seems capable, in the not too remote future, of leading to machine proofs of difficult new theorems. An easier preparatory task is to use machines to formalize proofs of known theorems. This line of work may also lead to mechanical checks of new mathematical results comparable to the debugging of a program.

Introduction

If we compare calculating with proving, four differences strike the eye: (1) Calculations deal with numbers; proofs, with propositions. (2) Rules of calculation are generally more exact than rules of proof. (3) Procedures of calculation are usually terminating (decidable, recursive) or can be made so by fairly well-developed methods of approximation. Procedures of proof, however, are often nonterminating (undecidable or nonrecursive, though recursively enumerable), indeed incomplete in the case of number theory or set theory, and we do not have a clear conception of approximate methods in theorem-proving. (4) We possess efficient calculating procedures, while with proofs it frequently happens that even in a decidable theory, the decision method is not practically feasible. Although short-cuts are the exception in calculations, they seem to be the rule with proofs in so far as intuition, insight, experience, and other vague and not easily imitable principles are applied. Since the proof procedures are so complex or lengthy, we simply cannot manage unless we somehow discover peculiar connections in each particular case.

Undoubtedly it is such differences that have discouraged responsible scientists from embarking on the enterprise of mechanizing significant portions of the activity of mathematical research. The writer, however, feels that the nature and the dimension of the difficulties have been misrepresented through uncontrolled speculation and exaggerated because of a lack of appreciation of the combined capabilities of mathematical logic and calculating machines.

Of the four differences, the first is taken care of either by quoting Gödel representations of expressions or by recalling the familiar fact that alphabetic information can be handled on numerical (digital) machines. The second difference has largely been removed by the achievements of mathematical logic in formalization during the past eighty years or so. Item (3) is not a difference that is essential to the task of proving theorems by machine. The immediate concern is not so much theoretical possibility as practical feasibility. Quite often a particular question in an undecidable domain is settled more easily than one in a decidable region, even mechanically. We do not and cannot set out to settle all questions of a given domain, decidable or not, when, as is usually the case, the domain includes infinitely many particular questions. In addition, it is not widely realized how large the decidable subdomains of an undecidable domain (e.g., the predicate calculus) are. Moreover, even in an undecidable area, the question of finding a proof for a proposition known to be a theorem, or formalizing a sketch into a detailed proof, is decidable theoretically. The state of affairs arising from the Gödel incompleteness is even less relevant to the sort of work envisaged here. The purpose here is at most to prove mathematical theorems of the usual kind, e.g., as exemplified by treatises on number theory, yet not a single "garden-variety" theorem of number theory has been found unprovable in the current axiom system of number theory. The concept of approximate proofs, though undeniably of a kind other than approximations in numerical calculations, is not incapable of more exact formulation in terms of, say, sketches of and gradual improvements toward a correct proof.

The last difference is perhaps the most fundamental. It is, however, easy to exaggerate the degree of complexity which is necessary, partly because abstract estimates are hardly realistic, partly because so far little attention has been paid to the question of choosing more efficient alternative procedures. There will soon be occasion to give illustrations to these two causes of exaggeration. The problem of introducing intuition and experience into machines is a bit slippery. Suffice it to say for the moment, however, that we have not realized that much of our basic strategies in searching for proofs is mechanizable, because we had little reason to be articulate on such matters until large, fast machines became available. We are in fact faced with a challenge to devise methods of buying originality with plodding, now that we are in possession of slaves which are such persistent plodders. In the more advanced areas of mathematics, we are not likely to succeed in making the machine imitate the man entirely. Instead of being discouraged by this, however, one should view it as a forceful reason for experimenting with mechanical mathematics. The human inability to command precisely any great mass of details sets an intrinsic limitation on the kind of thing that is done in mathematics and the manner in which it is done. The superiority of machines in this respect indicates that machines, while following the broad outline of paths drawn up by man, might yield surprising new results by making many new turns which man is not accustomed to taking.

It seems, therefore, that the general domain of algorithmic analysis can now begin to be enriched by the inclusion of inferential analysis as a younger companion to the fairly well established but still rapidly developing leg of numerical analysis.

The writer began to speculate on such possibilities in 1953, when he first came into contact with calculating machines. These vague thoughts were afterwards appended to a paper on Turing machines.[1] Undoubtedly many people have given thought to such questions. As far as the writer is aware, works more or less in this area include Burks-Warren-Wright,[2] Collins,[3] Davis,[4] Newell-Shaw-Simon, Gelernter.[5] Of these, the most extensively explained and most widely known is perhaps that of Newell-Shaw-Simon, a series of reports and articles[6] written since 1956. Their work is also most immediately relevant to the results to be reported in this paper. It will, therefore, not be out of place if we indicate the basic differences in the respective approaches and the specific advances beyond their work.

They report[7] that their program LT on JOHNNIAC was given the task of proving the first 52 theorems of *Principia Mathematica* of Whitehead and Russell: "Of the 52 theorems, proofs were found for a total 38. . . . In 14 cases LT failed to find a proof. Most of these unsuccessful attempts were terminated by time or space limitations. One of these 14 theorems we know LT cannot prove, and one other we believe it cannot prove." They also give as examples that a proof for *2.45 was found in 12 minutes and a report of failure to prove *2.31 was given after 23 minutes.

The writer wrote three programs last summer on an IBM 704. The first program provides a proof-decision procedure for the propositional calculus which prints out a proof or a disproof according as the given proposition is a theorem or not. It was found that the whole list of over 200 theorems of the first five chapters of *Principia Mathematica* were proved within about 37 minutes, and 12/13 of the time is used for read-in and print-out, so that the actual proving time for over 200 theorems was less than 3 minutes. The 52 theorems chosen by Newell-Shaw-Simon are among the easier ones and were proved in less than 5 minutes (or less than ½ minute if not counting input-output time). In particular, *2.45 was proved in about 3 seconds and *2.31 in about 6 seconds. The proofs for these two theorems and some more complex proofs are reproduced in Appendix I as they were printed out on the machine.

The other two programs deal with problems not considered by Newell-Shaw-Simon in their published works. The second program instructs the machine to form propositions of the propositional calculus from basic symbols and select nontrivial theorems. The speed was such that about 14,000 propositions were formed and tested in one hour, storing on tape about 1000 theorems. The result was disappointing in so far as too few theorems were excluded as being trivial, because the principles of triviality actually included in the program were too crude.

The third program was meant as part of a larger program for the whole predicate calculus with equality which the writer did not have time to complete during 1958. The predicate calculus with equality takes up the next five chapters of *Principia Mathematica* with a total of over 150 theorems. The third program as it stands can find and print out proofs for about 85% of these theorems in about an hour. The writer believes that slight modifications in the program will enable the machine to prove all these theorems within 80 minutes or so. The full program, as envisaged, will be needed only when we come to propositions of the predicate calculus which are much harder to prove or disprove than those in this part of *Principia Mathematica*.

It will naturally be objected that the comparison with the program of Newell-Shaw-Simon is unfair, since the approaches are basically different. The writer realizes this but cannot help feeling, all the same, that the comparison reveals a fundamental inadequacy in their approach. There is no need to kill a chicken with a butcher's knife. Yet the net impression is that Newell-Shaw-Simon failed even to kill the chicken with their butcher's knife. They do not wish to use standard algorithms such as the method of truth tables,[8] because "these procedures do not produce a proof in the meaning of Whitehead and Russell. One can invent 'automatic' procedures for producing proofs, and we will look at one briefly later, but these turn out to require computing times of the orders of thousands of years for the proof of *2.45." It is, however, hard to see why the proof of *2.45 produced by the algorithms to be described in this paper is less acceptable as a proof, yet the computing time for proving *2.45 is less than ¼ second by this algorithm. To argue the superiority of "heuristic" over algorithmic methods by choosing a particularly inefficient algorithm seems hardly just.

The word "heuristic" is said to be synonymous with "the art of discovery," yet often seems to mean nothing else than a partial method which offers no guarantees of solving a given problem. This ambiguity endows the word with some emotive meaning that could be misleading in further scientific endeavors. The familiar and less inspiring word "strategy" might fare better.

While the discussions by Newell-Shaw-Simon are highly suggestive, the writer prefers to avoid hypothetical considerations when possible. Even though one could illustrate how much more effective partial strategies can be if we had only a very dreadful general algorithm, it would appear desirable to postpone such considerations till we encounter a more realistic case where there is no general algorithm or no efficient general algorithm, e.g., in the whole predicate calculus or in number theory. As the interest is presumably in seeing how well a particular procedure can enable us to prove theorems on a machine, it would seem preferable to spend more effort on choosing the more efficient methods rather than on enunciating more or less familiar generalities. And it is felt that an emphasis on mathematical logic is unavoidable, because it is just as essential in this area as numerical analysis is for solving large sets of simultaneous numerical equations.

The logical methods used in this paper are along the general line of cut-free formalisms of the predicate calculus initiated by Herbrand[9] and Gentzen.[10] Ideas of Hilbert-Bernays,[11] Dreben,[12] Beth,[13] Hintikka,[14] Schütte[15] and many others on these formulations, as well as some from standard decision methods for subdomains of the predicate calculus presented by Church[16] and Quine,[17] are borrowed. The special formulations actually used seem to contain a few minor new features which facilitate the use on machines. Roughly speaking, a complete proof procedure for the predicate calculus with equality is given which becomes a proof-decision procedure when the proposition to be proved or disproved falls within the domain of the propositional calculus or that of the "*AE* predicate calculus"* which includes the monadic predicate calculus as a subdomain.

The treatment of the predicate calculus by Herbrand and Gentzen enables us to get rid of every "Umweg" (cut or *modus ponens*) so that we obtain a cut-free calculus in which, roughly speaking, for every proof each of the steps is no more complex than the conclusion. This naturally suggests that, given any formula in the predicate calculus, we can examine all the less complex formulae and decide whether it is provable. The reason that this does not yield a decision procedure for the whole predicate calculus is a rule of contraction which enables us to get rid of a repetition of the same formula. As a result, in searching for a proof or a disproof, we may fail in some case because we can get no proof, no matter how many repetitions we introduce. In such a case, the procedure can never come to an end, although we do not know this at any finite stage. While this situation does not preclude completeness, it does exclude a decision procedure.

Now if we are interested in decidable subdomains of the predicate calculus, we can usually give suitable reformulations in which the rule of contraction no longer occurs. A particularly simple case is the propositional calculus. Here we can get a simple system which is both a complete proof procedure and a complete decision procedure. The completeness receives a very direct proof, and as a decision procedure it has an advantage over usual procedures in that if the proposition tested is provable, we obtain a proof of it directly from the test. This procedure is coded in Program I. Moreover, it is possible to extend the system for the propositional calculus to get a proof-decision procedure for the *AE* predicate calculus, which has the remarkable feature that in searching for a proof for a given proposition in the "miniscope" form, we almost never need to introduce any premise which is longer than its conclusion. This procedure is coded in Program III.

A rather surprising discovery, which tends to indicate our general ignorance of the extensive range of decidable subdomains, is the absence of any theorem of the predicate calculus in *Principia* which does not fall within the simple decidable subdomain of the *AE* predicate calculus. More exactly, there is a systematic procedure of separating variables to bring a proposition into the "miniscope" form, a term to be explained below. Since this procedure can be easily carried out by hand or by machine for these particular theorems, every theorem in the predicate calculus part of *Principia* can then be proved by the fairly simple Program III.

*Those propositions which can be transformed into a form in which no existential quantifier governs any universal quantifier.

Originally the writer's interest was in formalizing proofs in more advanced domains, such as number theory and differential calculus. It soon became clear that for this purpose a pretty thorough mechanization of the underlying logic is a necessary preliminary step. Now that this part is near completion, the writer will discuss in the concluding part of this paper some further possibilities that he considers to be not too remote.

Work for this paper was done at the Poughkeepsie IBM Research Laboratory in the summer of 1958. The writer much appreciates the satisfactory working conditions, especially the stimulating suggestions of friends in the Laboratory and the easy access to a good calculating machine.

The propositional calculus (System *P*)

Since we are concerned with practical feasibility, it is preferable to use more logical connectives to begin with when we wish actually to apply the procedure to concrete cases. For this purpose we use the five usual logical constants $\sim$ (not), & (conjunction), $\vee$ (disjunction), $\supset$ (implication), $\equiv$ (biconditional), with their usual interpretations.

A propositional letter *P, Q, R, M* or *N*, et cetera, is a formula (and an "atomic formula"). If ϕ, ψ are formulae, then $\sim\phi$, $\phi \& \psi$, $\phi \vee \psi$, $\phi \supset \psi$, $\phi \equiv \psi$ are formulae. If π, ρ are strings of formulae (each, in particular, might be an empty string or a single formula) and ϕ is a formula, then π, ϕ, ρ is a string and $\pi \rightarrow \rho$ is a sequent which, intuitively speaking, is true if and only if either some formula in the string π (the "antecedent") is false or some formula in the string ρ (the "consequent") is true, i.e., the conjunction of all formulae in the antecedent implies the disjunction of all formulae in the consequent.

There are eleven rules of derivation. An initial rule states that a sequent with only atomic formulae (proposition letters) is a theorem if and only if a same formula occurs on both sides of the arrow. There are two rules for each of the five truth functions—one introducing it into the antecedent, one introducing it into the consequent. One need only reflect on the intuitive meaning of the truth functions and the arrow sign to be convinced that these rules are indeed correct. Later on, a proof will be given of their completeness, i.e., all intuitively valid sequents are provable, and of their consistency, i.e., all provable sequents are intuitively valid.

P1. Initial rule: if λ, ζ are strings of atomic formulae, then $\lambda \rightarrow \zeta$ is a theorem if some atomic formula occurs on both sides of the arrow.

In the ten rules listed below, λ and ζ are always strings (possibly empty) of atomic formulae. As a proof procedure in the usual sense, each proof begins with a finite set of cases of *P1* and continues with successive consequences obtained by the other rules. As will be explained below, a proof looks like a tree structure growing in the wrong direction. We shall, however, be chiefly interested in doing the steps backwards, thereby incorporating the process of searching for a proof.

The rules are so designed that given any sequent, we can find the first logical connective, i.e., the leftmost symbol in the whole sequent that is a connective, and apply the appropriate rule to eliminate it, thereby resulting in one or two premises which, taken together, are equivalent to the conclusion. This process can be repeated until we reach a finite set of sequents with atomic formulae only. Each connective-free sequent can then be tested for being a theorem or not, by the initial rule. If all of them are theorems, then the original sequent is a theorem and we obtain a proof; otherwise we get a counterexample and a disproof. Some simple samples will make this clear.

P2a. Rule $\rightarrow\sim$: If $\phi, \zeta \rightarrow \lambda, \rho$, then $\zeta \rightarrow \lambda, \sim\phi, \rho$.

P2b. Rule $\sim\rightarrow$: If $\lambda, \rho \rightarrow \pi, \phi$, then $\lambda, \sim\phi, \rho \rightarrow \pi$.

P3a. Rule $\rightarrow\&$: If $\zeta \rightarrow \lambda, \phi, \rho$ and $\zeta \rightarrow \lambda, \psi, \rho$, then $\zeta \rightarrow \lambda, \phi \& \psi, \rho$.

P3b. Rule $\&\rightarrow$: If $\lambda, \phi, \psi, \rho \rightarrow \pi$, then $\lambda, \phi \& \psi, \rho \rightarrow \pi$.

P4a. Rule $\rightarrow\vee$: If $\zeta \rightarrow \lambda, \phi, \psi, \rho$, then $\zeta \rightarrow \lambda, \phi \vee \psi, \rho$.

P4b. Rule $\vee\rightarrow$: If $\lambda, \phi, \rho \rightarrow \pi$ and $\lambda, \psi, \rho \rightarrow \pi$, then $\lambda, \phi \vee \psi, \rho \rightarrow \pi$.

P5a. Rule $\rightarrow\supset$: If $\zeta, \phi \rightarrow \lambda, \psi, \rho$, then $\zeta \rightarrow \lambda, \phi \supset \psi, \rho$.

P5b. Rule $\supset\rightarrow$: If $\lambda, \psi, \rho \rightarrow \pi$ and $\lambda, \rho \rightarrow \pi, \phi$, then $\lambda, \phi \supset \psi, \rho \rightarrow \pi$.

P6a. Rule $\rightarrow\equiv$: If $\phi, \zeta \rightarrow \lambda, \psi, \rho$ and $\psi, \zeta \rightarrow \lambda, \phi, \rho$, then $\zeta \rightarrow \lambda, \phi \equiv \psi, \rho$.

P6b. Rule $\equiv\rightarrow$: If $\phi, \psi, \lambda, \rho \rightarrow \pi$ and $\lambda, \rho \rightarrow \pi, \phi, \psi$, then $\lambda, \phi \equiv \psi, \rho \rightarrow \pi$.

For example, given any theorem of *Principia,* we can automatically prefix an arrow to it and apply the rules to look for a proof. When the main connective is $\supset$, it is simpler, though not necessary, to replace the main connective by an arrow and proceed. For example:

$*2.45.\ \ \vdash : \sim(P \vee Q) \cdot \supset \cdot \sim P,$

$*5.21:\ \ \vdash : \sim P \,\&\, \sim Q \cdot \supset \cdot P \equiv Q$

can be rewritten and proved as follows.

$*2.45$	$\sim(P \vee Q) \rightarrow \sim P$	(1)
	(1) $\rightarrow \sim P, P \vee Q$	(2)
	(2) $P \rightarrow P \vee Q$	(3)
	(3) $P \rightarrow P, Q$	
	VALID	
		QED

$*5.21.$	$\rightarrow \sim P \,\&\, \sim Q \cdot \supset \cdot P \equiv Q$	(1)
	(1) $\sim P \,\&\, \sim Q \rightarrow P \equiv Q$	(2)
	(2) $\sim P, \sim Q \rightarrow P \equiv Q$	(3)
	(3) $\sim Q \rightarrow P \equiv Q, P$	(4)
	(4) $\rightarrow P \equiv Q, P, Q$	(5)
	(5) $P \rightarrow Q, P, Q$	
	VALID	
	(5) $Q \rightarrow P, P, Q$	
	VALID	
		QED

These proofs should be self-explanatory. They are essentially the same as the proofs printed out by the machine, except that certain notational changes are made both to make the coding easier and to avoid symbols not available on the machine printer. The reader may wish to read the next section, which explains these changes, and then compare these with more examples of actual print-outs reproduced in Appendix I. It is believed that these concrete examples will greatly assist the understanding of the procedure if the reader is not familiar with mathematical logic.

- *Program I: The propositional calculus P*

There is very little in the program which is not straightforward. To reserve the dot for other purposes and to separate the numbering from the rest, we write, for example, 2*45/ instead of *2.45.

For the other symbols, we use the following dictionary:

–	$\rightarrow$
B	$\equiv$
C	&
D	$\vee$
F	$\sim$
I	$\supset$

Moreover, we use a modified Polish notation by putting, for example, *CFP . FQ* instead of $\sim P \,\&\, \sim Q$. By putting the connective at the beginning, we can more easily search for it. The use of dots for grouping makes it easier to determine the two halves governed by a binary connective. The reader will have no difficulty in remembering these notational changes if he compares the examples *2.45 and *5.21 with the corresponding proofs in the new notation given in Appendix I.

With the longer examples in Appendix I, the reader will observe that the numbers on the right serve to identify the lines, while the numbers on the left serve to identify the conclusions for which the numbered lines are premises. Essentially each proof is a tree structure. Since we have to arrange the lines in a one-dimensional array, there is a choice among various possible arrangements. The one chosen can be seen from the example 4*45 given in Appendix I. The tree structure would be:

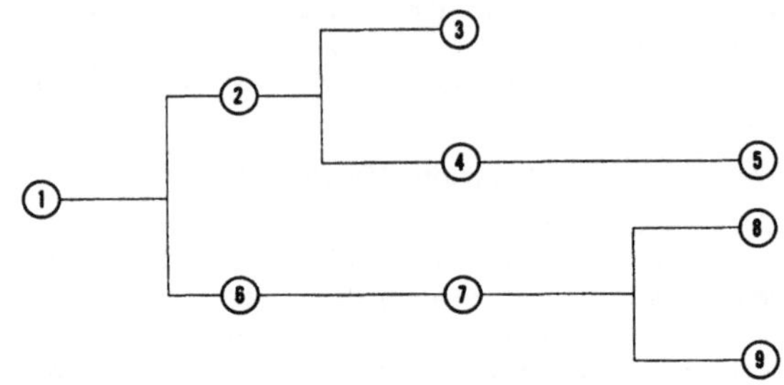

and the one-dimensional arrangement we use is:
(1), (1,2), (2,3), (2,4), (4,5), (1,6), (6,7), (7,8), (7,9).

The whole program has about 1000 lines. The length of the sequents to be tested is deliberately confined to 72 symbols, so that each sequent can be presented by a single punched card. Although this restriction can be removed, it makes the coding considerably easier and gives ample room for handling the problems on hand. Thus, for instance, the longest theorem of the propositional calculus in *Principia,* 5*24, has only 36 symbols. When presented with any punched card, the program enables the machine to proceed as follows.

Copy the card into the reserved core storage COL1 to COL72 (72 addresses in all) in the standard *BCD* notation, i.e., a conventional way of representing symbols by numbers, one symbol in each address. Append the number 1 at the last address, viz., COL72. Search for the arrow sign. If it does not occur, then the line is regarded as ordinary prose, printed out without comment, and the machine begins to study the next card. In particular, the machine stops if the card is blank. If the arrow sign occurs, then the machine marks all symbols before the arrow sign as negative and proceeds to find the earliest logical connective. According as it is *F, C, D, I,* or *B,* the machine turns to *RTNF, RTNC, RTND, RTNI,* or *RTNB.* In each case the proper rule is applied according as whether the connective is before or after the arrow.

After COL1 to COL72, 144 addresses are reserved for getting the one or two premises. As soon as the premises are found according to the proper rule, the original line is printed out and the first premise is shifted into COL1 to COL72 and gets the next number for its identification. If there is a second premise, it has to be shifted away to the idle section and wait for its turn. When the line in COL1 to COL72 contains no more logical connectives, the machine goes to a COMPARE routine to determine whether there is a formula occurring on both sides of the arrow or not and prints the line out with VALID or NOT VALID appended to it. Then it looks at the idle section to see whether any earlier premises remain there. If there is, it moves the first line there into COL1 to COL72 and pushes the remaining lines of the idle section to fill up the vacancy. If there is no more line left then it concludes, according as whether all final sequents are valid, that the original sequent is a theorem (QED) or not (NOT VALID). When the original sequent is not a theorem, the conjunction of all the resulting nonvalid connective-free sequents amounts to a conjunctive normal form of the original sequent.

Several alternatives are permitted by putting down suitable sense switches. If the interest is to determine merely whether the given sequent is a theorem, it is natural to stop as soon as a nonvalid, connective-free sequent is found. This is indeed taken as the normal procedure in the program. Another permissible choice is to omit the proofs or disproofs altogether but only print out the final answer. A third possible choice is to give the output by punching cards rather than by printing. The possibility of omitting the proof or the disproof enables us to separate the calculating time from the input-output time.

While the program is sufficiently fast for testing propositions we ordinarily encounter, it is not the most efficient testing procedure for more complex propositions. If the purpose is to find an isolated fast-test procedure just for the propositional calculus, and not to obtain at the same time a proof procedure which can be combined naturally with proofs in more advanced domains, it is possible to find much more efficient methods. For example, B. Dunham, R. Fridshal, and G. Sward† have one which is being coded by them.

On the other hand, the storage needed is not large. Not only is each theorem proved from scratch so that earlier theorems need not be kept in the store, but in each proof there is no need to keep all the intermediate lines. In fact, at any time the machine needs to keep in its store at most one line from each level of the proof in its tree form.

This can be made clear by the following example:

6*6/–*BBBBBQR . P . . BQR . . . BBPQ . . BP . BQR . . . . BP . . BQ . BQR*.

There are 13 occurrences of *B* (if and only if) in this theorem. Since every elimination of *B* gives two new branches, the complete proof consists of $2^{14}-1$, or about 16,000 lines. Yet at no time need the machine keep in the store for the idle section more than 13 lines. If we use one address for each symbol, we need 72×13 addresses for this; if we pack up these idle lines, we need only 12×13 addresses, a very small number for such a long proof. The length of the proof, incidentally, illustrates how inefficient the procedure can be for certain long propositions. While the complexity of the ordinary truth-table test is determined by the number m of distinct proposition letters (2^m rows) and the number n of distinct subformulae (n columns), the length of proof of the present approach is determined by the number k of occurrences of propositional connectives, i.e., $2^{k+1}-1$ lines or less, since in many cases a conclusion has only one instead of two premises.

The estimated running time for a complete proof of the above theorem with all steps printed out on line is about five hours (about 2500 of the 16,000 lines were printed out in 48 minutes or so). On the other hand, if the proof is not printed out, it takes the machine less than 30 minutes to get the answer. Hence, about 12/13 of the time is spent in reading and printing out. Since axioms and definitions are not used in this approach, they are taken as theorems to be proved, and one arrives at about 220 theorems from the first five chapters of *Principia*. These were proved in about 37 minutes. When only the theorems themselves and the answers (i.e., QED) are printed out, it takes 8 minutes. The actual calculating time, i.e., not counting the input-output time, is less than 3 minutes.

- *Program II: Selecting theorems in the propositional calculus*

A natural question to ask is "Even though the machine can prove theorems, can it select the theorems to be proved?" A very crude experiment in this direction was made with some quite preliminary results. These will be reported here, not for their intrinsic interest, but for suggesting further attempts on the same line. The motive is quite simple: by including suitable principles of triviality, the machine will only select and print out less trivial theorems. These may in turn suggest further principles of triviality; after a certain stage, one would either arrive at essentially the same theorems which have already been discovered and considered interesting, or find in addition a whole crowd of interesting new theorems.

The machine has been made to form a fairly large class of propositions (sequents) and select "interesting" theorems from them. At first all formulae with exactly six symbols containing at most the propositional letters *P, Q, R* are formed. These come to a total of 651, of which 289 are basic and 362 are trivial variants obtainable from the basic ones by renaming the propositional letters. Of these 651 formulae, 107 are theorems. The program enables the machine to form these formulae, one stored in a single address, and select the theorems among them, prefixing each with a minus sign.

Then the machine is to form all non-ordered pairs (π, ρ) such that either π or ρ is (or both are) among the 289 basic formulae and for each pair (π, ρ), the sequents $\pi \rightarrow \rho$, $\rho \rightarrow \pi$, $\pi, \rho \rightarrow$, $\rightarrow \pi, \rho$, when neither π nor ρ is a theorem. When π, ρ are the same, $\pi \rightarrow \rho$, $\rho \rightarrow \pi$, $\rightarrow \pi, \rho$ are not formed. These are the only principles of triviality which are included. Thus, the distinction between basic formulae and their variants avoids the necessity of testing that large number of sequents which are variants of other tested sequents. Moreover, if either π or ρ is a theorem, $\rightarrow \pi, \rho$ is a trivial consequence, $\pi, \rho \rightarrow$ is

†All at the IBM Poughkeepsie Research Laboratory.

a trivial variant of ρ, $\rho\rightarrow$ or π, $\pi\rightarrow$; moreover, then $\pi\rightarrow\rho$ (or $\rho\rightarrow\pi$) is a theorem, and a trivial one, if and only if ρ (or π) is a theorem. Finally, $\pi\rightarrow\pi$ is always a trivial theorem.

It was at first thought that these crude principles are sufficient to cut down the number of theorems to a degree that only a reasonably small number of theorems remain. It turns out that there are still too many theorems. The number of theorems printed out after running the machine for a few hours is so formidable that the writer has not even attempted to analyze the mass of data obtained. The number of sequents to be formed is about half a million, of which about 1/14 are theorems. To carry out the whole experiment would take about 40 machine hours.

The reason that such a high portion are theorems comes from the bias in our way of forming sequents. If we view an arbitrary truth table with n proposition letters, since the table gives a theorem if and only if every row gets the value true and there are 2^n rows, the probability of getting a theorem is $\frac{1}{2}^{(2^n)}$. In particular, if $n=3$, we get 1/256. However, the sequents $\pi\rightarrow\rho$, $\rho\rightarrow\pi$, $\rightarrow\pi,\rho$, $\pi,\rho\rightarrow$ amount to $\sim\pi\vee\rho$, $\sim\rho\vee\pi$, $\rho\vee\pi$, $\sim\rho\vee\sim\pi$, each being a disjunction. Hence, the probability is much higher since the probability of $\phi\vee\psi$ being true is ¾. If there are three proposition letters, we have $(\frac{3}{4})^8$, which is about 1/10. The few crude principles of triviality, besides cutting the sequents to be tested to less than half, reduces this percentage to about 1/14. It would seem clear that other principles of triviality should be devised and included, e.g. if $\pi\equiv\rho$ and $\Theta(\pi)$ are theorems, $\Theta(\rho)$ is a trivial consequence which need not be recorded.

In the actual program, input cards are used to assign the region of sequents to be formed and tested, since otherwise the machine would simply run continuously for about 40 hours until all sequents are formed and tested. Each sequent formed is retained in a reserved region of the core memory or simply thrown away, according as it has been found to be a theorem or a nontheorem. When all sequents required by the input card have been tested or when the reserved region has been filled up, the theorems obtained to date are transferred onto a tape which afterwards is printed out off-line. Just as a curiosity, a very small random consecutive sample of the print-out is reproduced in Appendix II.

• *Completeness and consistency of the systems P and P_s*

A simple proof for the consistency and the completeness of the system P is possible. Since, however, such considerations become even shorter if fewer truth-functional connectives are used, a system P_s based on the single-stroke | connective (not both) will be given and proved consistent, as well as complete. It will then be clear that a similar proof applies to the system P.

The formulae and sequents are specified as with the system P except that the clause on forming new formulae is now merely: if ϕ and ψ are formulae, then $\phi|\psi$ is a formula. There are only three rules:

P_s1. Same as $P1$.

If λ and ζ are (possibly empty) strings of atomic formulae, then:

P_s2. If $\phi, \psi, \zeta\rightarrow\lambda, \rho$, then $\zeta\rightarrow\lambda, \phi|\psi, \rho$.

P_s3. If $\lambda, \rho\rightarrow\pi, \phi$ and $\lambda, \rho\rightarrow\pi, \psi$, then $\lambda, \phi|\psi, \rho\rightarrow\pi$.

Consistency of the calculus P_s. One can easily verify by the intended interpretation of the arrow and the comma that P_s1 is valid, i.e., true in every interpretation of the atomic formulae, that the conclusion of P_s2 is valid if (and only if) its premise is, and that the conclusion of P_s3 is valid if (and only if) its premises are. It follows that every provable sequent is valid.

Completeness of the calculus P_s. We wish to prove that every sequent, if valid, is provable. Given any sequent, we find the earliest non-atomic formula, if any, in the antecedent, and apply P_s3 in reverse direction, thereby obtaining two premises, each with less occurrences of | . If there is no | in the antecedent, we find the earliest, if any, occurrence of | in the consequent and apply P_s2 in reverse direction. We then repeat the same procedure with the results thus obtained. This process will be continued with each sequent until | no longer occurs. Since there are only finitely many occurrences of | in each sequent, this process always comes to an end and then we have a finite class of sequents in which only atomic formulae occur. Now the original sequent is valid if and only if every sequent in the class is. But a sequent with only atomic formulae is valid if and only if it is a case of P_s1, and we can decide effectively in each case whether this is so. Hence, the calculus is complete and we have a decision procedure for provability which yields automatically a proof for each provable sequent.

The propositional calculus with equality (System P_e)

It is convenient, though not necessary, to add the equality sign = before introducing quantifiers. This procedure serves to stress the fact that equality is more elementary than quantifiers, even though customarily quantifiers are presented prior to equality. The changes needed to reach this system from the system P are rather slight. Variables X, Y, Z, S, T, U, V, W, et cetera, are now taken as terms, and the domain of atomic formulae are extended to include all expressions of the form $\alpha=\beta$ when α and β are terms.

The only additional rules necessary for equality are an extension of the initial rule $P1$ so that in addition to the rules $P1$ to $P6b$ of the system P, we now have: If λ, ζ are strings (possibly empty) of atomic formulae, then:

P7. $\lambda\rightarrow\zeta$ is a theorem if there is a term α such that $\alpha=\alpha$ occurs in ζ.

P8. $\lambda\rightarrow\zeta$ is a theorem if $\alpha=\beta$ occurs in λ and $\lambda\rightarrow\zeta'$ is a theorem, where ζ' is obtained from ζ by substituting α (or β) for some or all occurrences of β (or α).

It is quite easy to extend Program I to obtain a program for the system P_e. The writer, however, did not write a separate program for P_e but includes such a program as a part in Program III. This part enables the machine to proceed exactly as in Program I except that, in testing whether a sequent of atomic formulae is a theorem, the machine does not stop if the sequent is not a theorem by the initial rule *P1* but proceeds to determine whether the sequent can be shown to be a theorem by using the additional initial rules *P7* and *P8*. To distinguish sequents of atomic formulae which are valid truth-functionally from those which become valid only after applying *P7* or *P8*, the former case is marked with VA only, while the latter case is marked in addition by =. Examples of print-outs are given in Appendix III.

The longish example *13.3 is included to make a minor point. It has been suggested that it would be interesting if the machine discovers mistakes in *Principia*. This example may be said to reveal a mistake in *Principia* in the following sense. The authors of *Principia* proved this theorem by using *10.13 and *10.221 from the predicate calculus. From the discussion attached to the proof of this theorem, it seems clear that the authors considered this as a theorem which presupposes the predicate calculus. Yet in the proof printed out by the machine, no appeal to anything beyond the system *P*, i.e., no appeal even to the additional rules *P7* and *P8* is made. This is revealed by the fact that all the sequents of atomic formulae, viz., lines 6, 10, 14, 18, 24, 28, 32, 36 in the proof, are marked with VA without the additional = sign. At first the writer thought this indicates a mistake in the program. An examination of the theorem shows, however, that *13.3 is indeed a theorem of the propositional calculus. In fact, Program I alone would yield essentially the same proof.

- *Preliminaries to the predicate calculus*

Thus far an attempt has been made to avoid heavy technicalities from symbolic logic. A few more exact definitions seem necessary, however, when one comes to the predicate calculus.

Formulae, terms, and sequents of the full predicate calculus are specified as follows. Basic symbols are =; the five truth-functional connectives; the two quantification symbols for "all" and "some"; proposition letters *P, Q, R*, et cetera; predicate letters *G, H, J, K*, et cetera; variables *X, Y, Z*, et cetera; function symbols *f, g, h*, et cetera; numerals 1, 2, ..., 9, et cetera; dots or parentheses for grouping. Terms are: (i) a variable is a term; (ii) a numeral is a term; (iii) if α, β, et cetera are terms and σ a function symbol, then $\sigma\alpha$, $\sigma\alpha\beta$, et cetera, are terms. A variable or a numeral is a simple term; other terms are composite. The five truth-functional connectives and two quantification symbols are called logical constants. Atomic formulae are: (i) proposition letters; (ii) $\alpha=\beta$, when α, β are terms; (iii) $G\alpha$, $H\alpha\beta$, et cetera, where α, β, et cetera are terms. Formulae are: (i) atomic formulae are formulae; (ii) if ϕ, ψ are formulae, $\sim\phi$, $\phi\vee\psi$, $\phi\supset\psi$, $\phi\equiv\psi$, $\phi\&\psi$, $(E\alpha)\phi$, $(\alpha)\phi$ are formulae, where α is a variable. A string may be empty or a single formula, and if π, ρ are nonempty strings π, ρ is a string. Given any two strings π and ρ, $\pi\rightarrow\rho$ is a sequent.

Intuitively the scope of a logical constant is clear. In mechanical terms, the method of finding the scope of a logical constant in a given formula depends on the notation. According to the notation actually chosen for the machine, $\phi\vee\psi$, $\phi\&\psi$, $\phi\supset\psi$, $\phi\equiv\psi$ are written as $D\phi\_\psi$, $C\phi\_\psi$, $I\phi\_\psi$, $B\phi\_\psi$, with the blank filled in by a string of dots whose number is one larger than the longest string in ϕ and ψ except that no dot is used when both ϕ and ψ are proposition letters. These and related details in the notation can be understood easily from the Appendices. Given a sequent, the scope of a logical constant Γ standing at the beginning of a whole formula in the sequent is the entire formula minus Γ. If Γ is singulary, that is, $\sim$ or one of the two quantification symbols, the scope of the next logical constant Γ' in the formula, if any, is the whole remaining part of the formula minus Γ'. If Γ is binary, then its scope breaks into two parts at the longest string of dots in the scope and each part, if containing a logical constant at all, must begin with one, say Γ', whose scope is the whole part minus Γ'. This gives a mechanizable inductive definition of the scope of every logical constant in any sequent. A logical constant Γ is said to "govern" a logical constant Γ', if Γ' falls within the scope of Γ.

To avoid the explicit use of the prenex normal form, i.e., the form in which all quantifiers in a formula stand at its beginning, it is desirable to introduce, after Herbrand,[18] the sign of every quantifier in a sequent. Two simple preliminary operations will be performed on a given sequent before calculating the signs of the quantifiers in it. First, distinct quantifiers are to get distinct variables, even when one quantifier does not govern the other; moreover, the free variables in the sequent are not used as variables attached to explicit quantifiers. This simplifies the elimination of quantifiers afterwards. Second, all occurrences of $\equiv$ which govern any quantifiers at all are eliminated by either of two simple equivalences: $\phi\equiv\psi$ if and only if $(\phi\&\psi)\vee(\sim\phi\&\sim\psi)$, or, alternatively, $(\sim\phi\vee\psi)\ \&\ (\sim\psi\phi)$.

The positive and negative parts of any formula in the sequent are defined thus: (i) (an occurrence of) ϕ is a positive part of (the same occurrence of) ϕ; (ii) if ϕ is a positive (a negative) part of ψ, then ϕ is a negative (a positive) part of $\sim\psi$; (iii) if ϕ is a positive (a negative) part of ψ or of χ, then ϕ is a positive (a negative) part of $\psi\vee\chi$; (iv) similarly with ϕ and $\psi\&\chi$; (v) if ϕ is a positive (a negative) part of ψ, then ϕ is a positive (a negative) part of $(\alpha)\psi$; (vi) similarly with ϕ and $(E\alpha)\psi$; (vii) if ϕ is a positive (a negative) part of ψ, then ϕ is a positive (a negative) part of $\chi\supset\psi$, and a negative (a positive) part of $\psi\supset\chi$. Any formula ϕ in a sequent is a positive or negative part of the sequent according as (i) it is a positive (a negative) part of a whole formula in the consequent (the antecedent), or (ii) it is a negative (a positive) part of a whole formula in the consequent (the antecedent). Any quantifier (α) with the scope ϕ in a given sequent is positive (negative) in the sequent if and only if $(\alpha)\ \phi$ is a positive (a negative) part of the sequent; $(E\alpha)$ is positive (negative) if and only if $(E\alpha)\ \phi$ is a negative (a positive) part of the sequent; the different occurrences of a same free variable α in the sequent also make up a positive quantifier (as if (α) were put at the head of the whole sequent).

This involved definition can be illustrated by an example:

Ex.0. $(X)\ (GXY \supset (\sim GXX \,\&\, (EZ)\ HXZ)),$
$(W)((\sim GWW \,\&(EU)HWU) \supset GWY)$
$\rightarrow \sim (EV)HYV.$

In this example, (X) is a negative quantifier, (EZ) is positive, (W) is negative, (EU) is negative, (EV) is positive. For instance, $(EU)HWU$ is positive in $\sim GWW \,\&(EU)HWU$ but negative in $(\sim GWW \,\&(EU)HWU) \supset GWY$ and $(W)((\sim GWW \,\&\, (EU)HWU) \supset GWY)$, which is a whole formula in the antecedent. Hence, $(EU)HWU$ is positive in the sequent. Hence, (EU) is negative. The assignment of signs to quantifiers coincides with the result in a prenex normal form; positive for universal, negative for existential. Thus, one prenex form of Ex. 0 is:

$(Y)\ (EX)\ (EW)\ (EU)\ (Z)\ (V)\ \{[(GXY \supset$
$(\sim GXX \,\&\, HXZ))$
$\&((\sim GWW \,\&\, HWU) \supset GWY)] \supset \sim HYV\}.$

Another useful but involved concept is "miniscope" forms of a formula of the predicate calculus. It is in a sense the opposite of the prenex form, which generally gives every quantifier the maximum scope. Since the interweaving of quantifiers and variables is the main factor determining the complexity of a formula of the predicate calculus, it is not hard to see that separating variables and reducing the ranges of quantifiers may help to simplify the problem of determining whether a formula is a theorem. The unfortunate part is that sometimes it can be a very complicated process to get a formula into the miniscope form.

It is easier to explain the notion for a formula in which $\equiv$ and $\supset$ no longer occur (say, eliminated by usual definitions) so that the only truth-functional connectives are $\sim$, &, v. Such a formula is said to be in the miniscope form if and only if: (i) an atomic formula ϕ is in the miniscope form; (ii) if ϕ in $(\alpha)\phi$ (or $(E\alpha)\phi$) is a disjunction (a conjunction) of formulae, each of which is in the miniscope form and either contains α or contains no free variable at all, then $(\alpha)\phi$ (or $(E\alpha)\phi$) is in the miniscope form; (iii) if ϕ and ψ are in the miniscope form, so are $\sim\phi$, $\phi \vee \psi$, $\phi \,\&\, \psi$; (iv) if ϕ in $(\alpha)\phi$ (or $(E\alpha)\phi$) begins with $(E\beta)$ (or (β)) and is in the miniscope form, so is $(\alpha)\phi$ (or $E\alpha)\phi$); (v) a formula beginning with a string of quantifiers of the same kind is in the miniscope form if every formula obtained by permuting these quantifiers and then dropping the first, is in the miniscope form. One procedure for bringing a formula into the miniscope form is explained in detail by Quine.[19] In what follows, only parts of Quine's procedure will be used and explained, the machine will follow quite different procedures if the formulae in a given sequent are not easily brought into the miniscope form.

The system Qp and the *AE* predicate calculus

A specially simple decision procedure is available for many of those sequents not containing function symbols in which each formula is in the miniscope form and in the *AE* form, i.e., no positive quantifier is governed by a negative quantifier. The procedure can be extended by two preliminary steps and described as follows.

Step 1. Bring every formula into the miniscope form and at the same time apply the truth-functional rules *P2 - P6b*, whenever possible. In general, we obtain a finite set of sequents which all are theorems if and only if the original sequent is.

Step 2. Test each sequent and decide whether it is in the *AE* form. If this is so for all the sequents, then they and the original sequent all fall within the *AE* predicate calculus, and we proceed to decide each sequent by continuing with Step 3. If this is not so for some sequent, then the original sequent does not belong to the *AE* predicate calculus and has to be treated by appealing to a richer system Q to be described below.

Step 3. For a sequent in the *AE* predicate calculus, drop all quantifiers and replace all the variables attached to negative quantifiers by numerals, one numeral for each quantifier. The resulting sequent contains no more quantifiers.

Step 4. Apply the truth-functional rules to obtain a finite set of sequents which contain no more logical constants. Test each sequent by the initial rules and retain only the non-valid ones.

Step 5. List all the variables and numerals occurring in this last set of sequents of atomic formulae, make all possible substitutions of the variables for the numerals in the sequents, (substitute X for the numerals if no variables occur) and test each time whether the resulting sequents are all valid. The initial sequent of Step 3 is a theorem if there is a substitution which makes all the sequents in the finite set theorems.

Step 6. The original sequent is a theorem if all the sequents obtained by Step 1 are theorems by Steps 3 to 5.

This completes the description of Qp. It is possible to formulate this system more formally, as derived from the basic system P_c, by adding additional explicit rules. But the result would be rather lengthy and a bit artificial. Since the above less explicit formulation conforms to the general theoretical requirements of a formal system, we shall not give a formally more pleasing description. This remark applies also to the systems to be given below.

Here are a few simple examples which illustrate this procedure and some minor modifications in it:

*10.25. $(X)GX \to (EY)GY$ (1)

(1) $G1 \to G2$ NOT (2)

(1) $GX \to GX$ VA (2)

QED

*11.21. $\to (X)(Y)(Z)GXYZ \equiv (V)(U)(W)GUVW$ (1)

(1) $(X)(Y)(Z)GXYZ \to (V)(U)(W)GUVW$ (2)

(1) $(V)(U)(W)GUVW \to (X)(Y)(Z)GXYZ$ (3)

(2) $G123 \to GUVW$ NOT (4)

(2) $GUVW \to GUVW$ VA (4)

PQED

(3) $G213 \to GXYZ$ NOT (5)

(3) $GXYZ \to GXYZ$ VA (5)

QED

*11.57 $(X)GX \equiv (Y)(Z)(GY \& GZ)$ (1)

(1) $(X)GX \to (Y)(Z)(GY \& GZ)$ (2)

(1) $(Y)(Z)(GY \& GZ) \to (X)GX$ (3)

(3) $G1 \& G2 \to GX$ (4)

(4) $G1, G2 \to GX$ NOT (5)

(4) $GX, GX \to GX$ VA (5)

PQED

(2) $(X)GX \to (Y)GY \& (Z)GZ$ (6)

(6) $(X)GX \to (Y)GY$ (7)

(6) $(X)GX \to (Z)GZ$ (8)

(7) $G1 \to GY$ NOT (9)

(7) $GY \to GY$ VA (9)

PQED

(8) $G1 \to GZ$ NOT (10)

(8) $GZ \to GZ$ VA (10)

QED

*9.22. $(X)(GX \supset HX) \to (EY)GY \supset (EZ)HZ$ (1)

(1) $(X)(GX \supset HX), (EY)GY \to (EZ)HZ$ (2)

(2) $G1 \supset H1, GY \to H2$ (3)

(3) $H1, GY \to H2$ NOT (4)

(3) $GY \to H2, G1$ NOT (5)

(3) $GY \to HY, GY$ VA (5)

QED.

(3) $HY, GY \to HY$ VA (4)

These examples are intended to show several things. In the first place, for example, in line (5) of *11.21, the possible substitutions for (1,2,3) are (X,X,X), (X,X,Y), (X,X,Z), (X,Y,X), (X,Y,Y), (X,Y,Z), $(Y,X,X,)$ (Y,X,Y), (Y,X,Z), (Y,Y,X), (Y,Y,Y), (Y,Y,Z), et cetera, 27 in all. If one tries out the substitutions one by one, as was done in Program III, it will take some time before one reaches the correct substitution (Y,X,Z). It is, however, clear that an equally mechanizable procedure is to single out occurrences of the same predicate letter on both sides of the arrow and select the substitutions which would make all the sequents in question theorems. This is one minor change which will be made in Program III. Incidentally, this is also an instance of a simple strategy which improves the program.

In the second place, *9.22 shows that there is no need to eliminate $\supset$ because the definition of formulae in the miniscope form can easily be modified to include formulae containing $\supset$ in addition to $\sim$, v, &. All that is needed is to remember that $\phi \supset \psi$ is the same as $\sim \phi \vee \psi$.

In the third place, as will be proved later on, the Steps 3 to 5 given above are applicable to a sequent in the *AE* form which contains at most one positive quantifier even if it is not in the miniscope form. This is why in the proof of *11.57, (3) does not have to be transformed into the miniscope form, while (2) has to. Thus if the quantifiers in (2) are not separated as in line (6), one gets:

(2) $G1 \to GY \& GZ$ (6)

(6) $G1 \to GY$ NOT (7)

(6) $G1 \to GZ$ NOT (8)

No possible substitution can make both (7) and (8) theorems. Hence, although *11.57 is a theorem, no proof would be obtained in this way, unless a reduction to the miniscope form is made first. On the other hand, in an alternative procedure to be described below, Step 5 in the above procedure is replaced by a different substitution, performed before Step 4, so that $G1$ is replaced by GY,GZ in the above example and it is no longer necessary to have the sequent in the miniscope form to begin with. The relative merits of the two procedures will be compared below.

- *Program III: The AE predicate calculus*

This program was originally intended to embody the procedure *Qp*. But the preliminary part of bringing a formula into the miniscope form has not been debugged. It is now clear that just for the purpose of proving all the theorems of the predicate calculus in *Principia*, it is not necessary to include all the rules for bringing a formula into a miniscope form. Indeed, only about 5% of the theorems need such rules at all, and only rather simple ones.

There are, however, a few other differences between Program III and the procedure described above. Instead of eliminating all quantifiers at once according to their signs, quantifiers are treated on the same basis as the truth-functional connectives with two rules for each. If λ and ζ are (possibly empty) strings of atomic formulae and i is a new numeral, v is a new variable:

Rule $\to\forall$: If $\zeta \to \lambda, \phi\, v, \pi$, then $\zeta \to \lambda, (\alpha)\, \phi\, \alpha, \pi$.

Rule $\to\exists$: If $\zeta \to \lambda, \phi\, i, \pi$, then $\zeta \to \lambda, (E\,\alpha)\, \phi\, \alpha, \pi$.

Rule $\forall\to$: If $\lambda, \phi\, i, \rho \to \pi$, then $\lambda, (\alpha)\, \phi\, \alpha, \rho \to \pi$.

Rule $\exists\to$: If $\lambda, \phi\, v, \rho \to \pi$, then $\lambda, (E\,\alpha)\, \phi\, \alpha, \rho \to \pi$

These rules make for uniformity in the whole procedure except that precaution should be taken that the same quantifier, when recurring at different places on account of truth-functional reductions, should still be replaced by the same variables or numerals, although when the replacement is by a numeral, the difference is not vital.

When this precaution is not taken, it can happen that certain theorems of the *AE* predicate calculus do not get proofs. This in fact happened with Program III, which failed to yield proofs for *10.3, *10.51, *10.55, *10.56, *11.37, *11.52, *11.521, *11.61, for no other reason than this.

A less serious defect in Program III is that truth-functional reductions are not always made as often as possible before eliminating quantifiers. This has the defect that several separate problems are sometimes treated as one whole problem and the running time required for getting a proof becomes unnecessarily long. In four special cases, viz., *10.22, *10.29, *10.42, *10.43, this defect in fact results in the failure to get a proof, even though a proof for each can be found by Program III, if all possible truth-functional reductions are made before eliminating quantifiers. Both this and the preceding defects of Program III can easily be amended. The reason for dwelling so long on them is to illustrate how machines can assist mathematical research in revealing theoretical defects in preliminary formulations of general procedures.

Since the present methods do not use axioms and definitions, the axioms and definitions of *Principia* are rewritten as theorems. The resulting augmented list of theorems in *Principia* (*9 to *13) from the predicate calculus with equality, has a total number of 158 members. Of these, 139 can be proved by Program III as it stands, although some of them require unnecessarily long running time, e.g., *11.21 and *11.24. If we make the few minor modifications mentioned above, the running time for all becomes reasonably short and the 12 theorems listed in the last two paragraphs become provable. Altogether there are only 7 of the 158 theorems which stand in need of some preliminary simple steps to get the formulae into the miniscope form: *11.31, *11.391, *11.41, *11.57, *11.59, *11.7, *11.71. The rules needed to take care of these cases are three:

(i) Replace $(\alpha)(\phi\alpha \& \psi\alpha)$ by $(\alpha)\phi\alpha \& (\alpha)\psi\alpha$.

(ii) Replace $(E\alpha)(\phi\alpha \vee \alpha)$ by $(E\alpha)\phi\alpha \vee (E\alpha)\psi\alpha$.

(iii) Replace $(\alpha)(\chi\alpha \supset (\phi\alpha \& \psi\alpha))$ by
$(\alpha)(\chi\alpha \supset \phi\alpha) \& (\alpha)(\chi\alpha \supset \psi\alpha)$.

Hence, to summarize, Program III can be somewhat modified to prove all the 158 theorems of *Principia*, with the modified program doing the following. Given a sequent, see whether the rules (i), (ii), (iii) are applicable and apply them if so. Then make all truth-functional simplifications by the rules *P2 - P6b*. This in general yields a finite set of sequents. If every one is in the *AE* form and either contains no more than one positive quantifier or is in the miniscope form, then the original sequent is often decidable by the method; otherwise it is beyond the capacity of the method. If the former is the case, proceed to decide each sequent either by eliminating all quantifiers at once, as in Step 3 of the preceding section, or by mixing the application of the rules *P2 - P6b* with the rules $\rightarrow\forall$, $\rightarrow\exists$, $\forall\rightarrow$, $\exists\rightarrow$. Finally, use Steps 4 and 5 of the preceding section.

A sample of the print-outs by Program III without the modifications is given in Appendix IV. In this connection, it may be of interest to report an amusing phenomenon when Program III is in operation. The machine usually prints out the lines of a proof in quick succession, and then there is a long pause, as if it were thinking hard, before it prints out the substitution instance which makes all the nonvalid sequents of atomic formulae valid. For example, the first 15 lines of *11.501 in Appendix IV were printed out in quick succession, followed by a long pause of over 2 minutes, and then the remaining few lines were printed out.

• *Systems Qq and Qr: Alternative formulations of the AE predicate calculus*

As remarked in connection with the proof of *11.57, one can replace Step 5 by a different way of substitution and then the new method *Qq* becomes applicable even when the formulae in the sequent are not in the miniscope form, as long as the sequent is in the *AE* form.

More exactly, the new method of substitution is as follows. Consider the sequent obtained immediately after the elimination of quantifiers, and determine all the occurring variables and numerals. If $\alpha_1, \ldots, \alpha_n$ are the variables, a formula $\phi\, i$, where i is a numeral, is replaced by $\phi\alpha_1, \ldots, \phi\alpha_n$. For example, take (2) of *11.57:

	$(X)GX \rightarrow (Y)(Z)(GY \& GZ)$	(1)
(1)	$G1 \rightarrow GY \& GZ$	(2)
(2)	$GY, GZ \rightarrow GY \& GZ$	(3)
(3)	$GY, GZ \rightarrow GY$	VA (4)
(3)	$GY, GZ \rightarrow GZ$	VA (5)
	QED.	

What are the comparative merits of *Qp* and *Qq*? According to *Qq*, only one substitution is made, and at the beginning rather than at the end, but unlike *Qp*, the results obtained may be sequents longer than any previous sequents in the proof. To apply *Qq* it is not necessary that the formulae be first brought to the miniscope form—the procedure is applicable as long as the sequent is in the *AE* form. The method *QP*, however, has the compensating advantage that sometimes a sequent not in the *AE* form can be reduced to sequents in this form by bringing it into the miniscope form. For example, all sequents of the monadic predicate calculus are decidable by *Qp*, but not by *Qq*. Church's example[20] in Appendix VI is such a case. On the other hand, the same example also shows that the procedure of bringing a sequent into the miniscope form can get very involved, so that other methods become preferable. While both *Qp* and *Qq* can be incorporated in the system *Q* for the whole predicate calculus, to be described below, a third method *Qr* is closer to *Q* in spirit. Hence, *Q* will be presented as an extension of *Qr* rather than one of *Qp* or *Qq*.

The method *Qr* proceeds in the same way as *Qp* except that at the step of substitution, the disjunction of all substitution instances is tested for truth-functional validity. Thus, take (2) of *11.57 again:

$$(X)GX \rightarrow (Y)(Z)(GY \& GX) \quad (1)$$

(1) $G1 \rightarrow GY \& GZ$ (2)

(2) $G1 \rightarrow GY$ (3)

(2) $G1 \rightarrow GZ.$ (4)

Now we test whether the disjunction of the conjunction of $GY \rightarrow GY$ and $GY \rightarrow GZ$, and that of $GZ \rightarrow GY$ and $GZ \rightarrow GZ$, is truth-functionally valid. This is indeed so, because if the first disjunctant is false, then $GY \rightarrow GZ$ is false and then $GZ \rightarrow GY$ is true and therewith the second disjunctant is true. Alternatively, this disjunction can also be expressed as a conjunction of four clauses:

(i) $GY, GZ \rightarrow GY, GY$
(ii) $GY, GZ \rightarrow GY, GZ$
(iii) $GY, GZ \rightarrow GZ, GY$
(iv) $GY, GZ \rightarrow GZ, GZ.$

Now we have three alternative methods, *Qp, Qq, Qr* for the *AE* predicate calculus. In Appendix V, we give an example with its three disproofs.

How do we justify the methods *Qp, Qq,* and *Qr*? First, if there is no proof, then the original sequent is not valid. The proof for this is easiest for *Qq*. In the example in Appendix V, line (3) under method *Qq* is not valid if and only if the original sequent (1) is not valid in the domain $\{X, Z\}$. The fact that (4), (5), (6), (7) are not all valid shows that (3) is not valid. In fact, if GXX and GXZ are true but GZX and GZZ are false, (3) is not valid. Therefore, (1) is false under the particular interpretation and hence not valid. More exactly, we wish to find an interpretation under which $(EX)(Y)(GXY \vee GYX) \& \sim (Z)(EW)GZW$, or more simply, $(Y)\ (GXY \vee GYX)\ \&\ (W) \sim GZW$, is satisfiable in $\{X, Z\}$. That is to say, to find an interpretation under which the conjunction of $GXX \vee GXX$, $GXZ \vee GZX$, $\sim GZX$, $\sim GZZ$ is true, or its negation (3) is false. The interpretation of G given above serves this purpose. It is not hard to generalize the argument to all sequents in the *AE* form. Indeed, such considerations are familiar from standard decision procedures.[21]

The justification of *Qp* consists in the fact that if an *AE* sequent is in the miniscope form, then all the negative quantifiers essentially govern a disjunction in the consequent and a conjunction in the antecedent. As a result, the substitution at the end is often equivalent to that obtained by the method *Qq*. For example, in Appendix V, the disjunction under method *Qp* is of:

(i) $GXX \rightarrow GZX; GXX \rightarrow GZX$ or simply $GXX \vee GXX \rightarrow GZX$
(ii) $GXX \rightarrow GZZ; GXX \rightarrow GZZ$ or simply $GXX \vee GXX \rightarrow GZZ$
(iii) $GXZ \rightarrow GZX; GZX \rightarrow GZX$ or simply $GXZ \vee GZX \rightarrow GZX$
(iv) $GXZ \rightarrow GZX; GZX \rightarrow GZZ$ or simply $GXZ \vee GZX \rightarrow GZZ$

Hence, the disjunction is equivalent to:

$$\sim (GXX \vee GXX) \vee \sim (GXZ \vee GZX) \vee GZX \vee GZZ,$$

which, in turn, is equivalent to (3) under method *Qq*.

When a sequent has no more than one positive quantifier, whether in the miniscope form or not, it is quite obvious that a proof by *Qq* is obtainable if and only if one by *Qp* is, since in either method there is only a single possible substitution. In the general case, however, as Mr. Richard Goldberg† has pointed out in correspondence, there are *AE* sequents in the miniscope form which are provable by *Qq* but not by *Qp*. He gives the following valid sequent as example:

$$(X)GXU \rightarrow (EW)(GYW \& GZW).$$

It follows that *Qp* is applicable only to a subclass of *AE* sequents in the miniscope form. Hence, it would seem that the correct course is to modify Program III to embody the procedure *Qq* or the procedure *Qr*. The modifications needed are, fortunately, again not extensive.

With some care, it is possible to prove by an inductive argument that a proof by *Qq* is obtainable if and only if one by *Qr* is. This is so because the substitutions at the beginning of *Qq* give the same result as the taking of disjunctions at the end of *Qr*. Hence, it is true that all valid sequents of the *AE* predicate calculus are provable in *Qq* and in *Qr*.

It is easier to prove that if there is a proof by any of the methods, then the sequent is valid. In the case of *Qp*, one can simply replace all numerals throughout by the correct variables found at the end, and the result would be a quite ordinary proof which can easily be seen to yield only valid results. In the case of *Qq*, one can again make the replacement throughout, so that the result is a proof which, instead of $\rightarrow\forall$, $\rightarrow\exists$, $\forall\rightarrow$, $\exists\rightarrow$, uses $\rightarrow\forall$, $E\rightarrow$, and:

Rule $\rightarrow\exists^*$: If $\zeta \rightarrow \lambda, \phi\,\alpha_1, \ldots, \phi\,\alpha_n\,\pi$, then $\zeta \rightarrow \lambda,\ (E\alpha)\,\phi\,\alpha, \pi$.

Rule $\forall\rightarrow^*$: If $\lambda, \phi\,\alpha_1, \ldots, \phi\,\alpha_n, \rho \rightarrow \pi$, then $\lambda,\ (\alpha)\,\phi\,\alpha, \rho \rightarrow \pi$.

Finally, since there is a proof by *Qr* if and only if there is one by *Qq*, every theorem of *Qr* is also valid.

- *System Q: The whole predicate calculus with equality*

Thus far we have considered only *AE* sequents and have used no function symbols. Now we shall consider arbitrary sequents of the predicate calculus and make use of function symbols from time to time.

The method *Q*, an extension of *Qr*, can be explained as follows. It is desirable (for shorter running time) but not necessary, to make preliminary truth-functional reductions so that one problem is broken up into several simpler problems. For each problem, the following steps are used.

Step I. Eliminate all occurrences of $\equiv$ whose scopes contain quantifiers. Determine the positive quantifiers, the negative quantifiers, and for every positive quantifier governed by negative quantifiers, if there is any, the negative quantifiers which govern it. Use distinct variables for all the free variables and positive quantifiers.

† IBM Research Laboratory, Yorktown Heights, N. Y.

Step II. Drop all quantifiers and replace all variables attached to a negative quantifier by a distinct numeral, all variables attached to a positive quantifier governed by negative quantifiers by a function symbol, followed by the numerals for the governing negative quantifiers.

Step III. Make truth-functional simplifications until all logical constants are eliminated and a finite set of sequents of atomic formulae is obtained.

Step IV. Make all possible substitutions on these sequents obtaining results S_1, S_2, S_3, et cetera. The original sequent is a theorem if and only if there is a truth-functional tautology among S_1, $S_1 \vee S_2$, $S_1 \vee S_2 \vee S_3$, et cetera.

The substitutions to be made are a bit more complex than those in the *AE* predicate calculus. The basic terms consist of not only all the occurring variables, say X and Y, but also instances of the composite terms such as, say, fX, ffX or f^2X, fY, f^2Y, f^3X, f^3Y, f^4X, et cetera. The numerals are to be substituted by all possible selections from these basic terms. If no variables occur in the sequents, a single variable X is added.

We give a few simple examples. More complex examples are included in Appendices VI and VII.[22]

Ex. 1.	$\rightarrow (EY)(Z)(GXZ \supset GXY)$	(1)
(1)	$\rightarrow GXf1 \supset GX1$	(2)
(2)	$GXf1 \rightarrow GX1$	(3)
(2)	$GXfX \rightarrow GXX$	(4)
(2)	$GXf^2X \rightarrow GXfX$	(5)
(2)	$GXf^3X \rightarrow GXf^2X$	(6)
	and so on.	

Since the disjunction of (4) and (5) is already valid, this is a theorem. In this simple case, (4), (5), (6) are S_1, S_2, S_3, and S_1, $S_1 \vee S_2$, $S_1 \vee S_2 \vee S_3$ can be simply rewritten as S_1 and:

$$GXfX, GXf^2X \rightarrow GXX, GXfX$$
$$GXfX, GXf^2X, GXf^3X \rightarrow GXX, GXfX, GXf^2X.$$

Ex. 2	$(X)(EY)(GXY \rightarrow (EZ)(W)GZW$	(1)
(1)	$G1f1 \rightarrow G2g2$	(2)
	$GXfX \rightarrow GXgX$	(S_1)
	$GfXf^2X \rightarrow GXgX$	(S_2)
	$GXfX \rightarrow GfXgfX$	(S_3)
	$GfXf^2X \rightarrow GfXgfX$	(S_4)
	and so on.	

In this example, the basic terms are X, fX, gX, fgX, gfX, f^2X, g^2X, fg^2X, gf^2X, f^2gX g^2fX, f^3X, g^3X, et cetera. Intuitively it can be seen that, no matter what basic terms we substitute for 1 and 2, we can never arrive at a tautologous disjunction. Thus, no matter what we do, we can never get an antecedent of the form $G\alpha g\beta$, or a consequent of the form $G\alpha f\beta$. Hence, this is not a theorem of Q.

An alternative procedure Q', which is more directly related to the standard method initiated by Herbrand, is to make the substitutions immediately after the quantifiers are eliminated. In Appendix VI, a proof of Ex. 3 is given by this method. Clearly Q' is an extension of Qq, just as Q is an extension of Qr. While it is not clear whether Q or Q' is superior if the problem is done by hand, it is conjectured that Q is mechanically less cumbersome, especially if the final test procedure is programmed along the line suggested in Appendix VII.

To justify the procedures Q and Q', that is, to prove their correctness and completeness, one need introduce only slight modifications into standard arguments of Skolem and Herbrand. The equivalence of the two procedures is established by the same kind of argument as that for the equivalence of Qq and Qr. Hence, it suffices to prove the correctness and completeness of Q'.

The correctness, i.e., that every provable sequent is valid, is apparent. Given a proof of Q', i.e., a disjunction of substitution instances which is tautologous, we can derive a sequent which is equivalent to the original sequent to be proved, except that every whole formula in the sequent is in the prenex form. Take Ex. 1. The proof of Q' consists merely in replacing the lines (3)–(6) in the above proof by the tautologous line:

$$\rightarrow GXfX \supset GXX, GXf^2X \supset GXfX. \qquad \text{(i)}$$

From this line, we can by the usual rules of quantification infer:

$$\rightarrow GXfX \supset GXX, (Z)(GXZ \supset GXfX)$$
$$\rightarrow GXfX \supset GXX, (EY)(Z)(GXZ \supset GXY)$$
$$\rightarrow (Z)(GXZ \supset GXX), (EY)(Z)(GXZ \supset GXY)$$
$$\rightarrow (EY)(Z)(GXZ \supset GXX), (EY)(Z)GXZ \supset GXY)$$
$$\rightarrow (EY)(Z)(GXZ \supset GXX).$$

It may be remarked that since a new term for the universal quantifier is introduced every time, viz., fX, f^2X, et cetera, we can always reintroduce it successively without violating the restriction that the term to be replaced by Z is not free elsewhere. Strictly speaking, the terms fX, f^2X, et cetera should first be replaced by new variables, say U, V, et cetera. Then the line (i) remains valid truth-functionally, and the resulting proof of the original sequent of Ex. 1 would conform entirely to usual rules for quantifiers.[23]

The completeness of Q′ can be proved by using ideas familiar in mathematical logic.[24] We wish to prove that if there is no proof, then its negation is satisfiable in an enumerable domain, and hence the original sequent is not valid. Consider Ex. 2 for which none of S_1, $S_1 \vee S_2$, $S_1 \vee S_2 \vee S_3$, et cetera, is valid. In other words, for each disjunction, there are truth-value assignments which would make it false. Since every later disjunction contains all earlier disjunctions as part, a falsifying assignment of $S_1 \vee \ldots \vee S_n$ (call it D_n) also falsifies all D_i, $i < n$. In other words, there is a truth-assignment which falsifies D_1, and for every n, there exists a falsifying assignment for D_n which has an extension that falsifies D_{n+1}. In the simple example Ex. 2, we have:

D_1 is falsified if $GXfX$ is true but $GXgX$ is false,
D_2 is falsified if, in addition, $GfXf^2X$ is true,
D_3 is falsified if, in addition, $GfXgX$ is false,
D_4 is falsified if, in addition, $GfXf^2X$ is true,
and so on.

In general, we have an infinite tree structure such that there is a finite set of nodes falsifying D_1, of which at least one has extensions or nodes on the second level, which falsify D_2. Among the nodes on the second level, i.e., truth-value assignments which falsify D_2, at least one has extensions which falsify D_3, and so on. It then follows by the *Unendlichkeitslemma*[25] that there exists an infinite path, or an infinite truth-value assignment which falsifies D_1, $D_2, \cdots$ simultaneously. Thus, since each node originates only finitely many (possibly zero) immediate branches and there are infinitely many paths, there must be one node a_1 at the first level which occurs in infinitely many paths, and among the finitely many nodes of the second level joined to a_1, there must be at least one node a_2 which occurs in infinitely many paths. This is true for every level, and hence $a_1, a_2, a_3 \cdots$ determines an infinite path. This is no longer true generally when there may be infinitely many branches for a given node. For example, there is no infinite path in a "spread" in which there is one node of the first level (the origin) and a path from it of length n, for every n, all disjoint except for the origin.

An assignment which falsifies D_1, $D_2, \cdots$ simultaneously is a model of the negation of, say, Ex. 2:

$$(X)(EY)GXY \& \sim (EZ)(W)GZW,$$

or $$(X)[(EY)GXY \& (EW) \sim GXW]. \quad \text{(N)}$$

Thus the individuals are X, fX, gX, f^2X, g^2X, fgX, et cetera, and we have found an interpretation of G such that for every individual a, there is an individual b, viz. fa, and an individual c, viz. ga, such that $Gafa \& \sim Gaga$. Hence, in this domain with this G, the negation N of Ex. 2 is true, and Ex. 2 is not valid.

A program for the method Q or the method Q' has not yet been written. It seems clear that certain auxiliary procedures will be useful in reducing the running time and extending the range of application. For example, it seems desirable to separate scopes of different quantifiers when possible, although it is not immediately obvious whether always bringing a sequent into the miniscope form first is feasible on the whole. Other simplifications such as dropping tautologous or repetitive conjunctants could easily and profitably be included. In general, the practical limitation of the machine will necessarily impose certain restrictions on the solvable problems. The machine will have to concede defeat when the running time is too long or the easily available storage is exhausted. When such a situation arises, it seems desirable to try some alternative procedure before giving up the problem entirely.

An intrinsic limitation of the methods Q and Q' is the following. There are various sequents which amount to an axiom of infinity, i.e., a proposition satisfiable in an infinite domain but in no finite domain. If the machine is given the negation of such a sequent, the method Q or the method Q' will never give the desired negative answer since it is, being the negation of an axiom of infinity, valid in every finite domain, though not a theorem of the predicate calculus. Simple examples of this type are:

Ex. 6. $\rightarrow (EX)GXX, (EX)(Y) \sim GXY,$
$(EX)(EY)(EZ)(GXY \& GYZ \& \sim GXZ)$

Ex. 7. $\rightarrow (EX)GXX, (EX)(Y) \sim GXY,$
$(EX)(Y)(EZ)(GYZ \& \sim GXZ).$

This class of propositions may be of special interest if we wish to test whether a formal system is consistent. Most of the interesting formal systems are intended to be satisfiable only in infinite domains. Hence, if the system is consistent, then its negation, though not a theorem of the predicate calculus, is valid in every finite domain. Hence, even theoretically, the methods Q and Q' can at most discover contradictions in an inconsistent formal system but cannot ascertain that an interesting formal system is indeed consistent.

This suggests the desirability of adding special decision procedures which cover some propositions in this class. Such results are rather scarce. The only one seems to be Ackermann's, which is applicable only to a rather special subclass.[26]

A question concerning the efficiency of the methods Q and Q' is the obvious remark that if some of D_1, D_2, $\cdots$ is indeed a tautology, the human being often finds such a disjunction in the sequence without actually examining all the preceding disjunctions. Hence, it may be possible to include suitable strategies for choosing such disjunctions. The difficult problem here is to find suitable strategies of sufficient generality. This is in part related to the larger questions of making use of previously proved theorems. The methods considered in this paper so far all begin from scratch. When we get into more advanced disciplines, it seems unlikely that the machine can feasibly avoid reference to previously proved theorems. Yet there is the analogous situation in ordinary calculations where it is often faster for the machine to calculate known results on the spot rather than look them up in tables stored in some remote corner of the machine. On account of questions of storage and access time, some golden mean has to be struck between the knowledgeable pedant and the prodigy who turns out to be somewhat ignorant.

Conclusions

The original aim of the writer was to take mathematical textbooks such as Landau on the number system,[27] Hardy-Wright on number theory,[28] Hardy on the calculus,[29] Veblen-Young on projective geometry,[30] the volumes by Bourbaki, as outlines and to make the machine formalize all the proofs (fill in the gaps). The purpose of this paper is to report work done recently on the underlying logic, as a preliminary to that project.

The restricted objective has been met by a running program for the propositional calculus and a considerable portion of the predicate calculus. Methods for dealing with the whole predicate calculus by machine have been described fairly exactly. A summary of results and a comparison with previous work in this field were given in the introductory section and will not be repeated here.

The writer sees the main interest of the work reported here, not so much in getting a few specific results which in some ways are stronger than expected (e.g., the fast speed attained and the relatively small storage needed), as in illustrating the great potentiality of machines in an apparently wide area of research and development. Various problems of the same type come to mind.

Decision procedures for the intuitionistic and modal propositional calculi are available but often too lengthy to be done by hand.[31] It seems possible and desirable to code these procedures in a manner similar to the classical systems of logic. The intuitionistic predicate calculus with its decidable subdomains, such as all those propositions which are in the prenex form, may also be susceptible to analogous treatment. Since the efficiency of the proof-decision procedure in Program I depends on the elimination of *modus ponens* (rule of detachment), a related question of logic is to devise cut-free systems for various partial and alternative systems of the propositional calculus.

A good deal of work has been spent in constructing various systems of the propositional calculus and of modal logic. The questions of completeness and independence are often settled by methods which are largely mechanizable and even of no great complexity. This suggests that many of the results in this area, such as those reported by Prior,[32] can be obtained by mechanical means. Given a system, in order to determine the independence and completeness (i.e., nonindependence of all axioms of some given complete system) of its axioms, we may simultaneously grind out proofs and matrices used for independence proofs and stop when we have either obtained a derivation or a matrix that establishes the independence of the formula under consideration. It is true that Linial and Post[33] have proved the undecidability of this class of problems so that we cannot be sure that we can always settle the particular question in each case. Nonetheless, we may expect this procedure to work in a large number of cases. The only practical difficulty is that, in grinding out proofs, the rules of *modus ponens* makes the matter rather unwieldy. When equivalent cut-free formulations are available, this mechanical aid to such simple mathematical research would become more feasible. Alternatively, the strategies devised by Newell-Shaw-Simon may find here a less wasteful place of application.

A mathematically more interesting project is to have machines develop some easy number theory. Here there are two possible alternative approaches: use quantifiers or avoid quantifiers. It is known in mathematical logic that ordinary number theory can be developed largely without appeal to quantifiers. Thus, from the discussions in the body of the paper, it is clear that quantifiers serve essentially to replace an indeterminate class of function symbols. In number theory, these function symbols can usually be replaced by specific function symbols introduced by recursive definitions. Since these are more specific and often intuitively more familiar, it seems quite plausible that avoiding quantifiers would be an advantage. On the other hand, it may be better to use existential quantifiers but avoid mixing quantifiers of both kinds ("all" and "some"), since that is the main source of the complexity of the predicate calculus.

If one wishes to prove that the square root of 2 is not a rational number, this can be stated in the free-variable form as: $2Y^2 \neq X^2$, and a proof can be written out without use of quantifiers. On the other hand, if one wishes to prove that there are infinitely many primes, it seems natural to state the theorem as:

$(EX)(Y < X \& X$ is a prime).

Essentially, the usual proof gives us a simple recursive function f, such that

$Y < fY \& fY$ *is a prime*

is true. But before we get the proof and the required function, it is convenient to use the quantifier (EX) which serves to express the problem that a yet unknown function is being sought for.

In this connection, it may be of interest to make a few general remarks on the nature of expansive features in different proof procedures. The attractive feature of the system P as a proof procedure is that, given a sequent, all the lines in a proof for it are essentially parts of the sequent. As a result, the task of searching for a proof is restricted in advance so that, at least in theory, we can always decide whether a proof exists or not. This contrasts sharply with those proof procedures for the propositional calculus which make use of the *modus ponens*. There, given q, we wish to search for p, such that p and $p \supset q$ are theorems. There is no restriction on the length and complexity of p. The cut-free formulation achieves a method such that for every proof by the expansive method there is a corresponding proof in this method without expansion, and vice versa.

Since there is no decision procedure for the predicate calculus or current number theory, it follows that expansive features cannot be eliminated entirely from these disciplines. The cut-free formulation for the predicate calculus concentrates the expansive feature in one type of situation: viz., a conclusion $(EX)FX$ may come from $F1$ or $F2$ or et cetera. The method Q given above further throws together all such expansions for a given sequent to be proved or disproved at the end of the process. These devices have the advantage that, for more efficient partial methods or strategies, one may direct the search mainly to one specific region which contains the chief source of expansion.

If number theory is developed with no appeal to quantifiers, the above type of expansion is avoided. It is not possible, however, to avoid in general another type of expansion. Thus we can conclude $X=Y$ from $fX=fY$, but given X and Y, there are in general infinitely many candidates for the function f, so that trying to find an f which leads to $X=Y$ through $fX=fY$ is an expansive procedure. So much for different expansive features.

Other possibilities are set theory and the theory of functions. In these cases, it seems desirable to use a many-sorted predicate calculus[34] as the underlying logic. While this is in theory not necessary, it will presumably make for higher efficiency.

As is well known, all standard formal systems can be formulated within the framework of the predicate calculus. In general, if a theorem p is derived from the axioms $A_1, \cdots, A_n$, then the sequent $A_1, \cdots, A_n \rightarrow p$ is a theorem of the predicate calculus. In particular, if a system with finitely many axioms is inconsistent, the negation of the conjunction of all its axioms is a theorem of the predicate calculus. (The restriction on finitely many axioms is, incidentally, not essential since in most cases we can reformulate a formal system to use only finitely many axioms, with substantially the same theorems.) Specker has proved[35] that Quine's *New Foundations* plus the axiom of choice is inconsistent. Hence, the negation of the conjunction of these (finitely many) axioms is a theorem of the predicate calculus. If a sufficiently efficient program for the predicate calculus on a sufficiently large machine yields, unaided, a proof of this, we would be encouraged to try to see whether the system without the axiom of choice might also be inconsistent. If a system is indeed inconsistent, then there would be a chance that a proof of this fact can be achieved first by a machine.

So far little is said about specific strategies. In number theory, we are often faced with the problem of choosing a formula to make induction on. Here an obvious strategy would be to try first to use as the induction formula the whole conclusion, and then the various subformulae of the conclusion to be established. When faced with a conclusion $(EX)FX$, it seems usually advantageous to try terms occurring elsewhere in the known part of the proof, or their variants, in order to find α such that $F\alpha$. Polya's book[36] contains various suggestions on strategies for developing number theory which will presumably be useful when one gets deeper into the project of mechanizing number theory. Efficient auxiliary procedures such as the one already mentioned by Dunham-Fridshal-Sward for the propositional calculus will undoubtedly be of use in shortening running time, when one tries to formalize proofs or prove theorems in more advanced domains.

While formalizing known or conjectured proofs and proving new theorems are intimately related, it is reasonable to suppose that the first type of problem is much easier for the machine. That is why the writer believes that machines may become of practical use more quickly for mathematical research, not by proving new theorems, but by formalizing and checking outlines of proofs. This proof formalization could be developed, say, from textbooks to detailed formulations more rigorous than *Principia,* from technical papers to textbooks, or from abstracts to technical papers.

The selection of interesting conjectures or theorems and useful definitions is less easily mechanizable. For example, Program II described above gives only very crude results. It should be of interest to try to get better results along the same line. In more advanced domains, however, the question seems to have a complexity of a different order.

If we use a machine to grind out a large mass of proofs, then there seems to be some mechanical test as to the importance and centrality of concepts and theorems. If a same theorem or a same expression occurs frequently, then we may wish to consider the theorem interesting or introduce a definition for the expression. This is, however, a rather slippery criterion. The finite number of proofs printed out at a given time may form a class that is determined on the ground of some formal characteristic of an accidental nature. Unless there is some acceptable norm in advance for ordering the proofs to be obtained, one can hardly justify in this way the claim that certain theorems are interesting.

A more stable criterion may be this: A formula which is short but can be proved only by long proofs is a "deep" theorem. A short expression which is equivalent only to very long expressions is a "rich" concept.

In the normal situations, of course, we have less restricted objective guidance. There is a fixed body of concepts and theorems which is for good reasons regarded as of special interest (the "archive of mathematical knowledge built up by the cumulative effort of the human intellect"). For such a body it is theoretically possible to select important theorems and concepts mechanically, as well as to find elegant alternative proofs. However, even in this case, one is looking backwards. It is not easy to find a forward-looking mechanizable criterion for mathematical centrality. For example, the nice criterion of ranges of application is hard to render articulate.

In one special kind of mathematics, one discipline is developed from another. For example, theories of natural numbers and real numbers can be developed from set theory. If theorems are generated mechanically from set theory, then any set of theorems isomorphic with the axioms for real numbers (or natural numbers) determines expressions which may be taken as definitions for the basic concepts of the theory of real numbers (or natural numbers). In such a case, one can claim that machines can discover definitions too.

It has often been remarked that the machine can do only what it is told. While this is true, one might be misled by an ambiguity. Thus the machine can be told to make a calculation, find a proof, or choose a "deep" theorem, et cetera. The main problem of using rather than building machines is undoubtedly to express more things in mechanical terms.

The limitation of machines has been seen as revealed by its inability to write love letters. That depends on the quality of the love letters to be composed. If one takes the common sort of love letter taught in manuals of effective letter-writing, the machine can certainly write some useful love letters more quickly than it can prove an interesting theorem. If the image of Don Juan in some films is to be believed, the machine can surely be taught to repeat the few sentences of flattery to every woman.

If experimenting with a machine to see what it can do is compared with the usual type of scientific research, it seems more like engineering than physics, in so far as we are not dealing with natural objects but man-made gadgets, and we are applying rather than discovering theories. On the other hand, calculating machines are rather unique among man-made things in that their potentialities are far less clear to the maker than are other gadgets. In trying to determine what a machine can do, we are faced with almost the same kind of problem as in animal or human psychology. Or, to quote Dunham, we are almost trying to find out what a machine is.

The suspiciously aggressive term "mechanical mathematics" is not unattractive to a mathematical logician. A common complaint among mathematicians is that logicians, when engaged in formalization, are largely concerned with pointless hairsplitting. It is sufficient to know that proofs can be formalized. Why should one take all the trouble to show exactly how such formalizations are to be done, or even to carry out actual formalizations? Logicians are often hard put to give a very convincing justification of their occupation and preoccupation. One lame excuse which can be offered is that they are of such a temperament as to wish to tabulate all scores of all baseball players just to have a complete record in the archives. However, the machines seem to supply, more or less after the event, one good reason for formalization. While many mathematicians have never learned the predicate calculus, it seems hardly possible for the machine to do much mathematics without first dealing with the underlying logic in some explicit manner. While the human being gets bored and confused with too much rigour and rigidity, the machine requires entirely explicit instructions.

It seems as though logicians had worked with the fiction of man as a persistent and unimaginative beast who can only follow rules blindly, and then the fiction found its incarnation in the machine. Hence, the striving for inhuman exactness is not pointless, senseless, but gets direction and justification. One may even claim that a new life is given to the Hilbert program of the *Entscheidungsproblem* which von Neumann thought was thoroughly shattered by Gödel's discoveries. Although a universal decision procedure for all mathematical problems is not possible, formalization does seem to promise that machines will do a major portion of the work that takes up the time of research mathematicians today.

Note added in proof: Recently the writer has succeeded by improved methods to have the machine prove the 158 theorems of *9 to *13 in *Principia* in about four minutes. This line includes the time needed for writing tapes but uses the off-line printer. The output is about 47 pages of 60 lines each. (November 10, 1959)

Appendix I: A sample from print-outs by Program I

```
2*45/FDPQ-FP                1
1/-FP, DPQ                  2
2/P-DPQ                     3
3/P-P, Q                    4
   VALID                    4
                      QED

5*21/-ICFP.FQ..BPQ          1
1/CFP.FQ-BPQ                2
2/FP, FQ-BPQ                3
3/FQ-BPQ, P                 4
4/-BPQ, P, Q                5
5/P-Q, P, Q                 6
   VALID                    6
5/Q-P, P, Q                 7
   VALID                    7
                      QED

2*31/DP.DQR-DPQ, R          1
1/P-DPQ, R                  2
2/P-P, Q, R                 3
   VALID                    3
1/DQR-DPQ, R                4
4/Q-DPQ, R                  5
5/Q-P, Q, R                 6
   VALID                    6
4/R-DPQ, R                  7
7/R-P, Q, R                 8
   VALID                    8
                      QED

4*45/-BP..CP.DPQ            1
1/P-CP.DPQ                  2
2/P-P                       3
   VALID                    3
2/P-DPQ                     4
4/P-P, Q                    5
   VALID                    5
1/CP.DPQ-P                  6
6/P, DPQ-P                  7
7/P, P-P                    8
   VALID                    8
7/P, Q-P                    9
   VALID                    9
                      QED

5*22/-BFBPQ...DCP.FQ..CQ.FP    1
                      QED

5*23/-BBPQ...DCPQ..CFP.FQ      1
                      QED

5*24/-BFDCPQ..CFP.FQ...DCP.FQ..CQ.FP  1
                      QED

7. PRELIMINARY TO PREDICATE CALCULUS  1
7*1/IG2.H2, GX, KX-CH3.K3      1
1/H2, GX, KX-CH3.K3            2
2/H2, GX, KX-H3                3
   NOT VALID                   3
2/H2, GX, KX-K3                4
   NOT VALID                   4
1/GX, KX-CH3.K3, G2            5
5/GX, KX-H3, G2                6
   NOT VALID                   6
5/GX, KX-K3, G2                7
   NOT VALID                   7
                      NOT VALID
```

Appendix II. Sample from print-outs by Program II

/–BDPR.R, CDPQ.P
/CCPQ.R–BIPR.P
/CCPQ.P–BIPR.R
/CBPQ.Q–CBPP.Q
/CCPP.Q–CIPP.Q
/CCPP.R–CIPP.P
/CDPP.R–CDPP.P
/BBPP.P–DCPQ.P
/DCPQ.P–BBPP.P
/BIPP.P–DDPP.P
/DDPP.P–BIPP.P
/CIPR.P–BDPQ.P
/BIPQ.P–CDPR.P
/CCPR.R–BIPQ.Q
/CCPR.Q–BIPQ.R
/CCPR.P–BBPR.P
/CBPR.P–BCPR.P
/CIPQ.P–BDPR.P
/BIPR.P–CDPQ.P
/CCPQ.R–BIPR.Q
/CCPQ.Q–BIPR.R
/CCPQ.P–CBPP.P
/CBPQ.P–CCPP.P
/CDPP.P–CIPP.P
/CIPP.P–CDPP.P
/BBPP.R–DBPQ.R
/BCPP.Q–DBPQ.P
/–BDPP.R, DDPP.R
/BIPP.P–DDPP.Q
/BIPP.R–DCPP.R
/BBPQ.P–DCPP.Q
/BBPQ.Q–DCPP.P
/DCPP.P–BBPQ.Q
/–BDPQ.P, CIPR.Q
/BIPQ.P–CDPR.Q
/CDPR.P–BIPQ.Q
/CCPR.Q–BBPR.P
/CBPR.R–BBPR.R
/CBPR.Q–BCPR.P
/–BCPR.R, CIPQ.R
/CIPQ.P–BDPR.Q
/CDPQ.R–BDPR.R
/CCPQ.R–BIPR.R
/CCPQ.Q–CBPP.P
/CCPQ.P–CBPP.Q
/CBPQ.R–CBPP.R
/CBPQ.Q–CCPP.P
/CCPP.Q–CBPQ.P
/CBPQ.P–CCPP.Q
/CCPP.R–CIPP.R
/CDPP.Q–CIPP.P
/BBPP.Q–DCPQ.Q
/DCPQ.Q–BBPP.Q
/BCPP.Q–DBPQ.Q
/BIPP.P–DDPP.R
/BIPP.Q–DDPP.Q
/BBPQ.Q–DCPP.Q
/CIPR.Q–BDPQ.Q
/CIPR.P–BDPQ.R
/CDPR.Q–BIPQ.Q
/CCPR.R–BBPR.P
/CCPR.Q–BBPR.Q
/CCPR.P–BBPR.R
/CBPR.R–BCPR.P
/CBPR.P–BCPR.R
/–BDPR.P, CIPQ.R
/BIPR.P–CDPQ.R
/CDPQ.P–BIPR.R
/CCPQ.R–CBPP.P
/CCPQ.Q–CBPP.Q
/CCPP.Q–CBPQ.Q
/CBPQ.Q–CCPP.Q
/CDPP.Q–CIPP.Q
/CDPP.R–CIPP.P
/BBPP.P–DDPQ.P
/BCPP.Q–DBPQ.R
/BBPQ.Q–DCPP.R
/–BCPQ.P, DCPP.P
/CCPR.Q–BBPR.R
/CCPR.P–BCPR.P
/CBPR.Q–BCPR.R
/CBPR.P–BDPR.P
/CBPP.P–CDPQ.P
/CDPQ.P–CBPP.P
/CCPQ.R–CBPP.Q
/CCPQ.P–CCPP.P
/CBPQ.P–CDPP.P
/BBPP.P–DDPQ.Q
/BBPP.Q–DDPQ.Q
/BBPP.R–DCPQ.R
/BDPP.Q–DBPQ.P
/BIPP.R–DDPP.R
/BBPQ.P–DDPP.Q
/BBPQ.Q–DDPP.P
/DDPP.P–BBPQ.Q
/–BCPQ.P, DCPP.Q
/DCPP.P–BCPQ.Q
/CIPR.P–BIPQ.Q
/CCPR.R–BBPR.R
/CCPR.Q–BCPR.P
/CBPR.R–BCPR.R
/CBPR.Q–BDPR.P
/CIPQ.R–BDPR.R
/CDPQ.R–BIPR.R
/CCPQ.R–CBPP.R
/CCPQ.Q–CCPP.P
/CCPP.Q–CCPQ.P
/CCPQ.P–CCPP.Q
/CBPQ.Q–CDPP.P
/CDPP.Q–CBPQ.P
/CBPQ.P–CDPP.Q
/CDPP.R–CIPP.R
/BBPP.P–DDPQ.R
/BBPP.Q–DDPQ.Q
/BDPP.Q–BDPQ.Q
/BBPQ.Q–DDPP.Q
/–BCPQ.P, DCPP.R
/–BCPQ.Q, DCPP.Q
/CIPR.Q–BIPQ.Q
/BBPR.P–CDPR.R
/CDPR.R–BBPR.P

Appendix III: Sample of print-outs by Program III (no quantifiers)

*13. IDENTITY		1
13*1/=XY–IGX.GY		1
1/GX, =XY–GY	=VA	2
QED		
13*12/=XY–BGX.GY		1
1/GX, =XY–GY	=VA	2
1/GY, =XY–GX	=VA	3
QED		
13*13/GX, =XY–GY	=VA	1
QED		
13*14/GX, FGY–F=XY		1
1/GX–F=XY, GY		2
2/=XY, GX–GY	=VA	3
QED		
13*15/–=XX	=VA	1
QED		
13*16/–B=XY.=YX		1
1/=XY–=YX	=VA	2
1/=YX–=XY	=VA	3
QED		
13*17/=XY, =YZ–=XZ	=VA	1
QED		
13*171/=XY, =XZ–=YZ	=VA	1
QED		
13*172/=YX, =ZX–=YZ	=VA	1
QED		
13*18/=XY, F=XZ–F=YZ		1
1/=XY–F=YZ, =XZ		2
2/=YZ, =XY–=XZ	=VA	3
QED		

13*194/-BCGX.=XY...CGX..CGY.=XY		1
1/CGX.=XY-CGX..CGY.=XY		2
2/GX,=XY-CGX..CGQ.=XY		3
3/GX,=XY-GX	VA	4
3/GX,=XY-CGY.=XY		5
5/GX,=XY-GY	=VA	6
5/GX,=XY-=XY	VA	7
1/CGX..CGY.=XY-CGX.=XY		8
8/GX.CGY.=XY-CGX.=XY		9
9/GX,GY,=XY-CGX.=XY		10
10/GX,GY,=XY-GX	VA	11
10/GX,GY,=XY-=XY	VA	12
QED		

13*3/DGY.FGY-BDGX.FGX..D=XY.F=XY		1
1/GY-BDGX.FGX..D=XY.F=XY		2
2/DGX.FGX,GY-D=XY.F=XY		3
3/GX,GY-D=XY.F=XY		4
4/GX,GY-=XY,F=XY		5
5/=XY,GX,GY-=XY	VA	6
3/FGX,GY-D=XY.F=XY		7
7/GY-D=XY.F=XY,GX		8
8/GY-=XY,F=XY,GX		9
9/=XY,GY-=XY,GX	VA	10
2/D=XY.F=XY,GY-DGX.FGX		11
11/=XY,GY-DGX,FGX		12
12/=XY,GY-GX,FGX		13
13/GX,=XY,GY-GX	VA	14
11/F=XY,GY-DGX.FGX		15
15/GY-DGX.FGX,=XY		16
16/GY-GX,FGX,=XY		17
17/GX,GY-GX,=XY	VA	18
1/FGY-BDGX.FGX..D=XY.F=XY		19
19/-BDGX.FGX..D=XY.F=XY,GY		20
20/DGX.FGX-D=XY.F=XY,GY		21
21/GX-D=XY.F=XY,GY		22
22/GX-=XY,F=XY,GY		23
23/=XY,GX-=XY,GY	VA	24
21/FGX-D=XY.F=XY,GY		25
25/-D=XY.F=XY,GY,GX		26
26/-=XY,F=XY,GY,GX		27
27/=XY-=XY,GY,GX	VA	28
20/D=XY.F=XY-DGX.FGX,GY		29
29/=XY-DGX.FGX,GY		30
30/=XY-GX,FGX,GY		31
31/GX,=XY-GX,GY	VA	32
29/F=XY-DGX.FGX,GY		33
33/-DGX,FGX,GY,=XY		34
34/-GX,FGX,GY,=XY		35
35/GX-GX,GY,=XY	VA	36
QED		

Appendix IV: Sample of print-outs by Program III (the *AE* predicate calculus)

10*25/AXGX-EXGX		1
1/G1-EXGX		2
2/G1-G2	NOT	3
2/GS-GS	VA	3
QED		

11*26/EXAYGXY-AYEXGXY		1
1/AYGSY-AYEXGXY		2
2/GS1-AYEXGXY		3
3/GS1-EXGXT		4
4/GS1-G2T	NOT	5
4/GST-GST	VA	5
QED		

11*501/-BEXFAYGXY.EXEYFGXY		1
1/EXFAYGXY-EXEYFGXY		2
2/FAYGSY-EXEYFGXY		3
3/-EXEYFGXY,AYGSY		4
4/-EYFG1Y,AYGSY		5
5/-FG12,AYGSY		6
6/G12-AYGSY		7
7/G12-GST	NOT	8
1/EXEYFGXY-EXFAYGXY		9
9/EYFGUY-EXFAYGXY		10
10/FGUW-EXFAYGXY		11
11/-EXFAYGXY,GUW		12
12/-FAYG3Y,GUW		13
13/AYG3Y-GUW		14
14/G34-GUW	NOT	15
14/GUW-GUW	VA	15
7/GST-GST	VA	8
QED		

13*22/-BEZEWC=ZX..C=WY.GZW...GXY		1
1//EZEWC=ZX..C=WY.GZW-GXY		2
2/EWC=SX..C=WY.GSW-GXY		3
3/C=SX..C=TY.GST-GXY		4
4/=SX.C=TY.GST-GXY		5
5/=SX,=TY,GST-GXY	=VA	6
1/GXY-EZEWC=ZX..C=WY.GZW		7
7/GXY-EWC=1X..C=WY.G1W		8
8/GXY-C=1X..C=2Y.G12		9
9/GXY-=1X	NOT	10
9/GXY-C=2Y.G12		11
11/GXY-=2Y	NOT	12
11/GXY-G12	NOT	13
11/GXY-GXY	VA	13
11/GXY-=YY	=VA	12
9/GXY-=XX	=VA	10
QED		

Appendix V: Different methods for the *AE* predicate calculus

$(EX)(Y)(GXY \vee GYX) \rightarrow (Z)(EW)GZW$ (1)

Method Qp.

(1) $GX1 \vee G1X \rightarrow GZ2$ (2)

(2) $GX1 \rightarrow GZ2$ NOT (3)

(2) $G1X \rightarrow GZ2$ NOT (4)

This is not valid since (3) and (4) are not both valid no matter whether, for (1, 2), we substitute (X, X), (X, Z), (Z, X) or (Z, Z).

Method Qq.

(1) $GX1 \vee G1X \quad GZ2$ (2)

(2) $GXX \vee GXX, GXZ \vee GZX \rightarrow GZX, GZZ$ (3)

(3) $GXX, GXZ \rightarrow GZX, GZZ$ NOT (4)

(3) $GXX, GZX \rightarrow GZX, GZZ$ VA (5)

(3) $GXX, GXZ \rightarrow GZX, GZZ$ NOT (6)

(3) $GXX, GZX \rightarrow GZX, GZZ$ VA (7)

This immediately suggests a simplification since (4) and (6), (5) and (7) are the same. The sequent (1) is not a theorem since (4) and (6) are not valid.

Method Qr. Proceed as in *Qp*, and then test the disjunction of the two conjunctions which is equivalent to the conjunction of (4), (5), (6), (7). Hence, again, (1) is not valid.

Appendix VI: An example of Church

Ex. 3.

$(X)(EY)\{[JX \equiv (JY \supset GY)] \& [GX \equiv (JY \supset HY)] \& [HX \equiv ((JY \supset GY) \supset HY)]\} \rightarrow (Z)(JZ \& GZ \& HZ)$ (1)

This is a sequent in the monadic predicate calculus. It is not of the *AE* form but its miniscope forms must be in the *AE* form. As a result, this can be proved either by *Qp* or by *Q*, though not by *Qq* or *Qr*. It seems more tedious to use *Qp* than *Q*.

To bring (1) into the miniscope form, we have to eliminate the occurrences of $\equiv$. Then we have to bring the quantifier-free part of the antecedent into a disjunctive normal form, distribute (EY) and separate out those parts of the scope of every occurrence of (EY) which contains the variable X. Then we have to bring the new antecedent minus the initial (X) into a conjunctive normal form and distribute (X) in the same way. The reader may wish to convince himself how complex the whole procedure is. The separation of quantifiers in the consequent is, of course, easy. After the separation of quantifiers, the resulting sequent in the miniscope form is very long. The remaining steps, while easy, are also tedious.

On the other hand, if method *Q* is used instead, the proof is not so lengthy. By hand, it is even easier by *Q'*.

(1) $J1 \equiv (Jf1 \supset Gf1), G1 \equiv (Jf1 \supset Hf1),$
$H1 \equiv ((Hf1 \supset Gf1) \supset HF1) \rightarrow JX \& GX \& HX$ (2)

We may substitute for 1, X, fX, f^2X, f^3X, et cetera to get S_1, S_2, S_3, S_4, et cetera and test for validity S_1, $S_1 \vee S_2$, $S_1 \vee S_2 \vee S_3$, $S_1 \vee S_2 \vee S_3 \vee S_4$, et cetera. For this purpose it is sufficient to find a disjunction such that any interpretation (truth-value assignment) which makes the antecedents of all the disjunctants true, will make the consequents of all the disjunctants true. It turns out that the disjunction of S_1 to S_7 does this. If one is doing this by hand, it is easier to argue, for example from $JX \equiv (JfX \supset GfX)$ that if JX is false, then JfX must be true and GfX must be false. In this way, one would see that it is not possible to make all the antecedents of S_1 to S_7 true without also making all their consequents true.

A simple consequence of Ex. 1 may illustrate that sometimes *Qq* is preferable to *Qp*:

Ex. 4.

$(EY)(X)\{[JX \equiv (JY \supset GY)] \& [GX \equiv (JY \supset HY)] \& [HX \equiv ((JY \supset GY) \supset HY)]\} \rightarrow (Z)(JZ \& GZ \& HZ)$.

This is in the *AE* form, though not in the miniscope form. The proof of this is easier than Ex. 3 no matter which method we use. Nonetheless the proof by *Qq* seems considerably shorter than the proof by *Qp* which still includes similar (though somewhat simpler) steps, as with Ex. 3. The proof by *Qq* is quite easy since we need only drop quantifiers, substitute 1 for Z and verify that:

$JY \equiv (JY \supset GY), GY \equiv (JY \supset HY),$
$HY \equiv ((JY \supset GY) \supset HY), JZ \equiv (JY \supset GY),$
$GZ \equiv (JY \supset HY), HZ \equiv ((JY \supset GY) \supset HY) \rightarrow$
$JZ \& GZ \& HZ$.

The verification for this is quite easy. Thus, we wish to show that if all clauses of the antecedent are true, then JZ, GZ, HZ are all true. By the first clause, JY and GY are true. Hence, by the fourth clause, JZ is true; by the second clause, HY is true. Hence, by the fifth and the sixth clauses, GZ and HZ are true.

Appendix VII: An example of Quine

Ex. 5. $\rightarrow (EY)(Z)(EW)\{[GYX \& (GYW \& GWY)] \vee [\sim GYX \& \sim (GYZ \& GZY)]\}$ (1)

(1) $\rightarrow G1X \& (G12 \& G21), \sim G1X \& \sim (G1f1 \& Gf11)$ (2)

(2) $G1X \rightarrow G1X$ (3)

(2) $G1f1, Gf11 \rightarrow G1X$ (4)

(2) $G1X \rightarrow G12$ (5)

(2) $G1X \rightarrow G21$ (6)

(2) $G1f1, Gf11 \rightarrow G12$ (7)

(2) $G1f1, Gf11 \rightarrow G21$ (8)

Now we wish to make substitutions on (3)–(8). Since (3) is valid for all substitutions, it can be omitted. We obtain S_1, S_2, S_3, S_4 et cetera by substituting, for (1, 2), (X, X), (X, fX), (fX, X), (fX, fX), et cetera. In forming each substitution instance, valid sequents and repetitions of a same sequent can be omitted so that, for example, the conjunctive clauses of S_1, S_2, S_3 are as follows.

S_1: $GXfX, GfXX \rightarrow GXX.$

S_2: $GXfX, GfXX \rightarrow GXX; GXX \rightarrow GXfX; GXX \rightarrow GfXX.$

S_3: $GfXf^2X, Gf^2XfX \rightarrow GfXX; GfXX \rightarrow GfXX; GfXX \rightarrow GXfX; GfXf^2X, Gf^2XfX \rightarrow GXfX.$

It can be verified that $S_1 \vee S_2 \vee S_3$ is a tautology. If GXX is true, then S_1 is true. If GXX is false but $GXfX$ or $GfXX$ is false, then S_2 is true. If both $GXfX$ and $GfXX$ are true, then S_3 is true.

While the verification is easy here, it appears that as the number of conjunctions increases, the test for the disjunction of all conjunctions can get mechanically cumbersome. A presumably more manageable method of testing suggests itself.

Make two lists for S_1, one for the antecedents, one for the consequents:

(A_1) $GXfX, GfXX.$

(C_1) $GXX.$

Test whether every string in (C_1) is contained in some string in (A_1). If yes, a proof is obtained. If not, form two lists for S_2:

(A_2) $GXfX, GfXX; GXX.$

(C_2) $GXX; GXfX; GfXX.$

Test whether every string obtained by joining a string of (C_1) to one of (C_2) is contained in every string obtained by combining a string from (A_1) with a string from (A_2). If yes, a proof is obtained. Otherwise, form two lists for S_3 and continue.

It is not hard to convince oneself that this procedure is equivalent to the usual procedure for testing the validity of S_1, $S_1 \vee S_2$, $S_1 \vee S_2 \vee S_3$, et cetera.

References

1. H. Wang, "A Variant to Turing's Theory of Computing Machines," *Journal ACM*, **4**, 88-92 (January 1957).
2. A. W. Burks, D. W. Warren, and J. B. Wright, "An Analysis of a Logical Machine Using Parenthesis-Free Notation," *Mathematical Tables and Other Aids to Computation*, **8**, 53-57 (April, 1954).
3. G. E. Collins, "Tarski's Decision Method for Elementary Algebra," *Proceedings of the Summer Institute of Symbolic Logic at Cornell University*, p. 64 (1957).
4. M. Davis, "A Program for Presburger's Algorithm," *Ibid.*, p. 215.
5. H. Gelernter, "Theorem Proving by Machine," *Ibid.*, p. 305.
6. For example, A. Newell, J. C. Shaw, and H. A. Simon, "Empirical Explorations of the Logic Theory Machine: A Case Study in Heuristics," Report P-951, Rand Corporation, March, 1957, 48 pp.
7. *Ibid.*, p. 26 and p. 28.
8. *Ibid.*, p. 8 and p. 10.
9. J. Herbrand, *Recherches sur la Théorie de la Démonstration*, Traveaux de la Société des Sciences de Varsovie, No. 33, 1930, 128 pp.
10. G. Gentzen, "Untersuchungen über das Logische Schliessen," *Math. Zeitschrift*, **39**, 176-210, 405-431 (1934-35).
11. D. Hilbert and P. Bernays, *Grundlagen der Mathematik*, vol. II, Berlin, 1939
12. B. Dreben, "On the Completeness of Quantification Theory," *Proc. Nat. Acad. Sci. U.S.A.*, **38**, 1047-1052 (1952).
13. E. W. Beth, *La Crise de la Raison et la Logique*, Paris et Louvain, 1957.
14. K. J. J. Hintikka, "Two Papers on Symbolic Logic," *Acta Philos. Fennica*, **8**, 7-55 (1955).
15. K. Schütte, "Ein System des Verknupfenden Schliessens," *Archiv f. Math. Logik u. Grundlagenforschung*, **2**, 375-387 (1955).
16. A. Church, *Introduction to Mathematical Logic, I.* Princeton, 1956.
17. W. V. Quine, *Methods of Logic*, New York, 1950.
18. Herbrand, *op. cit.*, p. 21.
19. Quine, *op. cit.*, pp. 101-107.
20. Church, *op. cit.*, p. 262, 46.12 (3).
21. Compare, e.g., Church, *op. cit.* p. 249.
22. The example in Appendix VII is from Quine, *Mathematical Logic*, *180 and *181, pp. 129-130.
23. Compare Herbrand, *op. cit.*, Ch. V.
24. Compare, e.g., the references to Schütte and Beth.
25. D. König, *Theorie der Graphen*, Leipzig, 1936, p. 81.
26. Church, *op. cit.*, Case X, p. 257.
27. E. Landau, *Grundlagen der Analysis*, Leipzig, 1930.
28. G. H. Hardy and E. M. Wright, *Introduction to the Theory of Numbers*, Oxford, 1954.
29. G. H. Hardy, *A Course of Pure Mathematics*, various editions.
30. O. Veblen and J. W. Young, *Projective Geometry*, 1910.
31. See, e.g., G. Kreisel and H. Putnam, "Ein Unableitbarsbeweismethode," *Arkiv f. Math. Logik u. Grundlagenforschung*, **3**, 74-78 (1957).
32. A. N. Prior, *Formal Logic*, Oxford, 1954.
33. S. Linial and E. L. Post, "Recursive Unsolvability of Axioms Problems of the Propositional Calculus," *Bull. Am. Math. Soc.*, **55**, 50 (1949).
34. See, e.g., Church, *op. cit.*, p. 339.
35. E. P. Specker, "The Axiom of Choice in Quine's New Foundations for Mathematical Logic," *Proc. Nat. Acad. Sci. U.S.A.* **39**, 972-975 (1953).
36. G. Polya, *Mathematics and Plausible Reasoning*, Oxford, 1954.

Received December 22, 1958

1962

A Machine Program for Theorem Proving

M. Davis, G. Logemann, D. Loveland

The programming of a proof procedure is discussed in connection with trial runs and possible improvements.

In [1] is set forth an algorithm for proving theorems of quantification theory which is an improvement in certain respects over previously available algorithms such as that of [2]. The present paper deals with the programming of the algorithm of [1] for the New York University, Institute of Mathematical Sciences' IBM 704 computer, with some modifications in the algorithm suggested by this work, with the results obtained using the completed algorithm. Familiarity with [1] is assumed throughout.

Changes in the Algorithm and Programming Techniques Used

The algorithm of [1] consists of two interlocking parts. The first part, called the *QFl-Generator*, generates (from the formula whose proof is being attempted) a growing propositional calculus formula in conjunctive normal form, the "quantifier-free lines." The second part, the *Processor*, tests, at regular stages in its "growth," the consistency of this propositional calculus formula. An inconsistent set of quantifier-free lines constitutes a proof of the original formula.

The algorithm of [1] used in testing for consistency proceeded by successive elimination of atomic formulas, first eliminating *one-literal* clauses (one-literal clause rule), and then atomic formulas all of whose occurrences were positive or all of whose occurrences were negative (affirmative-negative rule). Finally, the remaining atomic formulas were to have been eliminated by the rule:

III. *Rule for Eliminating Atomic Formulas.* Let the given formula F be put into the form

$$(A \vee p) \ \& \ (B \vee \bar{p}) \ \& \ R$$

where A, B, and R are free of p. (This can be done simply by grouping together the clauses containing p and "factoring out" occurrences of p to obtain A, grouping the clauses containing $\bar{p}$ and "factoring out" $\bar{p}$ to obtain B, and grouping the remaining clauses to obtain R.) Then F is inconsistent if and only if $(A \vee B) \ \& \ R$ is inconsistent.

After programming the algorithm using this form of Rule III, it was decided to replace it by the following rule:

III*. *Splitting Rule.* Let the given formula F be put in the form

$$(A \vee p) \ \& \ (B \vee \bar{p}) \ \& \ R$$

where A, B, and R are free of p. Then F is inconsistent if and only if $A \ \& \ R$ and $B \ \& \ R$ are both inconsistent.

Justification of Rule III*. For[1] $p = 0$, $F = A \& R$; for $p = 1$, $F = B \ \& \ R$.

The forms of Rule III are interchangeable; although theoretically they are equivalent, in actual applications each has certain desirable features. We used Rule III* because of the fact that Rule III can easily increase the number and the lengths of the clauses in the expression rather quickly after several applications. This is prohibitive in a computer if ones fast access storage is limited. Also, it was observed that after performing Rule III, many duplicated and thus redundant clauses were present. Some success was obtained by causing the machine to systematically eliminate the redundancy; but the problem of total length increasing rapidly still remained when more complicated problems were attempted. Also use of Rule III can seldom yield new one-literal clauses, whereas use of Rule III* often will.

In programming Rule III*, we used auxiliary tape storage. The rest of the testing for consistency is carried out using only fast access storage. When the "Splitting Rule" is used one of the two formulas resulting is placed on tape. Tape memory records are organized in the cafeterial stack-of-plates scheme: the last record written is the first to be read.

In the program written for the IBM 704, the matrix and conjunction of quantifier-free lines are coded into cross-referenced associated (or linked) memory tables by the QFL-Generator and then analyzed by the Processor. In particular, the QFL-Generator is programmed to read in the matrix M in suitably coded Polish (i.e., "parenthesis-free") form. The conversion to a quantifier-free matrix in conjunctive normal form requires, of course, a certain amount of pencil work on the formula, which could have been done by the computer. In doing this, we departed from [1], by not using prenex normal form. The steps are:

(1) Write all truth-functional connectives in terms of $\sim$, &, $\vee$.

(2) Move all $\sim$'s inward successively (using de Morgan laws) until they either are cancelled (with another $\sim$) or acting on an atomic formula.

(3) Now, replace existential quantifiers by function symbols (cf. [1], p. 205), drop universal quantifiers, and place in conjunctive normal form. A simple one-to-one assembler was written to perform the final translation of the matrix M into octal numbers.

It will be recalled that the generation of quantifier-free lines is accomplished by successive substitutions of "constants" for the variables in the matrix M. In the program

The research reported in this document has been sponsored by the Mathematical Sciences Directorate, Air Force Office of Scientific Research, under Contract No. AF 49(638)-777.

[1] As in [1], 1 stands for "truth", and 0 for "falsehood".

Article from: Communications of the Association for Computing Machinery, Vol. 5, No. 7, July 1962.

the constants are represented by the successive positive integers.

For a matrix containing n individual variables, the n-tuples of positive integers are generated in a sequence of increasing norm such that all n-tuples with a given norm are in decreasing n-ary numerical order. Here we define the norm of $(j_1, \cdots, j_n) = j_1 + \cdots + j_n = \|j_i\|$. Other norms could have been used. For example, Gilmore [2] takes for $\|j_i\|$ the maximum of $j_i, \cdots, j_n$. In [1] a more complicated norm is indicated.

Substitutions of successive n-tuples into the matrix cause new constants to appear in the matrix. The program numbers constants in their order of appearance. Thus, the constants are ordered by the program in a manner depending upon the input data. By rearranging the clauses of a formula a different order would in general be created. In some cases, whether or not the program could actually prove the validity of a given formula (without running out of fast access storage) depended on how one shuffled the punched-data deck before reading it into the assembler! Thus, the variation in ordering of constants did affect by a factor of 10 (from 500 to 5000) the number of lines needed to prove the validity of:

$$(e)(Ed)(x)(y)[S(x, y, d) \rightarrow T(x, y, e)]$$
$$\rightarrow (e)(x)(Ed)(y)[S(x, y, d) \rightarrow T(x, y, e)]$$

(This valid formula may be thought of as asserting that uniform continuity implies continuity if we set:

$$S(x, y, d) \leftrightarrow |x - y| < d$$
$$T(x, y, e) \leftrightarrow |f(x) - f(y)| < e.)$$

In storing the quantifier-free lines, two tables are used. The first, called the *conjunction table*, is a literal image of the quantifier-free lines in which one machine word corresponds to one literal, i.e., to p or $\sim p$ where p is an atomic formula. The lines in the second, or *formula table* are themselves heads of two chain lists giving the occurrences of p and $\sim p$ respectively in the conjunction table. In addition, included for formula p in the formula table are counts of the number of clauses in which p and $\sim p$ occur and total number of all literals in these clauses; the formula table is itself a two-way linked list. A third short list of those literals is kept in which are entered all formulas to which the one-lateral clause and affirmative-negative rules apply; this is called the *ready list*. If the program tries to enter p and $\sim p$ into the ready list, an inconsistency has been found; the machine stops.

The totality of the processing rules requires only two basic operations: a subroutine to *delete* the occurrences of a literal p or $\sim p$ from the quantifier-free lines, and a routine to *eliminate* from them all the clauses in which p or $\sim p$ occur.

We may observe that only the deletion program can create new one-literal clauses, and likewise applications of the affirmative-negative rule can come only from the elimination program.

The machine thus performs the one literal-clause and affirmative-negative rules as directed by the ready list until the ready list is empty. It is possible that the choice of p to be eliminated first is quite critical in determining the length of computation required to reach a conclusion: a program to choose p is used, but no tests were made to vary this segment of the program beyond a random selection, namely the first entry in the formula table. To perform Rule III*, one saves the appropriate tables with some added reference information in a tape record, then performs an elimination on $\sim p$ and a deletion on p. At a consequent discovery of consistency, one must generate more quantifier free lines; the QFL-generator is recalled. Otherwise, at finding an inconsistency, the machine must check to see if there are any records on the Rule III* tape: if none, the quantifier-free lines were inconsistent; otherwise, it reads in the last record.

If one uses Rule III (which we did in an early version of our program), an entirely different code is needed. The problem is precisely that of mechanizing the application of the distributive law.

Results Obtained in Running the Program

At the time the programming of the algorithm was undertaken, we hoped that some mathematically meaningful and, perhaps nontrivial, theorems could be proved. The actual achievements in this direction were somewhat disappointing. However, the program performed as well as expected on the simple predicate calculus formulas offered as fare for a previous proof procedure program. (See Gilmore [1].) In particular, the well-formed formula

$$(Ex)(Ey)(z)\{F(x, y) \rightarrow (F(y, z) \;\&\; F(z, z)))$$
$$\&\; ((F(x, y) \;\&\; G(x, y)) \rightarrow (G(x, z) \;\&\; G(z, z)))\}$$

which was beyond the scope of Gilmore's program was proved in under two minutes with the present program. Gilmore's program was halted at the end of 7 "substitutions", (quantifier-free lines) after an elapsed period of about 21 minutes. It was necessary for the present program to generate approximately 60 quantifier-free lines before the inconsistency appeared.[2] Indeed, the "uniform continuity implies continuity" example mentioned above required over 500 quantifier-free lines to be generated and was shown to be valid in just over two minutes. This was accomplished by nearly filling the machine to capacity with generated quantifier-free lines (2000 lines in this case) before applying any of the rules of reduction.

Rather than describe further successes of the program, it will be instructive to consider in detail a theorem that the program was incapable of proving and to examine the cause for this. This particular example is one the authors originally had hoped the program could prove, an elementary group theory problem. In essence, it is to show that in a group a left inverse is also a right inverse.

[2] In [1], a hand calculation of this example using the present scheme showed inconsistency at 25 quantifier-free lines. The discrepancy is due to a different rule for generation of constants.

It is, in fact, quite easy to follow the behavior of the proof procedure on this particular example as it parallels the usual approach to the problem. The problem may be formulated as follows:

Axioms: 1. $e \cdot x = x$

2. $I(x) \cdot x = e$

3. $(x \cdot y) \cdot z = w \Rightarrow x \cdot (y \cdot z) = w$

4. $x \cdot (y \cdot z) = w \Rightarrow (x \cdot y) \cdot z = w$

Conclusion: $x \cdot I(x) = e$

The letter e is interpreted as the identity element and the function I as the inverse function. The associative law has been split into two clauses for convenience.

A proof is as follows:

1. $I(I(x)) \cdot I(x) = e$ by Axiom 2
2. $e \cdot x = x$ by Axiom 1
3. $I(x) \cdot x = e$ by Axiom 2
4. $I(I(x)) \cdot e = x$ by Axiom 3, taking $(I(I(x)), I(x), x)$ for (x, y, z)
5. $e \cdot I(x) = I(x)$ by Axiom 1
6. $I(I(x)) \cdot I(x) = e$ by Axiom 2
7. $I(I(x)) \cdot e = x$ step 4
8. $x \cdot I(x) = e$ by Axiom 4, taking $(I(I(x)), e, I(x))$ for (x, y, z)

Step 8 is the desired result.

To formalize this proof would require adjoining axioms of equality. To avoid this, one can introduce the predicate of three arguments $P(x, y, z)$, interpreted as $x \cdot y = z$. The theorem reformulated becomes:

Axioms: 1. $P(e, x, x)$

2. $P(I(x), x, e)$

3. $\sim P(x, y, u) \vee \sim P(u, z, w) \vee \sim P(y, z, v) \vee P(x, v, w)$

4. $\sim P(y, z, v) \vee \sim P(x, v, w) \vee \sim P(x, y, u) \vee P(u, z, w)$

Conclusion: $P(x, I(x), e)$.

The theorem to be proved valid is the implication of the conjunction of the four axioms with the conclusion, the universal quantifiers appearing outside the matrix.

To complete the preparation of the well-formed formula for encoding for the assembler, it is necessary to negate the conclusion. (cf. [1], p. 204.)

The single existential quantifier has no dependence on the universal quantifiers, hence leads to the constant function s when this existential quantifier is replaced by a function symbol. (cf. [1], p. 205.]

The conclusion then becomes

$$\sim P(s, I(s), e).$$

The conjunction of this with the four axioms gives the desired form.

As seen from the proof previously noted the quantifier-free clauses needed to produce the inconsistency are

1. $P(I(I(s)), I(s), e)$
2. $P(e, s, s)$
3. $P(I(s), s, e)$
4. $\sim P(I(I(s)), I(s), e) \vee \sim P(e, s, s) \vee \sim P(I(s), s, e) \vee P(I(I(s)), e, s)$
5. $P(e, I(s), I(s))$
6. $\sim P(e, I(s), I(s)) \vee \sim P(I(I(s)), I(s), e) \vee \sim P(I(I(s), e, s) \vee P(s, I(s), e)$
7. $\sim P(s, I(s), e)$

(It is quite clear in this case that successive applications of the one-literal clause rule reducing this set to

$$P(s, I(s), e) \& \sim P(s, (I(s), e).)$$

The question to be considered is: how many quantifier-free lines must be generated by the present program to realize these required lines? The constants are generated in the following:

1. e
2. s
3. $I(s)$
4. $I(e)$
5. $I(I(s))$

etc.

(The constants are identified directly with their index e.g. the 6-tuple (1, 1, 1, 1, 1, 1) represents (e, e, e, e, e, e). As this is the first substitution, the program assigns in order, reading the well-formed formula backwards and from the inside out for nesting functions: e, s, $I(s)$, $I(e)$, $I(I(s))$. The $I(I(s))$ appears when x is assigned $I(s)$, no new entries occurring until this time. Note that there are 6 free variables (u, v, w, x, y, z) in the matrix).

The program generates the needed n-tuples by producing all possible n-tuples of integers whose sum N of entries is fixed, $N = n, n+1, \cdots$. Thus it is only necessary to consider the n-tuple which has the maximum sum of entries. In this case, the substitution $u = s$, $v = I(s)$, $w = e$, $x = I(I(s))$, $y = e$, $z = I(s)$ (required for axiom 4 to produce the clause 6 in the "proof" above in a quantifier-free line) gives the n-tuple with maximum sum. The n-tuple is seen to be (2, 3, 1, 5, 1, 3), the sum equals 15. The combinatorial expression $\binom{N}{n}$ gives the total number of n-tuples of positive integers whose sum is less than or equal[3] to N.

[3] To see this, consider a sequence of $N+1$ ones. Flag n of these. The flag is to be interpreted "sum all 1's, including the flagged 1, to the next flag and consider this sum as an entry in the n-tuple". Placing an unflagged 1 on the extreme left, leaving it fixed, consider the possible permutations of all other symbols. The different sequences total $\binom{N}{n}$ and, regarding the set of 1's starting with the last flagged 1 as overflow, this is seen to represent precisely the desired n-tuples.

It is seen that to prove this theorem at least $\binom{14}{6} = 3003$ lines must be generated and that the inconsistency will be found on or before $\binom{15}{6} = 5005$ lines have been generated.[4]

The present program generated approximately 1300 quantifier-free lines. This number of quantifier-free lines was accomplished holding all major tables simultaneously in core memory, limited to 32,768 "words". (This was done to insure a reasonable time factor for any problem, within possible scope of the program. For this reason also, the entire program was coded in SAP with many time-saving devices employed.)

The authors believe that a reprogramming to make use of tape storage of tables might realize a factor of 4 for the total number of quantifier-free lines attainable before running time became prohibitive. This would be just sufficient for this problem. That realizing this extra capacity is really uninteresting is seen by noting that if the conclusion was placed before the axioms, altering the validity of the matrix not at all, the element $I(e)$ would be generated before $I(s)$ and the needed n-tuple would sum to 16. Then $\binom{16}{6} = 8008$ becomes the upper bound, beyond the capacity of the projected program. Other formulations of the same problem result in quite unapproachable figures for the number of quantifier-free lines needed. (For another example illustrating the same situation, see Prawitz [3].)

The existing program allows one to think of working with a capacity of 1000 or 2000 quantifier-free lines instead of a capacity of 10 or 20, the previous limit. The time required to generate additional quantifier-free lines is independent of the number of quantifier-free lines already existing. Against this linear growth of number of quantifier-free lines generated, there is, in a meaningful sense, an extreme nonlinear growth in the number of quantifier-free lines to be considered with increasingly more "difficult" problems. This is true of simple enumeration schemes of the nature considered here. It seems that the most fruitful future results will come from reducing the number of quantifier-free lines that need be considered, by excluding, in some sense, "irrelevant" quantifier-free lines. Some investigation in this area has already been done (see Prawitz [3]).

[4] If the rule for generating n-tuples had been, for each m, to generate all n-tuples possible such that each entry assumes a positive integral value less than or equal to m, it is clear that at least 4^6 (= 4096) quantifier-free lines would be needed and 5^6 (= 15625) lines would suffice to guarantee a solution. If no more information were available, one sees an intuitive advantage, in this case, for using the previous method. In general, the authors see no preference for either method, in contrast to some previous uggestions that the latter method seemed intuitively better.

REFERENCES

1. Davis, Martin, and Putman, Hilary. A computing procedure for quantification theory. *J. ACM* 7 (1960), 201–215.
2. Gilmore, P. C. A proof method for quantification theory. *IBM J. Res. Dev.* *4* (1960), 28–35.
3. Prawitz, Dag. An improved proof procedure. *Theoria* *26*, 2 (1960), 102–139.

Theorem Testing by Computer

B. Dunham, J. H. North

In order to test whether non-trivial logical formulae are theorems, we shall probably need computer programs which either manage very large truth-functional expressions or else simplify by an efficient look-ahead the Herbrand expansion of a quantificational formula. We have found a distinctive property of valid truth-functional formulae which should help greatly in both connections. The details of this result and some comments on the philosophy of our general approach are presented. In particular, we emphasize the need for ordering the methods of solution because of the large memory requirements.

THE ELIMINATION THEOREM

In two earlier papers,[1,2] we have presented theorem-testing techniques for large truth-functional expressions, having in mind the obvious application to quantificational formulae. In reference 2, we expressed the conviction that advantage could be taken of the *looseness* of the expanding truth-functional formula which results when quantifiers are eliminated by the Herbrand method. *** A valid formula in alternational normal form is loose if only a small percentage of its clauses are essential to validity. It turns out that, in the usual case, the great majority of the redundant clauses can be identified and eliminated.

In reference 3, Quine defined the operation of *consensus* to be one whereby a new clause is formed out of two other clauses which contradict each other at exactly one variable. By this process, for example, the clause 'prs' is obtained from 'pqr' and 'p$\bar{q}$s'. Quine used this process to generate the *prime implicants* of formulae for purposes of minimization; but it is clearly related to the question of validity, since the prime implicants of a valid formula are single-literal

The material in this paper was also presented in a public lecture at Harvard University in February, 1962.

*** For a description of the formation of an Herbrand expansion see B. Dreben, "On the Completeness of Quantification Theory," *Proc. Nat. Acad. Sci., USA*, Vol. 38, pp. 1048-1049 (1952).

clauses such as 'p' and '$\bar{p}$'. Let us suppose we invoke the process of consensus, but instead of forming new clauses, simply circle the contradicting literals of each pair of clauses which would normally interact to form a new clause; for example 'p$\textcircled{q}$r V p$\textcircled{\bar{q}}$s'. The *elimination theorem* is simply that, if we circle all possible literals of a formula ϕ in the above manner, any clause with an uncircled literal can be dropped without loss of validity. Put otherwise:

Every valid formula in alternational normal form contains a closed subset of clauses, each literal of which is circled by interaction with some other literal of that subset.

The converse is not true, as can be seen from the non-valid formula '$p\bar{q} \vee \bar{p}q \vee p\bar{r} \vee \bar{p}r \vee q\bar{r} \vee \bar{q}r$'.

To demonstrate the theorem, let us suppose that we circle all literals of a formula ϕ, and that the fourth clause of ϕ contains an uncircled literal α. We decide to test ϕ for validity by a treeing technique in which we progressively eliminate variables by branching. If β is the variable to be eliminated, we introduce 1's for all occurrences of β on one side and 0's for all occurrences on the other. The usual simplifications involving 1's and 0's are permitted, but no other. If any branch terminates in 0, ϕ is, of course, not valid. We decide that α will be the last variable eliminated. Clearly, the absence of the fourth clause will not affect which branches terminate in 1 before the final level, that is, before α's are eliminated. Nor, in fact, will its absence prevent validity there, since it is impossible that the α from the fourth clause can form an alternation with an $\bar{\alpha}$ from some other clause, say the tenth. If so, clauses four and ten would have interacted originally by consensus and the particular occurrence of α would have been circled.

THE THEOREM IN APPLICATION

There are two obvious modes of application for the elimination theorem. First, it may serve as the basis for a look-ahead mechanism which, by appropriate criticism, eliminates clauses and elects promising lines of attack. Such a look-ahead would most likely be based upon a criterion of probable relevance; and, of course, the theorem is directly pertinent to the relevance of clauses. Much can be said in this connection, but we shall not pursue the matter further in this paper.

A second mode of application is in the trimming down of an already generated expression, which may or may not have received prior criticism and pruning. The problem here, of course, is how one passes from the theorem itself to a reasonably efficient method for eliminating clauses.

Let us suppose we order the variables and then place all of the clauses on a partially ordered list based upon the variables. All clauses with the literal 'p', for example, will precede clauses with the literal '$\bar{p}$'; and these, in turn, will precede all clauses whose earliest literal is 'q', and so forth. Thus each clause will occur initially at only one position on the list. Suppose we then compare each of the clauses containing 'p' against all those containing '$\bar{p}$', circling the appropriate 'p' and '$\bar{p}$' literal occurrences. When this operation is completed, any clause with an uncircled 'p' or '$\bar{p}$' can be dropped at once. The surviving clauses can be managed in one of two different ways. Either they can be moved directly to their next point of occurrence on the list or else they can be kept in temporary cold storage until the appropriate part of the list has been reached, at which time they are inserted. When the entire list has been gone through, presumably some of the clauses will have dropped out and some will remain. We must then continue making passes through the reduced lists until either no clauses remain or else no clauses drop out in a given pass. If the latter, we will have reached the *closed residue* of the expression, which now, because of its smaller size, can be tested directly for validity by such methods as we have set forth in reference 2.

The reader has probably noted one obvious drawback to the above procedure. It is not difficult to obtain the ordered list, nor is it unreasonable that we should make a single pass through it, even though the problem is large. The comparison operation is simple, and the number of such operations is decently related to the number of clauses and the number of variables. For an appropriately selected variable, let k represent the number of undenied occurrence and ℓ the denied occurrences. Then, the total operations might approximate the product of k, ℓ, and n, where n represents the number of variables. The difficulty with the method, as described thus far, is that we must make many passes through the list, and each time we do so, we repeat operations already performed. This drawback can be avoided, however, if suitable records are kept, so that only one pass is necessary.

We thus modify the passage through the list in two ways. First, no clauses are dropped; secondly, a complete record is kept of all the consensus interactions between clauses. The latter information is totally provided, of course, by a single pass through the list, so that the only additional time required to form this *register*, as we shall call it, is for the requisite bookkeeping. Without embarking upon details, we simply note that this bookkeeping has been worked out and will not cause an unreasonable increase in the operating time of the program, particularly if a computer such as the IBM 7090 is available by which tapes can be positioned during computation and which integrates data transmission with computation. After the register has been formed and ordered, if we wish to find out which clauses can be dropped by the progressive elimination procedure, we can do so by actually fewer operations than were necessary to form the register, since, in the formation process, numerous clauses were compared which did not interact. In this elimination, numbers need not be dropped from the register; only the signs of the numbers need be changed. Hence, if we discount the bookkeeping, the total number of operations necessary to form the register and eliminate clauses will be less than twice the product of k, ℓ, and n mentioned earlier. Beyond this, should the truth-functional expression under test prove non-valid so that further clauses must be added, the operations already carried out in forming the register need not be repeated. Only the necessary comparisons involving new clauses must be made. In turn, work done upon closed residues (by the method set forth in reference 2) need not be duplicated when new larger residues are encountered, since the former will, in every case, be contained in the latter. Under these conditions, the test in reference 2 can be carried out without starting from scratch each time. The related problem, of when we should test a register by eliminating clauses, is not critical and we forego discussion of it at the present time.

SOME FURTHER OBSERVATIONS

In the interests of brevity, we have avoided burdening this paper with detail. It seems pertinent, however, to dwell briefly upon our general philosophy of approach, since this has strongly influenced our specific mode of attack. In particular, we feel that extensive use of tapes or other very large (and thus probably slow-access) memories, must be anticipated if the computer is to deal successfully with significant logical and mathematical formulae. We are not very much impressed by counter-arguments based upon success with simple problems. First of all, it is clear that many of the expressions to be tested are per se quite large. Secondly, as we have indicated in this paper and shown somewhat more completely in reference 2, a large-scale duplication of operations can often be avoided by judicious record keeping. Thirdly, with experience and study, our sophistication in handling the strategy of complicated problems should improve greatly. If so, when deciding what to do on a problem, we should be able to take very much better advantage of the intermediate results obtained as the solution progresses. In order to accomplish this, however, we shall need an adequate record of the process to date. For all of these reasons, therefore, we believe that a very large memory system will be required, much beyond the size of contemporary or envisaged fast-access memories.

When we make use of peripheral serial memories, we are defeated at once if we must constantly search for information stored at the wrong end of tapes. Hence, it is of paramount importance that we so order our methods of solution that we can reasonably anticipate the calls we must make upon storage and also store information in a workable order. Therein, apart from the problem of developing improved logical techniques and strategies, lies the whole trick. In this connection, the interested reader may wish to examine the variable elimination method outlined in reference 2, which is a rather detailed illustration of the point.

REFERENCES

1. B. Dunham, R. Fridshal, and G.L. Sward, "A Nonheuristic Program for Proving Elementary Logical Theorems," in *Proc. International Conf. on Information Processing* (Paris: UNESCO, 1959), pp. 282-284.

2. B. Dunham, R. Fridshal, and J.H. North, "Exploratory Mathematics by Machine," in *Recent Developments in Information and Decision Processes* (New York: Macmillan, 1962), pp. 149-160.

3. W.V. Quine, "A Way to Simplify Truth Functions," *American Mathematical Monthly*, Vol. 62 (1955), pp. 627-631.

BIBLIOGRAPHY

Some other papers not already cited, pertinent to the general problems herein discussed are the following.

H. Wang, "Toward Mechanical Mathematics," *IBM J. Res. and Dev.*, Vol. 4, No. 1, pp. 2-22 (January 1960).

H. Wang, "Proving Theorems by Pattern Recognition - I," *Comm. Assoc. Comp. Mach.*, Vol. 3, pp. 220-234 (1960).

H. Wang, "Proving Theorems by Pattern Recognition - II," *Bell Syst. Tech. J.*, Vol. XL, No. 1, pp. 1-41 (January 1961)

P.C. Gilmore, "A Program for the Production of Proofs for Theorems Derivable within the First Order Predicate Calculus from Axioms," *Proc. International Conference on Information Processing* (Paris: UNESCO, 1959), pp. 265-273.

P.C. Gilmore, "A Proof Method for Quantification Theory: Its Justification and Realization," *IBM J. Res. and Dev.*, Vol. 4, No. 1, pp. 28-35 (January 1960).

B. Dreben, "Systematic Treatment of the Decision Problem," *Summaries of Talks at Summer Institute of Symbolic Logic* (Cornell University, 1957), p. 363.

M. Davis and H. Putnam, "A Computing Procedure for Quantification Theory," *J. Assoc. Comp. Mach.*, Vol. 7, pp. 201-215 (1960).

M. Davis, G. Logemann, and D. Loveland, "A Machine Program for Theorem-Proving," submitted to the *Comm. Assoc. Comp. Mach.*

A. Newell, J.C. Shaw, and H.A. Simon, "Empirical Explorations of the Logic Theory Machine; A Case Study in Heuristics," *Proc. Western Joint Computer Conference*, 1957, pp. 218-230.

D. Prawitz, H. Prawitz, and N. Vogera, "A Mechanical Proof Procedure and Its Realization in an Electronic Computer," *J. Assoc. Comp. Mach.*, Vol. 7 (1960), pp. 102-128.

D. Prawitz, "An Improved Proof Procedure," *Theoria.*, Vol. 26, pp. 102-139 (1960).

Exploratory Mathematics by Machine

B. Dunham, R. Fridshal, J. H. North

This paper might better have been titled, "A Small Step Toward Exploratory Mathematics by Machine." Specifically, we present a way for testing the validity of an expanding truth-functional formula in alternational normal form. The relevance of such a test (or of the related consistency test for a formula in conjunctional normal form) to a more general theorem-testing program for logical and mathematical expressions has been extensively argued elsewhere [1, 2, 3]. Nevertheless, we present a few general remarks for the benefit of the reader unacquainted with the problem.

It is well known that elementary mathematical formulas can be reconstructed within the language of symbolic logic. Such a transformation is not accomplished, however, without considerable growth in the size of the formula. It is thus theoretically possible to investigate certain mathematical questions indirectly by testing purely logical formulas, though of a much larger size. The extent to which indirect investigations such as this can have true mathematical interest is not clear at the present time. Nevertheless, the development of large-scale computing machines gives added hope that, in the future, problems whose chief difficulty is sheer size can be managed more successfully. Hence, if effective techniques can be developed for bringing the full power of the computer to bear on purely logical problems, it is conceivable that some nontrivial mathematical questions can, in turn, be resolved. Many problems of interest primarily to the logician might also be solved. Suppose, however, computer techniques for managing purely logical problems prove insufficient, taken by themselves, in the realm of mathematics. Such techniques might still constitute a necessary part of the more expanded methods by which, hopefully, a computer might test significant mathematical formulas for "theoremhood." Hence, we have set ourselves the problem of developing efficient computer programs for testing nontrivial logical expressions.

Quantification theory, sometimes called the first-order predicate calculus, is perhaps the cornerstone within logic. Here a set of formulas is obtained from two basic kinds of logical particle: truth-functional and quantificational. Truth-functional expressions are composed of logical elements such as AND, OR, NOT, NEITHER-NOR. Quantificational expressions combine truth-functional particles with two additional logical operators – namely, the existential and universal quantifiers. The existential quantifier with respect to a variable x may be roughly translated by the phrase, "There is an x such that..."; and the universal quantifier, by the phrase, "For whatever x taken...." Quantifiers govern fragmentary formulas in which the variables they have specified reappear as parts of predicates which are, in turn, brought together truth functionally.

Generally speaking, quantificational languages without certain special restrictions are formally undecidable. This means that, if we are presented with a formula of the language, we have no finite procedure which will tell us in every case whether or not the formula in question is valid within the system. This does not mean, of course, that we cannot prove validity or nonvalidity in many cases. There is a fundamantal result due to Herbrand which enables, in large part, the validity test of a quantificational formula to be reduced to one for a truth-functional expression only. Specifically, he has shown that we may obtain from any quantificational formula an expanding truth-functional expression which at some point in the expansion, usually unknown, will become valid if the quantificational formula is itself valid (see, for example, [4]). Hence, one method for testing a quantificational formula is to examine the associated truth-functional expansion. Since the latter, even when simplified, becomes quite large by comparison with the quantificational formulas, which are, in turn, quite large by comparison with the mathematical expressions they may represent, it is important that computer techniques be available for handling truly vast truth-functional expressions. It is toward the latter objective that we are working in this paper, not only so as to have the techniques, but also as an experiment in how far we can go along these lines.

As noted above, the particular problem we have attacked is that of testing the validity of an expanding truth-functional formula in alternational normal form. An expression is in the latter state if denials are restricted to the individual truth-functional variables, no connectives other than AND and OR are used, and the connective AND never joins subexpressions containing OR. It is possible so to set up the transformation of a quantificational formula into an expanding truth-functional one that the latter will be continually presented in alternational normal form. As we shall see, we can then take good advantage of the normality of the expression.

THE BASIC METHOD

The basic method we propose is similar to the one set forth in our earlier paper with Sward, in that the fundamental technique is one of variable elimination [5]. Advantage is also taken of simplifications resulting from variables in partial state. As noted by Putnam and Davis, simplifications also follow from single literal clauses [3, p. 209]. Programming of the method on the IBM 7090 is not yet completed; but, because of the deadline imposed by the editors, we offer this, perhaps premature, account of it.

The technique set forth by Putnam and Davis,[1] although formidable, to our mind suffers certain disadvantages. First of all, the "consolidation" which takes place after each variable elimination will delay the unmasking of a nonvalid formula [3, pp. 210-211]. Second, after each addition of new clauses by the Herbrand expansion, or otherwise, the total expression must be tested from "scratch," [3, pp. 211-212]. Third, the computer is asked to carry out a variety of truth-functional operations which are probably "slow" computerwise.

We have found it difficult to avoid the above limitations without some loss elsewhere. We hope, however, to have gained by the transaction. Specifically, we eliminate variables throughout the entire solution in a specified order. This means that simplifications resulting from variables in partial state or from single literal clauses must wait until the variable in question is eliminated.

We do not anticipate undue space problems. Because of the generally systematic way in which the problem is handled, we believe the necessary use of tapes can be made on large problems without great loss of time. One final comment might be made before we present a rough statement of the method. Our planned approach is somewhat empirical. Certain questions are left open – for example, how the variables will be ordered initially, or how many clauses will be added between trials. These

---

[1]The problem they actually consider is that of an expression in conjunctional normal form. We have recast the argument to be applicable to the alternational case.

questions can best be resolved by trying out the program on proposed problems this way and that.

Suppose we are presented with the following matrix:

	a_1	a_2	a_3	a_4 . . . .
1	1	X	0	X
2	X	0	[1]	X
3	X	X	X	X
4	0	1	0	1
5	X	1	X	X
⋮				

The horizontal rows represent the clauses of the alternational expression. The columns represent in an obvious manner which variables, denied or undenied, occur in which clauses. The brackets around the positive occurrence of a_3 in clause 2 indicate a *terminal* occurrence, in that no variable occurs therein thereafter. The computer, in effect, passes through an abbreviated version of this matrix from top to bottom, column by column. Sometimes, as we shall see, it must return to a point already passed, but in a different connection.

Suppose our original matrix had 50 clauses. We use the information in the column under a_1 to channel clauses as we eliminate a_1. Consider the following:

┌[1, 2, 3, 4, 5, ... , 50]┐

[1, 2, 3, 5, . . .]   [2, 3, 4, 5, . . .]

We may conveniently call the bracketed collections of clauses *traps*. Clause 1 does not end up in the lower right trap, nor clause 4 in the left. In point of fact, clause 1 is different by the absence of a_1 in the lower trap from the upper; but this change need not be recorded. We still possess all the information necessary to detect demonstrably valid or nonvalid traps. A nonvalid trap is one which is empty, without clauses. A valid trap results from the elimination of a terminal variable. For example, suppose we were eliminating a_3 from the following trap:

[1, 2 . . . .]

We would normally obtain the two resultant traps [2, ...] and [1, ...] but the former of these we know to be valid, since a_3 is the terminal variable for clause 2, and in turn, as we shall say, *governs* the trap in question. It should be noted that clauses are transmitted to a lower trap either *actively* or *passively*. The latter occurs when the eliminated variable is absent from the clause. By detecting cases in which no clause is transmitted actively to a given trap, we can determine the occurrence of a variable in partial state and act accordingly.

Before continuing further, we note that such operations as the channeling of numbers are more speedily handled on the typical computer than familiar truth-functional operations. It is not apparent at this point in the discussion, however, how we shall avoid starting from "scratch" when new clauses are added by a Herbrand expansion, or otherwise. Suppose the expression proves nonvalid. New clauses are added. Without redoing the work already done, we might by a slightly more complicated procedure "update" the problem by channeling the new clauses to the point earlier reached at which the nonvalidity appeared, and then continue in the regular mode. The chief complication is the fact that some variables earlier treated as being in partial state may have lost that property through the addition of new clauses. As will be seen, however, this problem is easily managed by appropriate bookkeeping.

One final observation should be made before we present detailed rules of procedure. In addition to the clause-variable matrix, a record must be kept of the problem to date. We must know not to explore regions already resolved, and which avenues remain open. To keep the size of this record small, to facilitate bookkeeping, and, also, to expedite the determination of nonvalid formulas, we proceed in a basically serial manner. If each variable is taken as determining a *level* (beyond which it does not occur), we explore only one path at a time per level, and add the appropriate entry regarding the other path to the record. Such an entry is said to be *held*. Thus, at any one moment, the record will have no more entries than the number of levels above.

RULES OF PROCEDURE

In all, there are three modes of procedure: *descent*, *ascent*, and *updating*. Descent is the normal channeling, or bookkeeping, operation as we explore downward. Ascent occurs when the region under scrutiny has proved valid and we return to some earlier entry to be examined. We update after the formula has proved nonvalid and new clauses have been added. In effect, we carry these new clauses through the paths already explored. The entries held are of two kinds: *boxes* and *doors*. A box is a trap (a collection of clauses) which may possibly serve as the path of a future descent. A door is a barrier to future descent, since the area covered has already been shown valid.

The rules for descent, ascent, and updating are as follows:

Descent

CASE 1. There are no terminal variables and both paths receive clauses actively – for one path hold a *box one;* explore the other path.

CASE 2. There are no terminal variables but only one path receives clauses actively – explore the path which receives all clauses passively and hold a *box two* for the other path.

CASE 3. Neither path receives clauses actively – hold a *box three* for one path and explore the other.

CASE 4. There are no terminal variables, but one path yields an empty trap – hold a *box four* for the nonempty path, terminate descent and obtain new clauses (by the Herbrand expansion, or otherwise; updating will follow).

CASE 5. One path is governed by a terminal variable – hold a *door one* for that path and explore the other.

CASE 6. Both paths are governed by a terminal variable – terminate descent and commence ascent.

CASE 7. One path is governed by a terminal variable but the other yields an empty trap – hold a *door two* for the former path, terminate descent and obtain new clauses.

Ascent

We proceed backward up the chain of entries in the record until a box one is encountered. At this point, we hold a door one, all the lower entries being dropped, and recommence descent through the set of clauses contained in the box one. If no box one is encountered during ascent, the total expression is valid.

Updating

In the updating process, we *complete* a record entry by adding the appropriate new clauses. We *continue an active path* by channeling the appropriate new clauses down a path previously explored. We see, therefore, with the exception of a box three, record entries must also indicate which of the two possible paths was held.

CASE 1. With no terminal variables, a box one is met – complete and hold the box one and continue the active path.

CASE 2. As in Case 1 (no terminal variables), a box two is met – complete the box two; (a) if there are no clauses actively added to the active path, hold the box two and continue the active path; (b) otherwise, convert the box two to a box one and hold, continuing the active path.

CASE 3. As in Case 1, a box three is met – (a) if there is no active addition to either path, complete the box three, hold, and continue the active path; (b) if there is active addition to one path only, complete that path, hold as a box two, and continue the other path; (c) otherwise, complete and hold as a box one either of the paths and continue the other.

CASE 4. As in Case 1, a box four is met – complete the box four; (a) if clauses are added to the opposite path, but only passively, convert the box four to a box two, hold, terminate updating, and commence descent of the completed opposite path; (b) if clauses are added actively to the opposite path, convert the box four to a box one, hold, terminate updating, and commence descent of the completed opposite path; (c) otherwise, hold the completed box four, terminate updating, and obtain new clauses.

CASE 5. As in Case 1, a door one is met – hold the door one and continue the active path.

CASE 6. As in Case 1, a door two is met – (a) if no clauses are added to the opposite path, hold the door two, terminate updating, and obtain new clauses; (b) otherwise, convert the door two to a door one, hold, terminate updating, and commence descent of the completed opposite path.

CASE 7. One path is governed by a terminal variable, a box one is met – (a) if the variable governs the path held, convert to a door one, hold, and continue the active path; (b) otherwise, hold a door one for the governed path, terminate updating, and commence descent via the completed path previously held.

CASE 8. As in Case 7, a box two is met – proceed as in Case 7.

CASE 9. As in Case 7, a box three is met – hold a door one for the governed path, and continue the other path as the active path.

CASE 10. As in Case 7, a box four is met – (a) if the variable governs the path held and no clauses are added to the opposite path, convert the box four to a door two, hold, terminate updating, and obtain new clauses; (b) if the variable governs the path held and clauses are added to the opposite path, convert the box four to a door one, hold, terminate updating, and commence descent of the completed opposite path; (c) otherwise, proceed as in (b), Case 7.

CASE 11. As in Case 7, a door one is met – (a) if the variable governs the path held, hold the door one and continue the active path; (b) otherwise, terminate updating and commence ascent.

CASE 12. As in Case 7, a door two is met – (a) if the variable governs the path held and no clauses are added to the opposite path, hold the door two, terminate updating, and obtain new clauses; (b) if the variable governs the path held and clauses are added to the opposite path, convert the door two to a door one, hold, terminate updating, and commence descent of the completed opposite path; (c) otherwise, terminate updating and commence ascent.

CASE 13. Both paths are governed by a terminal variable – terminate updating and commence ascent.

CASE 14. No clauses remain to be channeled (a box four or door two has not yet been encountered) – terminate updating and obtain new clauses.

THE METHOD EXTENDED

In the present paper, we have not considered questions beyond the domain of truth-function theory. As is evident from our earlier general remarks, there is little doubt many such questions must receive effective treatment if the computer is to be used eventually as a powerful theorem tester for logical and mathematical formulas. Nevertheless, it is reasonable to suppose that for ultimate success we shall require powerful techniques for manipulating and testing very large truth-functional expressions. Because of the result we have cited, due to Herbrand, by which a quantificational formula is reduced to an expanding truth-functional one, the particular problem we have attacked here seems central, not only as an aid to future problem-solving, but also as a device which should help us in the search for greater problem-solving acumen.[2]

The method proposed represents the lineal descendant of our earlier Paris paper. What we have tried is to take advantage of the normality of the expression so as to handle not only very much larger cases but also cases which are not static, but expanding. In this connection, then, have we gone as far as we can go? Well, certainly, as suggested earlier, there are a number of open questions which we can perhaps best see how to handle by trial and error in using the method. Beyond such questions, however, is there something more we might do? Probably, yes. The expressions which result from a Herbrand expansion, censored or not, will most likely prove quite *loose* in the sense that, for valid formulas, a very small subset of the total clauses will be sufficient for the over-all validity of the formula. What we need to do, then, is to take proper advantage of the great looseness of the expanding alternational formulas we shall probably encounter.

To accomplish this aim, we have in mind two basic approaches. One of these is a nuts-and-bolts extension to the method as already explained, the rough details of which have been worked out, and which will be programmed in the immediate future. The other is a more speculative technique based upon the extended method proposed here, but, perhaps, very much more powerful. We cannot claim to have solved the basic problem necessary to make this speculative technique work, but it does not appear to us to be insurmountable.

---

[2]For other papers pertinent to the general problem see [6] through [12].

Let us consider first the nuts-and-bolts extension. As we enter upon a new problem, we probe downward through essentially unknown territory until a patent validity or nonvalidity causes us to ascend or seek new clauses, as the case may be. Because we have gone directly "to the bottom," so to speak, and, after each ascent, probe downward again, nonvalidities have an excellent opportunity to appear early. Suppose, however, we are working on a valid expression. Our various excursions downward reveal more and more about the structure of the problem. The presence of certain clauses at certain levels is enough to ensure validity. Consider, for example, the following expression:

$$\begin{matrix} [1 & 2 & 3 & 4 & 5 & 6\,] \\ a \vee & \bar{a} b c \vee & \bar{a} \bar{b} c \vee & \bar{a} \bar{c} e \vee & \bar{a} d \bar{e} \vee & \bar{a} \bar{d} \bar{e} \end{matrix}$$

If the clauses are numbered in the manner indicated and the variables are eliminated alphabetically, the following diagram shows the history of our initial descent:

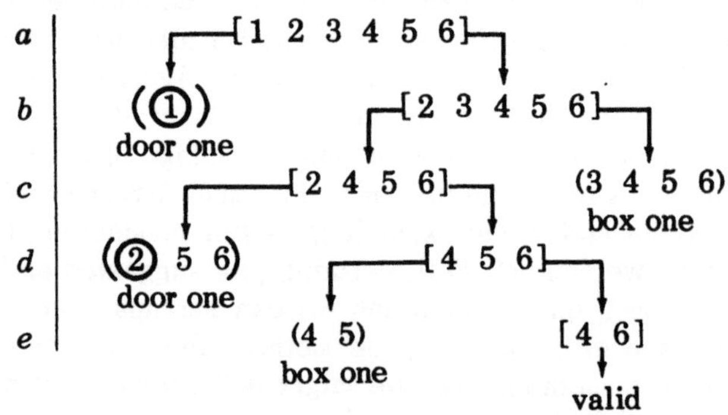

Normal parentheses enclose the entries held. Clearly, the presence of 4 and 6 at level *e* guarantees validity. As we continue working the problem, we obtain the following *register,* showing sufficient conditions for validity at the various levels.

a	(1, 2, 3, 4, 5, 6)
b	(2, 3, 4, 5, 6)
c	(2, 4, 5, 6) (3, 4, 5, 6)
d	(4, 5, 6)
e	(4, 5) (4, 6)

The register would be formed, of course, during ascent. Were the problem just considered part of a very much larger complex, we might, by proper use of the register, avoid exploring paths which contain known ingredients of validity. Although the register might become quite large, we would need to consider only the entries at the appropriate level. The comparison of traps with entries should not go slowly, since we would exclude an entry at the first recalcitrant clause. Let us now take a somewhat looser example, still eliminating variables in alphabetical order:

$$\begin{matrix}[\,1 & & 2 & & 3 & & 4 & & 5\,] \\ ab & \vee & \bar{a}\bar{b} & \vee & c & \vee & \bar{c} & \vee & c\end{matrix}$$

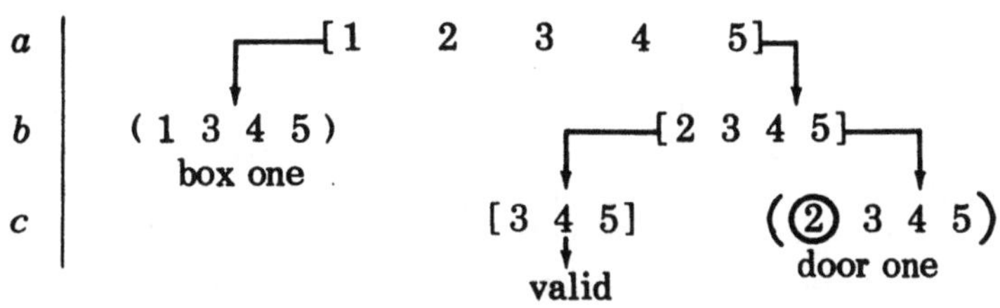

On reflection, we see that the following register is complete:

a	(3 – 5, 4)
b	
c	

By hyphenating clauses 3 and 5, we indicate that either is adequate without the other. Although, on first encounter, we do not recognize the patent validity inherent in clauses 3, 4, and 5 until the variable c is eliminated, the use of the register will prevent future tardiness.

The extent to which we may profitably employ registers (worked out in somewhat more elaborate detail) can best be determined by experiment. It may be they will prove useful only at specified points in the solution. Nevertheless, as we gauge the matter now, a rather heavy reliance on them is called for. In a loose problem, the clauses leading to validity at various stages should prove a small minority. There will probably be much repetition. The systematic method of solution with its "built-in" bookkeeping enables us to retain, without great effort, a running history of the problem, from which irrelevancies are largely eliminated. It seems reasonable that, when we explore regions having much in common with regions already examined, we should profit from our earlier experience. The use of a growing register seems a straightforward way of accomplishing this. Indeed, should our hopes in this connection prove justified, our speed of solution will increase as we move from the start of the problem to the end, and our over-all problem-solving machinery will possess a quite realistic self-improvement mechanism.

So much for the nuts-and-bolts extension. What of the more speculative procedure mentioned earlier? Suppose it generally happens that the transition from the nonvalid expanding expression to its valid resolution is abrupt and large, rather than gradual and small. That is to say, until certain clauses are added in the final appendage, the expression is not "near" validity. Were the truth table produced, it would contain many zeros. For extremely large examples, the method we have set forth should be able to unmask in a relatively short time nonvalid formulas "far" from validity. Suppose we proceed, then, adding new chunks of clauses and unmasking, until suddenly a transition occurs. The expression is no longer "far" from validity, as we determine by a series of probes. We might, for example, skip high up into the record and descend by a new path to ensure that our inability to uncover a non-validity is not due to a merely local phenomenon. We now suspect that the expression at hand is valid – although, if so, it will be much too large to be checked throughout. We believe it to be very loose, however. Our problem, then, earlier alluded to as unsolved, is to reduce the size of the expression without losing the ingredients necessary to validity. It does not seem impossible that this can be accomplished, especially if we can determine, as deletions are made, whether the expression has lost its "nearness" to validity. Once the expression has been reduced to a manageable size, we can test it outright. Here we are obviously within the realm of hopeful speculation. Still, if the problems to be dealt with are difficult, they are also interesting.

We wish to thank Professor Burton Dreben of Harvard, who has provided many helpful suggestions. In particular, Professor Dreben emphasized to us the fact that the expressions to be tested are both *expanding* and loose.

REFERENCES

1. Wang, H. "Proving Theorems by Pattern Recognition – I," *Communications of the Association for Computing Machines,* **3,** 220-234 (1960).
2. Herbrand, J. "Recherches sur la théorie de la demonstration," *Travaux de la Société des Sciences et des Lettres de Varsovie,* Classe III, No. 33, 1-128 (1930).
3. Davis, M., and Putnam, H. "A Computing Procedure for Quantification Theory," *Journal of the Association for Computing Machinery,* **7,** 31, 201-215 (July, 1960).
4. Dreben, B. "On the Completeness of Quantification Theory," *Proceedings of the National Academy of Sciences,* **38,** 1048-1049 (1952).

5. Dunham, B., R. Fridshal, and G. L. Sward. "A Nonheuristic Program for Proving Elementary Logical Theorems," *Proceedings of the International Conference on Information Processing* (Paris, 1959).
6. Wang, H. "Toward Mechanical Mathematics," *IBM Journal of Research and Development,* **4,** 1, 2-22 (January, 1960).
7. Wang, H. "Proving Theorems by Pattern Recognition – II," *The Bell System Technical Journal,* **50,** 1, 1-41 (January, 1961).
8. Gilmore, P. C. "A Proof Method for Quantification Theory: Its Justification and Realization," *IBM Journal of Research and Development,* **4,** 1, 28-35 (January, 1960).
9. Dreben, B. "Systematic Treatment of the Decision Problem," *Summaries of talks at Summer Institute of Symbolic Logic,* (Ithaca, N. Y.; 1957), p. 363.
10. Gilmore, P.C. "A Program for the Production of Proofs for Theorems Derivable within the First Order Predicate Calculus from Axioms," *Proceedings of the International Conference on Information Processing,* 265-273 (Paris, 1959).
11. Prawitz, D., et al. "A Mechanical Proof Procedure and its Realization in an Electronic Computer," *Journal of the Association for Computing Machinery,* **7,** 2, 102-128 (April, 1960).
12. Newell, A., et al. "Empirical Explorations of the Logic Theory Machine: A Case Study in Heuristics," *Proceedings of the Western Joint Computer Conference,* 218-230 (1957).

Machine-Generated Problem-Solving Graphs

H. Gelernter

Problems for which no well-defined solution procedures exist generally present the problem solver, at several stages during his search for a solution, with the necessity for choosing one path (hopefully, the best one) from among many alternative possibilities that present themselves at a given point. The process of solution may be formally represented by a tree-like structure called the problem-solving graph. When the problem-solving agent is a machine, rather than human, the generation, manipulation, and traversal of the problem-solving graph must be specified with precision and care. A properly constructed graph will not only favor the discovery of more elegant and efficient problem solutions, but will also guard the machine against certain pitfalls that humans seem to avoid naturally.

INTRODUCTION

It is the purpose of this note to present some of the problem-solving graphs generated by an IBM 704 computer program which is able to discover synthetic proofs in "high-school" elementary Euclidean plane geometry. The latter program (we have called it "the geometry machine") has been discussed elsewhere (see References 2, 4); we shall not describe its general features in any detail here. The results displayed in these earlier papers are, however, considerably elucidated by the corresponding problem-solving graphs, and the behavior of the geometry machine itself is certainly best understood and interpreted in the light of the material to be considered below. This report will also describe a rather curious and interesting result concerning the generation, manipulation, and traversal of problem-solving graphs by computer programs that do, in fact, produce such graphs in the course of their execution. While all of our specific examples are of necessity drawn from geometry, there are features of this work which are, we feel, of sufficient generality to be useful, or at least suggestive, outside the immediate area of their inception.

A problem-solving graph will be generated by a computer program whenever the programmed solution procedure makes use of a "trial and error" search to bridge the gap between the starting configuration for a problem (the premises) and the problem goal. Such programs generally, but not necessarily, incorporate heuristic processes as decision criteria in the generation and traversal of the graph.

When the computation has terminated, the problem-solving graph (henceforth, p-s graph) contains a complete record of the machine's activity in seeking and finding (or failing to find) the required problem solution. During the computation, the graph is the machine's guide to where it is going and where it has been, what it may do and what it has done. It would not be unrealistic to assert that the gross behavior of the geometry theorem proving machine is programmed by its p-s graph which, if you like, is first generated and then executed.

Although p-s graphs are generated for both analytic (goal to premises) and synthetic (premises to goal) solution procedures, our experience has been confined to the former method. This restriction should not be considered a serious one, since the analytic method seems the more suitable for exploitation in "trial and error" computer programs.

THE SIMPLE PROBLEM-SOLVING GRAPH

Let G_0 be the formal statement to be established by the computer. It will be called the problem goal. Then G_i is a subgoal of order i if there exists within the problem system a direct transformation taking G_i into some G_{i-1}. All G_j such that $j < i$ are higher subgoals than G_i. The problem-solving graph has as nodes the G_i, with each G_i joined to at least one G_{i-1} by a directed link. Each link represents a given transformation from G_i to G_{i-1}. The problem is solved when a G_k is generated which is the problem premise. Subgoal G_j is said to link G_k if G_j is a higher subgoal than G_k, and G_k is a subgoal for G_j as well as for G_0. The goal, of course, links all subgoals.

This definition of the simple p-s graph will supply our vocabulary for the discussion of the graphs generated by the geometry machine. It is not adequate, however, to describe fully the latter p-s graphs, for the following reason. Where a given subgoal is a conjunction of several expressions, it is convenient to consider each of the constituent expressions independently. Each such expression then becomes the head (or problem goal) of a new simple p-s graph. The subgoal in question is established only when each of the constituent expressions has been established. A completely general definition of the p-s graph will allow for the possibility of such "splitting" at each node of the graph. Since the ensuing discussion will be an informal one, we prefer to avoid the multi-index complexity of the general definition.

The problem-solving graph is a tree-like structure, exhibiting the tree's unhappy tendency to grow exponentially. It is clear, therefore, that vigorous pruning of all redundant and unlikely subgoals from the tree will yield rich dividends in decreased complexity of the p-s graph, and shorter traversal times. The geometry machine incorporates a number of important tree-pruning rules, some heuristic in nature, and some of which are well-defined algorithms. Each of these rules falls into one of three categories; some apply specifically to geometry, some are valid for any formal logistic system, and finally, some may reasonably be applied to any p-s graph, whatever the system that generated it.

As an example of the first class, consider the angle ABD in Fig. 2, which is identical with angle ABE in the diagram. In our formal system, however, "angle" is a predicate on three points, rather than a vertex and two rays, so that ∡ABD and ∡ABE are distinct expressions. In a strict proof, an additional step is required to establish the equality of the two angles, and hence, an additional node in the p-s graph. Also, for many of the subgoals referring to one of the angles, a redundant subgoal will appear containing the other. The geometry machine includes a procedure whereby all of the angle predicates defined by a vertex and two rays (such that more than one point lies on one or both rays) are treated as identical, ***provided that the collinearity of the vertex and the points on the ray is specified by problem's premise set.***

The rule of "syntactic symmetry", described in great detail in Reference 3, is a member of the second category. The latter rule enables the machine to recognize the formal equivalence of two distinct expressions with respect to a given logistic system when one of the expressions may be converted into the other by renaming the point variables, and when that same transformation applied to the set of premises transforms the premise set into itself, taking into consideration the symmetries of the predicates and connectives of the formal system.

Briefly, the inclusion of a syntactic symmetry recognizing procedure in the geometry machine offers two important advantages. First, if two expressions are syntactic conjugates (i. e., equivalent under the rule of syntactic symmetry), and one of the expressions has been established as valid by a proof sequence, the machine will establish the validity of the other expression by merely asserting that it is a syntactic conjugate of the first expression, just as the human mathematician makes use of the all-powerful term "similarly." Secondly, since the proof sequences of syntactically symmetric expressions are identical (except for the renaming of points, of course), syntactic conjugates are redundant subgoals in the sense of the explication below, and are thus excluded from the p-s graph.

The third category of tree-pruning rules is of most general interest. It is within this class that the simple-minded application of a few transparent and almost trivial maxims produced the curious result to which we have already alluded in the introduction above. These rules are intended to exclude redundant subgoals from the p-s graph, prevent a machine from generating circular proof sequences, and to force the machine to choose the shortest possible proof sequence from

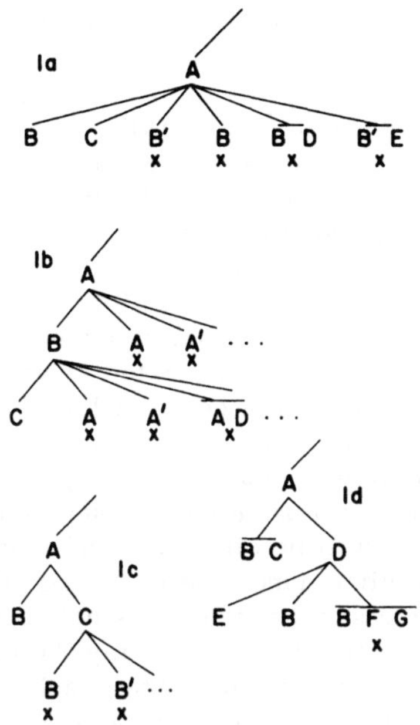

Fig. 1 Each letter represents a single expression. A horizontal bar above a sequence of letters indicates that the bracketed expressions are conjoined terms. A primed letter represents the syntactic conjugate of the unprimed expression. An "X" under a subgoal indicates that the subgoal is rejected by a problem-independent tree-pruning rule.

among sets of sequences containing the same initial sequence. The necessity for such procedures becomes apparent with the earliest, most tentative consideration of the role of the p-s graph in an "intelligent" computer program.

Let us examine, for example, the case where the machine has just generated a finite set of subgoals, $G_{i+1,k}$, for a given subgoal, G_i. If any of the G_{i+1} are identical, one will clearly want to exclude all of these but one. Where the problem domain is a logistic system, we will, in addition, want to keep only one of a subset of the $G_{i+1,k}$ that are syntactic conjugates, since the formal development of the subgraphs for each of the syntactically symmetric subgoals will be identical, except for the change in point names. And when, as is the case for the geometry machine, any of the G_{i+1} may be a conjunction of several expressions, each of which must be independently established, it frequently occurs that a given subgoal is included in, but smaller in extent than one or more of the others generated by G_i. In this case, it is clearly desirable to keep only the minimum subgoal of the latter set

(i. e., the subgoal containing the fewest terms in the conjunction), discarding the rest from the p-s graph. In determining whether a given subgoal is included in another, terms that are syntactically symmetric are considered to be identical (Fig. 1 (a)).

The foregoing procedures are intended to allow only one of a set of expressions that are in some way redundant to appear on the p-s graph as a subgoal for the given higher subgoal that generated them, and, in fact, to select the minimum expression from the set. These rules are easily seen to be equally valid for the set of all subgoals at a given level that are linked by the same higher subgoal, no matter how many levels intervene. A newly generated subgoal, having passed the test above, might, however, still be denied admission to the p-s graph. If it is identical to a subgoal already standing on the graph that links it (here and henceforth, syntactically symmetric *expressions* will be considered to be identical *subgoals*), the segment of proof sequence joining the two subgoals will be circular, and may be discarded, (Fig. 1(b)). If, on the other hand, the new subgoal is identical to an existing higher or lower subgoal that does not link it, but both these subgoals are linked by a third subgoal, only the higher of the identical subgoals will be allowed to remain on the graph, so that any proof in which the latter subgoals figure will be the shortest possible one, (Fig. 1(c)).

A further contingency that must be considered arises when a newly generated subgoal includes or is included in an existing higher or lower subgoal on the p-s graph. For the case where one of the latter pair links the other, it is reasonable to admit the new subgoal to the graph only if it is the smaller of the two (it is clear that in this case, the new subgoal will be a lower one than the existing one). When the "includes-included" subgoals do not link one another, then the newly generated subgoal may be either higher or lower on the graph than the existing one. In this event, the lower subgoal of the two is excluded if it is also the larger one, while both are allowed to remain if the lower subgoal is the minimum one (Fig. 1(d)).

It is well to emphasize at this point that the problem-independent third category tree-pruning rules above are formulated for the simple problem-solving graph. These procedures become considerably more complicated when subgoal splitting is permitted. The added difficulties are largely those of detailed genealogy tracing and the necessity for taking into consideration subgoals conjoined to all subgoals linking one or the other, but not both of the pair consisting of the newly generated subgoal and the existing one with which it is being compared. The detailed rules for tree-pruning in the general case are too tedious to be recorded here. It suffices to say that the principles governing the selection rules are the same for both cases, and that anyone faced with the necessity for using the more general rules will discover them quickly enough.

THE PROBLEM-SOLVING GRAPH OF THE GEOMETRY MACHINE

The proofs produced by the geometry machine are greatly illuminated by the corresponding problem-solving graphs. It is most instructive to follow the step-by-step behavior of the machine as it searches for a solution to a particular problem by examining the p-s graph at each stage of the search. We shall choose an extremely simple problem for our illustration; a problem, however, that contains the seeds of the paradox mentioned earlier.

Let us begin by reviewing the problem-solving procedure pursued by the machine in proving a theorem in elementary "high-school" geometry. We note, however, that questions of formal logic are beyond the scope of this paper, and will be dealt with briefly, or not at all. Also, the reader is called upon to consult the earlier papers to which we have already referred if he should care for a more detailed description of the internal structure of the theorem proving program.

In its initial configuration, the machine will contain a formal statement of the theorem to be proved together with a coordinate representation of the diagram for that theorem. It is assumed that the machine has at its disposal a library of all previously proved theorems, definitions of non-primitive symbols of the logistic system, and, of course, the complete set of axioms and postulates for Euclidean plane geometry. In operation, the machine executes the processes enumerated below.

1. With the theorem formulated as an implication, the consequent becomes the goal, G_0, on the p-s graph, and the terms of the antecedent (the premises) are placed on a "list of established formulas," labeled as established by hypothesis.

2. Premises containing non-primitive predicates or connectives are defined, and the terms of the definitions are placed on the list of established formulas, labeled as established by definition.

3. The problem goal is selected as the first generating subgoal. If the goal is a conjunction of expressions, the first term is taken as the generating subgoal.

4. The theorem library (which contains all previously proved theorems, axioms, and postulates that may be expressed as implications) is searched, and all theorems such that the consequent of the implication has the generating subgoal as a substitution instance are extracted.

5. A set of lower subgoals for the generating subgoals is constructed by making the required substitutions in the generating theorems. The diagram is used to find possible values for variables oc-

curring in the antecedent of the generating theorem, but not in the consequent.

6. Each of the generated subgoals is interpreted in the diagram. If the diagram does not contain a valid instance of the subgoal (i. e., the expression is false in the diagram), the subgoal is rejected as extremely unlikely to lead to a proof.

7. Each of the remaining subgoals is examined term by term as follows. First, each term is compared with the members of a list of fruitless, or "stuck" subgoals. If a term has previously been stuck, the subgoal containing it is rejected. Each term is then matched against the members of the list of established formulas. If it is discovered on the list, the term is labeled "established," with a reference to the occurrence of the formula on the list, which is in turn labeled with all pertinent information about the formula (i. e., how the validity of the formula was established, etc.). If the term has appeared on neither of the latter two lists, it is examined to see if the expression is a member of the limited class of those whose validity is permitted to be assumed, based on the diagram (cf., reference 1, p. 341). In this case the term is labeled as "established by assumption based on the diagram." Finally, if the expression is a member of a second limited class of formulas whose validity is often established by the direct application of a single special theorem (identities, for example), a quick, one-step proof of the term is attempted on the spot. If the attempt is successful, the term is labeled "established," and placed on the list of established formulas.

8. If, among the set of subgoals, there is one in which every term is established, then that subgoal is established, and by virtue of the generating theorem and modus ponens, so is the generating subgoal, which is placed on the list of established formulas together with a reference to the node of the p-s graph where it was established as valid. The procedure then continues at step (8a), below. If, on the other hand, no subgoal is established in every term, the procedure continues as at step (9).

8a. If the generating subgoal established in step (8) is a term conjoined to one or more terms that have yet to be established, the next non-established term is taken as a new generating subgoal, and the procedure continues at step (4). If there are no non-established terms conjoined to the generating subgoal, and there are no higher subgoals on the p-s graph, the problem has been solved, and the machine reconstructs a synthetic proof from the p-s graph. If there are higher subgoals on the p-s graph, the situation is exactly as at step (8), to which the machine returns.

9. The set of generated subgoals emerging from step (7) then

undergoes the problem-independent tree-pruning procedures described in detail above. The established terms of a subgoal are ignored in determining whether two subgoals are identical or whether one includes the other.

10. If any subgoals have survived to this point, they are placed on the p-s graph. One is chosen by the machine to work on, and its first term becomes the new generating subgoal for procedure (4), to which the machine returns.

11. If no subgoals remain after procedure (9), the generating subgoal is labeled stuck, and placed on the list of stuck subgoals. If an alternative to the generating subgoal exists (i. e., a subgoal generated by the same higher subgoal that produced the generating subgoal, with no stuck terms), it is chosen by the machine to work on, and its first term becomes the new generating subgoal for procedure (4), to which the machine returns. If no alternatives exist, and there are no higher subgoals, the machine has failed in its attempt to prove the theorem as it stands. A construction is suggested (reference 4, p. 145), and the machine returns to step (1). If higher subgoals do exist, the one that generated the stuck subgoal is labeled "stuck," and placed on the list of stuck subgoals. The situation is now exactly as at the beginning of step (11), to which the machine returns.

It should be kept in mind that throughout the discussion above, a subgoal is, in general, a conjunction of several terms. A single term of a subgoal becomes a subgoal itself only as the head of a problem-solving subgraph. A term becomes the goal of such a subgraph when it is chosen as a generating subgoal.

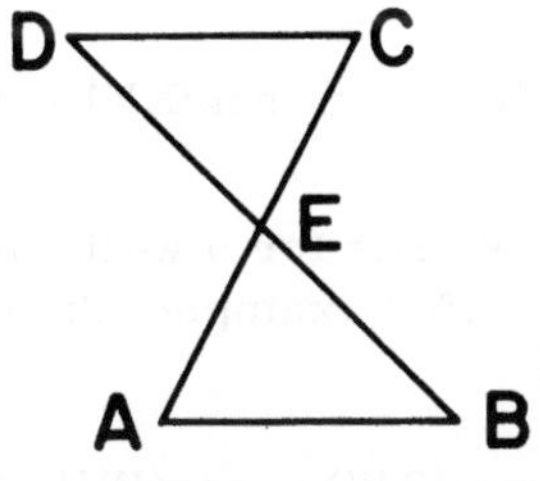

Figure 2

We can now proceed to the detailed consideration of our particular problem, which is simplicity itself. Referring to the diagram (Fig. 2), we are asked to prove that segment CD is both equal and parallel to segment AB, given point E as the midpoint of segments AC and BD. The formal statement of the problem, given as input to the machine, is the following:

$$\mathrm{pt(E)\,mdpt(AC)} \wedge \mathrm{pt(E)\,mdpt(BD)} \wedge \mathrm{seg(AB)} \wedge \mathrm{seg(CD)} \supset \mathrm{seg(AB)} = \mathrm{seg(CD)} \wedge \mathrm{seg(AB)} \parallel \mathrm{seg(CD)} \ .$$

The syntactic symmetry transformations for the theorem are the permutations: $(A \leftrightarrow C)(B \leftrightarrow D)$, and $(A \leftrightarrow B)(C \leftrightarrow D)$. These transformations are computed by the machine immediately upon acquiring the input data (cf., reference 3, p. 86).

The machine proceeds by setting up the initial configuration (step (1), above), placing the conjoined expressions seg(AB) = seg(CD) and seg(AB) ∥ seg(CD) on the problem-solving graph as G_0, the goal, and the four terms of the antecedent, the premises, on the list of established formulas. The machine, scanning the premise set, discovers two instances of the non-primitive predicate "midpoint," and, using the definition of that predicate,

$$\mathrm{pt(X)\,mdpt(YZ)} \supset \mathrm{p(YXZ)} \wedge \mathrm{seg(YX)} = \mathrm{seg(XZ)} \ ,$$

generates the expressions p(AEC), p(BED), seg(AE) = seg(EC), and seg(BE) = seg(ED), where the expression p(YXZ) has the interpretation "Y, X, and Z" are collinear in that order (step (2)). The definitions are placed on list of established formulas along with the premises.

The expression seg(AB) = seg(CD) is then withdrawn from the p-s graph to become the generating subgoal (step (3)), and from the theorem library all implications of the form

$$E_1 \wedge E_2 \wedge \ldots \supset \mathrm{seg(XY)} = \mathrm{seg(ZW)}$$

are extracted, where E_i stands for a well-formed formula of our logistic system (step (4)). For example, the following theorems will be among those extracted:

$$\mathrm{seg(XY)} = \mathrm{seg(UV)} \wedge \mathrm{seg(ZW)} = \mathrm{seg(UV)} \supset \mathrm{seg(XY)} = \mathrm{seg(ZW)} \ ,$$

$$\mathrm{parallelogram\ (XYZW)} \supset \mathrm{seg(XY)} = \mathrm{seg(ZW)} \ ,$$

$$\Delta\,\mathrm{(XYU)} \cong \Delta\,\mathrm{(ZWV)} \supset \mathrm{seg(XY)} = \mathrm{seg(ZW)} \ .$$

The number of such theorems withdrawn from the library will vary between one and thirty, depending upon the form of the generating subgoal. Twelve theorems probably represents a reasonable average for all cases. Each of the generating theorems will in turn spawn a great number of subgoals for the generating subgoal (step (5)). The exact number of "raw" subgoals produced depends upon the number of points named in the problem statement as well as on the form of the generating subgoal. This is true because, in general, each permutation of point names in the consequent of the generating theorem produces a different subgoal, and in fact, if unsubstituted variables remain in the antecedent, a single substitution in the consequent will generate a set of subgoals; one for each permutation of the remaining points of the unsubstituted variables. A special algorithm incorporated into the generating procedure immediately eliminates all but one of each set of subgoals produced by a given theorem which are identical except for symmetries within individual terms (i. e., the subgoals seg(AB) = seg(CD) and seg(CD) = seg(BA), for example). This is a trivial kind of syntactic symmetry that is independent of the premise set.

As subgoals are generated, they are interpreted in the diagram (step (6)). Consider, for example, the first of the representative generating theorems listed above. After equivalent forms as above have been reduced, only the following subgoal, containing the unsubstituted variables U and V, remains:

$$\text{seg}(AB) = \text{seg}(UV) \wedge \text{seg}(CD) = \text{seg}(UV) \ .$$

In general, the next step would be to substitute exhaustively the points named in the problem for the variables U and V, and then to check each of the resulting expressions for validity in the diagram. Special procedures are introduced, however, for certain frequently occurring forms, of which the above is an example. In this particular case, a routine is called upon to search the list of segments appearing in the diagram and to withdraw the names of those whose lengths satisfy the interpreted formula. The unsubstituted variables are then allowed to take only the values suggested by the special procedure.

Resulting from the first generating theorem, then, are the following two subgoals:

$$\text{seg}(AB) = \text{seg}(CD) \wedge \text{seg}(CD) = \text{seg}(CD) \ ,$$

and

$$\text{seg}(AB) = \text{seg}(AB) \wedge \text{seg}(AB) = \text{seg}(CD) \ .$$

The subgoal generating procedure, when applied to the second of the sample theorems produces the pair of subgoals:

$$\text{parallelogram}(ABCD) \ ,$$

and

$$\text{parallelogram(BACD)} .$$

Neither of these is valid in the diagram, and so both are rejected.

The third generating theorem produces the set:

$$\Delta(\text{ABU}) \cong \Delta(\text{CDV}),$$

and

$$\Delta(\text{BAU}) \cong \Delta(\text{CDV}),$$

containing the unsubstituted variables U and V. Here again, a special routine seeks acceptable values for U and V in the diagram, resulting in the single subgoal

$$\Delta(\text{ABE}) \cong \Delta(\text{CDE}) .$$

The process is repeated for each of the generating theorems withdrawn from the library during the earlier search. None of the subgoals thus produced, however, will survive the test of the diagram.

Of the acceptable subgoals generated above, none are found to contain terms that have been previously stuck or established, nor do they contain terms whose validity may be assumed from the diagram. Two of the subgoals, however, contain terms of the special class such that a one-step proof is attempted (the so-called "urgent subgoals"). These are the identities seg(CD) = seg(CD), and seg(AB) = seg(AB). The theorem seg(XY) ⊃ seg(XY) = seg(XY) is immediately withdrawn from the library to generate the subgoals for the identities seg(CD) and seg(AB), both of whose validities are assumed from the diagram. The identity terms are then labeled "established" (step (7)). No subgoal is found to be fully established in all of its terms, and so the problem-independent tree-pruning rules are applied to the generated subgoals before admitting them to the p-s graph (step (9)).

At this point, the subgoals remaining for consideration are the following:

$$\text{seg(AB)} = \text{seg(CD)} \wedge \text{seg(CD)} = \text{seg(CD)} ,$$

$$\text{seg(AB)} = \text{seg(AB)} \wedge \text{seg(AB)} = \text{seg(CD)} ,$$

and

$$\Delta(\text{ABE}) \cong \Delta(\text{CDE}) .$$

With the identities established, however, the first two subgoals become equivalent, so that one of these may be rejected. The term of the remaining one not yet established is seen to be identical with a higher subgoal that links it (the generating subgoal, in fact), and so it too is discarded. Only the last of the subgoals listed remains qualified for inclusion on the problem-solving graph, which at this point has the configuration shown in Fig. 3.

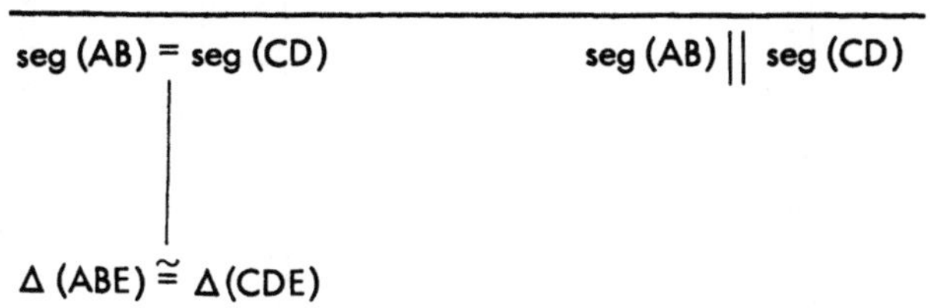

Fig. 3 p-s graph after first circuit through subgoal generating routine. Horizontal bracket indicates conjoined expressions, both of which must be established to establish subgoal. Each such expression is the head of a p-s subgraph.

For the next circuit, the expression $\Delta(ABE) \cong \Delta(CDE)$ is taken as the generating subgoal (step (10)), and the procedure is repeated. At the completion of the circuit, three new subgoals, generated by two distinct theorems, are admitted to the p-s graph. The configuration of the graph is now as in Fig. 4. The first two subgoals derive from the theorem "two triangles are congruent if side-angle-angle equals side-angle-angle," while the third derives from "side-angle-side equals side-angle-side." Every other subgoal generated has been rejected by some one of the tree-pruning rules.

Let us assume that the leftmost subgoal is chosen for the next circuit. The first two terms assert that the triangles in question are not degenerate. They are established by assumption based on the diagram. The third term of the subgoal will have been discovered on the list of established formulas, and will be so labeled. The fourth and fifth terms are syntactic conjugates, and so if one of the pair is established, the other may be merely asserted. By the same token, of course, if the machine gets stuck on one, the other too should be listed as stuck.

The next generating subgoal, then, is the expression

$$\measuredangle(EAB) = \measuredangle(ECD) \quad .$$

This time, the generating procedure produces only one subgoal that survives the tree-pruning algorithms, containing as a term the expression seg(AB) $\|$ seg(CD). Note that the latter subgoal is permitted on the p-s graph, even though it includes an expression that appears as a higher subgoal, for the earlier instance of the formula does not link it, nor are the two occurrences linked by a common higher subgoal. However, when seg(AB) $\|$ seg(CD) is taken as the generating subgoal for the next circuit, no new subgoals are produced that do not link it on the p-s graph, and so the situation of step (11) obtains. The formula seg(AB) $\|$ seg(CD) is labeled "stuck," and placed on the list of

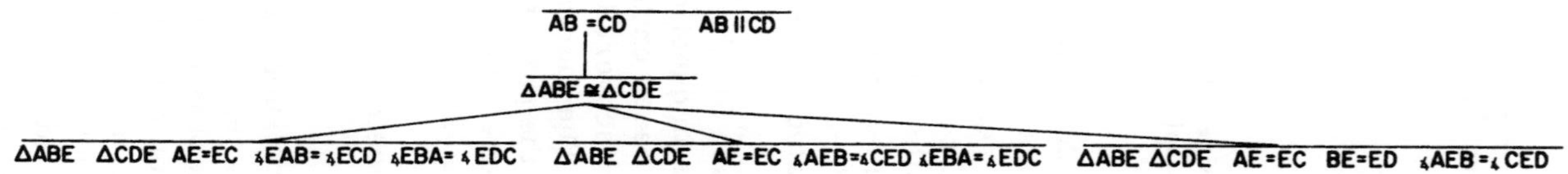

Fig. 4 p-s graph after two subgoal-generating circuits.

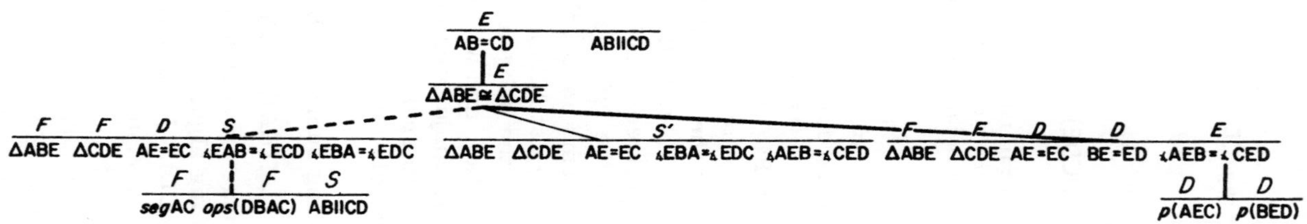

Fig. 5 p-s graph after the first term of the goal has been established. Note that the second term of the goal will at this point be listed as "stuck," unless corrective measures have been taken. Heavy lines trace out proof sequence, connecting a string of subgoals to the premise set. Broken line traces path the machine explored, but failed to establish. The letters E, S, F, and D standing above some of the subgoals indicate, respectively, that the corresponding term has been established, stuck, assumed from the figure, or established by definition of a premise. The primed "S" indicates that not that expression, but rather its syntactic conjugate has been previously stuck.

stuck formulas. Since there are no alternative subgoals to the latter, its generator, ∡(EAB) = ∡(ECD), is also labeled "stuck," and both it and its syntactic conjugate are banished to the stuck list.

The entire subgoal containing the generator as a term is now stuck, and so the machine seeks an alternative on the p-s graph. The middle subgoal is examined and rejected, because one of its terms, the syntactic conjugate of a term stuck earlier, is discovered on the list of stuck formulas. The third subgoal is then chosen, and easily established, for two of its terms are found to be premises on the list of established formulas, and the third is quickly established with a one-step proof. The configuration of the p-s graph is now as in Fig. 5. With the three terms of the third subgoal established, so is the entire subgoal, establishing its generator, as well (steps (8) and (8a) above). The same is true at the next higher level of the p-s graph, and so now the first term of the goal is established. When the machine takes the next term of the goal as a generating subgoal, however, it finds it stuck in the list of its previous failures! Here is the paradox to which our discussion has been leading, for the machine should have no difficulty proving segment AB parallel to segment CD by means of the alternate-interior angle theorem, since the angles are corresponding angles of two triangles already proved congruent.

Let us examine in somewhat greater detail the processes that brought on this result. The subgoal seg(AB) ∥ seg(CD) was stuck because every lower subgoal generated for it contained as terms the linking subgoal ∡(EAB) = ∡(ECD), or its syntactic conjugate. If the linking subgoal had been established by an alternative route, the lower subgoal that it blocked from further development towards a proof could have been established through that subgoal. In this case, the only alternative to seg(AB) ∥ seg(CD) generated for the subgoal ∡(EAB) = ∡(ECD) was the formula Δ(EAB) ≅ Δ(ECD), which was rejected because it, too, appears on the p-s graph as a higher linking subgoal.

The difficulty is resolved by noting that the latter expression was established through an alternative branch of the p-s graph. Once established, it no longer blocks its generator ∡(EAB) = ∡(ECD), but in fact supplies a proof sequence for that formula, so that it too is now established, releasing the subgoal seg(AB) ∥ seg(CD) that it had previously blocked. The procedure for traversal of the problem-solving graph must clearly be modified to allow for this contingency.

Referring back to the step-by-step description of the subgoal generating routine, the following amendments will correct the unsatisfactory condition above. In step (9), whenever a subgoal is discarded because it is identical to a higher linking subgoal, the instance of that subgoal on the p-s graph is labeled "blocking subgoal," with a reference to the lower subgoal that it blocks. The same expression may block several lower subgoals; there will be a reference added for each subgoal blocked. And in step (8), whenever the expression established is a blocking subgoal, every subgoal on the p-s graph that it blocks will be "released." That is, the subgoal will be removed from the list of stuck formulas although it remains stuck in the instance on the p-s graph referred to by the blocking subgoal, since we still wish to avoid circular proof sequences, or those that are not the shortest possible ones. Note that the release of a blocked subgoal releases every higher subgoal linking it to the expression that had blocked it. At this point, let the reader be assured that while the discussion above defies simplification, the procedure itself is not difficult to follow if the accompanying figures are referred to freely.

The problem we have just examined is a quite trivial one, and so the machine's behavior in solving it and the resulting p-s graph are not very profound. Figure 6 is the complete problem-solving graph for a somewhat more difficult problem. The proof synthesized by the machine from the problem-solving graph is displayed in Appendix I. This problem has been chosen for display because it contains a situation almost identical to that considered "raw" above. In particular, note that generating subgoal number 5 was "stuck" when every branch of its subgraph reached a dead end, with only higher linking subgoals generable at each terminal node. Five's generator, however, subgoal number 4, is subsequently established through an alternative branch, releasing, among several others, subgoal 5 from the list of stuck formulas. Later on, the same expression is generated by subgoal 37, to become generating subgoal 38. This time, three lower subgoals are produced, while the same formula produced only two lower subgoals as generating subgoal 5. Furthermore, the three subgoals generated by number 38 are all different from the two generated by number 5, and one of them leads directly to a solution for the problem!

The interested reader will find it most illuminating and rewarding to follow carefully the step-by-step progress of the machine in solving the problem displayed in Fig. 6. One cannot but be struck by the curious mixture of sophistication and naivete that stand side by side in the record of a computer's attempt at intellectual activity.

SOME FURTHER CONSIDERATIONS

The final p-s graph is an excellent indicator of how hard the machine had to work to solve a given problem. The same problem will generally result in different graphs when modifications are introduced or amendments are added to one section or another of the proof search procedure. A comparison of the final graphs for two such cases is a most effective tool for evaluating the impact of a given modification in the problem-solving system that differentiates the two procedures.

Figure 7 displays two p-s graphs in schematic form, both for the same problem, for which the proofs have been reproduced as

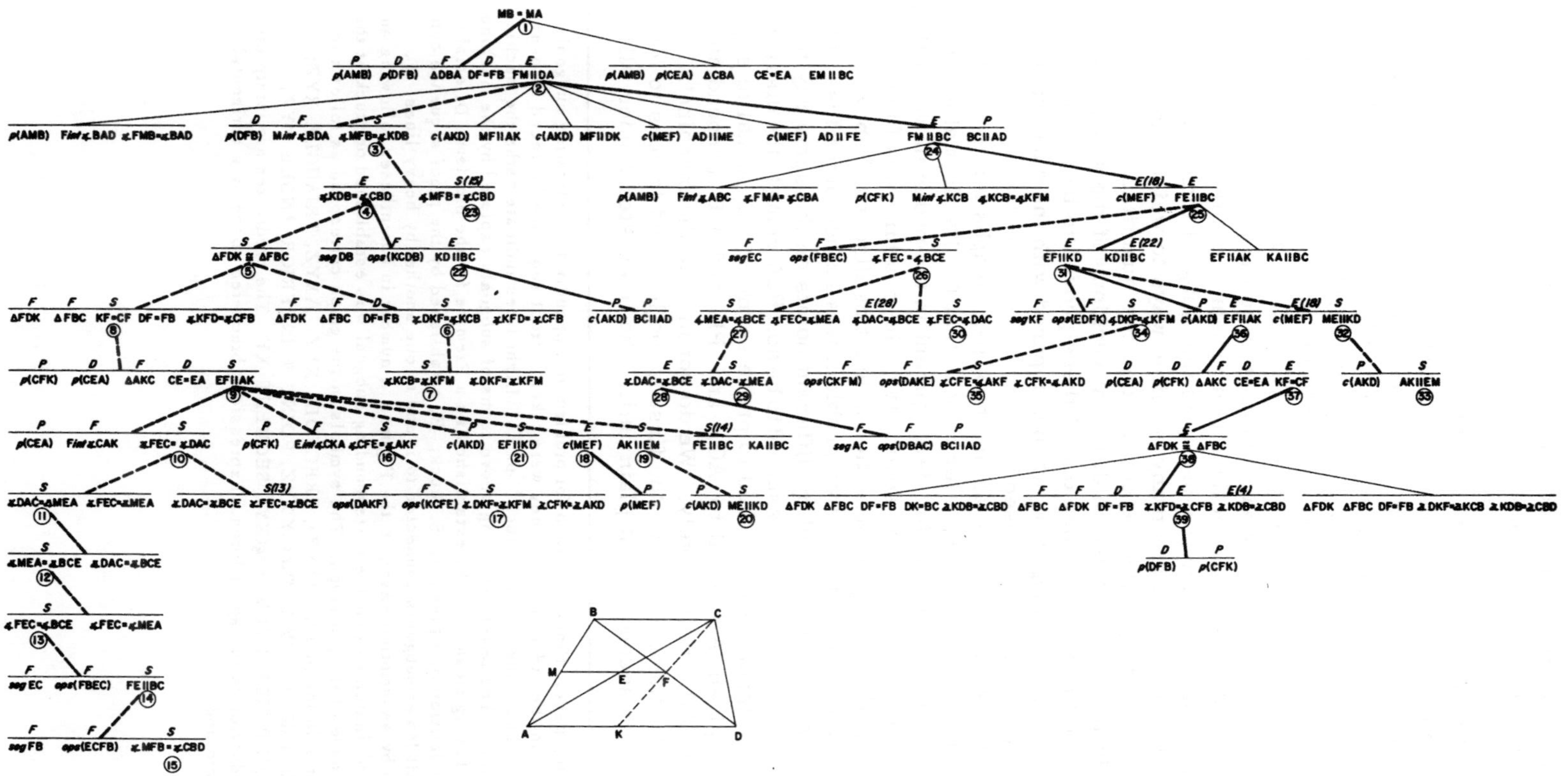

Fig. 6 Final p-s graph for problem in Appendix I.

Appendix II. In finding the second proof, the machine had at its disposal every technique described above except the ability to recognize the property of syntactic symmetry. The introduction of the latter feature resulted in the other rather elegant proof and problem-solving graph.

The geometry machine contains a heuristic procedure to assist it in selecting the subgoal most likely to succeed whenever it is faced with the necessity for making such a choice (in steps (8a) and (11) of the detailed procedure above.) Briefly, the machine chooses that subgoal that corresponds most closely to the premise set in points named, predicates, and connectives. The rule is an ad hoc heuristic; a subgoal is said to correspond most closely to the premises if the points, predicates, and connectives that appear most frequently in the premises appear more frequently in that subgoal than in any of the alternatives. There are many possible ways to weight the different factors in judging the correspondence quantitatively; several such possibilities were tried in different versions of the geometry machine. On the average, any one of these proved to be a superior method for selecting subgoals, when compared with the case where no selection rule at all was used.

Figure 8 displays, once again, two p-s graphs for the same problem. The two solutions differ only in that for its first attempt, the machine chose subgoals for development in the order in which they were generated, while for the second try, the machine was furnished with the best of the subgoal selection heuristics discussed above. The machine's formal system for this problem was rather more fussy than usual; identical angles named by different points on the same ray had to be proved equal, rather than assumed to be so. Referring to the diagram in Fig. 8, the machine was asked to prove segment BF equal to segment FC, given AB equal to AC, and BD and CE perpendicular to AC and AB, respectively. We do not include the machine's proofs for the theorem, since the first is too long and tedious to follow, while the second is just plain dull. The interested reader will

---

Fig. 6 (facing page) Final p-s graph for problem in Appendix I. Generating subgoals are numbered in the order in which they were taken. Proof sequence is traced by heavy lines, while broken lines trace "stuck" paths. Light lines indicate paths the machine did not explore. The letters standing above terms of subgoals explored by the machine have the following meanings: P, established as a Premise for the problem; D, established by Definition of a premise; S, Stuck; E, established by the proof sequence consisting of all lower subgoals connected to it and to one another by heavy lines; F, established by assumption based on the Figure. A number in parentheses following an "E" or an "S" indicates that the corresponding subgoal was established or stuck at the designated node of the p-s graph. The symbols in the graph correspond as follows to the predicates in the proof: p(XYZ), PRECEDES XYZ; ΔXYZ, TRIANGLE XYZ; c(XYZ), COLLINEAR XYZ; Wint∡XYZ, POINT W INTERIOR ANGLE XYZ; ops(XYUV), OPP-SIDE XYUV; seg(XY), SEGMENT XY. The predicates are interpreted in the introduction to the appendices. Expressions bracketed above by a horizontal line are conjoined.

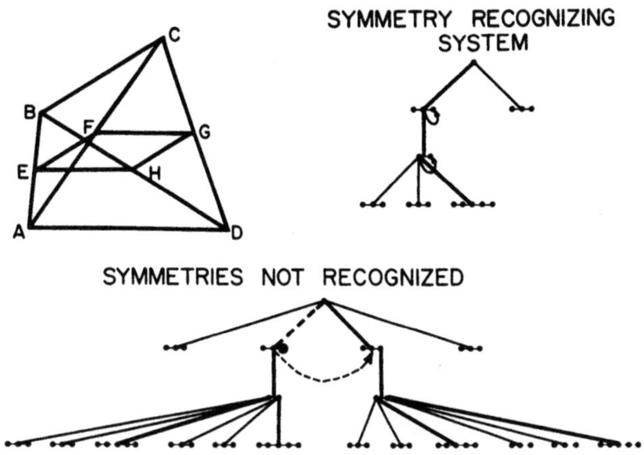

Fig. 7 Final p-s graphs for proofs in Appendix II. Each dot stands for a term of a subgoal. In the graph for the symmetry recognizing system, the arrows point to terms established as syntactic conjugates of the terms from which the arrows originate. In the graph for the system which does not recognize syntactic symmetry, the circled term was stuck. The term to the left of the circled one, which the machine established before failing on its right-hand neighbor, was found by the machine to be identical to the term indicated by the arrow. The machine therefore used that segment of proof at the second instance of the expression to complete the solution of the problem. As before, and for Figs. 8 and 9 as well, heavy solid lines trace established sequences, heavy broken lines trace paths the machine explored but failed to establish, and light lines trace unexplored alternatives.

find proofs with and without subgoal selection compared in reference 2.

A final problem-solving graph is reproduced in Fig. 9 to illustrate a point made in earlier papers on the geometry machine: the fact that the computer seems to find problems in inequalities relatively easier than people do. The problem (for which the machine's proof is reproduced in Appendix III) had to be invented by fusing two separate textbook problems, since the machine generally found the inequality problems given to high school students a snap. The final p-s graph will be seen to be quite uncluttered, and the machine's path to the solution almost a direct one, compared to the behavior the author has observed in human subjects solving the same problem. One is tempted to speculate that the computer might show a similar superiority to human mathematicians in the logic of non-Euclidean geometries.

The machine-generated problem-solving graph is a classic example of the kind of information for which computer list-processing procedures are almost mandatory. The author cannot conceive of the symbol-manipulation procedures described above being performed efficiently with any other organization of computer storage. The particu-

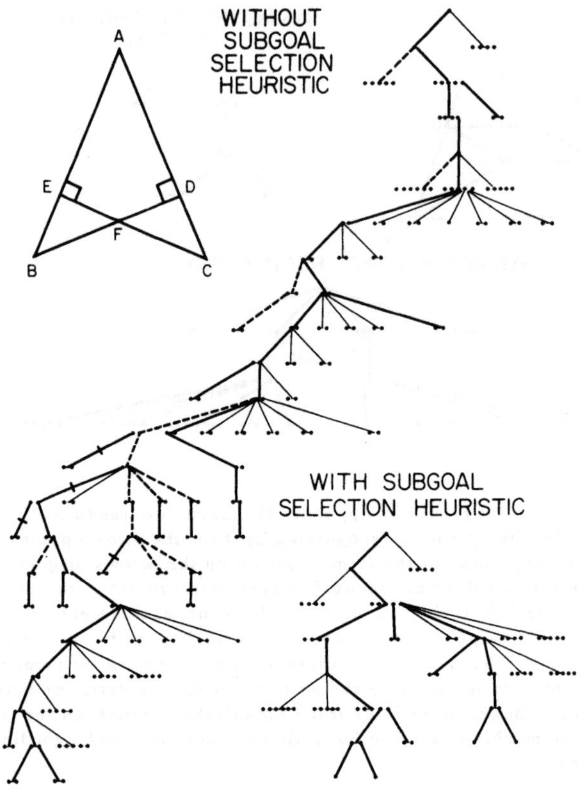

Fig. 8 Final p-s graphs for problem solved with and without subgoal selection heuristic. For the case without subgoal selection, the established sequences with a bar across the trace line were not used in the final proof. All other established sequences figured in the proof at one point or another.

lar list-processing language developed for the geometry program is described in detail in references 5 and 6. The language has since been further developed by groups in IBM France, Euratom at Ispra, Italy, and the University of Grenoble, in France. Of particular pertinence is the work of Tabory[7] on the outlines of a special purpose language for the manipulation of graphs.

It is a pleasure to acknowledge J. R. Hansen's important contribution to the sucess of the geometry machine. He is one of the rare computer programmers who can turn an idea into a working machine program overnight. D. W. Loveland and C. L. Gerberich contributed, too, toward the completion of our program.

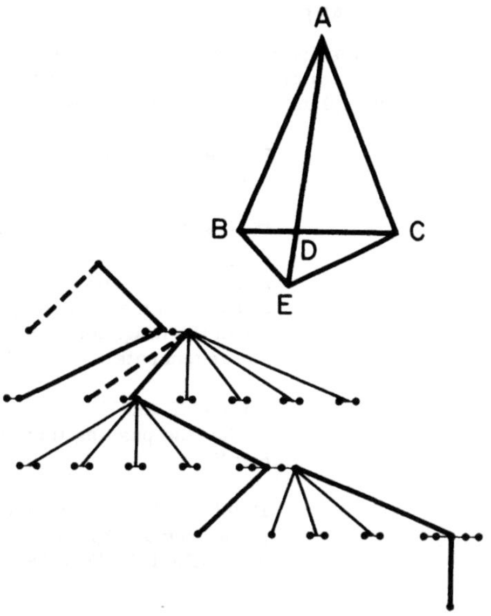

Fig. 9 Final p-s graph for inequality problem in Appendix III.

REFERENCES

1. H. Gelernter and N. Rochester, "Intelligent Behavior in Problem-Solving Machines," *IBM J. Res. and Devel.*, Vol. 2 (October 1958).

2. H. Gelernter, "Realization of a Geometry Theorem-Proving Machine," *Proc. First International Conf. on Information Processing* (Paris: UNESCO, 1959).

3. H. Gelernter, "A Note on Syntactic Symmetry and the Manipulation of Formal Systems by Machine," *Information and Control*, Vol. 2 (March 1959).

4. H. Gelernter, J.R. Hansen, and D.W. Loveland, "Empirical Explorations of the Geometry Theorem Machine," *Proc. Western Joint Computer Conference*,(San Fransisco), May 1960.

5. H. Gelernter, J.R. Hansen, and C.L. Gerberich, "A FORTRAN-Compiled List Processing Language," *J. Assoc. Comp. Mach.*, Vol. 7 (April 1960).

6. J.R. Hansen, "A Manual for the Use of the FORTRAN-Compiled List Processing Language," IBM Research Report RC-282, June 1960.

7. R. Tabory, "The Outlines of a Language for the Manipulation of Graphs," *Proc. Symp. on Symbolic Languages*, Rome, March 1962.

APPENDICES

The proofs exhibited in the following appendices are exactly as printed out by the geometry machine. Most of the predicates contained in these proofs have obvious interpretations; those that need further explanation are listed below.

Precedes XYZ	Points X, Y, and Z are collinear in that order.
Collinear XYZ	Points X, Y, and Z are collinear.
Triangle XYZ	Points X, Y, and Z are not collinear, and the segments joining the points exist in the diagram (i. e., the triangle is not degenerate).
Opp-Side XYUV	Points X and Y are on opposite sides of the line passing through points U and V.
Point W Interior Angle XYZ	Point W lies within that region of the plane bounded by rays YZ and YX, with ANGLE XYZ less than π.
Segment XY	The segment joining points X and Y exists in the diagram.

The proof in Appendix I and the first proof in Appendix II have been published previously in Reference 4. As an aid in interpreting the proofs, we note the following. In verbal form, the first theorem states that the line segment joining the midpoints of the diagonal of an arbitrary trapezoid bisects a side of the trapezoid when extended in either direction. The second theorem proves that in an arbitrary quadrilateral, if the midpoints of the diagonals and the midpoints of a pair of opposite sides are joined in alternating order, the figure produced is a parallelogram. In the third theorem, the apex of an isosceles triangle is joined to an arbitrary point not the midpoint on the base of the triangle, and extended an arbitrary distance. The terminal point of that segment is then joined to the endpoint of the base. The fourth theorem states that if the medians from two vertices of an arbitrary triangle are extended their own lengths through the base, the terminal points are collinear with the third vertex of the triangle.

Although the last problem is not a difficult one, it is included because the formula to be established is different from those usually encountered in high school texts.

APPENDIX I

```
   PREMISES
**************
QUAD-LATERAL ABCD
SEGMENT BC PARALLEL SEGMENT AD
POINT E MIDPOINT SEGMENT AC
POINT F MIDPOINT SEGMENT BD
PRECEDES MEF
PRECEDES AMB

   TO PROVE
**************
SEGMENT MB EQUALS SEGMENT MA

   NO SYNTACTIC SYMMETRIES
```

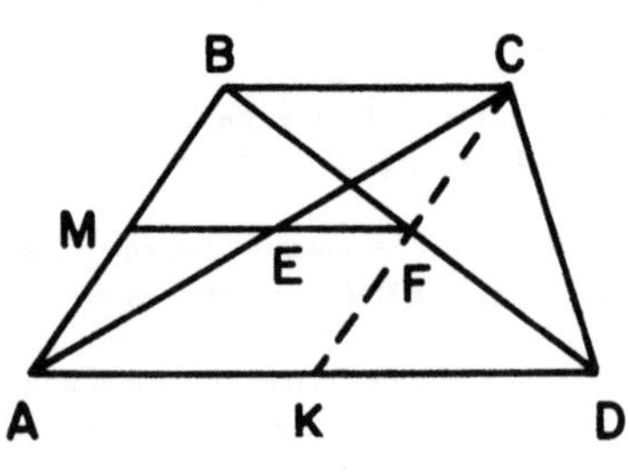

```
I AM STUCK, ELAPSED TIME   8.12 MINUTES
***************************************
CONSTRUCT SEGMENT CF
EXTEND SEGMENT CF TO INTERSECT SEGMENT AD IN POINT K

ADD TO PREMISES THE FOLLOWING STATEMENTS
****************************************
PRECEDES CFK
COLLINEAR AKD


   PROOF
***********
SEGMENT BC PARALLEL SEGMENT AD
      PREMISE
COLLINEAR AKD
      PREMISE
SEGMENT KD PARALLEL SEGMENT BC
      SEGMENTS COLLINEAR WITH PARALLEL SEGMENTS ARE PARALLEL
OPP-SIDE KCDB
      ASSUMPTION BASED ON DIAGRAM
SEGMENT DB
      ASSUMPTION BASED ON DIAGRAM
ANGLE KDB EQUALS ANGLE CBD
      ALTERNATE-INTERIOR ANGLES OF PARALLEL LINES ARE EQUAL
PRECEDES CFK
      PREMISE
PRECEDES DFB
      DEFINITION OF MIDPOINT
ANGLE KFD EQUALS ANGLE CFB
      VERTICAL ANGLES ARE EQUAL
SEGMENT DF EQUALS SEGMENT FB
      DEFINITION OF MIDPOINT
TRIANGLE FDK
      ASSUMPTION BASED ON DIAGRAM
TRIANGLE FBC
      ASSUMPTION BASED ON DIAGRAM
TRIANGLE FDK CONGRUENT TRIANGLE FBC
      TWO TRIANGLES ARE CONGRUENT IF ANGLE-SIDE-ANGLE EQUALS ANGLE-SIDE-ANGLE
SEGMENT KF EQUALS SEGMENT CF
      CORRESPONDING SEGMENTS OF CONGRUENT TRIANGLES ARE EQUAL
SEGMENT CE EQUALS SEGMENT EA
      DEFINITION OF MIDPOINT
TRIANGLE AKC
      ASSUMPTION BASED ON DIAGRAM
PRECEDES CEA
      DEFINITION OF MIDPOINT
SEGMENT EF PARALLEL SEGMENT AK
      SEGMENT JOINING MIDPOINTS OF SIDES OF TRIANGLE IS PARALLEL TO BASE
SEGMENT EF PARALLEL SEGMENT KD
      SEGMENTS COLLINEAR WITH PARALLEL SEGMENTS ARE PARALLEL
SEGMENT FE PARALLEL SEGMENT BC
      SEGMENTS PARALLEL TO THE SAME SEGMENT ARE PARALLEL
PRECEDES MEF
      PREMISE
COLLINEAR MEF
      ORDERED COLLINEAR POINTS ARE COLLINEAR
SEGMENT FM PARALLEL SEGMENT BC
      SEGMENTS COLLINEAR WITH PARALLEL SEGMENTS ARE PARALLEL
SEGMENT FM PARALLEL SEGMENT DA
      SEGMENTS PARALLEL TO THE SAME SEGMENT ARE PARALLEL
TRIANGLE DBA
      ASSUMPTION BASED ON DIAGRAM
PRECEDES AMB
      PREMISE
SEGMENT MB EQUALS SEGMENT MA
      LINE PARALLEL TO BASE OF TRIANGLE BISECTING ONE SIDE BISECTS OTHER SIDE

   TOTAL ELAPSED TIME  30.68 MINUTES
```

```
   PREMISES
***************
QUAD-LATERAL ABCD
POINT E MIDPOINT SEGMENT AB
POINT F MIDPOINT SEGMENT AC
POINT G MIDPOINT SEGMENT CD
POINT H MIDPOINT SEGMENT BD

   TO PROVE
***************
PARALELOGRAM EFGH

   SYNTACTIC SYMMETRIES
****************************
BA, AB, DC, CD, EE, HF, GG, FH,
CA, DB, AC, BD, GE, FF, EG, HH,
DA, CB, BC, AD, GE, HF, EG, FH,
```

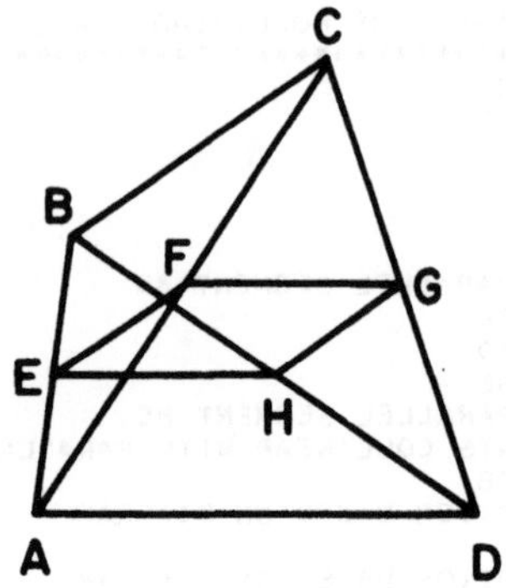


```
   PROOF
***********
SEGMENT DG EQUALS SEGMENT GC
      DEFINITION OF MIDPOINT
SEGMENT CF EQUALS SEGMENT FA
      DEFINITION OF MIDPOINT
TRIANGLE DCA
      ASSUMPTION BASED ON DIAGRAM
PRECEDES DGC
      DEFINITION OF MIDPOINT
PRECEDES CFA
      DEFINITION OF MIDPOINT
SEGMENT GF PARALLEL SEGMENT AD
      SEGMENT JOINING MIDPOINTS OF SIDES OF TRIANGLE IS PARALLEL TO BASE
SEGMENT HE PARALLEL SEGMENT AD
      SYNTACTIC CONJUGATE
SEGMENT GF PARALLEL SEGMENT EH
      SEGMENTS PARALLEL TO THE SAME SEGMENT ARE PARALLEL
SEGMENT HG PARALLEL SEGMENT FE
      SYNTACTIC CONJUGATE
QUAD-LATERAL HGFE
      ASSUMPTION BASED ON DIAGRAM
PARALELOGRAM EFGH
      QUADRILATERAL WITH OPPOSITE SIDES PARALLEL IS A PARALLELOGRAM

   TOTAL ELAPSED TIME   1.03 MINUTES
```

Proof of Above Theorem with Syntactic Symmetries Ignored.

```
   PROOF
***********
SEGMENT CF EQUALS SEGMENT FA
      DEFINITION OF MIDPOINT
SEGMENT BE EQUALS SEGMENT EA
      DEFINITION OF MIDPOINT
TRIANGLE CAB
      ASSUMPTION BASED ON DIAGRAM
PRECEDES CFA
      DEFINITION OF MIDPOINT
PRECEDES BEA
      DEFINITION OF MIDPOINT
SEGMENT FE PARALLEL SEGMENT BC
      SEGMENT JOINING MIDPOINTS OF SIDES OF TRIANGLE IS PARALLEL TO BASE
SEGMENT DG EQUALS SEGMENT GC
      DEFINITION OF MIDPOINT
SEGMENT DH EQUALS SEGMENT HB
      DEFINITION OF MIDPOINT
TRIANGLE CDB
      ASSUMPTION BASED ON DIAGRAM
```

```
PRECEDES DGC
      DEFINITION OF MIDPOINT
PRECEDES DHB
      DEFINITION OF MIDPOINT
SEGMENT HG PARALLEL SEGMENT BC
      SEGMENT JOINING MIDPOINTS OF SIDES OF TRIANGLE IS PARALLEL TO BASE
SEGMENT HG PARALLEL SEGMENT FE
      SEGMENTS PARALLEL TO THE SAME SEGMENT ARE PARALLEL
TRIANGLE DBA
      ASSUMPTION BASED ON DIAGRAM
SEGMENT HE PARALLEL SEGMENT AD
      SEGMENT JOINING MIDPOINTS OF SIDES OF TRIANGLE IS PARALLEL TO BASE
TRIANGLE DCA
      ASSUMPTION BASED ON DIAGRAM
SEGMENT GF PARALLEL SEGMENT AD
      SEGMENT JOINING MIDPOINTS OF SIDES OF TRIANGLE IS PARALLEL TO BASE
SEGMENT GF PARALLEL SEGMENT EH
      SEGMENTS PARALLEL TO THE SAME SEGMENT ARE PARALLEL
QUAD-LATERAL HGFE
      ASSUMPTION BASED ON DIAGRAM
PARALELOGRAM EFGH
      QUADRILATERAL WITH OPPOSITE SIDES PARALLEL IS A PARALLELOGRAM

   TOTAL ELAPSED TIME   2.63 MINUTES
```

APPENDIX III

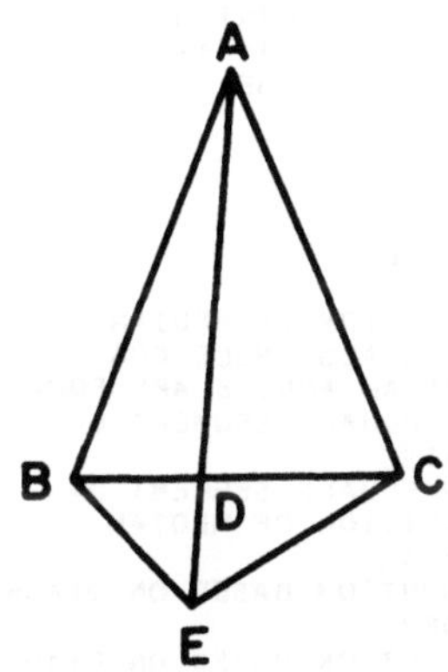


```
   PREMISES
***************
SEGMENT AB EQUALS SEGMENT AC
PRECEDES BDC
SEGMENT BD LESS-THAN SEGMENT DC
PRECEDES ADE

   TO PROVE
***************
ANGLE ACE LESS-THAN ANGLE ABE

   NO SYNTACTIC SYMMETRIES

   PROOF
***********
SEGMENT BD LESS-THAN SEGMENT DC
      PREMISE
SEGMENT AB EQUALS SEGMENT AC
      PREMISE
SEGMENT DA
      ASSUMPTION BASED ON DIAGRAM
SEGMENT DA EQUALS SEGMENT DA
      IDENTITY
TRIANGLE DAB
      ASSUMPTION BASED ON DIAGRAM
TRIANGLE DAC
      ASSUMPTION BASED ON DIAGRAM
ANGLE BAE LESS-THAN ANGLE CAE
      TWO TRIANGLES WITH TWO EQUAL SIDES AND THIRD SIDES UNEQUAL
SEGMENT EA
      ASSUMPTION BASED ON DIAGRAM
SEGMENT EA EQUALS SEGMENT EA
      IDENTITY
TRIANGLE EAC
      ASSUMPTION BASED ON DIAGRAM
TRIANGLE EAB
      ASSUMPTION BASED ON DIAGRAM
SEGMENT EB LESS-THAN SEGMENT EC
      TWO TRIANGLES WITH TWO EQUAL SIDES AND INCLUDED ANGLES UNEQUAL
TRIANGLE EBC
      ASSUMPTION BASED ON DIAGRAM
```

```
ANGLE ECD LESS-THAN ANGLE EBD
      IN TRIANGLE WITH UNEQUAL SIDES,SMALLER ANGLE IS OPPOSITE SMALLER SIDE
TRIANGLE CAB
      ASSUMPTION BASED ON DIAGRAM
ANGLE ACD EQUALS ANGLE ABC
      BASE ANGLES OF AN ISOSCELES TRIANGLE ARE EQUAL
POINT D INTERIOR ANGLE ABE
      ASSUMPTION BASED ON DIAGRAM
POINT D INTERIOR ANGLE ACE
      ASSUMPTION BASED ON DIAGRAM
ANGLE ACE LESS-THAN ANGLE ABE
      EQUALS ADDED TO UNEQUALS ARE UNEQUAL IN THE SAME ORDER

   TOTAL ELAPSED TIME  7.42 MINUTES
```

APPENDIX IV

```
   PREMISES
**************
SEGMENT CD MEDIAN TRIANGLE ABC
SEGMENT AE MEDIAN TRIANGLE ABC
PRECEDES AEG
SEGMENT AE EQUALS SEGMENT EG
PRECEDES CDF
SEGMENT CD EQUALS SEGMENT DF

   TO PROVE
**************
COLLINEAR FBG

   SYNTACTIC SYMMETRIES
***************************
CA,FG,AC,ED,DE,GF,BB,

   PROOF
***********
PRECEDES CDF
      PREMISE
PRECEDES BDA
      DEFINITION OF MEDIAN
ANGLE FDB EQUALS ANGLE CDA
      VERTICAL ANGLES ARE EQUAL
SEGMENT CD EQUALS SEGMENT DF
      PREMISE
SEGMENT BD EQUALS SEGMENT DA
      DEFINITION OF MEDIAN
TRIANGLE DCA
      ASSUMPTION BASED ON DIAGRAM
TRIANGLE DFB
      ASSUMPTION BASED ON DIAGRAM
TRIANGLE DFB CONGRUENT TRIANGLE DCA
      TWO TRIANGLES ARE CONGRUENT IF SIDE-ANGLE-SIDE EQUALS SIDE-ANGLE-SIDE
ANGLE CFB EQUALS ANGLE FCA
      CORRESPONDING ANGLES OF CONGRUENT TRIANGLES ARE EQUAL
OPP-SIDE BAFC
      ASSUMPTION BASED ON DIAGRAM
SEGMENT FC
      ASSUMPTION BASED ON DIAGRAM
SEGMENT BF PARALLEL SEGMENT AC
      SEGMENTS ARE PARALLEL IF ALTERNATE-INTERIOR ANGLES ARE EQUAL
SEGMENT BG PARALLEL SEGMENT AC
      SYNTACTIC CONJUGATE
COLLINEAR FBG
      ONLY ONE LINE MAY BE DRAWN PARALLEL TO ANOTHER THRU A GIVEN POINT

   TOTAL ELAPSED TIME  3.31 MINUTES
```

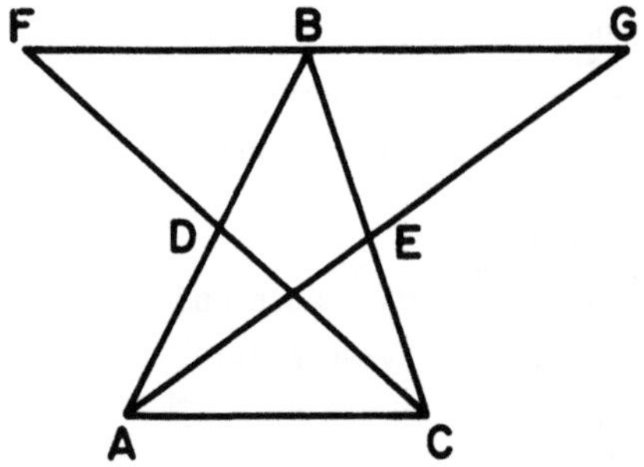

1963

Eliminating the Irrelevant from Mechanical Proofs

M. Davis

In mathematics, the exception *disproves* the rule. Hence a general statement of arithmetic (for example: Fermat's last theorem) could be disproved by exhibiting explicitly a single counter-example. When (as in the case of Fermat's last theorem) each special case of the general statement is decidable by use of a uniform algorithm, such a counter-example, if it exists, would eventually be found (assuming no limitations of time or storage capacity) by a computing machine programmed to search systematically through all possibilities. Is there a similar search procedure for proving (rather than disproving) such general propositions? If one completely specifies the axioms from which the proof is to proceed the answer is *yes*. Uniform search procedures can be formulated which will enable a suitably programmed machine to locate a proof of a given proposition from given axioms if such a proof exists. These search procedures are called *logical proof procedures*. There has recently been considerable interest in improving such procedures and in programming them for computers. Here, we shall give a general framework for the comparison of various extant proof procedures together with a discussion of their advantages and disadvantages. Finally, we shall describe a new improved proof procedure. We shall not discuss the work of Hao Wang, because it does not fit readily into our general framework. For Wang's work, as well as other relevant recent literature, cf. References.

1. **The logical language of mathematics.** In books or research articles in mathematics, assertions are made using ordinary English (or German, French, Russian, etc.) together with appropriate mathematical symbols. With the development of modern symbolic logic it has become possible to express mathematical assertions using only technical symbols with no English words. It is remarkable how economical this necessary "vocabulary" from symbolic logic is. Here we shall give a brief logical vocabulary which is adequate for mathematics, but is by no means as economical as possible. We shall group the logical symbols under the headings: *punctuation marks, truth-functional connectives, variables*, and *quantifiers*.

Punctuation marks: ( ) , . That is, parentheses and comma.

Truth-functional connectives: $\sim$ $\vee$ & $\rightarrow$ $\leftrightarrow$. These symbols stand for certain operations on statements which result in new statements. If p and q are given statements, then:

This research was supported in part by the Mathematical Sciences Directorate of the Air Force of Scientific Research under Contract No. AF 49 (638)–995, and in part by Bell Telephone Laboratories.

Reprinted from: Proceedings, Symposia of Applied Mathematics, Volume 15, pp. 15-30, by permission of the American Mathematical Society.

"$\sim p$" stands for the statement: "not p," i.e. "it is not the case that p."
"$p \vee q$" stands for the statement "p or q."
"$p \,\&\, q$" stands for the statement "p and q."
"$p \rightarrow q$" stands for the statement "If p, then q."
"$p \leftrightarrow q$" stands for the statement "p if and only if q."
However, the ambiguity and lack of precision of the English words is resolved in the case of the truth-functional connective by insisting:

(a) that $\sim p$, $p \vee q$, $p \,\&\, q$, $p \rightarrow q$, and $p \leftrightarrow q$ are to be regarded as well-defined statements whenever p and q are such, whether or not their contents are in any way related, and

(b) that the truth or falsity of these statements are completely determined by those of p and q according to the "truth" tables below (cf. Figure 1).

p	q	$\sim p$	$p \vee q$	$p \,\&\, q$	$p \rightarrow q$	$p \leftrightarrow q$
t	t	f	t	t	t	t
f	t	t	t	f	t	f
t	f	f	t	f	f	f
f	f	t	f	f	t	t

FIGURE 1

This insistence has the somewhat counter-intuitive consequence that we must regard as *true* such unpalatable assertions as:

If $2 + 2 = 5$, then snow is hot.

But this is just another instance of the general tendency in mathematics to increase domains of definition of operations as much as possible even if meaning is thereby given to what was formerly regarded as meaningless. (Consider, for example: $2 - 5$ or $\sqrt{-4}$.) For further discussion of this matter, the reader is referred to any standard book on logic, e.g. [**1**; **2**; **7**; or **10**].

Variables: These are symbols which are intended to be thought of as taking on arbitrary values from some initially preassigned set. Here we shall use as variables the letters $x\ y\ z\ u\ v$ and w with or without subscripts.

Quantifiers: An expression formed by placing a variable in parentheses [such as (x) or (y)] is called a *universal quantifier*. An expression such as $(x)\,A$ is intended to assert that A is true for all values of x. Similarly, an expression formed by placing in parentheses the letter E followed by a variable [e.g. (Ex) or (Ey)] is called an *existential quantifier*, and an expression such as $(Ex)\,A$ is intended to assert that A is true for at least one value of x.

In addition to the *logical symbols* just listed we must allow certain special *mathematical symbols*. Examples of such symbols are: 0, 1, $+$, $=$, $<$, $\in$. Which symbols occur will depend on the particular branch of mathematics under discussion. However, all of these symbols fall into three categories: *constant symbols*, *function symbols*, and *relation symbols*.

A *constant symbol* (0 and 1 are usual examples) is intended to stand for some definite constant. A *function symbol* (such as +) is intended to stand for a function of one or more arguments (e.g. + is ordinarily intended to stand for a function of two arguments). A *relation symbol* (such as =, <, or $\in$) stands for a relation among one or more arguments.

We include a table giving a translation into logical symbolism of various assertions from ordinary mathematics. (See Figure 2.)

Sentence in English	Translation into logical symbolism	Intended range of variables	Constant symbols	Function symbols	Relation symbols
(1) There are infinitely many primes.	$(x)(Ey)[y > x \& p(y)]$	the positive integers	NONE	NONE	p (..is a prime) $>$ (..is greater than..)
(2) Addition is commutative.	$(x)(y)(x + y = y + x)$	the members of some Abelian group	NONE	+	= (..is equal to..)
(3) A set of positive integers which contains 1, and contains the successor of each of its members, consists of all positive integers.	$(v)\{[1 \in v \& (x)(x \in v \rightarrow s(x) \in v)] \rightarrow (x)[I(x) \rightarrow x \in v)]\}$	the set of all positive integers and all sets of positive integers	1	s (successor of)	$\in$ (..is a member of..) I(..is a positive integer)
(4) There is always a prime between n and $2n$.	$(x)(Ey)[y > x \& x + x > y \& p(y)]$	like (1)	NONE	+	$>$ p (like (1))

FIGURE 2

We can now systematically survey all statements which may be written using this symbolism. First, beginning with the variable and constant symbols we may form *terms*, that is expressions built up from these using the various function symbols. In particular, if g is a function symbol which takes n arguments and $u_1, u_2, \cdots, u_n$ are already known to be terms, then $g(u_1, u_2, \cdots, u_n)$ is also a *term*. Custom sanctions a slight variant of this notation:

For g a function symbol which takes 2 arguments (like +), one writes $(u_g v)$ instead of $g(u, v)$.

Examples of terms are:

$$(x + y), \qquad (0 + (x + y)), \qquad s(0 + y).$$

If R is a relation symbol which takes n arguments, and $u_1, \cdots, u_n$ are terms, then the expression $R(u_1, \cdots, u_n)$ is called an *atomic formula*. (The above special notational conventions for function symbols taking 2 arguments is also used for relation symbols; e.g. one writes $(x = y)$ rather than $=(x, y)$.) Now, *the statements which we propose to consider are just those which can be built up from atomic formulas using truth-functional connectives and quantifiers.* It will readily be seen that the statements in Figure 2 all fit into this category. Moreover, we assert that *all propositions of mathematics are among the statements which can be expressed in this way.* This assertion sometimes gives rise to a certain amount of confusion. Namely, the modes of logical expression that we have permitted are often called "first-order" and it is asserted that "higher-order" logics are necessary for mathematics. However, these very "higher-order" logics can themselves be expressed (e.g. by using the $\in$ of set theory) within the present framework. In particular, the various formulations of axiomatic set theory can be expressed in the present framework (cf. [2]), and then, as is well known, the Peano-Dedekind program can be used to develop classical analysis.

2. **Herbrand proofs.** Let C be a statement expressed in the symbolism that has just been described. In C there will be logical symbols (namely: punctuation marks, truth-functional connectives, variables, and quantifiers) and mathematical symbols (namely: constant symbols, function symbols, and relation symbols). An *interpretation* of C is given by specifying "meanings" for the mathematical symbols. More accurately, an interpretation of C is given by specifying:

(1) a nonempty set u called the *universe*;

(2) for each constant symbol in C a definite element of u;

(3) for each function symbol which takes n arguments, a function of n variables from u into u (i.e. a mapping from u^n into u);

(4) for each relation symbol which takes n arguments, a relation among n elements of u.

When an interpretation of C is specified, C becomes a definite mathematical proposition, true or false. For example, let C be chosen as (2) from Figure 2:

$$(x)(y)(x + y = y + x).$$

Below are three interpretations of C:

(1) The universe is the set of rational numbers; $+$ is addition; $=$ is equality.

(2) The universe is the set of vectors in three-dimensional Euclidean space; $+$ is the cross-product; $=$ is equality.

(3) The universe is the set of positive integers; $+$ is multiplication; $=$ is "is less than."

Here C is true under interpretation (1) and false under interpretations (2) and (3). An interpretation which makes a statement C true is called a *model* of C. By a *model of the statements* $C_1, C_2, \cdots, C_n$ we mean a model for the single statement $C_1 \& C_2 \& \cdots \& C_n$; that is, such a model is an interpretation of the mathematical symbols of $C_1, \cdots, C_n$ which makes all of them true. For an example, consider the following set of statements, which we denote by $\mathfrak{G}_1$:

$$(x)(e \circ x = x),$$
$$(x)(I(x) \circ x = e),$$
$$(x)(y)(z)[(x \circ (y \circ z)) = ((x \circ y) \circ z)],$$

$$(x)(x = x),$$
$$(x)(y)[x = y \rightarrow y = x],$$
$$(x)(y)(z)[((x = y) \mathbin{\&} (y = z)) \rightarrow (x = z)],$$
$$(x)(y)[x = y \rightarrow I(x) = I(y)],$$
$$(x)(y)(u)(v)[((x = y) \mathbin{\&} (u = v)) \rightarrow (x \circ u = y \circ v)].$$

What is ordinarily called a *group* is simply a model of these statements in which "=" is interpreted as equality. The interpretations of e, $\circ$, and I are then the group identity element, group multiplication, and the function which maps each group element into its universe, respectively. In fact these statements are a formal system of *axioms for group theory*. In any model of these axioms = must be interpreted as an equivalence relation preserving the operations which are the interpretations of I and $\circ$. Then, the corresponding equivalence classes will form a group. (The group operation will naturally be that which multiplies the equivalence class containing a by that containing b to obtain the equivalence class containing $a \circ b$.)

Another set of statements which is of interest in connection with groups is the following which we call $\mathfrak{G}_2$:

$$(x)P(e,x,x)$$
$$(x)(y)P(I(x), x, e)$$
$$(x)(y)(z)(u)(v)(w)[(P(x,y,u) \mathbin{\&} P(u,z,w) \mathbin{\&} P(y,z,v)) \rightarrow P(x,v,w)]$$
$$(x)(y)(z)(u)(v)(w)[(P(y,z,v) \mathbin{\&} P(x,v,w) \mathbin{\&} P(x,y,u)) \rightarrow P(u,z,w)].$$

Namely, if we are given any group it can be used as a model for $\mathfrak{G}_2$ by interpreting e as the identity, I as the inverse function, and $P(x,y,z)$ as the relation $(x \circ y) = z$. We shall not discuss here the converse question of associating a group with each model of $\mathfrak{G}_2$, except to note that a model of $\mathfrak{G}_2$, in which the interpretation of P is such that for each x,y there is *exactly* one z for which $P(x,y,z)$ is true, is a group under the multiplication operation:

$x \circ y$ is the unique z for which $P(x,y,z)$ is true.

Now, suppose we wish to prove a statement T. The situation in which we are interested is that in which we have obtained a *finite* set of axioms $A_1, \cdots, A_m$ from which, as a starting point, we wish to demonstrate T. We take this to mean that we want to show that:

Every model of $A_1, \cdots, A_m$ is also a model of T.

For example, suppose we wish to prove the elementary theorem in group theory: $(x)(x \circ I(x) = e)$; i.e. the postulated left inverse is also a right inverse. What we wish to show is that this proposition is true in all groups. Since every group is a model of $\mathfrak{G}_1$ it suffices to show that the proposition is true in every model of $\mathfrak{G}_1$. Alternatively we may use the notation of $\mathfrak{G}_2$ and write the proposition as:

$$(x)P(x,\ I(x),\ e)$$

and seek to show that this last is true in every model of $\mathfrak{G}_2$. This point of view towards provability from axioms is usually characterized as *semantic* or *model-theoretic*. Textbooks of logic tend to emphasize a different, *syntactic* point of view in which the derivability of T from $A_1, \cdots, A_m$ is a matter of using a set of formal *rules of inference* to obtain T from $A_1, \cdots, A_m$. But the main result then turns out to be the Gödel completeness theorem which guarantees that the specified rules of inference are *complete* in the following sense:

T can be obtained from $A_1, \cdots, A_m$ by employing the specified rules of inference if and only if every model of $A_1, \cdots, A_m$ is also a model of T.

So, the syntactic and semantic points of view turn out to be equivalent. Here, we shall take the semantic point of view throughout. The reader who wishes more information about Gödel's completeness theorem is referred to [**1**; **2**; **7**; **10**].

Every interpretation of T must make T either true or false. Hence, every model of $A_1, \cdots, A_m$ is either a model of T or of $\sim T$. Thus, to say that every model of $A_1, \cdots, A_m$ is also a model of T is equivalent to saying that *no model of $A_1, \cdots, A_m$ is also a model of $\sim T$*. The proof procedures to be treated here may all be discussed from this point of view; that is, *one proposes to demonstrate that T is a consequence of $A_1, \cdots, A_m$ by showing that there is no model for the set of sentences $A_1, \cdots, A_m, \sim T$.*

We begin by subjecting the separate statements $A_1, \cdots, A_m, \sim T$ to preliminary processing as described in Steps 1–8 below. In general each step is to be performed over and over again as long as possible; when a given step can no longer be applied one goes on to the next step. The reader should note that:

(1) The steps could easily be made purely mechanical and hence capable of being programmed for a computer. (In practice it is as easy to carry out these preliminary steps by hand.)

(2) The set of statements has a model before one of the steps has been carried out if and only if it does after the step in question has been carried out.

Step 1. *Relabel variables.* If the same variable occurs in more than one quantifier in the same statement, use a new variable for one of them. For example, $(x)R(x) \vee (x)S(x)$ should be written as, say, $(x)R(x) \vee (y)S(y)$.

Step 2. *Eliminate* $\rightarrow$ *and* $\leftrightarrow$. Wherever $\rightarrow$ and $\leftrightarrow$ occur, make the replacements:

Replace $A \rightarrow B$ by $(\sim A) \vee B$.
Replace $A \leftrightarrow B$ by $(A \,\&\, B) \vee [(\sim A) \,\&\, (\sim B)]$.

Step 3. *Move* $\sim$ *inwards.* Where possible make the replacements:

Replace $\sim (x)M$ by $(Ex) \sim M$.
Replace $\sim (Ex)M$ by $(x) \sim M$.
Replace $\sim (M \,\&\, N)$ by $(\sim M) \vee (\sim N)$.
Replace $\sim (M \vee N)$ by $(\sim M) \,\&\, (\sim N)$.
Replace $\sim \sim M$ by M.

Ultimately the statements are obtained in a form where each $\sim$ occurs immediately preceding an atomic formula.

For atomic formulas we write $\bar{R}(U_1, \cdots, U_m)$ instead of $\sim R(U_1, \cdots, U_m)$; $\bar{R}(U_1, \cdots, U_m)$ and $R(U_1, \cdots, U_m)$ are both referred to as literals.

Thus beginning with

$$(x)\{R(x) \vee \sim (Ey)[S(x,y) \mathbin{\&} (R(x) \vee U(x))]\}$$

we obtain

$$(x)\{R(x) \vee (y)[\bar{S}(x,y) \vee (\bar{R}(x) \mathbin{\&} \bar{U}(x))]\}.$$

Step 4. *Eliminate existential quantifiers.* Cross out each existential quantifier, say (Ey). The corresponding variable (in this case y) is replaced by $g(x_1, \cdots, x_m)$, where g is a new function symbol, and $x_1, \cdots, x_m$ are all of the variables occurring in universal quantifiers to the left of the existential quantifier (in this case (Ey)).

In case, there are no universal quantifiers to the left of the existential quantifier, the variable is replaced by a new constant symbol g; this case may be included in the general case *by regarding a constant symbol as just a function symbol of* 0 *arguments.*

To see that Step 4 preserves the properties of having or not having a model, we note that if the statements had a model before one use of Step 4, for each $x_1, \cdots, x_m$ of the universe there would be one or more values of y making the statement being processed true. Hence we may take as the interpretation of g a function such that for each $x_1, \cdots, x_m$ the statement is true when $y = g(x_1, \cdots, x_m)$. (The tacit use of the axiom of choice here can be avoided; however we need not concern ourselves with this point.) Conversely, if the statements have a model after processing, then $y = g(x_1, \cdots, x_m)$ will make the statement true for all $x_1, \cdots, x_m$ so that the original existential condition likewise has a model.

Step 5. *Advance universal quantifiers.* Move all universal quantifiers to the left so that the statement has the form of a sequence of universal quantifiers followed by a quantifier-free expression.

Step 6. *Distribute, wherever possible, & over* $\vee$. I.e. replace $(A \mathbin{\&} B) \vee C$ by $(A \vee C) \mathbin{\&} (B \vee C)$.

After Steps 1–6 have been performed each statement must have the form

$$(x_1)(x_2) \cdots (x_n)[B_1 \mathbin{\&} B_2 \mathbin{\&} \cdots \mathbin{\&} B_k],$$

where, for each B_i

$$B_i = l_1^{(i)} \vee l_2^{(i)} \vee \cdots \vee l_{r_i}^{(i)},$$

and where each $l_j^{(i)}$ is a literal. In this expression each B_i is called a *clause*, and the $l_j^{(i)}$'s are called the *literals of* B_i.

Step 7. *Simplify.* If an atomic formula and its negation are both literals of a clause B_i, strike out the entire clause. If a literal occurs twice as a literal of the same clause B_i, strike out one occurrence. (E.g. the clause $l_1 \vee l_2 \vee l_1$ is to be replaced by $l_1 \vee l_2$.)

This completes the preliminary processing of the clauses. It must be emphasized that it is not being claimed that the statements resulting from this processing are equivalent in meaning to $A_1, \cdots A_m, \sim T$.

What is being claimed is that a model exists for the former if and only if it does for the latter.

We take advantage of the especially simple form into which the statements have now been brought to introduce certain abbreviations: We omit the universal quantifiers. We also omit explicit use of the symbols & and $\vee$. Thus, we write $l_1 l_2 \cdots l_r$ instead of $l_1 \vee l_2 \vee \cdots \vee l_r$. The conjunction of clauses is indicated by listing them in a single vertical column. The clauses coming from each of the statements $A_1, \cdots, A_m, \sim T$ may now be combined into a single list of clauses. This single list may be further simplified by using:

Step 8. *Eliminate redundant clauses.* Strike out any clause which another clause is part of (or all of). (For, the conjunction of all the clauses is true just when all of the clauses are true; but when a clause is true, any clause of which it is a part is likewise true.)

As an example, let $A_1, \cdots, A_4$ be the group theory axioms $\mathfrak{G}_2$ and let T be $(x)P(x, I(x), e)$. Applying Steps 1–8 to $\mathfrak{G}_2$ together with $\sim T$, we obtain the clauses:

(1) $P(e,x,x)$

(2) $P(I(x),x,e)$

(3) $\bar{P}(x,y,u)\bar{P}(u,z,w)\bar{P}(y,z,v)P(x,v,w)$

(4) $\bar{P}(y,z,v)\bar{P}(x,v,w)\bar{P}(x,y,u)P(u,z,w)$

(5) $\bar{P}(a,I(a),e)$,

where a is a constant symbol. To see how e.g. $\sim T$ was processed we begin with:

$$\sim (x)P(x, I(x), e).$$

Steps 1 and 2 do not apply. Step 3 yields

$$(Ex)\bar{P}(x, I(x), e).$$

Step 4 gives

$$\bar{P}(a, I(a), e)$$

and the remaining steps do not apply.

The combined list of clauses $C_1, C_2, \cdots, C_k$ may be thought of as the single quantifier-free expression

$$C_1 \,\&\, C_2 \,\&\, \cdots \,\&\, C_k.$$

Such an expression is in *conjunctive normal form*; this means that it is the conjunction of clauses to which the simplification of Step 7 has been applied. In general, an expression in conjunctive normal form (or, for that matter, any quantifier-free expression) is called a *truth-functional contradiction* if no matter how the truth values *truth* and *falsehood* are assigned to the atomic formulas of the expression, the expression is false. (Equivalently, we are here demanding that its negation be a tautology.)

With each complete list of clauses obtained from $A_1, \cdots, A_m, \sim T$ we associate a certain set H of terms which we call the *Herbrand universe* for that list. We first define H_0 to be the set of constant symbols and variables which occur in the list.

Next we take H_{n+1} to consist of the elements of H_n together with all terms that can be obtained by applying function symbols occurring in the list of clauses to elements of H_n. Finally, we set $H = \bigcup_n H_n$.

For the group theory example just considered $H_0 = \{e,a,x,y,z,u,v,w\}$,

$$H_1 = \{e,a,x,y,z,u,v,w,I(e),I(a),I(x),I(y),I(z),I(u),I(v),I(w)\}.$$

The search procedures for proofs which we are considering are all based on the following form of Herbrand's theorem:

Let $A_1, \cdots, A_m$ be given axioms and T a proposed theorem. Let $C_1, C_2, \cdots, C_k$ be the list of clauses obtained from $A_1, \cdots, A_m$ and $\sim T$ by using Steps 1–8 and let H be the resulting Herbrand universe. Then, T is a consequence of $A_1, \cdots, A_m$ if and only if there is a list of clauses $Q_1, \cdots, Q_r$, each of which is obtained from some C_i, $i = 1, 2, \cdots, k$, by replacing its variables by members of H, such that the formula Q_1 & Q_2 & $\cdots$ & Q_r is a truth-functional contradiction.

This theorem is proved as follows:

(1) Suppose such a Q_1 & Q_2 & $\cdots$ & Q_r is a truth-functional contradiction, but that nevertheless $A_1, \cdots, A_m, \sim T$ have a model. Hence $C_1, \cdots, C_k$ have a model with universe U. Assign to each variable and each constant symbol in H some fixed element of U. Then, each element of H is associated with a unique element of U. Under this interpretation each Q_i, $i = 1, 2, \cdots, r$ is true. So, Q_1 & Q_2 & $\cdots$ & Q_r cannot be a truth-functional contradiction.

(2) Conversely suppose $A_1, \cdots, A_m, \sim T$ has no model. Then $C_1, \cdots, C_k$ has no model. Let

$$Q_1, Q_2, Q_3, \cdots$$

be all possible clauses obtained from $C_1, \cdots, C_k$ by replacing variables by members of H. Then, we claim that no matter how truth values are assigned to the atomic formulas of $Q_1, Q_2, Q_3, \cdots$ at least one Q_i must be made false. Assuming this claim to be true for the moment, we are done, since by [**10**, pp. 254–256] (the "law of infinite conjunction") we will then have that Q_{s_1} & Q_{s_2} & $\cdots$ & Q_{s_r} is a truth-functional contradiction for some $s_1, s_2, \cdots, s_r$. Finally, that our claim is true can be seen by noting that if it were false we could construct a model for $C_1, \cdots, C_k$ with universe H. In this model, the interpretation of the constant and function symbols is automatic. And, given an assignment of truth values to the atomic formulas of $Q_1, Q_2, Q_3, \cdots$ which makes all of the Q_i true, we can interpret each relation symbol R by regarding $R(U_1, \cdots, U_m)$ for $U_1, \cdots, U_m \in H$ as true just in case the given assignment makes the atomic formula $R(U_1, \cdots, U_m)$ true. So, the theorem has been proved.

Using Herbrand's theorem, we may construct any number of complete search procedures for proofs along the following lines:

Given axioms $A_1, \cdots, A_m$ and the proposed Theorem T, we first apply Steps 1–7 to obtain the list of clauses $C_1, \cdots, C_k$. We then apply a *generation* procedure which generates the clauses $Q_1, Q_2, \cdots$ that can be obtained from $C_1, \cdots, C_k$ by replacing variables by members of the Herbrand universe. At various times we interrupt the generation procedure and test the conjunction $Q_1 \& Q_2 \& \cdots \& Q_N$ of clauses so far generated to see if it is a truth-functional contradiction. If it is, we have a proof. If not we generate more clauses. This general scheme is illustrated by the flow chart of Figure 3.

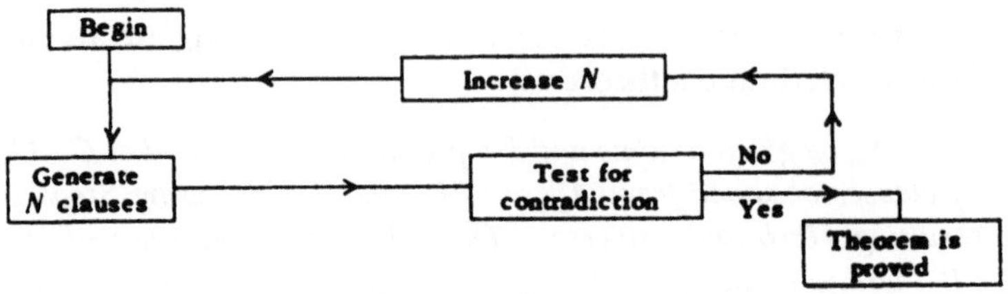

FIGURE 3

3. **Gilmore's procedure.** The procedure of Gilmore [6] may be described (with inessential modifications) as follows: Let the variables present in $C_1, \cdots, C_k$ be $x_1, \cdots, x_q$. Let $(t_1^{(j)}, \cdots, t_q^{(j)})$, $j = 1, 2, 3, \cdots$ be some definite enumeration of all q-tuples of elements H. Let $C_1^{(j)}, \cdots, C_k^{(j)}$ be the result of replacing $x_1, \cdots, x_q$ by $t_1^{(j)}, \cdots, t_q^{(j)}$, in $C_1, \cdots, C_k$, respectively. For each successive value of r, put in *disjunctive* normal form, the formula

$$(*)\qquad \begin{cases} C_1^{(1)} \& C_2^{(1)} \& \cdots C_k^{(1)} \\ \& C_1^{(2)} \& C_2^{(2)} \& \cdots C_k^{(2)} \\ \cdots \qquad\qquad \cdots \\ \& C_1^{(rN)} \& C_2^{(rN)} \& C_k^{(rN)}, \end{cases}$$

where N is some given integer (say 10).

To place (*) into disjunctive normal form means to iteratively distribute $\vee$ over & (that is to replace wherever possible $(A \vee B) \& C$ by $(A \& C) \vee (B \& C)$ until the formula has the form

$$V_1 \vee V_2 \vee \cdots \vee V_k,$$

where, for each V_i,

$$V_i = l_1^{(i)} \& l_2^{(i)} \& \cdots \& l_{r_i}^{(i)}.$$

Such a formula is a truth-functional contradiction if and only if in each V_i one of the literals occurring is the negation of the other. (Since otherwise, truth values can be chosen for the atomic formulas making that V_i, and hence the entire formula, true.)

The entire procedure can be represented by the flow diagram of Figure 4.

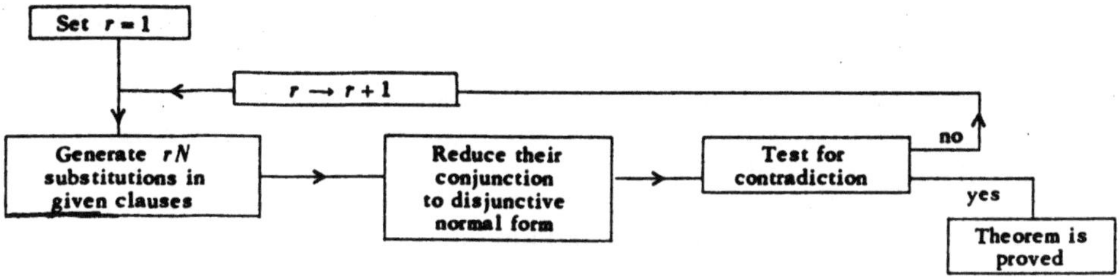

FIGURE 4

4. **The Davis-Putnam procedure.** The Gilmore procedure, when programmed for an IBM 704, failed to yield a proof for the following example:

$$(Ex)(Ey)(z)\{[F(x,y) \rightarrow (F(y,z)\&F(z,z))]$$
$$\& [(F(x,y)\&G(x,y)) \rightarrow (G(x,z)\&G(z,z))]\}.$$

Here this theorem is to be proved using no special mathematical axioms; i.e. it is what is called a *logical theorem*. Applying Steps 1–8 to the negation of this theorem (note that Step 8 eliminates 3 clauses) we obtain the list of clauses:

$$F(x,y),$$
$$\bar{F}(y, f(x,y))\bar{F}(f(x,y), f(x,y))G(x,y),$$
$$\bar{F}(y, f(x,y))\bar{F}(f(x,y), f(x,y))\bar{G}(x, f(x,y))\bar{G}(f(x,y), f(x,y)).$$

The Gilmore program was reported to have tried just 7 substitutions, i.e. $rN = 7$. It is easy to see that this involves a disjunctive normal form of 12^7 conjunctive clauses each consisting of 21 literals. The procedure of [3] (as well as the slightly improved version of [4]) is quite like the Gilmore procedure, but omits the step of reducing to disjunctive normal form. The procedure for testing a formula in conjunctive normal form for being a contradiction proceeds by using the rules:

1. ONE-LITERAL CLAUSE RULE.

(a) If the one-literal clauses p and $\bar{p}$ are both present the formula is a contradiction.

(b) If (a) does not apply and one-literal clause l is present then all occurrences of $\sim l$ are eliminated together with all clauses in which l occurs. (By $\sim l$ we understand $\bar{p}$ if l is an atomic formula p and p if l is a negated atomic formula $\bar{p}$.) If no clauses remain the formula is not a contradiction.

2. AFFIRMATIVE-NEGATIVE RULE. In case the literal l is present but $\sim l$ is not, then all clauses containing l may be eliminated. If no clauses remain the formula is not a contradiction.

3. SPLITTING RULE. Let p be an atomic formula such that p and $\bar{p}$ are both present. Let A_1 and A_2 be obtained from the original formula by suppressing all

clauses in which p or $\bar{p}$, respectively, occur. Then the formula is a contradiction if and only if A_1 and A_2 are both contradictions.

The Gilmore example is easily managed (even by hand) using this procedure (cf. [3]); but the group theory problem discussed above proves intractable (for a discussion of the difficulties, cf. [4]).

5. **The Prawitz procedure.** The Davis-Putnam procedure suffers from generation of far more clauses than are needed to accomplish a proof. In [9], Prawitz proposed an algorithm free of this defect.

Again, let the list of clauses be $C_1, \cdots, C_k$, and let the variables present be $x_1, \cdots, x_q$. Then, we let $C_1^{(j)}, \cdots, C_k^{(j)}$ be the result of replacing $x_1, \cdots, x_q$ by the new and distinct variables $x_{1j}, \cdots, x_{qj}$, respectively. Again we choose an integer N and for successive values of r, we put in *disjunctive normal form*, the formula (*) of §3. Then, seek substitutions of elements of the Herbrand universe H for variables which can bring two literals in the same (conjunctive) clause to be negations of one another. If no such substitutions exist for all clauses, increase r and try again.

As an example (given by Prawitz in [9]), consider the logical theorem

$$(x)(y)[P(x) \vee Q(y)] \rightarrow [(x)P(x) \vee (y)Q(y)].$$

Steps 1–8 yield the list of clauses:

$$P(x)Q(y)$$
$$\bar{P}(a)$$
$$\bar{Q}(b).$$

In disjunctive normal form this becomes

$$[P(x) \;\&\; \bar{P}(a) \;\&\; \bar{Q}(b)] \vee [Q(y) \;\&\; \bar{P}(a) \;\&\; \bar{Q}(b)].$$

This will be a contradiction just in case $x = a$ and $y = b$. Here $rN = 1$ already sufficed for a proof.

The Prawitz procedure has the great virtue of generating only the substitutions in $C_1 \;\&\; C_2 \;\&\cdots\&\; C_k$ actually needed for a contradiction. But the use of disjunctive normal form inevitably produces the same difficulties as Gilmore's procedure.

6. **A new procedure.** We now describe a new kind of procedure which seeks to combine the virtues of the Prawitz procedure and those of the Davis-Putnam procedure.

DEFINITION. *A formula in conjunctive normal form is called a linked conjunct if for each occurrence of a literal l in a given clause there is at least one occurrence of $\sim l$ in some other clause. Such an occurrence of $\sim l$ is said to be a mate for the given occurrence of l.*

THEOREM. *Let $C_1, \cdots, C_k$ be a list of (disjunctive) clauses whose conjunction is in conjunctive normal form and is a contradiction. Then there is a subset of $\{C_1, \cdots, C_k\}$ whose conjunction is a contradiction and a linked conjunct.*

Suppose a clause contains the literal l, and that no other clause contains $\sim l$. Then all clauses containing l can be deleted. For, setting l true will make all clauses containing l also true. Hence, any assignment to the remaining atomic formulas must make some clause not containing l false. (This remark is merely the justification for the affirmative-negative rule.) Repeating this process must lead to a linked conjunct which is also a contradiction.

This theorem on linked conjuncts (or rather its dual) was noted independently by Dunham and North (cf. [5]).

Hence, we can modify the Davis-Putnam procedure so as to search specifically for linked conjuncts; testing for a contradiction need only be done when one has obtained a linked conjunct. Let us find the appropriate linked conjunct in the group theory problem.

The desired linked conjunct must contain the clause (5), since it is the only trace of the theorem to be proved. Substitution in (1) or (2) can never produce (5). Hence a clause must be present which is obtained by substitution from (3) or (4). An attempt to use (3) in this way leads to a dead end, as the reader will readily verify. To use (4), we obtain:

$$\bar{P}(a, I(a), e),$$
$$\bar{P}(y, I(a), v)\bar{P}(x, v, e)\bar{P}(x, y, a)P(a, I(a), e).$$

By substitution in this last we must make the first three literals agree in form either with a substitution in (1) or (2) or with the final literal of (3) (or actually of another substitution in (4)). Of the various ways of accomplishing this, we choose to set $y = e$, $v = I(a)$. Then the first literal will have the form of (1), and if we further set $x = I(I(a))$, the second literal has the form of (2). This leads to the list of clauses:

$$\bar{P}(a, I(a), e)$$
$$\bar{P}(e, I(a), I(a))\bar{P}(I(I(a)), I(a), e)\bar{P}(I(I(a)), e, a)P(a, I(a), e)$$
$$P(e, I(a), I(a))$$
$$P(I(I(a)), I(a), e).$$

But, because the third literal of the second clause has no mate, this is not yet a linked conjunct. (1) and (2) are no help, so we try (3):

$$\bar{P}(I(I(a)), y, u)\bar{P}(u, z, a)\bar{P}(y, z, e)P(I(I(a)), e, a).$$

Here the first literal suggests setting $y = I(a)$, $u = e$, and then the second clause suggests $z = a$. We thus obtain the linked conjunct:

$$\bar{P}(a, I(a), e)$$
$$\bar{P}(e, I(a), I(a))\bar{P}(I(I(a)), I(a)e)\bar{P}(I(I(a)), e, a)P(a, I(a), e)$$
$$P(e, I(a), I(a))$$
$$P(I(I(a)), I(a), e)$$
$$\bar{P}(I(I(a)), I(a), e)\bar{P}(e, a, a)\bar{P}(I(a), a, e)P(I(I(a)), e, a)$$
$$P(e,a,a)$$
$$P(I(a), a, e).$$

Several applications of the one-literal clause rule suffice to show that this linked conjunct is a contradiction.

It remains to propose an exhaustive search procedure for finding linked conjuncts. We begin by noting that in a linked conjunct obtained from (1)–(5) we would expect more clauses coming from (1) and (2) than from (3) and (4). This is because each use of (3) or (4) requires us to locate mates for 3 literals of the form $\bar{P}(U,V,W)$. So we might guess that these will be about 3 times as many clauses coming from (1) and (2) as from (3) or (4). We can express this feeling in the form of the ratios:

$$3:3:1:1:1.$$

In general, beginning with clauses $C_1, C_2, \cdots, C_k$, let the integers $n_1, n_2, \cdots, n_k$ be assigned to them, where n_i/n_j is thought of as the ratio of expected clauses coming from C_i to those from C_j. For each fixed value of r, $r = 1, 2, 3, \cdots$, there are only a *finite number* of essentially different linked conjuncts containing no more than rn_i clauses coming from each C_i, $i = 1, 2, \cdots, k$. (Two linked conjuncts are not *essentially different* if one can be obtained from the other by suitable substitutions for variables.) These may be located systematically by seeking all possible ways of finding mates for the various literals.

D. McIlroy has proposed an interesting scheme for carrying out the search for all linked conjuncts with each fixed value of r. The scheme is similar to certain game-playing algorithms in that it is based on *scoring* of "positions," and on *moving* to the "position" with the best score.

A list of clauses is scored as follows: Each occurrence of a literal which does not yet have a mate gets as its score the total number of literals in the other clauses which could be used as mates for it (after suitable substitutions). The scoring for the group theory problem begins as follows:

(1) $$\overset{6}{P(e,x,x)}$$

(2) $$\overset{6}{P(I(x),x,e)}$$

(3) $$\overset{3}{\bar{P}(x,y,u)}\overset{3}{\bar{P}(u,z,w)}\overset{3}{\bar{P}(y,z,v)}\overset{3}{P(x,v,w)}$$

(4) $$\overset{3}{\bar{P}(y,z,v)}\overset{3}{\bar{P}(x,v,w)}\overset{3}{\bar{P}(x,y,u)}\overset{3}{P(u,z,w)}$$

(5) $$\overset{2}{\bar{P}(a,I(a),e)}.$$

The strategy is to find a mate first for the literal (or one of the literals) with the least score. If, as in this case, this least score is greater than 1, there still will be more than one possibility. Then our strategy is to tentatively try all of them ("look ahead") and to then rescore. The one we keep is one with smallest minimum score. In this case the two possibilities are:

$$\overset{5}{P(e,x,x)}$$

$$\overset{4}{P(I(x),\ x,\ e)}$$

$$\overset{1}{\bar{P}(a,y,u)}\overset{3}{\bar{P}(u,z,e)}\overset{2}{\bar{P}(y,z,I(a))}P(a,I(a),e)$$

$$\overset{3}{\bar{P}(y,z,v)}\overset{3}{\bar{P}(x,v,w)}\overset{3}{\bar{P}(x,y,u)}\overset{4}{P(u,z,w)}$$

$$\bar{P}(a,\ I(a),\ e)$$

with minimum score of 1, and

$$\overset{5}{P(e,x,x)}$$

$$\overset{5}{P(I(x),\ x,\ e)}$$

$$\overset{3}{\bar{P}(x,y,u)}\bar{P}(u,z,w)\overset{3}{\bar{P}(y,z,v)}\overset{4}{P(x,v,w)}$$

$$\overset{3}{\bar{P}(y,\ I(a),\ v)}\overset{3}{\bar{P}(x,v,e)}\overset{2}{\bar{P}(x,y,a)}P(a,\ I(a),\ e)$$

$$\bar{P}(a,\ I(a),\ e)$$

with a minimum score of 2.

Hence, we continue with the first possibility. Eventually, we may be forced to return to the second, either because no linked conjunct was obtained or because the one obtained was not a contradiction. In the present group theory problem, the linked conjunct found above is the first encountered.

A refinement on the proposed procedure which might possess certain practical advantages would be to attempt amalgamation of linked conjuncts by further substitutions so as to make such conjuncts share literals. This is not theoretically necessary, but might, in practice, lead to quicker proofs of certain theorems.

A proof procedure on these lines is currently being programmed for the IBM 7090 at Bell Telephone Laboratories, Murray Hill, N.J.

References

1. Alomzo Church, *Introduction to mathematical logic*, Vol. I, Princeton Univ. Press, Princeton, N. J., 1956.

2. Martin Davis, *Mathematical logic*, Mimeographed lecture notes, New York Univ. Inst. Math. Sciences, New York, 1960.

3. Martin Davis and Hilary Putnam, *A computing procedure for quantification theory*, J. Assoc. Comput. Mach. 7 (1960), 201–215.

4. Martin Davis, George Logemann, and Donald Loveland, *A machine program for theorem-proving*, Comm. ACM **5** (1962), 394–397.

5. B. Dunham and J. H. North, *Theorem testing by computer*, Proc. Sympos. Math. Theory Automata, Polytechnic Press, Brooklyn, N. Y. (to appear).

6. Paul C. Gilmore, *A proof method for quantification theory*, IBM J. Res. Develop. **4** (1960), 28–35.

7. David Hilbert and Wilhelm Ackermann, *Principles of mathematical logic*, Chelsea, New York, 1950.

8. Dag Prawitz, *An improved proof procedure*, Theoria **26** (1960), 102–139.

9. Dag Prawitz, Håkan Prawitz and Neri Voghera, *A mechanical proof procedure and its realization in an electronic computer*, J. Assoc. Comput. Mach. **7** (1960), 102–128.

10. Willard Van Orman Quine, *Methods of logic*, Henry Holt, New York, 1959.

11. Abraham Robinson, *Proving a theorem (as done by man, logician or machine)*, Summaries of talks at Cornell Summer 1957 Institute for Symbolic Logic, IDA, Princeton, N. J., 1960, pp. 350–352.

12. Hao Wang, *Towards mechanical mathematics*, IBM J. Res. Develop. **4** (1960), 2–22.

13. ———, *Proving theorems by pattern recognition*. I, Comm. ACM **3** (1960), 220–234.

14. ———, *Proving theorems by pattern recognition*. II, Bell System Tech. J. **40** (1961), 1–41.

A Semi-Decision Procedure for the Functional Calculus

J. Friedman

Abstract. This paper develops algorithms by which decision procedures for the functional calculus can be applied mechanically. Given any expression in Skolem normal form with prefix $(\exists y_1)(\exists y_2) \cdots (\exists y_m)(z_1)(z_2) \cdots (z_n)$ and matrix M, the algorithms lead to the construction of a matrix M^* such that $M \supset M^*$ is valid and such that the expressions $(\exists y_1)(\exists y_2) \cdots (\exists y_m)(z_1)(z_2) \cdots (z_n) \cdot M$ and $(\exists y_1)(\exists y_2) \cdots (\exists y_m)(z_1)(z_2) \cdots (z_n) \cdot M^*$ are interprovable. This procedure is thus a semi-decision procedure for the general Skolem case. For two special cases of this prefix, it is further proved that the method provides a solution to the decision problem in the sense that the given expression is a theorem if and only if M^* is tautologous. These cases are (1) a matrix M in which every elementary part contains at least one of the individual variables $z_1, z_2, \cdots, z_n$ or contains only one individual variable or contains both y_1 and y_2 and no other individual variables; and (2) a matrix M in which every elementary part contains at least one of the individual variables $z_1, z_2, \cdots, z_n$ or contains only one individual variable or contains all of $y_1, y_2, \cdots, y_m$.

Introduction

In this paper we describe algorithms for the application of a decision procedure derived from the Gödel-Herbrand decision table method for the functional calculus. We develop the algorithms in a way which is designed for implementation by computer, and in the course of doing so discover that additional previously unsolved cases of the decision problem are covered [7]. In this paper we report on the algorithms and the additional cases obtained. It is anticipated that future papers will discuss the realization of these algorithms on a computer.

The problem of mechanized theorem-proving in the functional calculus has received increasing attention in recent years. Hao Wang [13, 14, 15] has obtained many results on this problem and has independently obtained the result given in Part II. Other approaches to the problem have been presented by Gilmore [8] and by Prawitz, et al. [12].

The discussion here is concerned with formulas in Skolem normal form, $(\exists y_1)(\exists y_2) \cdots (\exists y_m)(z_1)(z_2) \cdots (z_n) \cdot M$, without free variables. Rather than dealing directly with the matrix M we consider the set S of systems of truth-values which falsify M. It is known from the work of Gödel and Herbrand that an expression $(Qx_1)(Qx_2) \cdots (Qx_k) \cdot M$ in prenex normal form is valid if and only if the set S of falsifying systems cannot be used to falsify every row of an infinite decision table.

It is clear (trivially) that if the decision table can be falsified by any subset of S, it can also be falsified by S. Our approach to implementation of this decision

Received June, 1962; revised September, 1962. The research here reported was done at Avion Division, ACF Industries, Inc., and was sponsored by the Air Force Cambridge Research Laboratories, Air Research and Development Command, under Contract AF 19(604)-1582, [1], [2].

Article from: Journal of the Association for Computing Machinery, Volume 10, Number 1, January 1963.

method is to develop algorithms for the formation of a particular subset S^* of S such that the converse is also true: if the decision table can be falsified by S it can also be falsified by S^*. Then, clearly, if the set S^* is empty, the table cannot be falsified and the given expression is a theorem. The procedures can thus be applied to all cases, including unsolvable ones, and provide a semi-decision procedure. For special solvable cases of the decision problem it is proved that if any falsifying systems remain in S^* then the table can be falsified and the given expression is not a theorem.

In summary we shall construct a matrix M^* such that $M \supset M^*$ and prove that: in every case

$$\vdash (Qx_1)(Qx_2) \cdots (Qx_k).\, M \Leftrightarrow \vdash (Qx_1)(Qx_2) \cdots (Qx_k).\, M^*$$

and in certain special cases

$$\vdash (Qx_1)(Qx_2) \cdots (Qx_k).\, M \Leftrightarrow M^* \text{ is tautologous.}$$

I. DECISION TABLE METHOD

The known decision procedures for expressions in prenex normal form have been outlined by Alonzo Church in his paper "Special Cases of the Decision Problem" [4]. We will use that paper as a starting point for our work. We first review some definitions and previous results which will be needed.

Well-Formed Formula. An elementary *well-formed formula* either consists of a propositional variable alone, or has the form $f(x_1, x_2, \cdots, x_n)$, where f is an n-adic functional variable, and $x_1, x_2, \cdots, x_n$ are individual variables (not necessarily all distinct). A well-formed formula (wff) is an expression built up by means of connectives of the propositional calculus and quantifiers from elementary well-formed formulas. And the elementary well-formed formulas thus used in building up a well-formed formula are called its *elementary parts*.

Valid; Satisfiable. If a particular non-empty domain D is taken as the range of the individual variables of a well-formed formula, then the range of the n-adic functional variables is to consist of all n-adic propositional functions over this domain, and the range of the propositional variables is to consist of the two truth-values, truth and falsehood. Upon taking the ranges of the variables in this way, a wff is said to be *valid in the domain D* if it has the value truth for every system of values of its free variables, *satisfiable in the domain D* if it has the value truth for at least one system of values of its free variables, *falsifiable in the domain D* if it has the value falsehood for at least one system of values of its free variables. A wff is said to be *valid* if it is valid in every non-empty domain, *satisfiable* if it is satisfiable in some non-empty domain.

It is obvious that a wff is valid in a domain D if and only if it is not falsifiable in D, or (what is really the same thing) if and only if its negation is not satisfiable in D. And a wff is valid, or not falsifiable, if and only if its negation is not satisfiable.

Prenex Normal Form. A well-formed formula C of the functional calculus is said to be in *prenex normal form* if C has the form $(Qa_1)(Qa_2) \cdots (Qa_n) \cdot M$, where $a_1, a_2, \cdots, a_n$ are distinct individual variables, each of the (Qa_i) is either (a_i) or $(\exists a_i)$, and where M is a quantifier-free formula in which each of the variables $a_1, a_2, \cdots, a_n$ actually occurs. M is called the *matrix* of C; $(Qa_1)(Qa_2) \cdots (Qa_n)$ is called the *prefix*. We shall observe the convention that M will always represent a quantifier-free wff.

It is well-known that every well-formed formula A of the functional calculus can be reduced to a wff C in prenex normal form which is equivalent to A in the sense that the wff $A \equiv C$ is valid, and hence further in the sense that A is valid if and only if C is valid. Moreover, if L is a formula which contains free variables $a_1, a_2, \cdots, a_k$, then $La_1a_2 \cdots a_k$ is valid if and only if $(a_1)(a_2) \cdots (a_k) La_1a_2 \cdots a_k$ is valid.

Skolem Normal Form. As was shown by Skolem, a wff A in prenex normal form without free variables and with arbitrary prefix can be reduced to a wff B in prenex normal form without free variables and with a prefix of the form $(\exists y_1)(\exists y_2) \cdots (\exists y_m)(z_1)(z_2) \cdots (z_n)$, in such a way that A is valid if and only if B is valid. B is said to be in *Skolem normal form*. For a statement of the reduction procedure see, for example, Church [5, §§42, 43]. In general, the reduction will increase both the number of existential quantifiers and the number of universal quantifiers in the prefix, as well as the number of functional variables in the matrix. But Kalmár [11] (see Ackermann [3]) has shown that, in the special case in which A has a prefix of the form $(x_1)(x_2) \cdots (x_k)(\exists y_1)(\exists y_2) \cdots (\exists y_m)(z_1)(z_2) \cdots (z_n)$, the reduction can be accomplished in such a way that the number m of existential quantifiers in the prefix is not increased. Therefore, without loss of generality, we shall consider in this paper only formulas in Skolem normal form.

Completeness. The functional calculus of first-order is to be taken as based on some definite system of axioms and primitive rules of inference, which for the purposes at hand it is not necessary to state in detail, but which may be any one of various known *complete* systems of axioms and rules for the first-order calculus. We then rely on the Gödel completeness theorem for assurance that the class of valid wffs and the class of theorems coincide. And we use the notation "$\vdash A$" to mean either that A is a theorem or that A is valid (in all non-empty domains), since it is not important for our purposes to distinguish the two things.

Decision Procedure. By a *decision procedure* for a special class of wffs is meant an effective test for deciding whether any given wff L of this class is valid (or equivalently, is a theorem). In calling a procedure *effective* we mean, roughly speaking, that there is assurance of reaching the desired result in every particular case by a faithful and purely mechanical following of fixed instructions, without the need for exercise of ingenuity.

The treatment of many of the classical solvable cases is unified by the use of a metatheorem which may be obtained, either by an extension of the same method which was used by Gödel in his original proof of the completeness metatheorem [9], or else as a corollary of the results established by Herbrand in his dissertation [10]. Rather than state the metatheorem directly, we state it as it applies to a decision table which we then describe:

Metatheorem. The wff $(Qa_1)(Qa_2) \cdots (Qa_n) \cdot M$ is a theorem if and only if it is not falsifiable in the decision table below; $(Qa_1)(Qa_2) \cdots (Qa_n) \cdot M$ is falsifiable if and only if M can be falsified in every row of the decision table.

Construction of a Decision Table. Given a formula $(Qa_1)(Qa_2) \cdots (Qa_m) \cdot M$, we construct a decision table for it as follows.[1] Let the elementary parts of M be $A_1, A_2, \cdots, A_n$. Then the headings of the columns of the table will be:

a_1	a_2	$\cdots$	a_m	A_1	A_2	$\cdots$	A_n

The first m columns correspond to individual variables; the remaining n columns correspond to elementary parts.

As entries in the first m columns we put natural numbers whose choice depends on the quantifiers of the prefix, as will be described. As entries in the remaining n columns of the table, corresponding to the elementary parts $A_1, A_2, \cdots, A_n$, we put certain expressions which we shall call *atoms* and which represent particular values of the elementary part that stands at the head of the column. In any row the atom which occurs as an entry under an elementary part is formed by replacing the individual variables of the elementary part by the entries in the columns corresponding to those individual variables. As a special case, if the elementary part is a propositional variable, the atom is the same as the elementary part itself.

It remains now to show how the natural numbers which occur as entries in the m columns corresponding to the individual variables of M will be chosen. The choice depends on the prefix of the formula to be decided. It is motivated by the fact that the table will be used to decide whether or not it is possible to satisfy the negation of $(Qa_1)(Qa_2) \cdots (Qa_n) \cdot M$. We consider here only formulas in Skolem normal form:

$$(\exists b_1)(\exists b_2) \cdots (\exists b_m)(c_1)(c_2) \cdots (c_n) \cdot M.$$

If this formula is falsifiable, we must be able to satisfy

$$(b_1)(b_2) \cdots (b_m)(\exists c_1)(\exists c_2) \cdots (\exists c_n) \cdot \bar{M}.$$

That is, for every possible assignment to $b_1, b_2, \cdots, b_m$ there must exist an assignment to $c_1, c_2, \cdots, c_n$ for which M can be falsified.

Let the ordered m-tuples of natural numbers be enumerated in such a way that one m-tuple comes later in the enumeration than another if it contains as member a natural number greater than the greatest natural number contained in the other, and the m-tuples having the same greatest natural number are arranged among themselves in lexicographic order. For the ith member of the kth m-tuple in this enumeration, use the notation $[mki]$.

[1] Tables I and II are examples of decision tables.

TABLE I. *Prefix* $(\exists x)(y_1)(y_2)\cdots(y_n)$

x	y_1	$\cdots$ y_n	$F_1(x)$	$F_2(x)$	$\cdots$	$F_N(x)$	$F_1(y_1)$	$\cdots$	$F_N(y_1)$	$\cdots$	$F_1(y_n)$	$\cdots$	$F_N(y_n)$
0	1	$\cdots$ n	$F_1(0)$	$F_2(0)$	$\cdots$	$F_N(0)$	$F_1(1)$	$\cdots$	$F_N(1)$	$\cdots$	$F_1(n)$	$\cdots$	$F_N(n)$
1	$(n+1)$	$\cdots$ $2n$	$F_1(1)$	$F_2(1)$	$\cdots$	$F_N(1)$	$F_1(n+1)$	$\cdots$	$F_N(n+1)$	$\cdots$	$F_1(2n)$	$\cdots$	$F_N(2n)$
2	$\cdots$		$F_1(2)$	$\cdots$									
$\vdots$			$\vdots$										
k	$kn+1$	$\cdots$ $(k+1)n$	$F_1(k)$	$F_2(k)$	$\cdots$	$F_N(k)$	$F_1(kn+1)$	$\cdots$	$F_N(kn+1)$	$\cdots$	$F_1((k+1)n)$	$\cdots$	$F_N((k+1)n)$
$k+1$	$(k+1)n+1$	$\cdots$ $(k+2)n$	$F_1(k+1)$										
	$\cdots$			$\cdots$									

TABLE II.

y_1	y_2	z_1	z_2	$F(y_1, y_1)$	$F(y_1, y_2)$	$F(y_1, z_1)$	$F(y_1, z_2)$	$F(y_2, y_1)$	$F(y_2, y_2)$	$(F\ y_2, z_1)$
0	0	1	2	F(0, 0)	F(0, 0)	F(0, 1)	F(0, 2)	F(0, 0)	F(0, 0)	F(0, 1)
0	1	3	4	...						
1	0	5	6	...						
1	1	7	8	...						
0	2	9	10	F(0, 0)	F(0, 2)	F(0, 0)	F(0, 10)	F(2, 0)	F(2, 2)	F(2, 9)
1	2	11	12	...						
2	0	13	14	F(2, 2)	F(2, 0)	...		F(0, 2)	F(0, 0)	...
2	1	15	16	...						
2	2	17	18	F(2, 2)	F(2, 2)	F(2, 17)	F(2, 18)	F(2, 2)	F(2, 2)	F(2, 17)
...				...						
0	9	163	164	F(0, 0)	F(0, 9)	...		F(9, 0)	F(9, 9)	...
...				...						
2	9	167	168	F(2, 2)	F(2, 9)	...		F(9, 2)	F(9, 9)	...
...				...						
9	0	181	182	F(9, 9)	F(9, 0)	...		F(0, 9)	F(0, 0)	...
...				...						
9	2	185	186	F(9, 9)	F(9, 2)	...		F(2, 9)	F(2, 2)	...
...				...						
9	9	199	200	F(9, 9)	F(9, 9)	F(9, 199)	F(9, 200)	F(9, 9)	F(9, 9)	F(9, 199)
0	10	201	202	F(0, 0)	F(0, 10)	...		F(10, 0)	F(10, 10)	...
...				...						
2	10	205	206	F(2, 2)	F(2, 10)	...		F(10, 2)	F(10, 10)	...
...				...						
9	10	219	220	F(9, 9)	F(9, 10)	...		F(10, 9)	F(10, 10)	...
10	0	221	222	F(10, 10)	F(10, 0)	...		F(0, 10)	F(0, 0)	...
...				...						
10	2	225	226	F(10, 10)	F(10, 2)	...		F(2, 10)	F(2, 2)	...
...				...						
10	9	239	240	F(10, 10)	F(10, 9)	...		F(9, 10)	F(9, 9)	...
10	10	241	242	F(10, 10)	F(10, 10)	F(10, 241)	F(10, 242)	F(10, 10)	F(10, 10)	F(10, 241)
...				...						

Prefix $(\exists y_1)\,(\exists y_2)\,(z_1)\,(z_2)$

$F(y_2, z_2)$	$F(z_1, y_1)$	$F(z_1, y_2)$	$F(z_1, z_1)$	$F(z_1, z_2)$	$F(z_2, y_1)$	$F(z_2, y_2)$	$F(z_2, z_1)$	$F(z_2, z_2)$
$F(0, 2)$	$F(1, 0)$	$F(1, 0)$	$F(1, 1)$	$F(1, 2)$	$F(2, 0)$	$F(2, 0)$	$F(2, 1)$	$F(2, 2)$
$F(2, 10)$	$F(9, 0)$	$F(9, 2)$	$F(9, 9)$	$F(9, 10)$	$F(10, 0)$	$F(10, 2)$	$F(10, 9)$	$F(10, 10)$
$F(2, 18)$	$F(17, 2)$	$F(17, 2)$	...		$F(18, 2)$	$F(18, 2)$	...	
$F(9, 200)$	$F(199, 9)$	$F(199, 9)$	...		$F(200, 9)$	$F(200, 9)$	...	
$F(10, 242)$	$F(241, 10)$	$F(241, 10)$	...		$F(242, 10)$	$F(242, 10)$	...	

Then, for the formula in Skolem normal form given above, the entries in the first columns of the table will be:

	b_1	b_2	$\cdots$	b_m	c_1	c_2	$\cdots$	c_n
1st row:	0	0	$\cdots$	0	1	2	$\cdots$	n
2nd row:	0	0	$\cdots$	1	$n+1$	$n+2$	$\cdots$	$2n$
$\vdots$								
kth row:	$[mk1]$	$[mk2]$	$\cdots$	$[mkm]$	$(k-1)n+1$	$(k-1)n+2$	$\cdots$	kn

Whenever a number is not required to be the same as a number previously assigned, we assign the lowest number not previously occurring in the table, since it is possible to do this without loss of generality. We note that the construction may be varied, without destroying the truth of the metatheorem, by using any other enumeration of the ordered m-tuples of natural numbers in place of the enumeration that we actually have used, and retaining the same instructions as before for the assignments to the universally quantified variables. And we may wish to make use of the freedom to vary the construction of the decision table in this way, if it appears that the solution of a particular case of the decision problem will be facilitated by so doing.

Use of the Decision Table. The decision table described above has one column for each individual variable occurring in the prefix, and one column for each elementary part occurring in the matrix. By an assignment of truth-values to the table, we mean an assignment of truth or falsehood to each atom occurring as an entry in a column headed by an elementary part. Each atom is to denote a fixed truth-value, and so must receive the same truth-value assignment throughout the table regardless of the number of times it occurs.

The metatheorem given above can now be restated.

Metatheorem. $(Qa_1)(Qa_2) \cdots (Qa_m) \cdot M$ is a theorem if and only if there is no assignment of truth-values to the atoms in the decision table which falsifies M in every row.

An assignment of truth-values which falsifies M will be called a *falsifying system* of truth-values. In order to decide a given formula, we must check to see if the matrix M can be falsified by such a system in every row of the table. But checking an infinite number of rows of a table to see whether M can be falsified in each does not constitute an effective decision procedure. If, however, for a special case we need only check a finite number K of the rows, we have an effective procedure for that case. Thus, for each of the solved cases of the decision problem a finite K can be specified, such that if we can falsify M in each of the first K rows of the table, then we can falsify M in every row of an infinite table.

The classical decision procedures determine a value K such that if the first K rows of the table can be filled, then the infinite table can be filled. The algorithms to be given in the remainder of this paper are based on the observation that there are only certain ways in which an attempt to fill the table can fail. The algorithms provide means for checking to see whether or not this failure will occur, without actually filling in the table.

II. A SIMPLE SOLVABLE CASE

We examine now a solvable case which is defined by prefix only, without restriction on the elementary parts which may occur in the matrix. The case is subsumed by the more general cases of Part III, and is intended to serve primarily as an introduction to our method.

In the decision table method outlined in Part I, *systems* of truth-values are used to falsify *rows* of a decision table. In what follows, the concept of the decision table will be used in formulating rules and proving that we have a decision procedure. However, the decision procedure itself will make no reference to rows of the decision table, but will deal only with the set of falsifying systems of truth-values.

The basic argument with which we will derive a decision procedure is the following: (1) Consider an arbitrary row of the decision table and suppose a system s_0 of truth-values is used to falsify it. Certain restrictions will be imposed by this system s_0 on the systems to be used in each of several other rows of the table. If systems consistent with these restrictions do not exist, then s_0 cannot be used at all. (2) Now consider any pair of systems s_1 and s_2 and suppose they are used in any two rows of the same table. Again there will be restrictions imposed on systems for other rows of the table. In this case, if the restrictions are not met, we conclude only that systems s_1 and s_2 cannot be used in the same table. For the solvable case of Part II we will have the result that from considerations of these two types we can construct a decision procedure. We prove that if restrictions of these two types can be met, the table can be filled; otherwise not. The more general cases of Part III will involve a generalization of these two types of rules.

To illustrate the method we shall use the case of prefix

$$(\exists y_1)(\exists y_2)(z_1)(z_2)\cdots(z_n).$$

We deal with a matrix containing N functional variables $F_1, F_2, \cdots, F_N$, and assume that all permissible elementary parts are included.[2] In this case the decision procedure consists of the repeated application of three rules. We start with the set S of falsifying systems of truth-values. Rules 1 and 2 reduce S to S' by eliminating systems which fail to meet restrictions of type (1) above. Rule 3 divides S' into subsets on the basis of a consideration of type (2) above. The set S^* resulting from iteration of these rules is shown to be empty if and only if the given formula is a theorem.

Notation. We use

$$\mathrm{S}^{c_1 c_2 \cdots c_n}_{C_1 C_2 \cdots C_n} D \mid$$

to indicate the expression which results by simultaneous substitution of

$$C_1, C_2, \cdots, C_n \text{ for } c_1, c_2, \cdots, c_n$$

[2] For example, if M does not contain the elementary part $F_i(z_j, y_1)$, then we replace M by the equivalent matrix $M\cdot[\bar{F}_i(z_j, y_1) \vee F_i(z_j, y_1)]$. It can be shown that this assumption is actually stronger than necessary; however, it simplifies the proof.

in the expression D. For example,

$$\mathrm{S}_{x,y}^{y,z}\, F(y, z, x) \mid \quad \text{is} \quad F(x, y, x).$$

The Decision Procedure

Three rules for reduction of the set S of falsifying systems to S^* will be stated. For each rule, we will prove as a lemma that if the set S can be used to fill the decision table, so can the reduced set S'. Also, an example will be given for each rule. After all three rules and lemmas have been given, we will prove that the repeated application of the rules does in fact constitute a decision procedure.

Rule 1. Let the elementary parts which contain the functional variable F_i and none of the individual variables $z_1, z_2, \cdots, z_n$ be $A_{i1}, A_{i2}, \cdots, A_{ih_i}$. For each possible set of truth-values $a_1, a_2, \cdots, a_N$, *if* there is no remaining system of S in which

a_1 is assigned to all of the elementary parts $A_{11}, A_{12}, \cdots, A_{1h_1}$,
a_2 is assigned to all of the elementary parts $A_{21}, A_{22}, \cdots, A_{2h_2}$,
$\vdots$ $\qquad\qquad\qquad\qquad \vdots \qquad\qquad \vdots$

and

a_N is assigned to all of the elementary parts $A_{N1}, A_{N2}, \cdots, A_{Nh_N}$,

and in which the assignment to any two elementary parts B_1, B_2 is the same whenever $\mathrm{S}_{y_1}^{y_2} B_1 \mid$ is identical with $\mathrm{S}_{y_1}^{y_2} B_2 \mid$, *then* delete from S all systems in which, for some individual variable u, $F_1(u, u, \cdots, u) = a_1$, $F_2(u, u, \cdots, u) = a_2$, $\cdots$, and $F_N(u, u, \cdots, u) = a_N$.

Lemma 1. *The decision table can be filled by S if and only if it can be filled by S', where S' is obtained from S by application of Rule* 1.

Proof. Consider the general row of the decision table for this prefix (see Table II[3]). It assigns to the individual variables $y_1, y_2, z_1, z_2, \cdots, z_n$ natural numbers, say, $k_1, k_2, m_1, m_2, \cdots, m_n$, respectively. Elsewhere in the table there will be "uniform" rows assigning to y_1, y_2 the numbers k_1, k_1; k_2, k_2; m_j, m_j $(j = 1, \cdots, n)$. If any system s_0 eliminated by Rule 1 were to be used in the general row, then at least one of these uniform rows would be impossible to fill. Hence if the table can be filled it can be filled without using systems s_0.

Example.[4] Consider the expression

$$(\exists y_1)(\exists y_2)(z) \cdot F(y_1, y_2)F(y_2, z) \not\equiv F(y_1, z) \supset \cdot F(y_1, z) \equiv F(z, y_1)$$
$$\supset \cdot F(y_1, z) \equiv F(y_2, z) \supset \cdot F(y_2, z)$$
$$\supset F(y_1, y_2) \equiv F(z, z)$$
$$\supset \cdot F(y_1, y_2) \equiv F(y_2, y_1) \equiv F(z, y_2)$$

[3] Table II shows selected parts of the decision table for the case of one dyadic functional variable only. However, the method and proof apply to the general n-adic case.

[4] This example is problem 46.14.4 of Church [5].

The set S consists of the four systems of truth-values[5] which falsify M:

	$F(y_1, y_2)$	$F(y_2, y_1)$	$F(y_1, z)$	$F(z, y_1)$	$F(y_2, z)$	$F(z, y_2)$	$F(z, z)$
(1)	1	1	0	0	0	0	1
(2)	0	0	1	1	1	0	1
(3)	0	1	1	1	1	1	0
(4)	1	0	1	1	1	1	0

For this example Rule 1 becomes:

If there is no system in which both $F(y_1, y_2)$ and $F(y_2, y_1)$ are assigned the truth-value a_1 and in which $F(y_1, z)$ and $F(y_2, z)$ are both assigned the same truth-value, and $F(z, y_1)$ and $F(z, y_2)$ are both assigned the same truth-value, then delete all systems in which $F(z, z)$ is assigned a_1.

There is no system in which both $F(y_1, y_2)$ and $F(y_2, y_1)$ are assigned 0, and in which $F(z, y_1)$ and $F(z, y_2)$ have the same assignment. We therefore delete systems (3) and (4) in which $F(z, z)$ is 0. By Lemma 1, if the decision table can be filled, it can be filled using only systems (1) and (2).

Rule 2. For each possible set of truth-values $a_{11}, a_{12}, \cdots, a_{Nh_N}$, if there is no remaining system of S in which the elementary parts $A_{11}, A_{12}, \cdots, A_{Nh_N}$ are assigned the respective truth-values $a_{11}, a_{12}, \cdots, a_{Nh_N}$, delete from S:

(a) all systems in which the elementary parts $\mathrm{S}_{y_2 y_1}^{y_1 y_2} A_{i\lambda} \mid$ are assigned $a_{i\lambda}$ $(i = 1, \cdots, N; \quad \lambda = 1, \cdots, h_i)$,

(b) and all systems in which for some j, $1 \leqq j \leqq n$, the elementary parts $\mathrm{S}_{z_j}^{y_2} A_{i\lambda} \mid$ are assigned $a_{i\lambda}$ $\quad (i = 1, \cdots, N; \quad \lambda = 1, \cdots, h_i)$,

(c) and all systems in which for some j, $1 \leqq j \leqq n$, the elementary parts $\mathrm{S}_{y_2 z_j}^{y_1 y_2} A_{i\lambda} \mid$ are assigned $a_{i\lambda}$ $\quad (i = 1, \cdots, N; \quad \lambda = 1, \cdots, h_i)$,

(d) and all systems in which for some j, k, $1 \leqq j < k \leqq n$, the elementary parts $\mathrm{S}_{z_j z_k}^{y_1 y_2} A_{i\lambda} \mid$ are assigned $a_{i\lambda}$ $\quad (i = 1, \cdots, N; \quad \lambda = 1, \cdots, h_i)$.

LEMMA 2. *The decision table can be filled by S if and only if it can be filled by S', where S' is obtained from S by application of Rule* 2.

PROOF. Consider again the general row of the decision table which assigns to the individual variables $y_1, y_2, z_1, z_2, \cdots, z_n$ the natural numbers $k_1, k_2, m_1, m_2, \cdots, m_n$, respectively. Elsewhere in the table there will be rows assigning to y_1, y_2 the number pairs k_2, k_1; k_1, m_j $(j = 1, \cdots, n)$; k_2, m_j $(j = 1, \cdots, n)$; and $m_j m_k$ $(1 \leqq j < k \leqq n)$. From these rows we see immediately that Rule 2 eliminates no systems which could be used. For, if Rule 2 were not applied and any one of the systems in S, but not in S', were assigned to row k_1, k_2, there would be in the decision table at least one row which could not be falsified.

Example. We continue the example begun above. The set S' resulting after application of Rule 1 was:

	$F(y_1, y_2)$	$F(y_2, y_1)$	$F(y_1, z)$	$F(z, y_1)$	$F(y_2, z)$	$F(z, y_2)$	$F(z, z)$
(1)	1	1	0	0	0	0	1
(2)	0	0	1	1	1	0	1

[5] We use "0" for falsehood and "1" for truth.

Rule 2 becomes:

If there is no system in which $F(y_1, y_2)$ and $F(y_2, y_1)$ are assigned the truth-values a_1, a_2, respectively, delete all systems in which (a) $F(y_2, y_1)$ and $F(y_1, y_2)$, or (b) $F(y_1, z)$ and $F(z, y_1)$, or (c) $F(y_2, z)$ and $F(z, y_2)$ are assigned a_1, a_2, respectively.

There is no system in which $F(y_1, y_2)$ and $F(y_2, y_1)$ are assigned 1, 0 respectively. Hence by Rule 2(c), system (2) in which $F(y_2, z)$, $F(z, y_2)$ are assigned 1, 0, respectively, can be eliminated. Then, there is no longer a system in which $F(y_1, y_2)$, $F(y_2, y_1)$ are assigned 0, 0; hence by 2(b) system (1) in which $F(y_1, z)$, $F(z, y_1)$ are assigned 0, 0 can be eliminated.

Thus $S^* \equiv 0$, and therefore by Lemmas 1 and 2 the given expression is a theorem. For this example no further rules are needed because Rules 1 and 2 reduce S to $S^* \equiv 0$. However, as we shall show, a third rule is needed before we can prove that if S^* is not empty, the expression is not a theorem.

Rule 3. If there is no remaining system in which the elementary parts $F_1(y_1, y_1, \cdots, y_1)$, $F_2(y_1, y_1, \cdots, y_1)$, $\cdots$, $F_N(y_1, y_1, \cdots, y_1)$ have the respective truth-values $a_1, a_2, \cdots, a_N$ and in which $F_1(y_2, y_2, \cdots, y_2)$, $F_2(y_2, y_2, \cdots, y_2)$, $\cdots$, $F_N(y_2, y_2, \cdots, y_2)$ have the truth-values $b_1, b_2, \cdots, b_N$, then form, from S, two sets such that: (1) the first contains all systems of S except those in which the elementary parts $F_1(u, u, \cdots, u)$, $F_2(u, u, \cdots, u)$, $\cdots$, $F_N(u, u, \cdots, u)$ are assigned $a_1, a_2, \cdots, a_N$ for some individual variable u; and (2) the second contains all systems of S except those in which the elementary parts $F_1(t, t, \cdots, t)$, $F_2(t, t, \cdots, t)$, $\cdots$, $F_N(t, t, \cdots, t)$ are assigned $b_1, b_2, \cdots, b_N$ for some individual variable t.

LEMMA 3. *The decision table can be filled by S if and only if it can be filled by at least one of the subsets S_i formed from S by Rule* 3.

PROOF. From a consideration of any two rows of the table, we see that if Rule 3 were applied, the use in the table of two systems not both in a single S_i would lead to a row which could not be falsified. For if a system in which for all i $F_i(u, u, \cdots, u) = a_i$ is used in a row in which the individual variable u is

		$F_1(y_1)$	$F_2(y_1)$	$F_1(y_2)$	$F_2(y_2)$	$F_1(z)$	$F_2(z)$
S:	(1)	0	0	0	0	1	0
	(2)	0	0	0	1	1	0
	(3)	0	0	1	0	0	1
	(4)	0	0	1	1	0	1
	(5)	0	1	0	0	1	1
	(6)	0	1	0	1	1	1
	(7)	0	1	1	0	0	0
	(8)	0	1	1	1	0	0
	(9)	1	0	0	0	0	1
	(10)	1	0	0	1	0	0
	(11)	1	0	1	0	0	1
	(12)	1	1	0	0	0	1
	(13)	1	1	0	1	0	0
	(14)	1	1	1	1	0	0

FIG. 1. Assignments of truth-values

assigned a natural number k, and a system in which for all i $F_i(t, t, \cdots, t) = b_i$ is used in a row in which the individual variable t is assigned a natural number m, then the row $y_1 = k$, $y_2 = m$ can be falsified only by a system in which for all i $F_i(y_1, y_1, \cdots, y_1) = a_i$ and $F_i(y_2, y_2, \cdots, y_2) = b_i$.

Example. To show the necessity of Rule 3, consider the expression with prefix $(\exists y_1)(\exists y_2)(z)$, elementary parts $F_1(y_1)$, $F_2(y_1)$, $F_1(y_2)$, $F_2(y_2)$, $F_1(z)$, $F_2(z)$ and with the falsifying assignments of truth-values[6] shown in Figure 1.

The first application of Rule 1 fails to eliminate any of these systems; likewise, Rule 2 eliminates none of them. However, by Rule 3, as there is no system

$F_1(y_1)$	$F_2(y_1)$	$F_1(y_2)$	$F_2(y_2)$	$F_1(z)$	$F_2(z)$
1	0	1	1	—	—

we form two subsets S_1 and S_2, such that S_1 contains all systems except those in which for some individual variable u, $F_1(u) = 1$ and $F_2(u) = 0$ and S_2 contains all systems except those in which for some individual variable t, $F_1(t) = 1$ and $F_2(t) = 1$. Thus, we have:

		$F_1(y_1)$	$F_2(y_1)$	$F_1(y_2)$	$F_2(y_2)$	$F_1(z)$	$F_2(z)$
S_1 :	(4)	0	0	1	1	0	1
	(5)	0	1	0	0	1	1
	(6)	0	1	0	1	1	1
	(8)	0	1	1	1	0	0
	(12)	1	1	0	0	0	1
	(13)	1	1	0	1	0	0
	(14)	1	1	1	1	0	0

and

		$F_1(y_1)$	$F_2(y_1)$	$F_1(y_2)$	$F_2(y_2)$	$F_1(z)$	$F_2(z)$
S_2 :	(1)	0	0	0	0	1	0
	(2)	0	0	0	1	1	0
	(3)	0	0	1	0	0	1
	(7)	0	1	1	0	0	0
	(9)	1	0	0	0	0	1
	(10)	1	0	0	1	0	0
	(11)	1	0	1	0	0	1

Consider now the set S_1. There is no system:

$$0 \quad 0 \quad 0 \quad 0 \quad — \quad —.$$

Therefore by Rule 1 all systems except (6) are deleted. There is then no system:

$$1 \quad 1 \quad 1 \quad 1 \quad — \quad —.$$

So, by Rule 1, system (6) is eliminated and S_1^* is empty.

[6] These falsifying systems determine the matrix as:

$$(F_1(y_1) \vee [F_1(z) \equiv F_1(y_2)] \vee F_2(y_1) \equiv [F_2(z) \equiv F_1(y_2)])$$
$$\cdot(\bar{F}_1(y_1) \vee F_1(z) \vee [F_2(y_2) \equiv F_2(z)] \vee F_1(y_2)[F_2(y_1) \not\equiv F_2(y_2)]).$$

Similarly, in S_2 there is no system:

$$0 \quad 1 \quad 0 \quad 1 \quad — \quad —.$$

So, by Rule 1, all systems except (1) are eliminated. There is then no system:

$$1 \quad 0 \quad 1 \quad 0 \quad — \quad —.$$

So, by Rule 1, system (1) is eliminated and S_2^* is empty.

Thus, S^* is empty, and the given expression is shown to be a theorem.

Application of the Rules

Rules 1, 2 and 3 are to be repeated until no further reduction is possible. Rules 1 and 2 must be applied to each of the sets formed by Rule 3. When no further reduction by these rules is possible, the remaining sets are S_1^*, S_2^*, $\cdots$ and their sum is S^*. The corresponding matrices are M_1^*, M_2^*, $\cdots$ and their logical product is M^*. We now prove that the application of the rules in this way constitutes a decision procedure for formulas in this prefix case.

THEOREM 1. *$\vdash (\exists y_1)(\exists y_2)(z_1)(z_2) \cdots (z_n) \cdot M$ if and only if S^* is empty, where S^* is formed from the set S of falsifying systems for M by Rules 1, 2 and 3, above.*

An alternative statement of Theorem 1 in terms of the matrices M and M^* is

$$\vdash (\exists y_1)(\exists y_2)(z_1)(z_2) \cdots (z_n) \cdot M \Leftrightarrow M^* \textit{ is tautologous.}$$

PROOF. From Lemmas 1, 2 and 3 it follows immediately that if S^* is empty the decision table cannot be falsified by S. Thus, we need prove only that if S^* is non-empty, the decision table can be falsified.

If S^* is non-empty, then at least one of the S_i^*, say S_α^*, is non-empty. We show by induction that the table can be falsified by the systems of S_α^*.

The first row of the table assigns the natural number 0 to y_1 and y_2, and assigns the natural number β to z_β, $\beta = 1, \cdots, n$. The atoms corresponding to the elementary parts $A_{i1}, A_{i2}, \cdots, A_{ih_i}$, (that is, to the elementary parts of F_i which contain only y_1 or y_2 or both) are all $F_i(0, 0, \cdots, 0)$. If two elementary parts B_1 and B_2 are such that

$$\mathrm{S}_{y_1}^{y_2} B_1 \mid \equiv \mathrm{S}_{y_1}^{y_2} B_2 \mid$$

then the corresponding atoms in the first row are the same. Otherwise, the atoms of this row are distinct. By Rule 1, S_α^* contains a system which will falsify this row.

Suppose now we have falsified by systems of S_α^* all rows preceding the row in which y_1, y_2 are assigned the natural numbers k_1, k_2. Either (1) k_1 and k_2 have not occurred together in any previous row, or (2) they have both occurred in the same row.

1. If k_1 and k_2 have not occurred in the same row before, then by the construction of the table $k_1 < k_2$.[7] Further, k_1 has occurred as an assignment to y_1 in

[7] This proof depends in several places on the order of the rows as determined by the construction of the table. As described above, the order is determined by the pair of natural numbers assigned to y_1, y_2. Pair k_1, k_2 precedes pair k_3, k_4 if $\max(k_1, k_2) < \max(k_3, k_4)$. If $\max(k_1, k_2) = \max(k_3, k_4)$, then the pairs occur in lexicographic order.

the earlier row $y_1 = k_1$, $y_2 = 0$, where the atoms $F_i(k_1, k_1, \cdots, k_1)$ received some assignments, say a_i; k_2 has occurred as an assignment to some z_j in a row in which the atoms $F_i(k_2, k_2, \cdots, k_2)$ received assignments, say b_i. As k_1 and k_2 have not occurred together before, atoms containing both k_1 and k_2 have not been assigned truth-values. The values assigned to $z_1, z_2, \cdots, z_n$ are new and distinct and therefore atoms containing $z_1, z_2, \cdots$, or z_n have not previously been assigned truth-values. Thus the only restriction on row k_1, k_2 is that $F_i(k_1, k_1, \cdots, k_1)$ and $F_i(k_2, k_2, \cdots, k_2)$ be assigned a_i and b_i $(i = 1, \cdots N)$ respectively. There is in S^* a system in which the $F_i(y_1, y_1, \cdots, y_1) = a_i$ and the $F_i(y_2, \cdots, y_2) = b_i$, for otherwise by Rule 3 the system in which the $F_i(y_1, y_1, \cdots, y_1) = a_i$ and the system in which the $F_i(z_j, z_j, \cdots, z_j) = b_i$ could not both be in S_α^*.

2. If k_1 and k_2 have occurred together in some previous row, there are three cases to be considered:

$$\text{(a)}\quad k_1 = k_2, \qquad \text{(b)}\quad k_1 < k_2, \qquad \text{(c)}\quad k_1 > k_2.$$

(a) $k_1 = k_2$. Then a row $y_1 = k_1$, $y_2 = 0$ has occurred and by Rule 1 row k_1, k_1 can be falsified.

(b) $k_1 < k_2$. Then either a row $y_1 = k_1$, $z_j = k_2$ (for some j, $1 \leqq j \leqq n$) has occurred and Rule 2b is sufficient; or a row $y_2 = k_1$, $z_j = k_2$ $(1 \leqq j \leqq n)$ has occurred and Rule 2c is sufficient; or a row $z_j = k_1$, $z_k = k_2$ $(1 \leqq j < k \leqq n)$ has occurred and Rule 2d is sufficient.

(c) $k_1 > k_2$. Then $y_1 = k_2$, $y_2 = k_1$ has occurred and Rule 2a is sufficient.

III. THE SEMI-DECISION PROCEDURE

We now extend the method presented above to the class of formulas in Skolem normal form. The Skolem normal form is a prenex normal form with prefix $(\exists y_1)(\exists y_2) \cdots (\exists y_m)(z_1)(z_2) \cdots (z_n)$ and without free variables. It is well known that given any formula of the functional calculus one can construct another formula in Skolem normal form such that either both of the formulas are theorems or neither is. The decision problem for formulas in Skolem normal form is thus equivalent to the decision problem for the functional calculus, and therefore unsolvable. It is thus clear that there is no way to extend the method to give a decision procedure for the class. However, we shall show that it can be extended to provide a decision procedure for most of the previously solved subcases, and for two new subcases. For the remaining cases it gives a semi-decision procedure, which may provide an answer, but will not always do so.

As in the case above, we give rules for the formation of a subset S^* of S, with corresponding matrix M^* such that $M \supset M^*$, and prove that

$$\vdash (\exists y_1)(\exists y_2) \cdots (\exists y_m)(z_1)(z_2) \cdots (z_n) \cdot M^*$$
$$\Leftrightarrow \vdash (\exists y_1)(\exists y_2) \cdots (\exists y_m)(z_1)(z_2) \cdots (z_n) \cdot M.$$

Further, for a matrix M suitably restricted, we shall prove that the method provides a solution to the decision problem

$$M^* \text{ is tautologous} \Leftrightarrow \; \vdash (\exists y_1)(\exists y_2) \cdots (\exists y_m)(z_1)(z_2) \cdots (z_n) \cdot M.$$

Rules for Formation of S^*

We give four rules which can be applied in the formation of S^* in this case. The rules are analogous to those for the simple case described above. Rules 4 and 7 are of a form similar to that of Rules 1 and 2; Rules 5 and 6 extend Rule 3. We assume in every case that all permissible elementary parts are present.

Rule 4. Let $A_1, A_2, \cdots, A_N$ be any set of elementary parts of M which contain none of the individual variables $z_1, z_2, \cdots, z_n$.

Let $\alpha_1\alpha_2 \cdots \alpha_m$ be any one of the $(m + n)^m$ possible arrays of $y_1, y_2, \cdots, y_m, z_1, z_2, \cdots, z_n$ taken m at a time. Let $t_1, t_2, \cdots, t_n$ be distinct individual variables not occurring in M. For each array $\alpha_1\alpha_2 \cdots \alpha_m$ and for each possible set of truth values, $a_1, a_2, \cdots, a_N$, *if* there is no remaining system of S in which

(a) A_1 is assigned a_1
A_2 is assigned a_2
⋮
A_N is assigned a_N

and in which

(b)[8] for any two elementary parts B_1 and B_2 if $S^{y_1y_2\cdots y_mz_1z_2\cdots z_n}_{\alpha_1\alpha_2\cdots\alpha_mt_1t_2\cdots t_n} B_1 \mid$ is the same as $S^{y_1y_2\cdots y_mz_1z_2\cdots z_n}_{\alpha_1\alpha_2\cdots\alpha_mt_1t_2\cdots t_n} B_2 \mid$, B_1 and B_2 are assigned the same truth-value, *then* delete from S all systems in which

$S^{y_1y_2\cdots y_m}_{\alpha_1\alpha_2\cdots\alpha_m} A_1 \mid$ is assigned a_1,
$S^{y_1y_2\cdots y_m}_{\alpha_1\alpha_2\cdots\alpha_m} A_2 \mid$ is assigned a_2,
⋮
$S^{y_1y_2\cdots y_m}_{\alpha_1\alpha_2\cdots\alpha_m} A_N \mid$ is assigned a_N.

Rule 5. If there is no remaining system of S in which the elementary parts $F_1(y_1, y_1, \cdots, y_1)$, $F_2(y_1, y_1, \cdots, y_1)$, $\cdots$, $F_N(y_1, y_1, \cdots, y_1)$ have the truth-values $a_{11}, a_{12}, \cdots, a_{1N}$ respectively, and in which $F_1(y_2, y_2, \cdots, y_2)$, $F_2(y_2, y_2, \cdots, y_2)$, $\cdots$, $F_N(y_2, y_2, \cdots, y_2)$ have the truth-values $a_{21}, a_{22}, \cdots, a_{2N}$ respectively, etc., and in which $F_1(y_m, y_m, \cdots, y_m)$, $F_2(y_m, y_m, \cdots, y_m)$, $\cdots$, $F_N(y_m, y_m,, \cdots y_m)$ have the truth-values $a_{m1}, a_{m2}, \cdots, a_{mN}$ respectively, then form, from S, m sets, $S_1, S_2, \cdots, S_m$, such that S_i contains all remaining systems except those in which $F_1(t, t, \cdots, t)$, $F_2(t, t, \cdots, t)$, $\cdots$, $F_N(t, t, \cdots, t)$ have the truth-values $a_{i1}, a_{i2}, \cdots, a_{iN}$ respectively for some individual variable t.

Rule 6. Let s be an integer, $1 \leqq s < m$. Let $A_1, A_2, \cdots, A_K$ be the elementary parts of M which contain none of the individual variables $y_{s+1}, y_{s+2}, \cdots, y_m, z_1, z_2, \cdots, z_n$. Let $B_1, B_2, \cdots, B_L$ be the elementary parts of M which contain none of the individual variables $y_1, y_2, \cdots, y_s, z_1, z_2, \cdots, z_n$.

[8] It is clear that (b) imposes a restriction on the system only if $\alpha_1, \alpha_2, \cdots, \alpha_m$ are not all distinct.

If there is no remaining system of S in which the elementary parts A_1, A_2, $\cdots$, A_K, B_1, B_2, $\cdots$, B_L have the truth-values a_1, a_2, $\cdots$, a_K, b_1, b_2, $\cdots$, b_L respectively, then form from S two sets, S_1 and S_2, such that S_1 contains all remaining systems except those in which the elementary parts A_1, A_2, $\cdots$, A_K have the truth-values a_1, a_2, $\cdots$, a_K respectively and S_2 contains all remaining systems except those in which the elementary parts B_1, B_2, $\cdots$, B_L have the truth-values b_1, b_2, $\cdots$, b_L respectively.

Rule 7. Let the elementary parts of M which contain none of the variables z_1, z_2, $\cdots$, z_n be A_1, A_2, $\cdots$, A_K, $\cdots$, A_L, and let the subset of these which contains only one of the individual variables y_1, $\cdots$, y_s be A_1, A_2, $\cdots$, A_K. Let $\alpha_1\alpha_2 \cdots \alpha_m$ be an array of the individual variables y_1, y_2, $\cdots$, y_m such that α_1, α_2, $\cdots$, α_m are not all different. Let γ be a system of S which assigns the truth-value a_i to A_i $(i \leqq K)$ and in which $a_i = a_j$ if

$$\mathrm{S}^{y_1 y_2 \cdots y_m}_{\alpha_1 \alpha_2 \cdots \alpha_m} A_i \,| = \mathrm{S}^{y_1 y_2 \cdots y_m}_{\alpha_1 \alpha_2 \cdots \alpha_m} A_j \,| \qquad (i, j \leqq K).$$

Then there must also be in S a set of $m!$ systems $\beta_{12\cdots(m-1)m}$, $\beta_{12\cdots m(m-1)}$, $\cdots$, $\beta_{e_1 e_2 \cdots e_m}$, $\cdots$, $\beta_{m(m-1)\cdots 21}$ (not necessarily all different) with the following properties:

In system $\beta_{12\cdots m}$, A_i is assigned the truth-value a_i (a_1, a_2, $\cdots$, a_K have been determined by system γ; a_{K+1}, $\cdots$, a_L are arbitrary) and for any two elementary parts B_1, B_2 if

$$\mathrm{S}^{y_1 \cdots y_m z_1 z_2 \cdots z_n}_{\alpha_1 \cdots \alpha_m t_1 t_2 \cdots t_n} B_1 \,| = \mathrm{S}^{y_1 \cdots y_m z_1 z_2 \cdots z_n}_{\alpha_1 \cdots \alpha_m t_1 t_2 \cdots t_n} B_2 \,|$$

then B_1 and B_2 are assigned the same truth-value.

For each permutation e_1, e_2, $\cdots$, e_m of the integers 1, 2, $\cdots$, m, the system $\beta_{e_1 \cdots e_m}$ assigns a_i to

$$\mathrm{S}^{y_1 \cdots y_m}_{y_{e_1} \cdots y_{e_m}} A_i \,| \qquad (i = 1, 2, \cdots, L)$$

and assigns the same truth-value to any two elementary parts B_1, B_2 such that

$$\mathrm{S}^{y_1 y_2 \cdots y_m}_{\alpha_{e_1} \alpha_{e_2} \cdots \alpha_{e_m}} B_1 \,| = \mathrm{S}^{y_1 y_2 \cdots y_m}_{\alpha_{e_1} \alpha_{e_2} \cdots \alpha_{e_m}} B_2 \,|.$$

If there is no such set of systems $\beta_{12\cdots m}$, $\beta_{12\cdots m(m-1)}$, $\cdots$, $\beta_{m(m-1)\cdots 21}$, then system γ is to be deleted from S.

We note that when system γ is deleted it will then be possible by Rules 4 and 5 to delete many other systems and to further subdivide the set S.

Rules 4, 5, 6 and 7 are to be repeated and applied to each set S_i until no further reduction is possible. When no further reduction by these rules is possible, the remaining sets are S_1^*, S_2^*, $\cdots$, and their sum is S^*. The corresponding matrices are M_1^*, M_2^*, $\cdots$, and their logical product is M^*.

THEOREM 2. *For arbitrary matrix M, the set S will falsify the decision table if and only if at least one of the S_i^* will falsify the decision table. That is* $\vdash (\exists y_1)(\exists y_2) \cdots (\exists y_m)(z_1)(z_2) \cdots (z_n) \cdot M$ *if and only if for all* i $\vdash (\exists y_1)(\exists y_2) \cdots (\exists y_m)(z_1)(z_2) \cdots (z_n) \cdot M_i^*$.

COROLLARY. *For arbitrary matrix* M, $S^* \equiv 0 \Rightarrow \;\vdash (\exists y_1)(\exists y_2) \cdots (\exists y_m)(z_1)(z_2) \cdots (z_n) \cdot M$.

PROOF. If S_i^* will falsify the decision table, so will S, because S_i^* is a subset of S.

We now prove that if S will falsify the decision table, then at least one of the S_i^* will falsify the decision table.

1. *Rule 4* eliminates only systems which cannot be used in the decision table. Consider an arbitrary row, R, of the decision table. Row R assigns to $y_1, y_2, \cdots, y_m, z_1, z_2, \cdots, z_n$ natural numbers, say $k_1, k_2, \cdots, k_m, m_1, m_2, \cdots, m_n$ respectively. Elsewhere in the table there will be rows assigning to $y_1, y_2, \cdots, y_m$ every one of the $(m+n)^m$ possible sets of these natural numbers taken m at a time. Consider the row, R', in which $y_1, y_2, \cdots, y_m$ are assigned respectively the natural numbers which were assigned to $\alpha_1, \alpha_2, \cdots, \alpha_m$ in row R. Then the atoms in R' corresponding to the elementary parts A_{ij} (which contain as arguments only $y_1, y_2, \cdots, y_m$) will be assigned the same truth-values as those atoms were assigned in row R. That is, the atom corresponding to the elementary part A_{ij} will be given the same assignment in R' as $\mathrm{S}^{y_1 y_2 \cdots y_m}_{\alpha_1 \alpha_2 \cdots \alpha_m} A_{ij} \mid$ was given in row R. Furthermore, for any two elementary parts B_1 and B_2 such that $\mathrm{S}^{y_1 y_2 \cdots y_m z_1 z_2 \cdots z_n}_{\alpha_1 \alpha_2 \cdots \alpha_m t_1 t_2 \cdots t_n} B_1 \mid$ is the same as $\mathrm{S}^{y_1 y_2 \cdots y_m z_1 z_2 \cdots z_n}_{\alpha_1 \alpha_2 \cdots \alpha_m t_1 t_2 \cdots t_n} B_2 \mid$ the corresponding atoms in row R' will be identical and must therefore have the same truth-value in any system to be used in row R'. Thus, each system deleted by Rule 4 is such that if it were to be used in a row R of the table there would be in the table a row R' which could not be falsified.

2. Falsifying systems which by *Rule 5* cannot all occur in the same set S_i^* cannot all be used in the same decision table. For, suppose there were no system meeting the condition of Rule 5 and that assignments in which for the natural numbers $k_1, k_2, \cdots, k_m$ the atoms $F_j(k_1, k_1, \cdots, k_1)$, $F_j(k_2, k_2, \cdots, k_2)$, $\cdots$, $F_j(k_m, k_m, \cdots, k_m)$ were assigned the truth-values $a_{1j}, a_{2j}, \cdots, a_{mj}$ $(j = 1, 2, \cdots, N)$ respectively, were used in the table. Then there would be no system which would falsify the row in which the individual variables $y_1, y_2, \cdots, y_m$ were assigned the natural numbers $k_1, k_2, \cdots, k_m$ respectively.

3. If two falsifying systems are such that by *Rule 6* they cannot both occur in the same set S_i^* then they cannot both be used in the same decision table. For suppose a system in which the atoms corresponding to $A_1, A_2, \cdots, A_K$ are assigned $a_1, a_2, \cdots, a_K$ is used in a row in which $y_1, y_2, \cdots, y_s$ are assigned $k_1, k_2, \cdots, k_s$; and a system in which the atoms corresponding to $B_1, B_2, \cdots, B_L$ are assigned $b_1, b_2, \cdots, b_L$ is used in a row in which $y_{s+1}, y_{s+2}, \cdots, y_m$ are assigned $k_{s+1}, k_{s+2}, \cdots, k_m$. Then the row in which $y_1, y_2, \cdots, y_s, y_{s+1}, \cdots, y_m$ are assigned $k_1, k_2, \cdots, k_s, k_{s+1}, \cdots, k_m$ can be filled only by a system in which the atoms corresponding to $A_1, A_2, \cdots, A_K, B_1, B_2, \cdots, B_L$ are assigned $a_1, a_2, \cdots, a_K, b_1, b_2, \cdots, b_L$ respectively. But by hypothesis no such system exists.

4. Suppose the system γ described in *Rule 7* is used in the decision table in a row R_1. Somewhere in the table there will also be a row R_2 in which each y_i is assigned the natural number given to α_i in row R_1. There will also be rows in which the assignments to the y_i exhaust all permutations of the assignments to

R_2. These rows are related to R_2 in the way that $\beta_{12\cdots m(m-1)}, \cdots, \beta_{m(m-1)\cdots 21}$ of Rule 7 are related to system $\beta_{12\cdots m}$. Therefore, if no such set of systems exists, the use of system γ in the table will mean that some rows cannot be falsified. Thus, system γ can be deleted from S.

THEOREM 3. *For the special case prefix* $(\exists y_1)(\exists y_2) \cdots (\exists y_m)(z_1)(z_2) \cdots (z_n)$ *and a matrix in which every elementary part contains at least one of the individual variables* $z_1, z_2, \cdots, z_n$ *or contains only one individual variable or contains both* y_1 and y_2 *and no other individual variables,* $\vdash (\exists y_1)(\exists y_2) \cdots (\exists y_m)(z_1)(z_2) \cdots (z_n) \cdot M$ *if and only if* S^* *is empty, where* S^* *is formed from* S *by Rules* 4, 5 *and* 6.

PROOF. By Theorem 2, it remains only to prove that if S^* is non-empty then the decision table can be falsified and thus the given expression is not a theorem.

If S^* is non-empty, then at least one of the S_i^*, say S_α^*, is non-empty. We show by induction that the decision table can be falsified by the systems of S_α^*.

The first row of the table assigns the natural number 0 to all of $y_1, y_2, \cdots, y_m$ and assigns the natural number i to z_i, $i = 1, 2, \cdots, n$. If any two elementary parts B_1 and B_2 are such that

$$\mathrm{S}_{y_1 y_1 \cdots y_1}^{y_1 y_2 \cdots y_m} B_1 | = \mathrm{S}_{y_1 y_1 \cdots y_1}^{y_1 y_2 \cdots y_m} B_2 |$$

then the corresponding atoms of the first row are the same. Otherwise, the atoms of this row are distinct. By Rule 4, if S_α^* is non-empty it contains a system which will falsify this first row.

Suppose now that we have falsified by systems of S_α^* all rows preceding a row R in which the individual variables $y_1, y_2, \cdots, y_m$ are assigned the natural numbers $k_1, k_2, \cdots, k_m$ respectively. There are two cases to be considered, $k_1 = k_2$ and $k_1 \neq k_2$.

Case 1. $k_1 = k_2$. Each atom of row R except those corresponding to elementary parts containing at least one of $z_1, z_2, \cdots, z_n$ has previously occurred in the table (corresponding, i.e. to some elementary part $F_j(t, t, \cdots, t)$ in some earlier row). By Rule 5, there is then a system β in S_α^* which assigns the correct truth-values to the atoms of R that correspond to the elementary parts $F_j(y_i, y_i, \cdots, y_i)$, $(i = 1, 2, \cdots, m$, and $j = 1, 2, \cdots, N)$. But then by Rule 4, letting α_i be the first individual variable of R which is assigned the natural number k_i, there is also in S_α^* a system γ which will falsify row R. For otherwise system β could not be in S_α^*.

Case 2. $k_1 \neq k_2$. Either (A) the atoms containing both k_1 and k_2 have not previously occurred in the table, or (B) those atoms have occurred in some preceding row.

If (A), then each atom of row R which has previously occurred contains precisely one natural number, and corresponds to an elementary part that contains only one individual variable. By Rule 5, there is in S_α^* a system β which assigns to these atoms of R the previously assigned truth-values. But by Rule 4, (letting α_i be the first individual variable of R which is assigned k_i) system β cannot be in S_α^* unless there is also in S_α^* a system γ which assigns to these atoms the same values as system β and which in addition assigns the same truth-value to any two identical atoms. System γ will thus falsify row R.

If (B), then, in some previous row, k_1, k_2 were assigned to a subset α_1, α_2 of individual variables including either both y_1 and y_2 or at least one of $z_1, z_2, \cdots, z_n$. Then, by Rule 4, there is a system β_1 in S_α^* which will falsify the row in which y_1, y_2 are assigned k_1, k_2 and $y_3, y_4, \cdots, y_m$ are all assigned k_1. There is also in S_α^* a system β_2 which was used to falsify the previous row in which y_1 and y_2 were both assigned 0 and $y_3, y_4, \cdots, y_m$ were assigned $k_3, k_4, \cdots, k_m$, respectively.[9] But, by Rule 6 since systems β_1 and β_2 are in S_α^* there must also be in S_α^* a system β_3 which assigns to all previously occurring atoms of row R their correct values. And then, by Rule 4, (again letting α_i be the first individual variable of R which is assigned k_i), β_3 could not be in S_α^* unless there were also a system γ of S_α^* which will falsify row R.

We have now shown that if S_α^* is non-empty we can falsify the first row of the table and that if we can falsify all rows preceding row R we can also falsify row R. Thus, by induction, we can falsify the entire decision table.

THEOREM 4. *For the special case prefix* $(\exists y_1)(\exists y_2) \cdots (\exists y_m)(z_1)(z_2) \cdots (z_n)$ *and a matrix in which every elementary part contains at least one of the individual variables* $z_1, z_2, \cdots, z_n$ *or contains only one individual variable or contains all of* $y_1, y_2, \cdots, y_m$, $\vdash (\exists y_1)(\exists y_2) \cdots (\exists y_m)(z_1)(z_2) \cdots (z_n) \cdot M$ *if and only if* S^* *is empty, where* S^* *is formed from* S *by Rules 4, 5 and 7.*

PROOF. By Theorem 2, it remains only to prove that if S^* is non-empty then the decision table can be falsified and thus the given expression is not a theorem.

If S^* is non-empty, then at least one of the S_i^*, say S_α^*, is non-empty. We show by induction that the decision table can be falsified by the systems of S^* using the following rule:

Rule for Filling the Decision Table. If the natural numbers $k_1, k_2, \cdots, k_m$ assigned to $y_1, y_2, \cdots, y_m$ in a row R are all the same (Case 1) or if $k_1, k_2, \cdots, k_m$ are all distinct (Case 2) or if the atoms containing all of $k_1, k_2, \cdots, k_m$ have previously occurred in the table (Case 4) then assign to row R any system of S_α^* which will falsify R. If $k_1, k_2, \cdots, k_m$ are neither all the same nor all distinct and the atoms containing all of them have not previously occurred in the table (Case 3), then assign to row R a system β of S_α^* such that:

β assigns to the elementary part A_i (containing none of the individual variables $z_1, z_2, \cdots, z_n$) the truth-value a_i $(i = 1, \cdots, L)$ and for each permutation $e_1, e_2, \cdots, e_m$ of the integers $1, 2, \cdots, m$ there is in S_α^* a system $\beta_{e_1e_2\cdots e_m}$ which assigns a_i to

$$\mathrm{S}^{y_1y_2\cdots y_m}_{y_{e_1}y_{e_2}\cdots y_{e_m}} A_i \,| \qquad i = 1, 2, \cdots, L$$

and assigns the same truth-value to any two elementary parts B_1, B_2 such that

$$\mathrm{S}^{y_1y_2\cdots y_m}_{k_{e_1}k_{e_2}\cdots k_{e_m}} B_1 \,| = \mathrm{S}^{y_1y_2\cdots y_m}_{k_{e_1}k_{e_2}\cdots k_{e_m}} B_2 \,| .$$

[9] The order of the rows for this prefix case is the order of the m-tuples of natural numbers enumerated in such a way that one m-tuple comes later in the enumeration than another if it contains as a member a natural number greater than the greatest natural number contained in the other, and the m-tuples having the same greatest natural number are arranged among themselves in lexicographic order. By this rule $0 \cdots 0\, k_{s+1}k_{s+2} \cdots k_m$ must precede the m-tuple $k_1k_2 \cdots k_sk_{s+1}k_{s+2} \cdots k_m$.

INDUCTION: FIRST ROW. The first row of the table assigns the natural number 0 to all of $y_1, y_2, \cdots, y_m$ and assigns the natural number i to z_i, $i = 1, 2, \cdots, n$. If any two elementary parts B_1 and B_2 are such that

$$\mathrm{S}_{y_1 y_1 \cdots y_1}^{y_1 y_2 \cdots y_m} B_1 \mid = \mathrm{S}_{y_1 y_1 \cdots y_1}^{y_1 y_2 \cdots y_m} B_2 \mid$$

then the corresponding atoms of the first row are the same. Otherwise the atoms of this row are distinct. By Rule 4, if $S_\alpha{}^*$ is non-empty it contains a system which will falsify this first row.

INDUCTION: ROW R. Suppose now that we have falsified according to the above rule all rows preceding a row R in which the individual variables $y_1, y_2, \cdots, y_m$ are assigned the natural number $k_1, k_2, \cdots, k_m$ respectively. There are four cases to be considered:

Case 1. $k_1, k_2, \cdots, k_m$ are all the same.

Case 2. $k_1, k_2, \cdots, k_m$ are all different.

Case 3. $k_1, k_2, \cdots, k_m$ are neither all the same nor all different and the atoms containing all of $k_1, k_2, \cdots, k_m$ have not previously occurred in the table.

Case 4. $k_1, k_2, \cdots, k_m$ are neither all the same nor all different and the atoms containing all of $k_1, k_2, \cdots, k_m$ have previously occurred in the table.

We consider these cases separately.

Case 1. $k_1, k_2, \cdots, k_m$ are all the same.

Each atom of row R except those corresponding to elementary parts containing at least one of $z_1, z_2, \cdots, z_n$ has previously occurred in the table (corresponding to some elementary part $F_j(t, t, \cdots, t)$ in some earlier row). By Rule 5, there is then a system β in $S_\alpha{}^*$ which assigns the correct truth-values to the atoms of R that correspond to the elementary parts

$$F_j(y_i, y_i, \cdots, y_i), \quad (i = 1, 2, \cdots, m; \text{and } j = 1, 2, \cdots, N).$$

But then, by Rule 4, letting α_i be the left-most individual variable of R which is assigned the natural number k_i, there is also in $S_\alpha{}^*$ a system γ which will falsify row R. For otherwise system β could not be in $S_\alpha{}^*$.

Case 2. $k_1, k_2, \cdots, k_m$ are all distinct.

If atoms containing all of $k_1, k_2, \cdots, k_m$ have not previously occurred then each atom of row R which has occurred corresponds to an elementary part containing just one individual variable and so contains precisely one natural number. Then, by Rule 5, there is a system in $S_\alpha{}^*$ which will falsify row R.

If the atoms containing all of the distinct natural numbers $k_1, k_2, \cdots, k_m$ have occurred in a preceding row R_1, then by Rule 4 (letting α_i be the individual variable to which k_i was assigned in row R_1) there is a system in $S_\alpha{}^*$ which will falsify row R.

Case 3. $k_1, k_2, \cdots, k_m$ are neither all the same nor all different and the atoms containing all of $k_1, k_2, \cdots, k_m$ have not previously occurred in the table.

By Rule 5, there is at least one system γ of $S_\alpha{}^*$ which would falsify row R. But by Rule 7, γ cannot be in $S_\alpha{}^*$ unless there is also in $S_\alpha{}^*$ a system β which will falsify row R and which has the additional property required by the rule for filling the table.

Case 4. $k_1, k_2, \cdots, k_m$ are neither all the same nor all distinct and the atoms containing all of $k_1, k_2, \cdots, k_m$ have previously occurred in the table.

If these atoms have occurred in a row R_2 in which at least one of the natural numbers $k_1, k_2, \cdots, k_m$ was assigned to one of the individual variables $z_1, z_2, \cdots, z_n$, then by Rule 4 (letting α_i be the first individual variable to which k_i was assigned in row R_2) there must be a system which will falsify R.

If no such row R_2 has occurred, then in at least one previous row the natural numbers $k_1, k_2, \cdots, k_m$ were assigned to a permutation of $y_1, y_2, \cdots, y_m$. Let the first such row be R_3. By the rule for filling the table R_3 was assigned a system β such that there is also in S_α^* a system β' which will falsify row R.

We have now proved that if S_α^* is non-empty we can falsify the first row of the table and that if we can falsify all rows preceding row R we can also falsify row R. Thus, by induction, we can falsify the entire decision table.

THEOREM 5. $\vdash (\exists y_1)(\exists y_2) \cdots (\exists y_m). M$ *if and only if* S^* *is empty, where* S^* *is formed from* S *by Rule* 4.

PROOF. If S^* is empty, by Theorem 2 the expression is not a theorem. If S^* is non-empty, then by Rule 4 it contains a system which will falsify the first row of the table. And for this special case of the decision problem the decision table can be falsified if and only if the first row can be falsified.

THEOREM 6. $\vdash (x_1)(x_2) \cdots (x_n). M$ *if and only if* S^* *is empty, where* S^* *is* S.

PROOF. The given expression is a theorem if and only if M is tautologous. That is, if and only if $S \equiv 0$.

CONCLUSIONS

It is of interest to examine the relationship between the cases solved by the above method and the previously solved cases of the decision problem. Using Kalmár's reduction to Skolem normal form, cited above, we find that the cases solved relate to the known solved cases of the decision problem,[10] as follows:

Case 1, prefix $(x_1)(x_2) \cdots (x_k)(\exists y_1)(\exists y_2) \cdots (\exists y_m)$, is solved by Theorem 5, above.

Case 2, prefix $(x_1)(x_2) \cdots (x_k)(\exists y)(z_1)(z_2) \cdots (z_n)$, is a subcase of the cases solved by Theorems 3 and 4.

Case 3, prefix $(x_1)(x_2) \cdots (x_k)(\exists y_1)(\exists y_2)(z_1)(z_2) \cdots (z_n)$, is solved by Theorem 1 above, and is a subcase of the cases solved by Theorems 3 and 4.

Case 4, prefix $(x_1)(x_2) \cdots (x_k)(\exists y_1)(\exists y_2) \cdots (\exists y_m)(z_1)(z_2) \cdots (z_n)$ and a matrix in which every elementary part that contains individual variables other than $x_1, x_2, \cdots, x_k$ contains either all of the variables $y_1, y_2, \cdots, y_m$ or at least

[10] The list of solved cases is that given by Church [4]. The cases are those in which the special class of wff's is characterized by conditions on the prefix and matrix of the wff in prenex normal form. References to the original solutions of these cases are given by Church. Dreben [6] has announced a general method for all classes of quantificational schemata whose decision problems are known from the literature to have affirmative solutions and also for several (unspecified) classes not previously known to have affirmative solutions.

one of the variables $z_1, z_2, \cdots, z_n$ is, for $k = 0$, a subcase of the case solved by Theorem 4, above. For $k > 0$, the reduction by the theorem of Kalmár may introduce elementary parts which are excluded by Theorem 4, so that the solution does not extend automatically.

Case 5, prefix beginning with $(x_1)(x_2)\cdots(x_k)$ and ending with $(z_1)(z_2)\cdots(z_n)$, and a matrix in which every elementary part that contains individual variables other than $x_1, x_2, \cdots, x_k$ contains at least one of the variables $z_1, z_2, \cdots, z_n$, and Case 6, there is at most one falsifying system of truth-values for the elementary parts of the matrix, apply to more general prefixes and are thus not in general solved above. For the Skolem prefix, however, Case 5 is a subcase of Case 4 and Case 6 can be solved by the application of Rule 4 alone.

Case 7a, prefix $(\exists y_1)(\exists y_2)\cdots(\exists y_m)(z)$, where $m \leqq 4$, and a matrix of the form $N \supset F(y_1, z)$, where N does not contain the variable z and does not contain any functional variable other than F, contains only a finite number of formulas, but has not been solved by the above methods.

In all cases, Theorem 2 provides a sufficient condition for the validity of the given formula.

Thus, we have the result that the rules provide a method, suitable for programming, which can be applied as a decision procedure for most solved cases of the decision problem. For the general unsolvable Skolem case, the method is a semidecision procedure: it will terminate, and may provide an answer, but will not always do so.

* * *

Acknowledgment. I should like to thank Professor Alonzo Church for his guidance and his many helpful suggestions.

REFERENCES

1. Avion Division, ACF Industries, Inc. Decision procedures for the functional calculus. Scientific Report No. AFCRC-TN-56-387, Air Force Cambridge Research Center, Air Research and Development Command, 1956.
2. Avion Division, ACF Industries, Inc. Decision procedures for the functional calculus (Supplementary report). Scientific Report No. AFCRC-TN-56-785, Air Force Cambridge Research Center, Air Research and Development Command, 1956.
3. Ackermann, W. *Solvable Cases of the Decision Problem.* North-Holland Publishing Co., Amsterdam, 1954, pp. 67–70.
4. Church, Alonzo. Special cases of the decision problem. *Revue Philosophique de Louvain 49*(1951), 203–221. A correction, *ibid. 50*(1952), 270–272.
5. Church, Alonzo. *Introduction to Mathematical Logic, Vol. 1.* Princeton University Press, Princeton, N. J., 1956.
6. Dreben, Burton. Systematic treatment of the decision problem. Summer Institute for Symbolic Logic, Cornell University, 1957, p. 363. (abstract)
7. Friedman, Joyce. Extensions of two solvable cases of the decision problem. *J. Symbolic Logic 22*(1957), 108. (abstract)
8. Gilmore, P. C. A proof method for quantification theory; its justification and realization. *IBM J. Res. Dev. 4*(1960), 28–35.
9. Gödel, Kurt. Die Vollständigkeit der Axiome des logischen Funktionenkalküls *Monat. Math. Physik 37*(1930), 349–360.

10. Herbrand, Jacques. Recherches sur la Théorie de la Démonstration. *Traveaux de la Société des Sciences et des Lettres de Varsovie*, Cl. 3 math.-phys., *33*(1930).

11. Kalmár, László. Über die Erfüllbarkeit derjenigen Zählausdrücke, welche in der Normalform zwei benachbarte Allzeichen enthalten. *Math. Ann. 108*(1933), 466–484.

12. Prawitz, D; Prawitz, H.; Voghera, N. A mechanical proof procedure and its realization in an electronic computer. *J. ACM 7*(1960), 102–128.

13. Wang, Hao. Proving theorems by pattern recognition {1}, *Comm. ACM 3*(1960), 220–234.

14. Wang, Hao. Proving theorems by pattern recognition {2}, *Bell System Tech. J. 40* (1960), 1–41.

15. Wang, Hao. Toward mechanical mathematics, *IBM J. Res. Dev. 4*(1960), 2–22.

A Computer Program for a Solvable Case of the Decision Problem

J. Friedman

Abstract. In a previous paper [4] by the same author a semi-decision procedure for the functional calculus was described. A part of the procedure has now been programmed for the 7090 computer. In this paper a FORTRAN program is described which decides expressions with prefix $(\exists y_1)(\exists y_2)(z_1) \cdots (z_n)$ and matrix containing only dyadic functional variables. The program has been run for some standard examples and takes less than seven seconds for the longest one.

Introduction

In a previous paper [4] we described and proved a semi-decision procedure for certain classes of formulas of mathematical logic. The procedure was developed with a view toward possible implementation on a digital computer. We have now constructed a FORTRAN program which implements part of the procedure on the 7090. The program provides a decision procedure for a special class of well-formed formulas. This is the class of formulas in closed prenex normal form with prefix $(\exists y_1)(\exists y_2)(z_1) \cdots (z_n)$ and matrix M containing dyadic functional variables only.

The impossibility of constructing a decision procedure for all of the first-order predicate calculus was proved by Church in 1936. Faced with this negative result, logicians have found three directions in which its effects can be mitigated: (1) restriction of the class of expressions to be considered so that *solvable sub-cases* are obtained, (2) construction of *proof procedures* which will eventually recognize any theorem, but may run on indefinitely for a non-theorem, (3) construction of *semi-decision procedures* which include decision procedures for known solvable subcases and which, when confronted with expressions outside these cases, may recognize a theorem or may fail to reach any decision. The procedure which we have developed is of this third type; the present program treats a solvable subcase.

The decision procedure discussed here does not construct a proof from axioms as do some theorem-proving programs. Instead, the procedure arrives at the result "theorem" if and only if there are no counter-examples. Then, by the completeness of the first-order functional calculus, we conclude that a formal proof exists. We are not concerned with constructing it.

Received February, 1963. The research reported in this paper was sponsored by the Air Force Electronic Systems Division, Air Force Systems Command, under contract AF33(600)39852. This paper is identified as ESD Technical Documentary Report No. ESD-TDR-63-153. Further reproduction is authorized to satisfy needs of the U. S. Government.

Article from: Journal of the Association for Computing Machinery. Volume 10, Number 3, July 1963.

Procedure

The decision procedure operates on the set S of falsifying systems for the matrix M (corresponding to the disjuncts of $\sim M$). There are two groups of rules which are applied.

The core of the procedure is the application of reduction rules which reduce the set S to a set S^* by deleting systems and by dividing S into smaller sets S_1 and S_2 which are in turn reduced.

If the reduction procedure terminates with all systems of S eliminated, the given expression is a theorem. If, on the other hand, the procedure terminates with S^* nonempty, then an additional set of rules, called "amplification rules," must be applied. These rules are needed in order to avoid initially saturating the formula with all possible elementary parts. The amplification rules operate on the list of elementary parts and determine if there is any new part which if added might lead to further reductions. If so, the part is added (in such a way that the truth-value of the matrix is unchanged) and the process continues. If not, the program terminates with the result "not theorem."

Reduction Rules. Let S be the set of systems of truth-values which falsify the matrix M. Let $A_1, A_2, \cdots, A_h$ be the elementary parts of M which contain the functional variable F and none of the individual variables $z_1, z_2, \cdots, z_n$. A formula with N functional variables is treated as having only one functional variable with 2^N truth-values.

Rule 1. A system in which the truth-value assigned to any two elementary parts B_1 and B_2 is the same whenever $\mathrm{S}_{y_1}^{y_2}B_1|$ is identical with $\mathrm{S}_{y_1}^{y_2}B_2|$ will be called a *uniform system*. For each truth-value a $(0 \leqq a \leqq 2^N - 1)$, if there is no uniform system in which a is assigned to all of the elementary parts A_1, $A_2, \cdots, A_h$, then delete from S all systems in which for some individual variable u, $F(u, u) = a$. If there is no uniform system in S, delete all of S.

Rule 2. Let $\alpha_1\alpha_2$ be in turn each of the arrays y_1z_j, y_2y_1, y_2z_j, z_jy_1, z_jy_2, z_jz_i $(1 \leqq j \leqq n, 1 \leqq i \leqq n, i \neq j)$. For each such array, let

$$B_i = \mathrm{S}_{\alpha_1\alpha_2}^{y_1y_2} A_i| \quad (i = 1, \cdots, h)$$

and let $B_1', B_2', \cdots, B_r'$ $(r \leqq h)$ be the subset of the B_i which occur in the formula. Let $A_1', A_2', \cdots, A_r'$ be the corresponding A_i.[1] Then for each possible set of truth-values $a_1, a_2, \cdots, a_r$, if there is no system in which A_i' is assigned a_i for all $i = 1, \cdots, r$, then delete all systems in which B_i' is assigned a_i for all $i = 1, \cdots, r$.

Rule 3. If there is no remaining system of S in which $F(y_1, y_1)$ has the truth-value a and $F(y_2, y_2)$ has the truth-value b, then form, from S, two sets S_1 and S_2 such that: (1) S_1 contains all systems of S except those in which for some individual variable u the elementary part $F(u, u)$ is assigned a; (2) S_2 con-

[1] *Example*. If the elementary parts are $F(y_1, y_1)$, $F(y_1, y_2)$, $F(y_2, y_1)$, $F(z, y_1)$, then $A_1 = F(y_1, y_1)$, $A_2 = F(y_1, y_2)$, $A_3 = F(y_2, y_1)$. For the array $\alpha_1\alpha_2 = y_1z$, $B_1 = F(y_1, y_1)$, $B_2 = F(y_1, z)$, $B_3 = F(z, y_1)$. Then $B_1' = F(y_1, y_1)$, $B_2' = F(z, y_1)$ and $A_1' = F(y_1, y_1)$, $A_2' = F(y_2, y_1)$.

TABLE I. AMPLIFICATION A1

	1	2	3	4	5	6	7	8	9	10	11	12	13	14	15	16	17	18
Fy_1y_1	1	1	1	0	0	0	0	0	0	1	1	1	1	1	1	0	0	0
Fy_1z_i	0	0	0	1	1	1	0	0	0	1	1	1	0	0	0	1	1	1
Fz_iy_1	0	0	0	0	0	0	1	1	1	0	0	0	1	1	1	1	1	1
Fy_2y_2	0	0	0	0	1	1	0	1	1	1	0	0	1	0	0	1	1	1
Fy_2z_i	1	0	1	0	0	0	1	0	1	0	1	0	1	1	1	1	0	0
Fz_iy_2	0	1	1	1	0	1	0	0	0	1	1	1	0	0	1	0	0	1
	$F(y_2, y_2)$			$F(y_2, z_1)$			$F(z_i, y_2)$			$F(z_i, y_1)$			$F(y_1, z_i)$			$F(y_1, y_1)$		

tains all systems except those in which for some individual variable u the elementary part $F(u, u)$ is assigned b.

Rules 1, 2 and 3 are to be repeated until no further reduction is possible. Rules 1 and 2 must be applied to each of the sets formed by Rule 3. When no further reduction is possible, the remaining sets are S_1^*, S_2^*, $\cdots$ and their logical sum is S^*. If S^* is empty, then $(\exists y_1)(\exists y_2)(z_1)(z_2) \cdots (z_n)\cdot M$ is a theorem. If S^* is not empty, then the amplification rules are applied.

AMPLIFICATION RULES. In stating these rules we use Fuv to stand for the proposition *The elementary part* $F(u, v)$ *occurs*. If $\sim(Fy_1y_2 \vee Fy_2y_1)$, no amplification is necessary. Assume then that $(Fy_1y_2 \vee Fy_2y_1)$.

Rule A1. If one of the patterns of Table I occurs for some z_i, then amplify by the elementary part shown below. In Table I the value 1 is given if Fuv is true, 0 if false. For example, if the elementary parts $F(z, y_1)$ and $F(y_2, y_2)$ occur but none of $F(y_1, y_1)$, $F(y_1, z)$, $F(y_2, z)$, $F(z, y_2)$ occurs, amplify with $F(z, y_2)$.

Rule A2. If $(Fy_1y_1 \vee Fy_2y_2)$ and if there exist distinct individual variables u, w, and z_i for which

$$(Fuz_i \vee Fz_iu)(Fwz_i \vee Fz_iw)(\bar{F}z_iz_i),$$

then amplify with the elementary part $F(z_i, z_i)$.

Rule A3. If $Fy_1y_1Fz_iz_i\bar{F}y_2y_2(Fy_1y_2Fy_2y_1 \vee Fy_1y_2Fy_2z_i \vee Fy_2y_1Fz_iy_2)$, then amplify with the elementary part $F(y_2, y_2)$. Symmetrically, if

$$Fy_2y_2Fz_iz_i\bar{F}y_1y_1(Fy_1y_2Fy_2y_1 \vee Fy_2y_1Fy_1z_i \vee Fy_1y_2Fz_iy_1),$$

then amplify with $F(y_1, y_1)$.

Rule A4. If $Fy_1y_1Fy_1y_2\bar{F}y_2y_1Fy_2y_2$ and for some u and z_i, $Fuz_i\ Fz_iu\ Fz_iz_i$, then amplify with $F(y_2, y_1)$. Symmetrically for $\bar{F}y_1y_2\ Fy_2y_1$.

After the formula has been amplified with a new elementary part the reduction rules are applied again. If S^* is then not empty, the next amplification rule is applied. If all necessary amplification has occurred and S^* is still not empty, then the given formula is not a theorem.

The proof for the decision procedure is given in [5], available on request. The proof is omitted here because it is hoped that it will soon be subsumed by a proof for a more general case.

Program

The program was written for the 7090 computer in FORTRAN. The choice of computer was governed by availability. The choice of FORTRAN as the programming language was made because of ease of modification and, in particular, ease of expansion. There is, of course, a sacrifice in running time in using FORTRAN and we were prepared to use FAP for any lengthy subroutines. But the FORTRAN program is fast enough. That is, time can now be better spent in trying to increase the class of formulas which the program will accept.

The program now consists of a CONTROL routine, subroutines RULE1, RULE2, RULE3, and subroutines CONSAT and SATRAT (which apply rules A1, A2, A3 and A4 and amplify the matrix as required). A flow chart of the CONTROL routine is given as Figure 1. The program (including data storage) uses approximately 8000 computer words.

Inputs

A weak point in the present program is the need for manually preprocessed inputs. This is a weakness of the program and not of the method, for the preprocessing could clearly be done by computer. An outline of a program for preprocessing has been prepared by Robert R. Fenichel [3], and we hope to incorporate such a program in the future.

At present the preprocessing of the inputs consists of four steps:

(1) The problem is brought into closed prenex normal form

(2) The falsifying systems (the systems of truth-value assignments under which the matrix M is false) are found

(3) The variables are coded by numbers starting with 1 for the first variable of the prefix

(4) The information is then entered on cards. For each problem the program calls for a "header" card giving the number of falsifying systems (NFS), the number of existential quantifiers (NOY), the number of universal quantifiers (NOZ), the number of elementary parts (NCOL) and the number of functional variables (NFV). This is followed by a card with the problem name, a card with column headings, and finally NFS cards with the actual falsifying systems.

Examples

Consider example 46.14.2 of Church [1]:

$$(\exists x)(\exists y)(z)\cdot(Fxz \equiv Fzy)(Fzy \equiv Fzz)(Fxy \equiv Fyx) \supset (Fxy \equiv Fxz)$$

The matrix is false for two systems of truth-values:

Fxy	Fxz	Fyx	Fzy	Fzz
t	f	t	f	f
f	t	f	t	t

Coding the variables in the order in which they appear in the prefix, and using

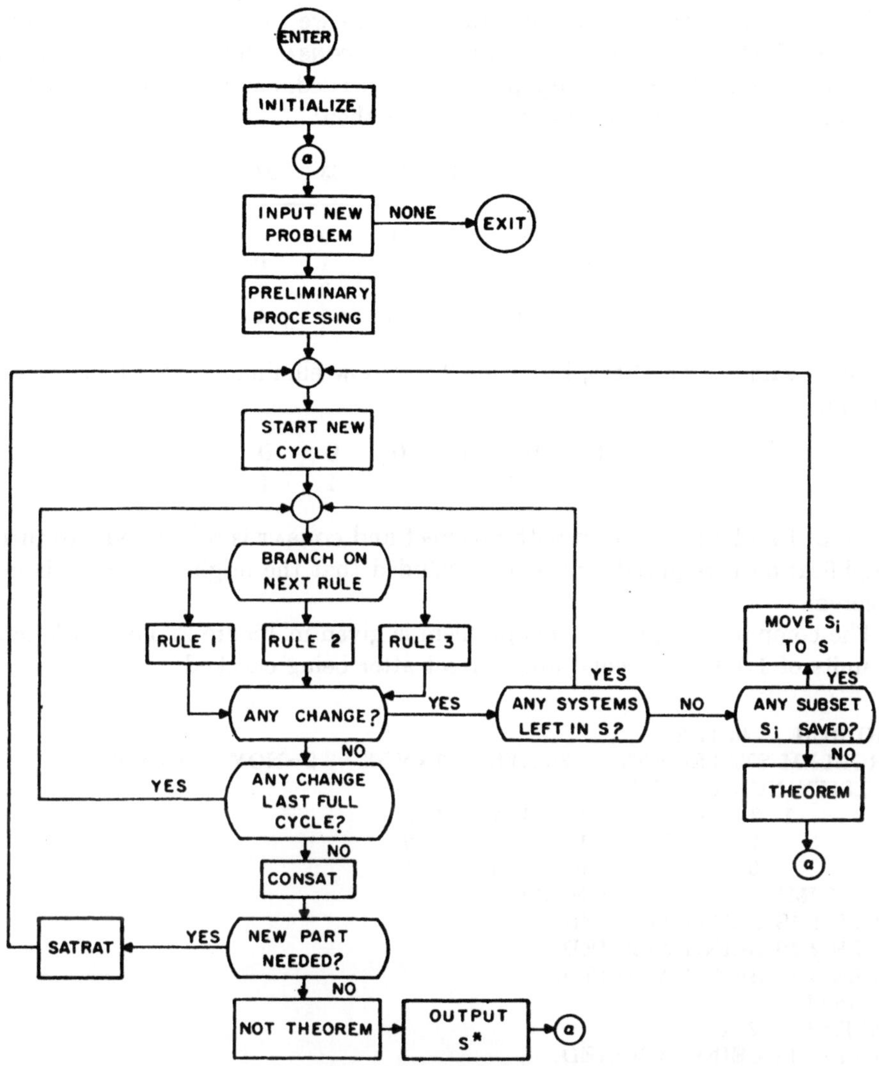

FIG. 1. CONTROL routine

0 and 1 for *f* and *t*, we obtain the input cards:

Identifiers:	NFS	NOX	NOY	NOZ	NCOL	NFV
Card 1:	2	0	2	1	5	1
Problem Name:						
Card 2:	CHURCH 46.14.2					
Column heads:	*Fxy*	*Fxz*	*Fyx*	*Fzy*	*Fzz*	
Card 3:	1 2	1 3	2 1	3 2	3 3	
Falsifying Systems:						
Card 4:	1	0	1	0	0	
Card 5:	0	1	0	1	1	

Following the example through the program we find that Rules 1, 2 and 3 have no effect when first applied. Subroutine CONSAT determines (by Rule A1, Column 6) that the elementary part Fy_2z is needed. SATRAT adds the part, and the falsifying systems for the problem are then:

12	13	21	32	33	23
1	0	1	0	0	0
0	1	0	1	1	0
1	0	1	0	0	1
0	1	0	1	1	1

Rule 1 again has no effect, but Rule 2 (y_2z)now eliminates two of the systems, leaving

1	0	1	0	0	0
0	1	0	1	1	1

Rules 3, 1 and 2 are applied with no effect and CONSAT is re-entered. No further amplification is required, and it is concluded that the original formula is not a theorem.

The computer output for this problem is given in Figure 2. The clock counts in units of 1/60 of a second and is reset after being checked.

```
CHURCH    46.14.2
PROBLEM STATEMENT    2=NFS    0=NOX    2=NOY    1=NOZ
FALSIFYING SYSTEMS
          1 2   1 3   2 1   3 2   3 3
            1     0     1     0     0
            0     1     0     1     1
      TIME =      8 SECONDS/60
RULE 1 IS BEING APPLIED.
RULE 2 IS BEING APPLIED.
RULE 3 IS BEING APPLIED.
CONSAT
SATRAT    2  3
RULE 1 IS BEING APPLIED.
RULE 2 IS BEING APPLIED
RULE 3 IS BEING APPLIED.
RULE 1 IS BEING APPLIED.
RULE 2 IS BEING APPLIED.
CONSAT
NOT A THEOREM.    FALSIFYING SYSTEMS ARE
      TIME =     16 SECONDS/60
          1 2   1 3   2 1   3 2   3 3   2 3
            1     0     1     0     0     0
            0     1     0     1     1     1
      TIME =      3 SECONDS/60
```

FIG. 2. Output for example 46.14.2

In the above example, if we stop prior to entering CONSAT, there are two falsifying systems and we conclude that the formula is not a theorem. Since this is in fact the correct answer, the need for amplification should perhaps be demonstrated. To do this we take as a second example a problem which differs from the first only in the values of a single column. For this example the additional elementary part is required to enable us to reach the correct conclusion that the corresponding formula is a theorem. The identifiers and the column heads are the same as above, but the two systems are

12	13	21	32	33
1	0	1	1	0
0	1	0	0	1

As before, Rules 1, 2 and 3 have no effect. CONSAT then adds the Fy_2z-column giving:

12	13	21	32	33	23
1	0	1	1	0	0
0	1	0	0	1	0
1	0	1	1	0	1
0	1	0	0	1	1

Again Rule 1 has no effect and Rule 2 (y_2z) eliminates two systems, leaving this time:

0	1	0	0	1	0
1	0	1	1	0	1

Rule 3 has no effect, but now Rule 1 eliminates both systems because there is no uniform system. The conclusion in this case is thus "theorem," a result which would not have been reached without amplification.

Results

Some of the examples which have been run were taken from Church; others were generated to test the program. In all, seventy problems have been run. The total running time for these is less than forty-five seconds. The most difficult problems for the program are the following ones given by Church in Section 46.14:

(1) $(\exists x)(\exists y)(z)\cdot FxxFyyFxz \supset Fzy$

(2) $(\exists x)(\exists y)(z)\cdot(Fxz \equiv Fzy)(Fzy \equiv Fzz)(Fxy \equiv Fyx) \supset (Fxy \equiv Fxz)$

(3) $(\exists x)(\exists y)(z)\cdot FxzFyz(Fxy \equiv Fzz)(Fyx \vee Fzz) \supset (Fzx \vee Fzy)$

(4) $(\exists x)(\exists y)(z)\cdot(FxyFyx \not\equiv Fxz)(Fxz \equiv Fzx)(Fxz \equiv Fyz)$
$((Fyx \supset Fxy) \equiv Fzz) \supset ((Fxy \equiv Fyx) \equiv Fzy)$

(5) $(\exists x)(\exists y)(z)\cdot(Fxy \supset FyzFzz)(FxyGxy \supset GxzGzz)$

(6) $(\exists x)(\exists y)(z)\cdot(Fxy \supset (Fxz \equiv Gyz))(Fxy \equiv (Fzz \supset Gzz)) \supset (Gxy \equiv Gzz)$

(7) $(\exists x)(\exists y)(z)\cdot Fxz((Fzz \supset Gzz) \equiv Fxy)((Gzz \supset Fzz) \equiv Gxy)$
$((Gxy \supset Fyx) \equiv Gyz) \supset (Fzy \equiv Fyx)$

TABLE II. Running Times for Church's Examples

Example	Time (in seconds) *Input*	*Process*	*Output*	*Total*	*Result*
46.14.1	.12	.10	.03	.24	not theorem
.2	.13	.27	.05	.45	not theorem
.3	.17	.08	—	.25	theorem
.4	.18	.10	—	.28	theorem
.5	2.10	.28	—	2.38	theorem
.6	.85	.22	—	1.07	theorem
.7	2.62	3.33	.38	6.33	not theorem

Running times for these examples are given in Table II. The breakdown into input and processing times shows the effect of the lack of a preprocessing routine. The input time is roughly proportional to the number of falsifying systems, and includes time for writing a copy of the input as output. For non-theorems the falsifying systems of S^* are output.

Some of these problems have been discussed earlier in the literature. Gilmore [6] was unable to do problem (2) and manually stopped his attempt at problem (5) after 21 minutes on the 704. Davis, Logemann and Loveland [2] are able to do problem (5) in less than two minutes on the 704, but do not handle non-theorems such as (2). However, it should be pointed out that both programs handle a wider range of problems than does our present program. Wang [7] discusses a method similar to ours in connection with these problems, but has not constructed the corresponding program.

A Correction

Reference [4] contains an error in Rule 7 which requires modification of the proof of Theorem 4. The correction to be made is, in essence, the following.

In *Rule 7* the array $\alpha_1\alpha_2 \cdots \alpha_m$ is to be specifically $y_1 \cdots y_1 y_{q+1} \cdots y_m$, $1 < q < m$, and for each array $e_1e_2 \cdots e_m$ of all and only the numbers 1, $q+1$, $\cdots$, m there is to be a system $\beta_{e_1 \cdots e_m}$ as described. Further, if

$$\mathrm{S}^{y_1 \cdots y_m}_{y_{e_1} \cdots y_{e_m}} A_j| = \mathrm{S}^{y_1 \cdots y_m}_{y_{e_1'} \cdots y_{e_m'}} A_k| \quad (j, k \leq L),$$

then $\beta_{e_1 \cdots e_m}$ must assign to A_j the same value that $\beta_{e_1' \cdots e_m'}$ assigns to A_k.

In the proof of *Theorem 4*, row R of Case 3 is to be filled by $\beta_{1 \cdots 1 (q+1) \cdots m}$ of the set of systems corresponding to γ by Rule 7. All subsequent rows which assign to $y_1, \cdots, y_m$ combinations of all and only the distinct numbers assigned to $y_1, \cdots, y_m$ in R can then be filled by the corresponding systems of this same set.

Conclusions

The results obtained with this program show that theorem proving can be done much faster than previously indicated. The increase in speed is intrinsic to the method used rather than to the particular program. The method has been extended in theory well beyond the rigid limitations of the present program. The direction of development will now be toward extending the program to more general cases.

Acknowledgment. I am grateful to Hao Wang for his interest and encouragement.

REFERENCES

1. CHURCH, ALONZO. *Introduction to Mathematical Logic, Vol. I.* Princeton University Press, Princeton, New Jersey, 1956.
2. DAVIS, MARTIN; LOGEMANN, GEORGE; LOVELAND, DONALD. A machine program for theorem proving. *Comm. ACM 5* (1962), 394–397.
3. FENICHEL, ROBERT R. Preliminary description of the Gödel preprocessor. Unpublished paper, 1962.
4. FRIEDMAN, JOYCE. A semi-decision procedure for the functional calculus. *J. ACM 10* (1963), 1–24.
5. ——, A computer program for a solvable case of the decision problem. MITRE SR-88, 1963.
6. GILMORE, P. C. A proof method for quantification theory. *IBM J. Res. Develop. 4* (1960), 28–35.
7. WANG, HAO. Proving theorems by pattern recognition. Part I: *Comm. ACM 3* (1960), 220–234; Part II: *Bell System Tech. J. 40* (1961), 1–41.

A Simplified Proof Method for Elementary Logic

S. Kanger

It is a well-known fact that there are complete effective proof methods for elementary logic. In other words, there are effective methods M such that whenever a formula F in the language of elementary logic happens to be a logical consequence of a sequence Γ of such formulas, this fact may be demonstrated by means of the method M. And since M is effective, we may — in principle at least — program a computer to carry out the demonstrations. Thus there is a possibility to use a computer to prove theorems in any elementary axiomatic theory. In order to prove a theorem F in a theory with the axioms Γ, we have the computer show that F is a logical consequence of Γ.

However, every effective proof method for elementary logic available today seems to be much too time- or space-consuming to fit even the largest and fastest computer, except when applied to some special cases.

The least suitable method, from the point of view of the computer, so to speak, is the so-called British museum method. Suppose a formula F ınd a finite formula sequence Γ are given and suppose we wish to show that F is a logical consequence of Γ. We then order the class of finite formula sequences beginning with Γ and ending with F and check these sequences one by one. If F happens to be a logical consequence of Γ, we will sooner or later reach a sequence which is a deduction of F from Γ in, say, the Hilbert-Ackermann calculus of elementary logic. We readily see that this method very soon exhausts the capacity of any computer.

Better methods may be obtained if we add various "heuristic" devices which sometimes give short cuts in the demonstrations but sometimes do not. The use of heuristic devices has been suggested by Newell and Simon [6] and by Gelernter [2]. The introduction of heuristics may yield considerable simplifications of a given proof method, but I have the impression that it would be wise to postpone the heuristics until we have a satisfactory proof method to start with.

The most suitable methods are perhaps those based on a Gentzen-type calculus, for instance the calculus given in Kanger [4] or Beth's semantic tableaux. Proof methods of this kind — or essentially of this kind — suitable for programming have been given by Gilmore [3], by Wang [9], by Prawitz and Vogera [7] and by others. But as far as I know, these methods are still too time-consuming to be of any practical interest. In the Prawitz-Vogera program, for instance, we may easily construct very simple cases of logical consequence such that the computer would need an astronomic number of years to provide the demonstration.

Thus, what we need is a radical simplification of the ordinary routine, so to speak, of the Gentzen-type proof methods.

In this paper I shall suggest a simplification of these methods. The simplified method I describe is identical with the proof method for elementary logic given in my "Handbok i logik" and is similar to a method independently given by Prawitz [8].

First I shall outline a Gentzen-type calculus.

Let us use the letters F and G to denote formulas of elementary logic — or to be more precise: formulas of the first order predicate logic with symbols for identity and operations from individuals to individuals. It is convenient in this connection to assume that bound variables are typographically distinct from free variables, or parameters, as we may also call them. We shall use the letter x to denote bound individual variables and the letters c and d to denote individual parameters or individual constants or terms formed by means of individual parameters or constants and operation symbols. We shall use Greek capital letters Γ, Δ, Z and Λ to denote finite (and possibly empty) sequences of formulas. The notation Γ^c_d shall denote the result of replacing each occurrence of c in each formula of Γ by occurrences of d. We shall use expressions of the form $\Gamma \Rightarrow Z$ to express the fact that Z is a logical consequence of Γ. Expressions of this form we call sequents. (We say that Z is a logical consequence of Γ, if for each non-empty domain of individuals and each choice of a possible interpretation with respect to this domain of the non-logical symbols for individuals, operations, classes and relations, some formula of Z is true or some formula of Γ is false.)

We may assume that the symbols $=$ (identity), $\sim$ (negation), & (conjunction), $\vee$ (disjunction), $\supset$ (implication), $\equiv$ (equivalence), **U** (universal quantification) and **E** (existential quantification) are the only logical symbols, and we may adopt two axiom schemes and sixteen rules of inference as postulates in the calculus:

P.1 $$\Gamma, F, \Delta \Rightarrow Z, F, \Lambda$$

P.2 $$\Gamma \Rightarrow Z, (c = c), \Lambda$$

P.3 $$\frac{\Gamma^{c}_{d}, (c = d), \Delta^{c}_{d} \Rightarrow Z^{c}_{d}}{\Gamma, (c = d), \Delta \Rightarrow Z}$$

P.4 $$\frac{\Gamma^{c}_{d}, (d = c), \Delta^{c}_{d} \Rightarrow Z^{c}_{d}}{\Gamma, (d = c), \Delta \Rightarrow Z}$$

P.5 $$\frac{F, \Gamma \Rightarrow Z, \Lambda}{\Gamma \Rightarrow Z, \sim F, \Lambda}$$

P.6 $$\frac{\Gamma, \Delta \Rightarrow F, Z}{\Gamma, \sim F, \Delta \Rightarrow Z}$$

P.7 $$\frac{\Gamma \Rightarrow Z, F, \Lambda \qquad \Gamma \Rightarrow Z, G, \Lambda}{\Gamma \Rightarrow Z, (F \,\&\, G), \Lambda}$$

P.8 $$\frac{\Gamma, F, G, \Delta \Rightarrow Z}{\Gamma, (F \,\&\, G), \Delta \Rightarrow Z}$$

P.9 $$\frac{\Gamma \Rightarrow Z, F, G, \Lambda}{\Gamma \Rightarrow Z, (F \vee G), \Lambda}$$

P.10 $$\frac{\Gamma, F, \Delta \Rightarrow Z \qquad \Gamma, G, \Delta \Rightarrow Z}{\Gamma, (F \vee G), \Delta \Rightarrow Z}$$

P.11 $$\frac{F, \Gamma \Rightarrow Z, G, \Lambda}{\Gamma \Rightarrow Z, (F \supset G), \Lambda}$$

P.12 $$\frac{\Gamma, \Delta \Rightarrow F, Z \qquad \Gamma, G, \Delta \Rightarrow Z}{\Gamma, (F \supset G), \Delta \Rightarrow Z}$$

P.13 $$\frac{F, \Gamma \Rightarrow Z, G, \Lambda \qquad G, \Gamma \Rightarrow Z, F, \Lambda}{\Gamma \Rightarrow Z, (F \equiv G), \Lambda}$$

P.14 $$\frac{\Gamma, \Delta \Rightarrow F, G, Z \qquad \Gamma, F, G, \Delta \Rightarrow Z}{\Gamma, (F \equiv G), \Delta \Rightarrow Z}$$

P.15 $$\frac{\Gamma \Rightarrow Z, F^{x}_{i}, \Lambda}{\Gamma \Rightarrow Z, \mathrm{U}xFx, \Lambda}$$

where i is a parameter not occurring in the conclusion

P.16 $$\frac{\Gamma, F_c^x, \mathrm{U}xFx, \Delta \Rightarrow Z}{\Gamma, \mathrm{U}xFx, \Delta \Rightarrow Z}$$

P.17 $$\frac{\Gamma \Rightarrow Z, F_c^x, \mathrm{E}xFx, \Lambda}{\Gamma \Rightarrow Z, \mathrm{E}xFx, \Lambda}$$

P.18 $$\frac{\Gamma, F_i^x, \Delta \Rightarrow Z}{\Gamma, \mathrm{E}xFx, \Delta \Rightarrow Z}$$

where i is a parameter not occurring in the conclusion.

Now, when we wish to give a demonstration of a sequent $\Gamma \Rightarrow Z$, we start from below with this sequent and construct a tree of sequents above it by means of backwards applications of the rules of inference. If we succeed in constructing a tree in which every branch has an instance of P.1 or P.2 at its top, the demonstration has succeeded. And if $\Gamma \Rightarrow Z$, we will always be able to construct such a tree, provided we follow a certain routine (and provided of course that we have a sufficient amount of time and space at our disposal). Thus, if we supply such a routine we have an effective proof method for elementary logic.

The main difficulty is this: Suppose we have reached a sequent at a certain level in a branch of the tree. Then it frequently happens that we may continue the branch in more than one way. And it may happen that some of these ways are more favorable than others from the viewpoint of simplicity. Thus our routine ought to involve some devices for choosing the favorable ways of continuing the branches. Without good devices of this sort the proof method will be much too time-consuming. The lack of such devices was the source of trouble with the Prawitz-Vogera program.

In order to give a routine which involves devices of this kind we shall make some modifications of the rules.

Let us say that an individual parameter or constant is a term of rank zero, and let us say that a term $f(c_1,\ldots,c_n)$ is a term of rank $r+1$, if r is the maximum rank of the terms $c_1,\ldots,c_n$. Now we restrict the rules P.3 and P.4 for identity by the requirement that $\mathrm{rank}(c) \geq \mathrm{rank}(d)$ in P.3 and $\mathrm{rank}(c) > \mathrm{rank}(d)$ in P.4. Thus, in a backwards application of P.3 or P.4 we never replace a term c with a term d of higher rank.

We say that a sequent $\Gamma \Rightarrow Z$ is directly demonstrable if it is demonstrable by means of the restricted postulates P.1–P.4 only. We note that we can always decide if a given sequent is directly demonstrable or not.

We also restrict the rules P.16 and P.17 by the requirement that the term c shall occur in the conclusion below the line or — if there are no terms in the conclusion — c shall be the alphabetically first individual parameter. We shall also change the formulation of the rule. When we apply the rule we shall not have to choose the term c immediately. Instead we replace x by a dummy α and make a note in the margin that α stands for one of the terms in the conclusion. Thus the rules are given the following form:

P.16 $$\frac{\Gamma, F\overset{x}{\alpha}, \mathrm{U}xFx, \Delta \Rightarrow Z}{\Gamma, \mathrm{U}xFx, \Delta \Rightarrow Z} \qquad \alpha/c_1, \ldots, c_n$$

where $c_1, \ldots, c_n$ are the terms occurring in the conclusion; if there are no such terms, α is the first individual parameter.

P.17 $$\frac{\Gamma \Rightarrow Z, F\overset{x}{\alpha}, \mathrm{E}xFx, \Lambda}{\Gamma \Rightarrow Z, \mathrm{E}xFx, \Lambda} \qquad \alpha/c_1, \ldots, c_n$$

where $c_1, \ldots, c_n$ are the terms occurring in the conclusion; if there are no such terms, α is the first individual parameter.

The note $\alpha/c_1, \ldots, c_n$ we may call a substitution list for the dummy α and the terms $c_1, \ldots, c_n$ we call the values of α.[1])

I shall now describe the routine of the proof method. Suppose we wish to demonstrate that $\Gamma \Rightarrow Z$. We start from below with $\Gamma \Rightarrow Z$ and going upwards we construct a tree of sequents by applications of the rules. We divide the construction of the tree into stages. Within each stage we apply only the rules for truth-functions and quantification, i.e. rules P5–P18. At the end of each stage we test the top sequents of each branch of the stage. If the test gives a positive result we stop the construction of the tree, otherwise we construct a new stage.

Within each stage we prefer applications of rule P.15 and P.18 to applications of the truth-functional rules, i.e. rules P.5–P.14, and we prefer applications of truth-functional rules to applications of P.16 or P.17. When we apply P.16 and P.17 and when more than one term occurs in the conclusion, we shall always introduce a new dummy which is not yet introduced in the tree and we shall always give the substitution list for the dummy. Moreover, when we apply P.16 and P.17 we shall prefer to split a formula G rather than a formula which has been split more times than G in previous application of P.16 or P.17 in the branch

1) Note that dummies (other than α) may occur in the list $c_1, \ldots, c_n$.

in question. When we apply P.15 or P.18, the parameter introduced shall be new and of course different from the values of the dummies in the conclusion.

At the end of each stage we stop for a moment and check whether we can choose values for the dummies from their substitution lists in such a way that all top sequents will be directly demonstrable when we replace the dummies by their values.

If there is such a choice, our demonstration has succeeded. If there is no such choice, we keep the dummies and construct a new stage of the tree by continuing the construction of each branch which is not completed in the sense that its top sequent is directly demonstrable for every value of the dummies.

It remains to fix the extension of the stages. We let a stage be completed when each branch of the stage has a top sequent $\Delta \Rightarrow \Lambda$ such that (1) every formula in Δ is either an atomic formula or a formula beginning with a universal quantifier, and every formula in Λ is either an atomic formula or a formula beginning with an existential quantifier, and (2) every non-atomic formula in Δ and Λ has been split equally many times by previous applications of P.16 or P.17 in the branch.

This concludes the description of the proof method. To illustrate the method, I shall give a demonstration of the sequent $\Rightarrow \mathbf{E}x\mathbf{U}y((x = f(x)) \supset (f(f(y)) = y))$. The formula in this sequent we may abbreviate as $\mathbf{E}xF$. The demonstration shall be read from below.

direct demonstration:

$$(j = f(j)), (i = f(i)) \Rightarrow (j = j), (f(f(k)) = k), \mathbf{E}xF$$
$$\overline{} \quad \text{P.4}$$
$$(j = f(j)), (i = f(i)) \Rightarrow (f(j) = j), (f(f(k)) = k), \mathbf{E}xF$$
$$\overline{} \quad \text{P.4}$$
$$(j = f(j)), (i = f(i)) \Rightarrow (f(f(j)) = j), (f(f(k)) = k), \mathbf{E}xF$$

choice of values: $\alpha/_j$

stage 3:

$$(\alpha = f(\alpha)), (i = f(i)) \Rightarrow (f(f(j)) = j), (f(f(k)) = k), \mathbf{E}xF$$
$$\overline{} \quad \text{P.11}$$
$$(i = f(i)) \Rightarrow (f(f(j)) = j), ((\alpha = f(\alpha)) \supset (f(f(k)) = k)), \mathbf{E}xF$$
$$\overline{} \quad \text{P.15}$$
$$(i = f(i)) \Rightarrow (f(f(j)) = j), \mathbf{U}y((\alpha = f(\alpha)) \supset (f(f(y)) = y)), \mathbf{E}xF$$
$$\overline{} \quad \text{P.17} \quad \alpha/i, f(i), j, f(j), f(f(j))$$

stage 2:

$$(i = f(i)) \Rightarrow (f(f(j)) = j), \mathbf{E}x\mathbf{U}y((x = f(x)) \supset (f(f(y)) = y))$$
$$\overline{} \quad \text{P.11}$$
$$\Rightarrow ((i = f(i)) \supset (f(f(j)) = j)), \mathbf{E}xF$$
$$\overline{} \quad \text{P.15}$$
$$\Rightarrow \mathbf{U}y((i = f(i)) \supset (f(f(y)) = y)), \mathbf{E}xF$$
$$\overline{} \quad \text{P.17}$$

stage 1:

$$\Rightarrow \mathbf{E}x\mathbf{U}y((x = f(x)) \supset (f(f(y)) = y))$$

We shall compare this demonstration with the demonstrations of the sequents $\Rightarrow UyEx((x = f(x)) \supset (f(f(y)) = y))$ and $\Rightarrow (Ux(x = f(x)) \supset Uy(f(f(y)) = y))$. The formulas here are logically equivalent with the formula in the example, but the demonstrations will be simpler. We need only two stages, since th econstant j which we substitute for y will now be available as a dummy value already at the first application of P.17 or P.16. The simplest demonstration is that of the last sequent:

$$\text{direct demonstration}\left\{\begin{array}{ll} \dfrac{\dfrac{(j = f(j)), Ux(x = f(x)) \Rightarrow (j = j)}{(j = f(j)), Ux(x = f(x)) \Rightarrow (f(j) = j)}}{(j = f(j)), Ux(x = f(x)) \Rightarrow (f(f(j)) = j)} & \begin{array}{l}\text{P.4}\\ \text{P.4}\end{array} \end{array}\right.$$

choice of values: α/j

$$\begin{array}{ll} \text{stage 2}\ \{ & \dfrac{(\alpha = f(\alpha)),\ Ux(x = f(x)) \Rightarrow (f(f(j)) = j)}{Ux(x = f(x)) \Rightarrow (f(f(j)) = j)} \quad \text{P.16} \quad \alpha/j, f(j), f(f(j)) \\ \text{stage 1} & \dfrac{}{Ux(x = f(x)) \Rightarrow Uy(f(f(y)) = y)} \quad \text{P.15} \\ & \dfrac{}{\Rightarrow (Ux(x = f(x)) \supset Uy(f(f(y)) = y))} \quad \text{P.11} \end{array}$$

Thus, to obtain simple demonstrations it usually pays not to extend the scopes of the quantifiers more than necessary.

REFERENCES

[1] BETH, E. W., *Formal methods.* Dordrecht 1962.

[2] GELERNTER, H. and ROCHESTER, N., Intelligent behavior in problem-solving machines. *IBM Journal of Research and Development*, vol. 2 (1958), pp. 336–345.

[3] GILMORE, P. C., A proof method for quantification theory: its justification and realization. *IBM Journal of Research and Development*, vol. 4 (1960), pp. 28–35.

[4] KANGER, S., *Provability in logic.* Stockholm 1957.

[5] —, *Handbok i logik.* Stockholm 1959 (Mimeographed).

[6] NEWELL, A. and SIMON, H., The logic theory machine. *IRE Transactions on Information Theory*, vol. IT-2, no. 3 (1956), pp. 61–79.

[7] PRAWITZ, D., PRAWITZ, H. and VOGERA, N., A mechanical proof procedure and its realization in an electronic computer. *The Journal of the Association for ComputingMachinery*, vol. 7 (1960), pp. 102–128.

[8] PRAWITZ, D., An improved proof procedure. *Theoria*, vol. 26 (1960), pp. 102–139.

[9] WANG, H., Toward mechanical mathematics. *IBM Journal of Research and Development*, vol. 4 (1960), pp. 2–22.

[10] —, Proving theorems by pattern recognition. Part I, *Communications of the Association for Computing Machinery*, vol. 3 (1960), pp. 220–234; Part II Bell System Technical Journal, vol. 40 (1961) pp. 1–41.

Theorem-Proving on the Computer

J. A. Robinson

There are excellent explanations in the literature of the formulation, in quantification theory, of problems in which a *proof* is to be found, if one exists, for a given *conclusion* from a set of given *premises*. In particular, in [1] and [3] it is shown how to transform the original problem into a standard form which contains no *quantifiers* and which consists of a *conjunction* of *disjunctions*, each disjunct being an *open atomic sentence-form* or the *negation* of one. We assume familiarity with these methods of formulation and preliminary transformation, and provide just those definitions of our working terminology which will be required for the immediate purposes of the paper.

The paper discusses the "combinatorial explosion" difficulties encountered by computer programs embodying proof-construction procedures. A program developed at Argonne National Laboratory is described in which these difficulties are somewhat alleviated in two ways. The first way, which although very useful in practice is less intellectually satisfying than the second way, consists essentially in incorporating the mathematician-user of the program into the search-loop. Several examples of proofs obtained by this means are exhibited and discussed, one interesting feature of them being that they are "reasonably nontrivial" mathematical exercises. The second way involves a complete proof procedure which seems to be new to the literature but which has not yet been programmed and tested.

Following [1] and [3], then, consider logical expressions

$$\begin{pmatrix} L_{11} & L_{12} & \cdots & L_{1m_1} \\ L_{21} & L_{22} & \cdots & L_{2m_2} \\ \vdots & & & \vdots \\ L_{k1} & L_{k2} & \cdots & L_{km_k} \end{pmatrix} \tag{1}$$

in which each L_{ij} is a *literal*. A *literal* is any expression of the form

$$P^n(A_1, A_2, \cdots, A_n) \tag{2}$$

or of the form

$$\sim P^n(A_1, A_2, \cdots, A_n) \tag{3}$$

where P is a *predicate letter*; n (>0), its *degree*; $\sim$, the *negation sign*; and A_1, $A_2, \cdots, A_n$ are *arguments*. An *argument* is one of three kinds of expression: (a) an *individual variable*, (b) an *individual constant*, or (c) an expression of the form

$$f^n(A_1, A_2, \cdots, A_n) \tag{4}$$

Received November, 1962. Work was performed under the auspices of the U. S. Atomic Energy Commission.

† Department of Philosophy.

Article from: Journal of the Association for Computing Machinery, Volume 10, Number 2, April 1963.

where f is a *function letter*; n (>0), its *degree*; and $A_1, A_2, \cdots, A_n$ are individual variables.

In (1) each row is a *disjunction* of its literals, and the entire expression (1) is a *conjunction* of its rows.

The individual constants, if any, occurring in (1) are said to be *primitive*, or *of level zero*. For each system (1), we define a set H of constant expressions, called the *Herbrand universe* (following [3]) of the system (1), as follows: H_0 is the set of constants of level zero,[1] and H_{r+1} is the set of all expressions of the form

$$f^n(K_1, K_2, \cdots, K_n) \tag{5}$$

in which f^n is a function letter of degree n occurring in (1), and each K_i is a member of one of the sets $H_0, H_1, \cdots, H_r$, with at least one K_i being actually a member of H_r. Members of H_r are said to be (*compound*) *constants of level r*. Finally, H is defined to be $\bigcup_{r=0}^{\infty} H_r$. For later convenience, we also define, for each r, the set H_r^* as the "partial sum"

$$H_r^* = \bigcup_{j=0}^{j=r} H_j \tag{6}$$

so that H_r^* contains all constants of level r or less. In later examples we shall omit superscripts on predicate letters and function letters. An expression is said to be an *instance* of a row of (1) if it is obtained from that row by replacement of each individual variable in the row by a member of H (each occurrence of the same individual variable being replaced by an occurrence of the same member of H). We shall also say that an expression is an instance of (1) if it is an instance of some row of (1). An instance of (1) is said to be *of level r* if r is the highest level of any member of H appearing in it. The elements of an instance will also be called *literals*, it being understood that, in this context, arguments can now be any member of H. Two literals are *complementary* if they are exactly alike except for one possessing the negation sign and the other lacking it.

A finite set $Q_1, Q_2, \cdots, Q_p$ of instances is *inconsistent* if there is no way of assigning the numbers 0 and 1 to the literals of $Q_1, Q_2, \cdots, Q_p$ so that complementary literals are assigned different numbers and each Q_i, $1 \leqq i \leqq p$, contains at least one literal receiving the number 1. A system of form (1) which has an inconsistent set of instances is said to be *inconsistent*.

The fundamental problem of automatic proof-construction is then: *to design a procedure which for any inconsistent system of the form (1) will find an inconsistent set of instances.*

Procedures satisfying this requirement have been given in [1] and [8]. In outline they are based on the idea that if an inconsistent set of instances exists it must, being finite, contain an instance of highest level, say q. Hence by examining successively the sets $S_0, S_1, S_2, \cdots$, where S_i is the set of *all* instances obtainable by using the constants of H_i^* in all possible ways, an inconsistent set will be found, since S_q certainly is inconsistent. [Note that if a set of instances is inconsistent so is any set which includes it.]

[1] If no primitive constants occur in (1), then H_0 is to contain a single constant.

The drawback to these procedures is that the size of the sets S_i increases explosively with i for most systems of any interest, so that for such systems it is completely infeasible to carry the procedure beyond very small values of i. There are interesting enough exceptions to this situation, however, and it is worthwhile pursuing the matter in greater detail.

The size of S_i depends on two factors: (a) the number of elements in H_i^*, and (b) the numbers $v_1, v_2, \cdots, v_k$ of distinct free variables in the respective rows of (1). Indeed, putting h_i as the size of H_i^*, there are

$$w_i = h_i^{v_1} + h_i^{v_2} + \cdots + h_i^{v_k} \tag{7}$$

instances in the set S_i.

In turn, h_i depends on the numbers $u_1, u_2, \cdots, u_t$ of distinct function letters of degrees $1, 2, \cdots, t$ (t being the largest degree occurring) in (1), together with the number u_0 of distinct primitive constants in (1). (We take $u_0 = 1$ if (1) contains no primitive constants.) Indeed, we have $h_0 = u_0$ and

$$h_{n+1} = u_0 + u_1 h_n + u_2 h_n^2 + \cdots + u_t h_n^t. \tag{8}$$

In practice, one rarely considers problems for which $t > 2$ and

$$\max [v_1, v_2, \cdots, v_k] > 6.$$

(The former is so because the familiar algebraic structures involve only binary and singular operations; the latter is so because in dealing with the same structures the condition requiring the most free variables seems always to be associativity, which takes 6. Of course, one can easily invent problems in which t and the v_i's are arbitrarily large.)

In [1], Davis and Putnam consider the system

$$\begin{pmatrix} F(x, y) & & & \\ \sim F(y, f(x, y)) & \sim F(f(x, y), f(x, y)) & G(x, y) & \\ \sim F(y, f(x, y)) & \sim F(f(x, y), f(x, y)) & \sim G(x, f(x, y)) & \sim G(f(x, y), f(x, y)) \end{pmatrix} \tag{9}$$

in which $v_1 = v_2 = v_3 = 2$, $u_0 = 1$, $u_1 = 0$, $u_2 = 1$, $t = 2$, so that $(h_0, h_1, \cdots) = (1, 2, 5, 26, 677, \cdots)$ and $(w_0, w_1, w_2, \cdots) = (3, 12, 75, 2028, 1374987, \cdots)$. For this problem $q = 2$, as it turns out; hence at most 75 instances need to be generated before an inconsistent system is obtained. By testing the growing set of instances more frequently than at the end of each level, one can in practice often detect the inconsistency before the level is exhausted, as was shown with respect to this example in [1]. Had q been 3, the problem would still have been feasible (though hardly by hand); but already if q had been 4 the limits of feasibility would have been left far behind. One notes that in this problem the only members of H that are actually needed in order to produce an inconsistent set of instances of (9) are:

$$a, \quad f(a, a), \quad f(a, f(a, a)), \quad f(f(a, a), \quad f(a, a)) \tag{10}$$

where a is the sole member of H_0 supplied in lieu of any primitive constants actually occurring in (9).

A subset P of H is called an *inconsistency set* of constants if the result of instantiating (1) completely over P is an inconsistent set of instances. If P is an inconsistency set but no proper subset of P is, then P is called a *proof set* of constants. We see that (10), for example, is a proof set for the problem (9). Clearly, given a proof set, it is a trivial matter to generate the corresponding set of instances and demonstrate its inconsistency (provided that $p^{v_1} + p^{v_2} + \cdots + p^{v_k}$, the size of this set, is not too large, where p is the number of constants in the proof set).

We define q to be the *level* of a proof set, in the obvious way, if q is the highest level of any constant in it, and remark that it is generally the case that p is very much smaller than h_q. This is shown by the following example, which arises from seeking to prove the existence of a right identity element in any algebra closed under a binary associative operation having left and right solutions x and y for all equations $x \cdot a = b$ and $a \cdot y = b$ whose coefficients a and b are in the algebra. Putting "$P(x, y, z)$" to represent that z is the result of operating on x and y, we get the system

$$\begin{bmatrix} \sim P(x, y, u) \sim P(y, z, v) \sim P(x, v, w) P(u, z, w) \\ \sim P(x, y, u) \sim P(y, z, v) \sim P(u, z, w) P(x, v, w) \\ P(g(x, y), x, y) \\ P(x, h(x, y), y) \\ P(x, y, f(x, y)) \\ \sim P(k(x), x, k(x)) \end{bmatrix} \tag{11}$$

in which $v_1 = 6$, $v_2 = 6$, $v_3 = 2$, $v_4 = 2$, $v_5 = 2$, $v_6 = 1$; $u_0 = 1$, $u_1 = 1$, $u_2 = 3$, $t = 2$. Using equations (7) and (8), we therefore have

$$(h_0, h_1, h_2, h_3, \cdots) = (1, 5, 81, 19765, \cdots) \tag{12}$$

and

$$(w_0, w_1, w_2, \cdots) = (6, 31330, 564859092726, \cdots). \tag{13}$$

Now a proof set for (11) is the following:

$$a, \quad h(a, a), \quad k(h(a, a)), \quad g(a, k(h(a, a))), \tag{14}$$

with a entering as before in lieu of any primitive constants occurring in (11) itself. It seems sad to contemplate generating the vertiginously large number w_3 of instances of (11) simply to locate the set (14). Indeed, the situation is even sadder; for by no means all of the instances of (11) over the set (14) are required for an inconsistent set of instances—the following, in fact, is an inconsistent set:

$$\begin{bmatrix} P(a, h(a, a), a) \\ \sim P(k(h(a, a)), h(a, a), k(h(a, a))) \\ P(g(a, k(h(a, a))), a, k(h(a, a))) \\ \sim P(g(a, k(h(a, a))), a, k(h(a, a))) \sim P(a, h(a, a), a) \\ \qquad \sim P(g(a, k(h(a, a))), a, k(h(a, a))) \\ \qquad P(k(h(a, a)), h(a, a), k(h(a, a))) \end{bmatrix} \tag{15}$$

It was therefore very exciting when Prawitz [4] announced a procedure for calculating a system like (15) *directly* from a system like (11). One notes in this example, following Prawitz, that 1, 0, 1, 1, 0, 1, instances, respectively, of the rows of (11), appear in (15). If we somehow knew these numbers ahead of time, we could set out to examine the different ways in which one instance each of the first, third, fourth and sixth rows of (11) could be obtained *using only compound constants actually generated during this process itself*. Each such set of instances would be examined for inconsistency, and among them would be (15)!

Unfortunately, the concrete process suggested by Prawitz for actually carrying out this beautiful idea is not at all feasible in general, as Davis has pointed out already in [3]. The trouble lies in Prawitz' reliance upon successive transformations from one normal form to another in such a way that at each such step the size of the arrays obtained increases explosively. His method actually suffices for the system (11) above but not for the system (9), and not, in general, for any but very special cases.

Davis [3] has therefore proposed to find a way of exploiting Prawitz' powerful idea while avoiding Prawitz' disastrous use of normal forms—in much the same way that the techniques of Davis and Putnam [1] avoid the use of normal forms which caused Gilmore's program [8] to be unable even to solve the system (9). From the few remarks on this matter at the end of [3] it does not yet seem clear just how Davis will proceed, and one awaits with great interest reports of his further researches along these lines.

Meanwhile, at the Argonne National Laboratory a sort of interim program has been developed and coded for the 704 based on the earlier remarks concerning proof sets. In designing this program, the main emphasis has been placed on the actual production by the machine of solutions in the form of annotated proofs; provision has been made for the program to accept guesses, by the mathematician using the program, as to what the proof set might contain. It had been noticed that very often it is fairly easy to guess some or all of the members of the proof set without yet being able to write down an actual proof for the problem in question. Indeed, "spotting the proof set" seems to be the really "creative" part of the act of proof-construction—the part that we know the least well how to make the computer *efficiently* perform.

Accordingly, the program makes no attempt to restrict the number of instances of each row of (1) that might be required for an inconsistent set, but simply considers every instance of (1) obtainable by instantiation over the alleged proof set. These instances are not written out *in extenso*: rather an attempt is made to infer outright as many *literals* as possible, by operations corresponding to repeated applications of Rule I of the Davis and Putnam procedure for testing truth-functions (cf. [1]). In many cases this suffices for obtaining a contradiction. Only if this strategem fails do we actually write out any instances of (1); even then we write out only those none of whose literals have been inferred true (for they would be dropped by the Davis and Putnam Rule II). And we omit, from any instance, occurrences of literals already proved false. The result is, in general, a very much smaller set of instances to be tested for inconsistency than would otherwise be the case.

In short, with this program, the user guesses at a proof set as shrewdly as he can, and if he is right the computer prints out a completely detailed and annotated proof as a reward. If he is wrong, he simply learns that he was wrong—a piece of information, at any rate, that he did not possess beforehand.

In trying out various examples with this program, it was interesting to discover that there are a number of proof construction problems with proof sets *of level zero*, but which are of a reasonably nontrivial nature. One such example occurs as a starred exercise for the reader in Birkhoff and Maclane [13, p. 130, exercise *11]: to prove that any group each of whose elements satisfies $x^2 = e$, where e is the group identity, is commutative. Setting up this problem in the standard form (1), we obtain the system (with $P(x, y, z)$ meaning the same as in (11), and e standing for the identity element):

$$\begin{bmatrix} \sim P(x,y,u) & \sim P(y,z,v) & \sim P(x,v,w) & P(u,z,w) \\ \sim P(x,y,u) & \sim P(y,z,v) & \sim P(u,z,w) & P(x,v,w) \\ P(e,x,x) & & & \\ P(x,e,x) & & & \\ P(h(x),x,e) & & & \\ P(x,h(x),e) & & & \\ P(x,x,e) & & & \\ P(a,b,c) & & & \\ \sim P(b,a,c) & & & \end{bmatrix} \tag{16}$$

If given no proof set at all, the program will use the primitive constants occurring in the input system, here the set:

$$e, a, b, c. \tag{17}$$

This was in fact what happened and the following rather pleasant proof was printed out. (The numeration of the lines carries on from the rows of (16), these being labelled 1 through 9 for reference):

(10) $P(a, a, e)$ from (7), a/x. (I.e., replacing x by a.)
(11) $P(b, b, e)$ from (7), b/x.
(12) $P(c, c, e)$ from (7), c/x.
(13) $P(a, e, a)$ from (4), a/x.
(14) $P(e, b, b)$ from (3), b/x.
(15) $P(e, c, c)$ from (3), c/x.
(16) $P(a, c, b)$ from (8), (10), (14) and (1), putting a/x, a/y, b/z, e/u, c/v, b/w in (1).
(17) $\sim P(b, c, a)$ from (9), (11), (15) and (1), putting b/x, b/y, c/z, e/u, a/v, c/w in (1).
(18) $\sim P(a, c, b) \sim P(c, c, e) \sim P(a, e, a)\ P(b, c, a)$ from (2), putting a/x, c/y, c/z, b/u, e/v, a/w.

But (18) is contradicted by (16), (12), (13) and (17). Q.E.D.

This is also an example of a proof obtainable by repeated applications of the "one literal" rule of Davis and Putnam; hence our program at no time wrote out any instances, merely examining each one singly before throwing it away. This contrasts nicely with the 8212 instances which would otherwise have had to be written out.

A problem set by Birkhoff as an exercise in elementary lattice theory [14, p. 20, exercise 1] likewise turned out to possess a zero-level proof. (The number of primitive constants occurring in the standard formulation is ten.) The reader is asked by Birkhoff to prove that in any lattice one always has the general inequality:

$$(a \cup b) \cap (c \cup d) \geqq (a \cap c) \cup (b \cap d). \tag{18}$$

The ten primitive constants arise from denying (18), for this comes to asserting simultaneously the seven statements:

$$\begin{array}{llll} (a \cup b) = g & (g \cap h) = k & (b \cap d) = n & \sim(k \geqq p) \\ (c \cup d) = h & (a \cap c) = m & (m \cup n) = p & \end{array} \tag{19}$$

The machine proof is the straightforward one in which "$k \geqq p$" is deduced from the first six lines of (19), using the definitions of l.u.b. and g.l.b. in terms of $\geqq$, and using the transitivity of $\geqq$. Again the "one-literal" rule suffices, and the proof is made without recording instances in extenso.

Unlike other proof-construction programs, this program is able to accept highly redundant sets of premises without much loss of efficiency. Premises which are not needed for a proof are no particular burden (and of course in general one does not know in advance that a given premise will *not* be used). We give here a further example, in which a great many premises were supplied, including lemmas which we knew were relevant to the theorem. The theorem is: *If a is a prime number, then the square root of a is irrational.*

To set up the problem, we denied the theorem, asserting that:

1.	$P(a)$	a is prime
2.	$Q(b, b, d)$	the square of b is d
3.	$Q(c, c, g)$	the square of c is g
4.	$Q(a, g, d)$	the product of a with g is d
5.	$\sim D(x, b) \sim D(x, c)$	b and c have no common divisor

and then supplied the following premises:

6.	$\sim D(x, y)\ Q(x, h(x, y), y)$	if x divides y then y is the product of x and some integer.
7.	$\sim Q(x, y, z)\ D(x, z)$	if z is the product of x and y then x divides z.
8.	$Q(x, y, f(x, y))$	existence of products.
9.	$\sim P(x) \sim Q(y, y, z) \sim D(x, z)\ D(x, y)$	a lemma: *If a prime x divides y^2, it divides y.*
10.	$\sim Q(x, y, u) \sim Q(x, z, u)\ E(y, z)$	cancellation law: if $xy = xz$, then $y = z$
11.	$\sim Q(x, y, u) \sim Q(y, z, v) \sim Q(x, v, w)\ Q(u, z, w)$	associativity of multiplication
12.	$\sim Q(x, y, u) \sim Q(y, z, v) \sim Q(u, z, w)\ Q(x, v, w)$	associativity of multiplication
13.	$\sim Q(x, y, z) \sim E(z, u)\ Q(x, y, u)$	substitutivity of equality
14.	$\sim Q(x, y, z) \sim E(y, u)\ Q(x, u, z)$	substitutivity of equality
15.	$\sim Q(x, y, z) \sim E(x, u)\ Q(u, y, z)$	substitutivity of equality
16.	$\sim D(x, y) \sim E(x, z)\ D(z, y)$	substitutivity of equality
17.	$\sim D(x, y) \sim E(z, y)\ D(x, z)$	substitutivity of equality
18.	$\sim P(x) \sim E(x, y)\ P(y)$	substitutivity of equality
19.	$\sim Q(x, y, z) \sim Q(x, y, u)\ E(z, u)$	uniqueness of products
20.	$E(x, x)$	reflexivity of equality
21.	$\sim E(x, y)\ E(y, x)$	symmetry of equality
22.	$\sim E(x, y) \sim E(y, z)\ E(x, z)$	transitivity of equality

23. $\sim Q(x, y, z)\, Q(y, x, z)$ commutativity of multiplication
24. $\sim D(x, y) \sim D(y, z)\, D(x, z)$ transitivity of "is a divisor of"

This is quite a large system. (We label it (20).) For a proof set, the program was given, in addition to the primitive constants already present in the first five premises: $h(a, b)$, $f(h(a, b), b)$ which we knew in advance, because of familiarity with the theorem, would suffice. It did. The machine printed out this proof:

25. $D(a, d)$ from (4) and (7); putting a/x, g/y, d/z in (7).
26. $D(a, b)$ from (1), (2), and (9); putting a/x, b/y, d/z in (9).
27. $Q(h(a, b), b, f(h(a, b), b))$ from (8); putting $h(a, b)/x$, b/y.
28. $Q(b, h(a, b), f(h(a, b), b))$ from (27) and (23); putting $h(a, b)/x$, b/y, $f(h(a, b), b)/z$ in (23).
29. $\sim D(a, c)$ from (26) and (5); putting a/x in (5).
30. $Q(a, h(a, b), b)$ from (26) and (6); putting a/x, b/y in (6).
31. $D(b, f(h(a, b), b))$ from (28) and (7); putting b/x, $h(a, b)/y$, $f(h(a, b), b/z$ in (7).
32. $\sim D(a, g)$ from (1), (3), (29), and (9); putting a/x, c/y, g/z in (9).
33. $Q(a, f(h(a, b), b), d)$ from (30), (27), (2), and (12); putting a/x, $h(a, b)/y$, b/z, b/u, $f(h(a, b), b)/v$, d/w in (12).
34. $D(a, f(h(a, b), b))$ from (26), (31), and (24); putting a/x, b/y, $f(h(a, b), b)/z$ in (24).
35. $\sim E(g, f(h(a, b), b))$ from (32), (34), and (17); putting a/x, $f(h(a, b), b)/y$, g/z in (17).
36. $\sim E(f(h(a, b), b), g)$ from (35) and (21); putting $f(h(a, b), b)/x$, g/y in (21).
37. $\sim Q(a, f(h(a, b), b), d) \sim Q(a, g, d)\, E(f(h(a, b), b), g)$
from (10); putting a/x, $f(h(a, b), b)/y$, d/u, g/z.

But (37) is contradicted by (33), (4), and (36). Q.E.D.

If the reader will follow the steps of this proof he will see that it is essentially the famous piece of reasoning which was known to the Pythagoreans—to demonstrate the irrationality of the square root of 2, simply generalized to the case of any prime. The reader will also note that premises (11), (13), (14), (15), (16), (18), (19), (20) and (22) were not used in the proof, although they are plausible enough candidates for inclusion in the list of premises. Finally, we remark that this proof, like the earlier examples, required, besides instantiation, only the "one-literal" rule of inference. In other programs discussed in the literature, the six free variables in premises (11) and (12) alone would have required the generation of $2 \times 7^6 = 235298$ instances in extenso.

One final example is given to illustrate the kind of proof that is produced when the program is given a problem for which the "one-literal" rule does not suffice. The subject matter chosen for this example is the problem of the constructibility of "Graeco-Latin" squares of order $4t + 2$ for $t = 0$ and $t = 1$. As is well-known, Euler's conjecture that Graeco-Latin squares were impossible for all $t \geqq 0$ was recently disproved (cf. [16] and [17]), and indeed the only two cases for which Euler was right are those cited above. The case $t = 0$ is very easy to see merely by a few moments of paper-and-pencil experimentation; the case $t = 1$ is very difficult to see. We obtained a machine proof of zero level for the case $t = 0$ and had to abandon the machine's attempt at the case $t = 1$, along similar lines, pending rewriting our program for speed. (Our current program, a "pilot-plant" code, contains several features making for slowness.) We set up the $t = 0$ case as follows (the marginal comments should make clear the intended meanings of the postulates):

1.	$\sim E(a, b)$	The elements a and b are different.
2.	$E(x, x)$	Equality is reflexive, and symmetric.
3.	$\sim E(x, y)\ E(y, x)$	
4.	$\sim L(x, y, z) \sim L(x, y, u)\ E(z, u)$	The "Latin" element in each cell is unique, occurs at most once in each column, and at most once in each row.
5.	$\sim L(x, y, z) \sim L(x, u, z)\ E(y, u)$	
6.	$\sim L(x, y, z) \sim L(u, y, z)\ E(u, x)$	
7.	$\sim G(x, y, z) \sim G(x, y, u)\ E(u, z)$	The "Greek" element in each cell is unique, occurs at most once in each column, and at most once in each row.
8.	$\sim G(x, y, z) \sim G(x, u, z)\ E(u, y)$	
9.	$\sim G(x, y, z) \sim G(u, y, z)\ E(u, x)$	
10.	$L(x, y, a)\ L(x, y, b)$	In each cell, the Latin element is either a or b.
11.	$L(x, a, y)\ L(x, b, y)$	Each Latin element occurs at least once in each row and at least once in each column.
12.	$L(a, x, y)\ L(b, x, y)$	
13.	$G(x, y, a)\ G(x, y, b)$	Likewise for Greek elements.
14.	$G(x, a, y)\ G(x, b, y)$	
15.	$G(a, x, y)\ G(b, x, y)$	
16.	$\sim L(x, y, z) \sim G(x, y, u) \sim L(v, w, z) \sim G(v, w, u)\ E(x, v)\ E(y, w)$	It is not permitted that two different cells have the same Latin element and the same Greek element.

(We remark here that for the case $t = 1$ we simply postulate *six* distinct elements instead of two, and modify premises (10) through (15) accordingly.)

The machine printed out the following proof, which as the reader will notice contains several inferences arising from applications of the Davis and Putnam Rule III (the "affirmative-negative" rule).

17.	$E(b, a)$	from (1) and (3); a/x, b/y in (3).
18.	$\sim L(a, a, a) \sim L(b, a, a)$	from (17) and (6); a/x, a/y, a/z, b/u in (6).
19.	$\sim L(a, a, b) \sim L(b, a, b)$	from (17) and (6); a/x, a/y, b/z, b/u in (6).
20.	$\sim G(a, a, a) \sim G(a, a, b)$	from (17) and (7); a/x, a/y, a/z, b/u in (7).
21.	$\sim G(a, a, a) \sim G(a, b, a)$	from (17) and (8); a/x, a/y, a/z, b/u in (8).
22.	$\sim G(a, a, a) \sim G(b, a, a)$	from (17) and (9); a/x, a/y, a/z, b/u in (9).
23.	$\sim G(a, a, b) \sim G(b, a, b)$	from (17) and (9); a/x, a/y, b/z, b/u in (9).
24.	$\sim G(a, b, b) \sim G(b, b, b)$	from (17) and (9); a/x, b/y, b/z, b/u in (9).
25.	$L(b, a, a)\ L(b, a, b)$	from (10); b/x, a/y.
26.	$L(a, a, a)\ L(a, b, a)$	from (11); a/x, a/y.
27.	$L(a, a, b)\ L(a, b, b)$	from (11); a/x, b/y.
28.	$G(a, a, a)\ G(a, b, a)$	from (14); a/x, a/y.
29.	$G(a, a, b)\ G(a, b, b)$	from (14); a/x, b/y.
30.	$G(a, a, a)\ G(b, a, a)$	from (15); a/x, a/y.
31.	$G(a, a, b)\ G(b, a, b)$	from (15); a/x, b/y.
32.	$G(a, b, b)\ G(b, b, b)$	from (15); b/x, b/y.
33.	$\sim L(a, b, a) \sim L(b, a, a) \sim G(a, b, a) \sim G(b, a, a)$	from (1), (17), and (16); putting a/x, b/y, a/z, a/u, b/v, a/w in (16).
34.	$\sim L(a, b, b) \sim L(b, a, b) \sim G(a, b, a) \sim G(b, a, a)$	from (1), (17), and (16); putting a/x, b/y, b/z, a/u, b/v, a/w in (16).
35.	$\sim L(a, b, a) \sim L(b, a, a) \sim G(a, b, b) \sim G(b, a, b)$	from (1), (17), and (16); putting a/x, b/y, z/a b/u, b/v, a/w in (16).

36. $\sim L(a, b, b) \sim L(b, a, b) \sim G(a, b, b) \sim G(b, a, b)$ from (1), (17), and (16); putting a/x, b/y, b/z, b/v, b/u, a/w, in (16).

37. $L(a, b, a) \sim L(b, a, a)$ from (18) and (26).

38. $L(a, b, b) \sim L(b, a, b)$ from (19) and (27).

39. $\sim L(b, a, a) \sim G(a, b, a) \sim G(b, a, a)$ from (33) and (37).

40. $\sim L(b, a, a) \sim G(a, b, b) \sim G(b, a, b)$ from (35) and (37).

41. $\sim L(b, a, b) \sim G(a, b, a) \sim G(b, a, a)$ from (34) and (38).

42. $\sim L(b, a, b) \sim G(a, b, b) \sim G(b, a, b)$ from (36) and (38).

43. $L(b, a, b) \sim G(a, b, a) \sim G(b, a, a)$ from (25) and (39).

44. $L(b, a, b) \sim G(a, b, b) \sim G(b, a, b)$ from (25) and (40).

45. $\sim G(a, b, a) \sim G(b, a, a)$ from (41) and (43).

46. $\sim G(a, b, b) \sim G(b, a, b)$ from (42) and (44).

47. $\sim G(a, a, b)\ G(a, b, a)$ from (20) and (28).

48. $G(a, b, a) \sim G(b, a, a)$ from (22) and (28).

49. $\sim G(a, b, a)\ G(b, a, a)$ from (21) and (30).

50. $G(a, b, b) \sim G(b, a, b)$ from (23) and (29).

51. $G(a, b, a)\ G(b, a, b)$ from (31) and (47).

52. $\sim G(b, a, a)$ from (45) and (48).

53. $G(b, a, b)$ from (49), (51), and (52).

54. $G(b, b, b)$ from (32), (46), and (53).

55. $\sim G(b, b, b)$ from (24), (50), and (53).

But (54) and (55) are contradictory. Q.E.D.

The proof for the interesting case $t = 1$ is much the same as this one just given, but of course is a good deal longer. This proof was completely new to us: for obvious reasons it was not feasible to try to check the consistency of premises (1) through (16) by hand. However, we were confident that we had completely characterized a Graeco-Latin square of order 2 and we were also confident that this structure is logically impossible; so we simply "turned the program loose" to see what would happen. We intend to try many similar problems, i.e. those which involve proving that finite structures satisfying certain conditions are logically impossible. An easy example would be the impossibility of a noncommutative group of order 3; or again, the impossibility of a finite field of some small order not a prime power. The reader will think of many more such examples.

By these examples we have illustrated the capability of a proof-construction program which expects to be supplied with a proof set of compound constants with which to supplement the primitive constants contained in the standard formulation of the problem. We now ask: is there some way whereby we can restore the machine's autonomy and have *it* decide what proof sets to try? We must be careful, in providing such a procedure, to avoid the "combinatorial explosiveness" which has plagued earlier procedures—or at least to slow it down or postpone it. There does indeed appear to be a rather neat way of examining in turn all possible proof sets; let us discuss it.

The reader may have noticed already that the proof sets in our examples satisfy a certain closure property: *if a constant C is in a proof set and the constant D is one of its constituents, then D is also in the proof set.* This is not accidental; an examination of Prawitz' method will suffice to show that the only way a compound constant can enter a proof is by virtue of its having been "manufactured" through instantiation, out of its immediate constituents, at some step of the proof. But we can better set the scene for what follows by remarking that for *any* finite subset P of elements of an infinite Herbrand universe H there is always a *single element* p of H *which contains every member of* P *as a constituent provided only that the problem contains at least one function letter of degree two or higher*. If the function letters in the problem are all of degree one, this "packing" property does not hold; such problems therefore are not amenable to the method suggested below and would have to be handled by one of the other autonomous methods which are known. To obtain P we simply disassemble p level by level. For instance, the proof set for example (9) is given by the constant $f(f(a, f(a,a)), f(f(a, a), f(a, a)))$; for example (11), by the constant $g(a, k(h(a, a)))$; for example (20), by the constant $f(h(a, b), b)$. As example (20) shows, we of course include in the proof set all the primitive constants of the problem, so that they need not actually be packed into the "key constant" explicitly; but it does not matter if some of them are.

The suggested procedure, then, is: first, try H_0 itself as a proof set; but if it is not one, take each member of H_1 in turn, using it and all its constituents together with the members of H_0 as possible proof sets. Then do the same with H_2 , H_3 , and so on. If we proceed in this way every possible proof set will eventually have been tried, so that if one exists we will have found it.

The advantage of this procedure is that we never have to instantiate with too many distinct constants at any one time, even though we are probing quite high levels of the Herbrand universe. Of course, as we have seen, the populations of successive levels increase with horrible speed in many problems and we are still faced with trying each member of each level, one at a time.

A further question bearing upon feasibility of this method is: how rapidly can our program decide, given an alleged proof set, that it is or is not one? Since this is the "inner loop" of the entire proof procedure, its speed is crucial. As was stated earlier, our present version of this inner loop program is quite slow compared with the improvements we already know how to effect by various programming tricks. Furthermore, it is written in FORTRAN for the 704. We feel that by careful hand-coding on a next-generation machine we shall obtain a very rapid program to do this part of the task. Work is therefore currently being started on a code for the CDC 3600.

Finally, however, we wish to stress the usefulness of a "man-machine" proof-construction system, in which the creative insight of the human mathematician is exploited to fullest advantage. It is felt that via the principle of channeling the creative aspect of proof-construction into the selection of possible proof-sets, with the machine taking over the rest of the task, we have provided a flexible and potentially powerful means of exploiting computers in problems of deduction.

* * *

Acknowledgments. I wish to thank the Applied Mathematics Division of the Argonne National Laboratory for its support of this work in the summers of 1961 and 1962. I am particularly indebted to Dr. W. F. Miller, director of the Division, and to Dr. G. A. Robinson, head of its Programming Development Section, for many hours of valuable critical discussion.

REFERENCES

1. Davis, Martin, and Putnam, Hilary. A computing procedure for quantification theory. *J. ACM 7* (1960), 201–215.
2. Davis, Martin, Logemann, George, and Loveland, Donald. A machine program for theorem-proving. *Comm. ACM 5*, 7 (July 1962), 394–397.
3. Davis, Martin. Eliminating the irrelevant from mechanical proofs. Paper, Symp. Exp. Arith., A.M.S. Meeting no. 589, Apr. 1962 (Mimeo.).
4. Prawitz, Dag. An improved proof procedure. *Theoria 26* (1960), 102–139.
5. Wang, Hao. Towards mechanical mathematics. IBM J. Res. Develop. *4* (1960), 2–22.
6. Wang, Hao. Proving theorems by pattern recognition, I. *Comm. ACM 3* (1960), 220–234.
7. Wang, Hao. Proving theorems by pattern recognition, II. *Bell System Tech. J. 40* (1961), 1–41.
8. Gilmore, Paul C. A proof method for quantification theory. IBM J. Res. Develop. *4* (1960), 28–35.
9. Prawitz, Dag, Prawitz, Hakan, and Vogera, Neri. A mechanical proof procedure and its realization in an electronic computer. *J. ACM 7* (1960), 102–128.
10. Robinson, J. A. A general theorem-proving program for the IBM 704. Argonne Nat. Lab. Rep. 6447, Nov. 1961.
11. Quine, W. V. *Methods of Logic*. Henry Holt, New York, rev. ed. (1959).
12. Quine, W. V. A proof procedure for quantification theory. J. Symbolic Logic *20* (1955), 141–149.
13. Birkhoff, Garrett and MacLane, Saunders. *A Survey of Modern Algebra*. Macmillan, rev. ed. (1953).
14. Birkhoff, Garrett. *Lattice Theory*. AMS Colloq. Publ., Vol. 25, rev. ed. (1948).
15. Beth, Evert W. *Formal Methods*. D. Reidel, Holland (1962).
16. Bose, R. C., and Shrikhande, S. S. On the falsity of Euler's conjecture about the nonexistence of two orthogonal latin squares of order $4t + 2$. Proc. Nat. Acad. Sci. *45* (1959), 734–737.
17. Parker, E. T. Orthogonal Latin squares. Proc. Nat. Acad. Sci. *45* (1959), 859–862.

1964

The Unit Preference Strategy in Theorem Proving

L. Wos, D. Carson, G. A. Robinson

Unit Preference and Set of Support Strategies

The theorems, axioms, etc., to which the algorithm and strategies described in this paper are applied are stated in a normal form defined as follows: A *literal* is formed by prefixing a predicate letter to an appropriate number of arguments (constants, variables, or expressions formed with the aid of function symbols) and then perhaps writing a negation sign (−) before the predicate letter. For example:

P(b,x) −P(b,x) Q(y) R(a,b,x,z,c) S

are all literals if P, Q, R, and S are two-, one-, five-, and zero-place predicate letters, respectively. The predicate letter is usually thought of as standing for some n-place relation. Then the literal P(a,b), for example, is thought of as saying that the ordered pair (a,b) has the property P. The literal −P(a,b) is thought of as saying that (a,b) does not have the property P.

One may build a *clause* by writing a sequence of literals separated by disjunction (logical "or") signs. Logical "or" will be symbolized by a small letter "v" (distinguishable from possible uses of "v" as a variable from context). Where it is desired to indicate dependence of a particular argument of a predicate letter upon one or more other variables, function symbols are employed. The clause

P(x,y,z) v Q(x,f(x,y))

thus says that either the ordered triple (x,y,z) has the property P or the ordered pair (x,f(x,y)) has the property Q (or both). Each of the variables in a clause is then thought of as being universally quantified (that is, if the variable x occurs in a clause, the clause is assumed to be preceded by an implicit quantifier, "for each x." Functional expressions such as f(x,y) are then treated as existentially quantified variables. Roughly speaking, f(x,y) in the example above stands for an element (depending on x and y) which forms, when used as the second element with x as the first element, an ordered pair which has the property Q A *unit clause* is a clause composed of a single literal.

Finally, one may consider a sequence of clauses implicitly joined by logical "and" (conjunction). Such a sequence of clauses will be said to be in *(conjunctive) normal form.*

Instantiation, as applied to this normal form, can be thought of as the forming of a possibly less general (more specific) *instance* of a clause by performing a systematic replacement of variables by constants, new variables, or by expressions formed with the aid of function symbols. Substituting b for x and f(d,u) for y in the clause

P(x,y) v Q(b,y)

would yield a less general instance of that clause:

P(b,f(d,u)) v Q(b,f(d,u)).

Work performed under the auspices of the U. S. Atomic Energy Commission.

The latter is less general in that, while it can be deduced from the former, the former cannot be deduced from the latter.

Early computer-oriented theorem-proving efforts employed search techniques involving instantiation of a given set of logical formulae. In these methods, successively larger sets of constants were generated. As each set was generated, all permissible substitutions of such constants for variables were made in the original formulae, producing successively larger sets of instances. These instances differed from the original formulae in that they included no logical quantifiers: "For each x," "For some x," etc. For sets of such quantifier-free formulae, straightforward techniques were known for determining validity or inconsistency. P. C. Gilmore[2] described an IBM 704 program using the technique of conversion of sets of quantifier-free formulae from conjunctive normal form to disjunctive normal form. Davis and Putnam introduced a substantial improvement in the method of testing such sets of quantifier-free formulae. A 704 program[4] incorporating this improvement proved to be several orders of magnitude faster than Gilmore's program for the same machine.

These instantiation techniques employed the argument forms *existential instantiation* and *universal instantiation* (see for example Quine[5]) to infer the quantifier-free instances from the original formulae. All permissible substitutions were made in a systematic, exhaustive manner, guaranteeing that if a proof existed of the desired theorem, it would be captured in the steadily expanding sets of instances. The disastrous rate of growth of these sets, due to the inclusion of numerous unprofitable inferences, spelled the doom of exhaustive instantiation. Study of the nature of the instances which could be expected to result from such a program led, however, to the suspicion that the logical completeness of the method could be retained but the combinatoric explosion substantially reduced by considering the generalizations represented by classes of similar instances. This led to the formulation of computer-oriented rules of inference which suppressed *universal instantiation* and retained universally quantified variables in their more general form. These rules were codified and put on a sound logical basis by J. A. Robinson[3] in his paper on the resolution principle. As a computer algorithm, he proposed that the original set of formulae be completely "resolved" in the sense that all possible applications of the inference rule *resolution* be made in a systematic, exhaustive manner.

Where exhaustive instantiation methods met their downfall was in blindly forming all possible maximally specific instances from a given set of clauses; once formed, such instances were never to be discarded. Only then was consideration given to what interactions might occur between the instances (specifically, to whether the set thus generated was inconsistent). Furthermore, in most interesting cases, many new objects to be substituted were manufactured, resulting in a combinatoric explosion of further instances.

Resolution, as proposed by J. A. Robinson, curtails the number of instances produced, in that it produces a new clause only when it can be determined in advance that two existing clauses will, when each is instantiated, yield a pair of instances that will interact, forming a clause which could not have been inferred from either parent clause taken by itself. Specifically, in resolution a pair of clauses (each of which is called a resolvend) is examined to see if there is a substitution which will transform the clauses into a pair of the form

$$L \vee L_1 \vee L_2 \vee \ldots \vee L_m, \quad -L \vee K_1 \vee K_2 \vee \ldots \vee K_n$$

From such a pair, the clause

$$L_1 \vee L_2 \vee \ldots \vee L_m \vee K_1 \vee K_2 \vee \ldots K_n$$

can be inferred. This clause (called the resolvent) is then added to the set of clauses which have been accumulated from previous inferences. Furthermore, a substitution is considered only if it is the most general that could be made, thus maintaining the maximum degree of generality in the result. Another closely related method of inference is *factoring:* A substitution is sought such that (a) two or more of the literals of a clause will collapse into a single literal, and (b) no more general substitution would have the same effect. The resulting clause is called a *factor.*

Substitution of these rules of inference for the earlier instantiation inference rules was shown to produce a reduction of the combinatoric explosion by a factor in excess of 10^{50}, leading to some rather spectacular achievements when compared to the instantiation techniques. Nevertheless, the search algorithm employed a more or less random generation of resolvents inferred by means of the resolution principle. Using only these techniques, previously established bench-mark problems required prohibitive amounts of machine time. It seemed that a change of emphasis would be profitable.

One approach would be to concentrate upon the strategies of search. The current paper considers one such strategy, the *unit preference strategy*. This strategy has been implemented in a theorem-proving program now successfully running on the Control Data 3600. We will describe the search algorithm employed in the program, prove its soundness and completeness, consider some examples, and describe additional search strategies which can be employed to effect further improvement.

The principal search strategy arises from the fact that the object of the resolution principle is the generation through inference of two unit clauses which are manifestly contradictory. Here and elsewhere in this paper, *contradict* is used as in mathematics in the somewhat broader sense of conflict, rather than in the narrower sense frequently used in logic. Two clauses will be called contradictory if they are mutually exclusive (contrary) clauses. We do not require them to bear the relation of a clause to its negation. With this in mind, it seemed worthwhile to orient the program to produce shorter and still shorter clauses in preference to other possible inferences.

The Unit Preference Strategy

The data for consideration consists of a set of clauses in the normal form. The clauses correspond to a given set of axioms and the denial of a theorem to be proved from these axioms, for it is to be remembered that the program finds proofs and not theorems. Remember that all variables that occur are treated as universally quantified; all existentially quantified variables have been replaced by constants or functions.

The algorithm is divided into two sections, a unit section and a non-unit section. Where a j-clause is a clause of length j (i.e., has j literals), and the j-list is the list of j-clauses, the logic of the unit section is as follows, starting with $j = 1$:

1. Search the unit list and the j-list for a pair of elements C and D such that C is a unit, D is of length j, and for some literal m of D resolution has not been attempted for C and D on m; if no such pair exists, execute step 2; if such a pair exists, execute step 3.

(The search in the program proceeds by taking the first unit and the first j-clause and examining each of its literals; the procedure is repeated with the same j-clause and the next unit, and continues until the unit list is exhausted. Then the process is applied to the next j-clause.)

2. If j is the length of the longest clause present, enter the non-unit section; if not, increase j by 1 and return to 1.

3. Resolve C and D on m; if no resolvent is generated, return to 1 and resume the search; if a resolvent of length $i > 0$ is generated, add the resolvent to the i-list, set $j = i$, and return to 1; if the empty resolvent is generated, execute the proof recovery. (The generation of the empty resolvent is equivalent to finding that the clauses C and D are manifestly contradictory.)

In order to avoid the possibility of being caught in an infinite loop, there is a constraint placed on step 3. The constraint is formulated in terms of the concept of the level of a clause. Let S_0 be the original set of clauses; define S_i for $i > 0$ to be the set of resolvents of S_{i-1} together with S_{i-1}. The level k of a clause C is 0 if C is input, is that of B if C is obtained by factoring B, and is 1 greater than the maximum of the levels of A and B if C is obtained by resolving A and B. The constraint imposed on 3 is: if the resolvent of C and D is a non-unit whose level is a specified bound k_0, or a unit whose level exceeds k_0, then it is not added to the corresponding list, and the pair is treated as if no resolvent were generated. To illustrate the difficulty thus avoided consider the set consisting of the clauses $P(a)$, $-P(x) \vee P(f(x))$, which correspond to a subset of the Peano axioms, and some clause of length 3. Without the level bound, the program would generate P(f(a)), P(f(f(a))), P(f(f(f(a)))), . . ., *ad infinitum.* We would be caught in an infinite loop which would continually present and execute the task of resolving a new unit with the same 2-clause instead of either passing to the proof recovery or to the non-unit section.

The non-unit section beings by setting j = 2, then:

1. Search the j-list for a clause D with literals l and m on which factoring has not been attempted; if one exists, execute 2; if not, execute 3.
2. Factor D on l and m; if no factor is generated, return to 1; if a factor C of length i is generated, add C to the i-list, set j = i, and return to step 1 of the unit section.
3. If j = 2, increase j by 1 and return to step 1 of the non-unit section; if $j \neq 2$, execute 4.
4. Set h = 2, replace j by j-1, and execute 5.
5. Search the h-list and the j-list for a pair C and D such that C is an h-clause with a literal l, D is a j-clause with a literal m, and resolution has not been attempted for C and D on l and m; if no such pair exists, execute 6; if such a pair exists, execute 7.
6. If $h \geqq$ the maximum length of all clauses present, the program stops with the conclusion that no proof exists within level k_0; if h < this maximum and $h + 1 \geq j$, replace j by h + j and return to 1 of the non-unit section; if h + 1 < j, replace h by h + 1 and j by j − 1 and return to 5.
7. Resolve C and D on l and m; if no resolvent is generated, return to 5; if a resolvent B of length i is generated, then add B to the i-list, set j = i, and return to 1 of the unit section. Again the level constraint imposed on Step 3 of the unit section applies.

Soundness and Completeness

The soundness of the procedure follows from the fact that if C is in S_i or a factor of some D in S_i for any i, C is implied by S_0, i.e., C is a consequence of S_0. So, if C and D are elements of the unit list which conflict (are manifestly contradictory), S_0 is inconsistent.

The argument for completeness is as follows: J. A. Robinson[3] proved in effect that if S_0 is an inconsistent set, there exists a k such that S_k contains the empty resolvent; this implies that S_{k-1} contains two clauses which are units or have unit factors which conflict. The claim is that when the level bound k_0 of the program is equal to this k, whether k_0 is immediately set equal to k or set equal to 1,2, . . . k, the desired unit conflict will be obtained. At any given time the set of clauses which occur on any list is contained in S_k, and is therefore finite. The search through any list for potential factors or through any pair of lists for potential resolvents is a finite process since all lists are finite. The number of searches through any list is limited by the size of the input and the number of clauses which are adjoined and is, therefore, finite also. So any clause which occurs on any list will do so after a finite number of steps. In a similar fashion we can see that all clauses on all lists will be factored and, within the level bound, pairwise resolved in a finite number of steps. Consider the case where k_0 is sufficiently large to have S_{k0} contain the empty resolvent. If $k_0 = 1$, S_0 contains the clauses which yield the conflict, and the previous argument shows that the program will examine the pair in question. If $k_0 = 2$, the same argument shows that S_1 will be contained in the lists of clauses, and again the conflict will be obtained. Applying the argument k_0 times shows S_{k0-1} will be contained in the lists of clauses. So we have proven the following lemma.

Lemma: Using the search algorithm with level bound k_0 such that S_{k0} contains the empty resolvent, the program finds a proof if and only if S_0 is unsatisfiable.

Subsidiary Strategies

The most important subsidiary strategy is based on the concept of a chain. With the appropriate grouping, any clause which occurs on any list is expressible in terms of the elements of S_0 together with the operators of resolution and factoring. Such an expression is called a *chain,* and the number of resolutions which occur therein is its *length.* The elements of S_0 and the factors of those elements are chains of length 0; those of S_1 have chains of length 0 and 1; those of S_2 have chains of length of 0, 1, 2 and 3.

If T is a nonempty subset of S_0 and C is a chain of length 0 whose single element is in T or is a factor of an element in T, we say C is *supported by T* or C has *T-support.* Chains C of length greater than 0 have T-support if, for every resolution, that occurs in C, at least one of the resolvends has T-support. The clause B is said to be *derived from T,* if its chain (its expression or derivation in S_0) has T-support.

The strategy employed here is to choose T and generate only those elements of S_i for $i > 0$ which are derived from T. The choice of T obviously has a profound effect on the number of clauses generated during the search for a proof. The question remains as to which available choices of T preserve completeness of the procedure. By the lemma of the previous section, $T = S_0$ is admissible under the appropriate conditions. For a given S_0 the clauses might be divided into 3 categories; those which correspond to the basic axioms of the theory under study; those which correspond to the special hypotheses of the theorem under consideration, and those which correspond to the denial of the conclusion of the theorem. For example, consider the theorem: in a group, if the square of every element is the identity, the group is commutative. The first category would consist of a set of axioms which characterize groups; the second would consist of the axiom, for every x in the group $x^2 = e$; the third would be the denial of commutativity, there exist a and b such that $ab \neq ba$. (This example will be considered later under various conditions.) It appears that, where T is the join of the last two categories, this choice of T for set of support preserves completeness and is most valuable in obtaining proofs in a reasonable amount of time.

Among the other strategies which can be employed are: deletion of the unit clause B upon generation if the unit list contains a clause A such that every instance of B is also one of A; deletion upon generation of a clause if it contains two literals which have opposite sign but are otherwise identical; deletion of B if the unit A is a resolvent of B with some C such that all instances of some literal k of B are contained in the set of instances of A.

EXAMPLE 1: In an associative system with left and right solutions, there is a right identity element.

Basic Axioms:

A1. $-P(x,y,u) \vee -P(y,z,v) \vee -P(x,v,w) \vee P(u,z,w)$
A2. $-P(x,y,u) \vee -P(y,z,v) \vee -P(u,z,w) \vee P(x,v,w)$
(A1, A2: associativity)
A3. $P(g(x,y),x,y)$ (left solution)
A4. $P(x,h(x,y),y)$ (right solution)
A5. $P(x,y,f(x,y))$ (closure)

Negation of Conclusion:

A6. $-P(j(x),x,j(x))$ (no right identity)

When A6 was taken as the only member of the set of support, the computer generated the following proof in 35 milliseconds with 11 clauses in memory at the time the proof was detected:

1. $-P(x,y,u) \vee -P(y,z,v) \vee -P(x,v,w) \vee P(u,z,w)$ (A1)
2. $P(g(x,y),x,y)$ (A3)
3. $P(x,h(x,y),y)$ (A4)
4. $-P(j(x),x,j(x))$ (A6)
5. $-P(x,y,j(z)) \vee -P(y,z,v) \vee -P(x,v,j(z))$ (from 4 and 1)
6. $-P(y,z,v) \vee -P(g(y,j(z)),v,j(z))$ (from 2 and 5)
7. $-P(v,z,v)$ (from 2 and 6)

Since 3 and 7 are manifestly contradictory, the proof is complete.

EXAMPLE 2: In an associative system with an identity element, if the square of every element is the identity, the system is commutative.

Basic Axioms:

A1. $P(x,e,x)$ (right identity)
A2. $P(e,x,x)$ (left identity)
A3. $-P(x,y,u) \vee -P(y,z,v) \vee -P(u,z,w) \vee P(x,v,w)$
A4. $-P(x,y,u) \vee -P(y,z,v) \vee -P(x,v,w) \vee P(u,z,w)$
(A3, A4: associativity)

Special Hypothesis:

A5. P(x,x,e)
(The square of every element is the identity.)

Negation of Conclusion:

A6. P(a,b,c)

A7. −P(b,a,c)
(There are elements a and b which do not commute.)

When A1 through A7 were used as the set of support, the machine generated the following proof in 1.124 seconds with 119 clauses in memory at the time the proof was detected.

1. P(x,e,x) (A1)
2. P(e,x,x) (A2)
3. −P(x,y,u) v −P(y,z,v) v −P(u,z,w) v P(x,v,w) (A3)
4. −P(x,y,u) v −P(y,z,v) v −P(x,v,w) v P(u,z,w) (A4)
5. P(x,x,e) (A5)
6. P(a,b,c) (A6)
7. −P(b,a,c) (A7)
8. −P(x,y,e) v −P(y,w,v) v P(x,v,w) (from 2 and 3)
9. −P(y,w,v) v P(y,v,w) (from 5 and 8)
10. −P(w,y,u) v −P(y,z,e) v P(u,z,w) (from 1 and 4)
11. −P(w,z,u) v P(u,z,w) (from 5 and 10)
12. P(c,b,a) (from 6 and 11)
13. P(c,a,b) (from 12 and 9)
14. −P(c,a,b) (from 7 and 11)

Since 13 and 14 are manifestly contradictory, the proof is complete.

When, instead, only A5, A6, and A7 were used as the set of support, the machine generated the following proof in the faster time of 538 milliseconds with 72 clauses in memory at the time the proof was detected:

1. P(x,e,x) (A1)
2. P(e,x,x) (A2)
3. −P(x,y,u) v −P(y,z,v) v −P(u,z,w) v P(x,v,w) (A3)
4. −P(x,y,u) v −P(y,z,v) v −P(x,v,w) v P(u,z,w) (A4)
5. P(x,x,e) (A5)
6. P(a,b,c) (A6)
7. −P(b,a,c) (A7)
8. −P(y,z,v) v −P(e,z,w) v P(y,v,w) (from 5 and 3)
9. −P(y,w,v) v P(y,v,w) (from 2 and 8)
10. −P(x,z,u) v −P(x,e,w) v P(u,z,w) (from 5 and 4)
11. −P(w,z,u) v P(u,z,w) (from 1 and 10)
12. P(c,b,a) (from 6 and 11)
13. P(c,a,b) (from 12 and 9)
14. −P(c,a,b) (from 7 and 11)

Since 13 and 14 are manifestly contradictory, the proof is complete.

EXAMPLE 3: In a group, if the square of every element is the identity, the group is commutative.

Basic Axioms:

A1. P(x,y,f(x,y)) (closure)

A2. P(e,x,x)
A3. P(x,e,x)
(existence of identity)

A4. P(x,g(x),e)
A5. P(g(x),x,e)
(existence of inverse)

A6. R(x,x) (reflexivity of =)

A7. −R(x,y) v R(y,x) (symmetry of =)

A8. −R(x,y) v −R(y,z) v R(x,z) (transitivity of =)

A9. −P(x,y,u) v −P(x,y,v) v R(u,v)
(multiplication is well-defined)

A10. −P(x,y,u) v −P(y,z,v) v −P(u,z,w) v P(x,v,w)
A11. −P(x,y,u) v −P(y,z,v) v −P(x,v,w) v P(u,z,w)
(associativity)

A12. −R(u,v) v −P(x,y,u) v P(x,y,v)
A13. −R(u,v) v −P(x,u,y) v P(x,v,y)
A14. −R(u,v) v −P(u,x,y) v P(v,x,y)
A15. −R(u,v) v R(f(x,u),f(x,v))
A16. −R(u,v) v R(f(u,y),f(v,y))
A17. −R(u,v) v R(g(u),g(v))
(substitution for =)

Special Hypothesis:

A18. P(x,x,e)
(The square of every element is the identity.)

Negation of Conclusion:

A19. P(a,b,c)

A20. −P(b,a,c)
(There are elements a and b which do not commute.)

With A18 through A20 used as the set of support, the computer generated the following proof in 54 seconds with 563 clauses in memory at the time the proof was detected:

1.	P(x,y,f(x,y))	(A1)
2.	P(e,x,x)	(A2)
3.	P(x,e,x)	(A3)
4.	−P(x,y,u) v −P(y,z,v) v −P(u,z,w) v P(x,v,w)	(A10)
5.	−P(x,y,u) v −P(y,z,v) v −P(x,v,w) v P(u,z,w)	(A11)
6.	P(x,x,e)	(A18)
7.	P(a,b,c)	(A19)
8.	−P(b,a,c)	(A20)
9.	−P(y,z,v) v −P(e,z,w) v P(y,v,w)	(from 6 and 4)
10.	−P(e,z,w) v P(y,f(y,z),w)	(from 1 and 9)
11.	P(y,f(y,w),w)	(from 2 and 10)
12.	−P(x,y,z) v −P(y,z,v) v P(x,v,e)	(from 6 and 4)
13.	−P(x,y,z) v P(x,f(y,z),e)	(from 1 and 12)
14.	P(a,f(b,c),e)	(from 7 and 13)
15.	−P(b,y,u) v −P(y,z,a) v −P(u,z,c)	(from 8 and 4)
16.	−P(e,z,a) v −P(b,z,c)	(from 3 and 15)
17.	−P(e,f(b,c),a)	(from 11 and 16)
18.	−P(y,z,v) v −P(y,v,w) v P(e,z,w)	(from 6 and 5)
19.	−P(w,z,e) v P(e,z,w)	(from 3 and 18)
20.	P(e,f(b,c),a)	(from 14 and 19)

Since 17 and 20 are manifestly contradictory, the proof is complete.

Discussion

In order to study the importance of the various strategies discussed above, Example 2 was run without the aid of the *unit preference strategy* or the *set of support strategy*. At the end of 30 minutes the program had just finished generating S_2, having also generated S_1, without obtaining a proof. By comparison, with the aid of these strategies, the proof was obtained in about .5 seconds.

Example 3 serves to illustrate the comparative difficulty encountered when attacking a mathematical theorem without *a priori* knowledge, such as that employed in Example 2, as to which subset of the basic axioms of the theory is relevant. The proof of Example 2 was completed in about .5 seconds, while Example 3 required 54 seconds although the same strategy was applied to both. It should be noted that, without the *set of support strategy*, Example 3 was beyond the range of a 65,000 word machine.

BIBLIOGRAPHY

1. DAVIS, M., and PUTNAM, H., "A computing procedure for quantification theory," J. ACM 7, 201–215 (1960).
2. GILMORE, P. C., "A proof method for quantification theory," IBM J. Res. Develop. 4, 28–35 (1960).
3. ROBINSON, J. A., "A machine oriented logic based on the resolution principle." To be published.
4. ROBINSON, J. A., "GAMMA I, a general theorem-proving program for the IBM 704," Argonne National Laboratory Report ANL–6447. November, 1961.
5. QUINE, W. V. O., *Methods of Logic*, (Holt, Rinehart, and Winston, New York, 1959), Revised Ed.

1965

A Machine-Oriented Logic Based on the Resolution Principle

J. A. Robinson

Abstract. Theorem-proving on the computer, using procedures based on the fundamental theorem of Herbrand concerning the first-order predicate calculus, is examined with a view towards improving the efficiency and widening the range of practical applicability of these procedures. A close analysis of the process of substitution (of terms for variables), and the process of truth-functional analysis of the results of such substitutions, reveals that both processes can be combined into a single new process (called *resolution*), iterating which is vastly more efficient than the older cyclic procedures consisting of substitution stages alternating with truth-functional analysis stages.

The theory of the resolution process is presented in the form of a system of first-order logic with just one inference principle (the resolution principle). The completeness of the system is proved; the simplest proof-procedure based on the system is then the direct implementation of the proof of completeness. However, this procedure is quite inefficient, and the paper concludes with a discussion of several principles (called search principles) which are applicable to the design of efficient proof-procedures employing resolution as the basic logical process.

1. *Introduction*

Presented in this paper is a formulation of first-order logic which is specifically designed for use as the basic theoretical instrument of a computer theorem-proving program. Earlier theorem-proving programs have been based on systems of first-order logic which were originally devised for other purposes. A prominent feature of those systems of logic, which is lacking in the system described in this paper, is the relative *simplicity* of their inference principles.

Traditionally, a single step in a deduction has been required, for pragmatic and psychological reasons, to be simple enough, broadly speaking, to be apprehended as correct by a human being in a single intellectual act. No doubt this custom originates in the desire that each single step of a deduction should be indubitable, even though the deduction as a whole may consist of a long chain of such steps. The ultimate conclusion of a deduction, if the deduction is correct, follows logically from the premisses used in the deduction; but the human mind may well find the unmediated transition from the premisses to the conclusion surprising, hence (psychologically) dubitable. Part of the point, then, of the logical analysis of deductive reasoning has been to reduce complex inferences, which are beyond the capacity of the human mind to grasp as single steps, to chains of simpler inferences, each of which is within the capacity of the human mind to grasp as a single transaction.

Work performed under the auspices of the U. S. Atomic Energy Commission.

* Argonne, Illinois.

† Present address: Rice University, Houston, Texas.

Article from: Journal of the Association for Computing Machinery, Volume 12, Number 1, January 1965.

From the theoretical point of view, however, an inference principle need only be *sound* (i.e., allow only logical consequences of premisses to be deduced from them) and *effective* (i.e., it must be algorithmically decidable whether an alleged application of the inference principle is indeed an application of it). When the agent carrying out the application of an inference principle is a modern computing machine, the traditional limitation on the complexity of inference principles is no longer very appropriate. More powerful principles, involving perhaps a much greater amount of combinatorial information-processing for a single application, become a possibility.

In the system described in this paper, one such inference principle is used. It is called the *resolution principle*, and it is machine-oriented, rather than human-oriented, in the sense of the preceding remarks. The resolution principle is quite powerful, both in the psychological sense that it condones single inferences which are often beyond the ability of the human to grasp (other than discursively), and in the theoretical sense that it alone, as sole inference principle, forms a complete system of first-order logic. While this latter property is of no great importance, it is interesting that (as far as the author is aware) no other complete system of first-order logic has consisted of just one inference principle, if one construes the device of introducing a logical axiom, given outright, or by a schema, as a (premiss-free) inference principle.

The main advantage of the resolution principle lies in its ability to allow us to avoid one of the major combinatorial obstacles to efficiency which have plagued earlier theorem-proving procedures.

In Section 2 the syntax and semantics of the particular formalism which is used in this paper are explained.

2. *Formal Preliminaries*

The formalism used in this paper is based upon the notions of unsatisfiability and refutation rather than upon the notions of validity and proof. It is well known (cf. [2] and [5]) that in order to determine whether a finite set S of sentences of first-order logic is satisfiable, it is sufficient to assume that each sentence in S is in prenex form with no existential quantifiers in the prefix; moreover the matrix of each sentence in S can be assumed to be a disjunction of formulas each of which is either an atomic formula or the negation of an atomic formula. Therefore our syntax is set up so that the natural syntactical unit is a finite set S of sentences in this special form. The quantifier prefix is omitted from each sentence, since it consists just of universal quantifiers binding each variable in the sentence; furthermore the matrix of each sentence is regarded simply as the set of its disjuncts, on the grounds that the order and multiplicity of the disjuncts in a disjunction are immaterial.

Accordingly we introduce the following definitions (following in part the nomenclature of [2] and [5]):

2.1 *Variables.* The following symbols, in alphabetical order, are variables:

$$u, v, w, x, y, z, u_1, v_1, w_1, x_1, y_1, z_1, u_2, \cdots, \text{etc.}$$

2.2 *Function symbols.* The following symbols, in alphabetical order, are function symbols of degree n, for each $n \geqq 0$:

$$a^n, b^n, c^n, d^n, e^n, f^n, g^n, h^n, k^n, a_1{}^n, b_1{}^n, \cdots, \text{etc.}$$

When $n = 0$, the superscript may be omitted. Function symbols of degree 0 are *individual constants.*

2.3 *Predicate symbols.* The following symbols, in alphabetical order, are predicate symbols of degree n, for each $n \geqq 0$:

$$P^n, Q^n, R^n, P_1{}^n, Q_1{}^n, R_1{}^n, P_2{}^n, \cdots, \text{etc.}$$

The superscript may be omitted when n is 0.

2.4 *The negation symbol.* The following symbol is the negation symbol: $\sim$

2.5 *Alphabetical order of symbols.* The set of all symbols is well ordered in alphabetical order by adding to the above ordering conventions the rule that variables precede function symbols, function symbols precede predicate symbols, predicate symbols precede the negation symbol, function symbols of lower degree precede function symbols of higher degree, and predicate symbols of lower degree precede predicate symbols of higher degree.

2.6 *Terms.* A variable is a term, and a string of symbols consisting of a function symbol of degree $n \geqq 0$ followed by n terms is a term.

2.7 *Atomic formulas.* A string of symbols consisting of a predicate symbol of degree $n \geqq 0$ followed by n terms is an atomic formula.

2.8 *Literals.* An atomic formula is a literal; and if A is an atomic formula then $\sim A$ is a literal.

2.9 *Complements.* If A is an atomic formula, then the two literals A and $\sim A$ are said to be each other's complements, and to form, in either order, a complementary pair.

2.10 *Clauses.* A finite set (possibly empty) of literals is called a clause. The empty clause is denoted by: $\square$

2.11 *Ground literals.* A literal which contains no variables is called a ground literal.

2.12 *Ground clauses.* A clause, each member of which is a ground literal, is called a ground clause. In particular $\square$ is a ground clause.

2.13 *Well-formed expressions.* Terms and literals are (the only) well formed expressions.

2.14 *Lexical order of well-formed expressions.* The set of all well formed expressions is well ordered in lexical order by the rule that A precedes B just in case that A is shorter than B or, if A and B are of equal length, then A has the alphabetically earlier symbol in the first symbol position at which A and B have distinct symbols.

In writing well-formed expressions for illustrative purposes, we follow the more readable plan of enclosing the n terms following a function symbol or predicate symbol of degree n by a pair of parentheses, separating the terms, if there are two or more, by commas. We can then unambiguously omit all superscripts from symbols. In writing finite sets, we follow the usual convention of enclosing the members in a pair of braces and of separating the members by commas, with the understanding that the order of writing the members is immaterial.

2.15 *Herbrand universes.* With any set S of clauses there is associated a set of ground terms called the Herbrand universe of S, as follows: let F be the set of all function symbols which occur in S. If F contains any function symbols of degree 0, the functional vocabulary of S is F; otherwise it is the set $\{a\} \cup F$. The Herbrand universe of S is then the set of all ground terms in which there occur only symbols in the functional vocabulary of S.

2.16 *Saturation.* If S is any set of clauses and P is any set of terms, then by $P(S)$ we denote the saturation of S over P, which is the set of all ground clauses obtainable from members of S by replacing variables with members of P—occurrences of the same variable in any one clause being replaced by occurrences of the same term.

2.17 *Models.* A set of ground literals which does not include a complementary pair is called a model. If M is a model and S is a set of ground clauses, then M is a model of S if, for all C in S, C contains a member of M. Then, in general, if S is any set of clauses, and H is the Herbrand universe of S, we say that M is a model of S just in case that M is a model of $H(S)$.

2.18 *Satisfiability.* A set S of clauses is satisfiable if there is a model of S; otherwise S is unsatisfiable.

From the definition of satisfiability, it is clear that any set of clauses which contains $\square$ is unsatisfiable, and that the empty set of clauses is satisfiable. These two circumstances will appear quite natural as the development of our system proceeds. It is also clear that according to our semantic definitions each nonempty clause is interpreted, as explained in the informal remarks at the beginning of this section, as the universal closure of the disjunction of the literals which it contains.

2.19 *Ground resolvents.* If C and D are two ground clauses, and $L \subseteq C$, $M \subseteq D$ are two singletons (unit sets) whose respective members form a complementary pair, then the ground clause: $(C - L) \cup (D - M)$ is called a ground resolvent of C and D.

Evidently any model of $\{C, D\}$ is also a model of $\{C, D, R\}$, where R is a ground resolvent of C and D. Not all pairs of ground clauses have ground resolvents, and some have more than one; but in no case, as is clear from the definition, can two ground clauses have more than a finite number of ground resolvents.

2.20 *Ground resolution.* If S is any set of ground clauses, then the ground resolution of S, denoted by $\mathcal{R}(S)$, is the set of ground clauses consisting of the members of S together with all ground resolvents of all pairs of members of S.

2.21 *N-th ground resolution.* If S is any set of ground clauses, then the nth ground resolution of S, denoted by $\mathcal{R}^n(S)$, is defined for each $n \geqq 0$ as follows: $\mathcal{R}^0(S) = S$; and for $n \geqq 0$, $\mathcal{R}^{n+1} = \mathcal{R}(\mathcal{R}^n(S))$.

This completes the first batch of definitions. The next sections are concerned with the various forms that Herbrand's Theorem takes on in our system. To each such form, there is a type of refutation procedure which that form suggests and justifies. The basic version is stated as follows (cf. [2, 4]):

HERBRAND'S THEOREM. *If S is any finite set of clauses and H its Herbrand universe, then S is unsatisfiable if and only if some finite subset of $H(S)$ is unsatisfiable.*

3. *Saturation Procedures*

Is was noted in an earlier paper [5] that one can express Herbrand's Theorem in the following form:

THEOREM 1. *If S is any finite set of clauses, then S is unsatisfiable if and only if, for some finite subset P of the Herbrand universe of S, $P(S)$ is unsatisfiable.*

This version of Herbrand's Theorem suggests the following sort of refutation procedure, which we call a *saturation procedure*: given a finite set S of clauses, select a sequence $P_0, P_1, P_2, \cdots$, of finite subsets of the Herbrand universe H of S, such that $P_j \subseteq P_{j+1}$ for each $j \geqq 0$, and such that $\bigcup_{j=0}^{\infty} P_j = H$. Then examine in turn the sets $P_0(S), P_1(S), P_2(S), \cdots$, for satisfiability. Evidently, for any finite subset P of H, $P \subseteq P_j$ for some j, and therefore $P(S) \subseteq P_j(S)$. Therefore, by Theorem 1, if S is unsatisfiable then, for some j, $P_j(S)$ is unsatisfiable.

Of course, any specific procedure of this sort must make the selection of P_0, $P_1, P_2, \cdots$, uniformly for all finite sets of clauses. A particularly natural way of doing this is to use the so-called levels $H_0, H_1, H_2, \cdots$, of the Herbrand universe H; where H_0 consists of all the individual constants in H, and H_{n+1}, for $n \geqq 0$, consists of all the terms in H which are in H_n, or whose arguments are in H_n. In [5] we called procedures using this method *level-saturation procedures*. It was there remarked that essentially the procedures of Gilmore [4] and Davis-Putnam [2] are level-saturation procedures.

The major combinatorial obstacle to efficiency for level-saturation procedures is the enormous rate of growth of the finite sets H_j and $H_j(S)$ as j increases, at least for most interesting sets S. These growth rates were analyzed in some detail in [5], and some examples were there given of some quite simple unsatisfiable S for which the earliest unsatisfiable $H_j(S)$ is so large as to be absolutely beyond the limits of feasibility.

An interesting heuristic remark is that, for every finite set S of clauses which is unsatisfiable and which has a refutation one could possibly construct, there is at least one reasonably small finite subset of the Herbrand universe of S such that $P(S)$ is unsatisfiable and such that P is *minimal* in the sense that $Q(S)$ is satisfiable for each proper subset Q of P. Such a P was called a *proof set for* S in [5]. If only, then, a benevolent and omniscient demon were available who could provide us, in reasonable time, with a proof set P for each unsatisfiable finite set S of clauses that we considered, we could simply arrange to saturate S over P and then extract a suitable refutation of S from the resulting finite unsatisfiable set $P(S)$ of ground clauses. This was in fact the underlying scheme of a computer program reported in [5], in which the part of the demon is played, as best his ingenuity allows, by the mathematician using the program. What is really wanted, to be sure, is a simulation of the proof set demon on the computer; but this would appear, intuitively, to be out of the question.

It turns out that it is not completely out of the question. In fact, the method developed in the remainder of this paper seems to come quite close to supplying the required demon as a computing process. In Section 4 we take the first major step towards the development of this method by proving more versions of Herbrand's Theorem. We also give a preliminary informal account of the rest of the argument, pending a rigorous treatment in succeeding sections.

4. *The Resolution Theorems and the Basic Lemma*

As a specific method for testing a finite set of ground clauses for satisfiability, the method of Davis-Putnam [4] would be hard to improve on from the point of view of efficiency. However, we now give another method, far less efficient than theirs, which plays only a theoretical role in our development, and which is much simpler to state: given the finite set S of ground clauses, form successively the sets S, $\mathcal{R}(S)$, $\mathcal{R}^2(S)$, $\cdots$, until either some $\mathcal{R}^n(S)$ contains $\square$, or does not contain $\square$ but is equal to $\mathcal{R}^{n+1}(S)$. In the former case, S is unsatisfiable; in the latter case, S is satisfiable. One or other of these two terminating conditions must eventually occur, since the number of distinct clauses formable from the finite set of literals which occur in S is finite, and hence in the nested infinite sequence:

4.1 $S \subseteq \mathcal{R}(S) \subseteq \mathcal{R}^2(S) \subseteq \cdots \subseteq \mathcal{R}^n(S) \subseteq \cdots$,

not all of the inclusions are proper, since resolution introduces no new literals.

In view of the finite termination of the described process we can prove its correctness, as stated above, in the form of the ground resolution theorem.

Ground Resolution Theorem. *If S is any finite set of ground clauses, then S is unsatisfiable if and only if $\mathcal{R}^n(S)$ contains $\square$, for some $n \geqq 0$.*

Proof. The "if" part is immediate. To prove the "only if" part, let T be the terminating set $\mathcal{R}^n(S)$ in the sequence (4.1) above, so that T is closed under ground resolution. We need only show that if T does not contain $\square$, then T is satisfiable, and hence S is satisfiable since $S \subseteq T$. Let $L_1, \cdots, L_k$ be all the distinct atomic formulas which occur in T or whose complements occur in T. Let M be the model defined as follows: M_0 is the empty set; and for $0 < j \leqq k$, M_j is the set $M_{j-1} \cup \{L_j\}$, unless some clause in T consists entirely of complements of literals in the set $M_{j-1} \cup \{L_j\}$; in which case M_j is the set $M_{j-1} \cup \{\sim L_j\}$. Finally, M is M_k. Now if T does not contain $\square$, M satisfies T. For otherwise there is a least j, $0 < j \leqq k$, such that some clause (say, C) in T consists entirely of complements of literals in the set M_j. By the definition of M_j, therefore, M_j is $M_{j-1} \cup \{\sim L_j\}$. Hence by the leastness of j, C contains L_j. But since M_j is $M_{j-1} \cup \{\sim L_j\}$, there is some clause (say, D) in T which consists entirely of complements of literals in the set $M_{j-1} \cup \{L_j\}$. Hence by the leastness of j, D contains $\sim L_j$. Then the clause $B = (C - \{L_j\}) \cup (D - \{\sim L_j\})$ consists entirely of complements of literals in the set M_{j-1}, unless B is $\square$. But B is a ground resolvent of C and D, hence is in T, hence is not $\square$. Thus the leastness of j is contradicted and the theorem is proved.

The Ground Resolution Theorem now allows us to state a more specific form of Theorem 1, namely,

THEOREM 2. *If S is any finite set of clauses, then S is unsatisfiable if and only if, for some finite subset P of the Herbrand aniverse of S, and some $n \geqq 0$, $\mathcal{R}^n(P(S))$ contains $\square$.*

It is now possible to state informally the essential steps of the remaining part of the development. We are going to generalize the notions of ground resolvent and ground resolution, respectively, to the notions of resolvent and resolution. by removing the restriction that the clauses involved be only ground clauses. Any two clauses will then have zero, one or more clauses as their resolvents, but in no case more than finitely many. In the special case that C and D *are* ground clauses, their resolvents, if any, are precisely their ground resolvents as already defined. Similarly, the notations $\mathcal{R}(S)$, $\mathcal{R}^n(S)$ will be retained, with S allowed to be any set of clauses. $\mathcal{R}(S)$ will then denote the resolution of S, which is the set of clauses consisting of all members of S together with all resolvents of all pairs of members of S. Again, $\mathcal{R}(S)$ is precisely the ground resolution of S, already defined, whenever S happens to be a set of ground clauses.

The details of how this generalization is done must await the formal definitions in Section 5. However, an informal grasp of the general notion of resolution is obtainable now, prior to its exact treatment, from simply contemplating the fundamental property which it will be shown to possess: *resolution is semicommutative with saturation.* More exactly, this property is as stated in the following basic Lemma, which is proved in Section 5:

LEMMA. *If S is any set of clauses, and P is any subset of the Herbrand universe of S, then: $\mathcal{R}(P(S)) \subseteq P(\mathcal{R}(S))$.*

The fact is, as will be shown here, that any ground clause which can be obtained by *first* instantiating over P a pair C, D of clauses in S, and *then* forming a ground resolvent of the two resulting instances, can also be obtained by instantiating over P one of the finitely many resolvents of C and D.

It is an easy corollary of the basic Lemma that nth resolutions are also semicommutative with saturation:

COROLLARY. *If S is any set of clauses and P is any subset of the Herbrand universe of S, then: $\mathcal{R}^n(P(S)) \subseteq P(\mathcal{R}^n(S))$ for all $n \geqq 0$.*

PROOF. By induction on n. $\mathcal{R}^0(P(S)) = P(S) = P(\mathcal{R}^0(S))$, so that the case $n = 0$ is trivial. And if, for $n \geqq 0$, $\mathcal{R}^n(P(S)) \subseteq P(\mathcal{R}^n(S))$, then:

$$
\begin{aligned}
\mathcal{R}^{n+1}(P(S)) &= \mathcal{R}(\mathcal{R}^n(P(S))) && \text{by definition of } \mathcal{R}^{n+1}, \\
&\subseteq \mathcal{R}(P(\mathcal{R}^n(S))) && \text{by the induction hypothesis, as } \mathcal{R} \text{ preserves inclusion}, \\
&\subseteq P(\mathcal{R}(\mathcal{R}^n(S))) && \text{by the Lemma}, \\
&= P(\mathcal{R}^{n+1}(S)) && \text{by definition of } \mathcal{R}^{n+1},
\end{aligned}
$$

and the Corollary is proved.

Now by the above Corollary to the basic Lemma we may immediately obtain a third version of Herbrand's Theorem from Theorem 2:

THEOREM 3. *If S is any finite set of clauses, then S is unsatisfiable if and only if, for some finite subset P of the Herbrand universe of S, and some $n \geqq 0$, $P(\mathcal{R}^n(S))$ contains $\Box$.*

Here, the order of the saturation and nth resolution operations is reversed. Now a rather surprising simplification of Theorem 3 can be made, on the basis of the remark that mere replacement of variables by terms cannot produce $\Box$ from a nonempty clause. Hence $P(\mathcal{R}^n(S))$ will contain $\Box$ if and only if $\mathcal{R}^n(S)$ contains $\Box$. From Theorem 3, therefore, we immediately obtain our final version of Herbrand's Theorem, which is the main result of this paper, and which we call:

RESOLUTION THEOREM. *If S is any finite set of clauses, then S is unsatisfiable if and only if $\mathcal{R}^n(S)$ contains $\Box$, for some $n \geqq 0$.*

The statement of the Resolution Theorem is just that of the Ground Resolution Theorem with the word "ground" omitted. Apart, therefore, from the somewhat more complex way in which the resolvents of two clauses are computed (described in Section 5) the method suggested by the Resolution Theorem for testing a finite set S of clauses for unsatisfiability is exactly like that given earlier for the case that S is a set of ground clauses, and indeed it automatically reverts to that method when it is applied to a finite set of ground clauses. However, it is no longer the case in general that the nested sequence

$$S \subseteq \mathcal{R}(S) \subseteq \mathcal{R}^2(S) \subseteq \cdots \subseteq \mathcal{R}^n(S) \subseteq \cdots$$

must terminate for all finite S. By Church's Theorem this could not be so, for otherwise we would have a decision procedure for satisfiability for our formulation of first-order logic.

Consider now the "proof set demon" discussed in Section 3. We there supposed that if we were given a proof set P for an unsatisfiable set S of clauses, all we would have to do would be to compute until we encountered the first $\mathcal{R}^n(P(S))$ which contains $\Box$, in order to obtain from it a formal refutation of S. But the Resolution Theorem assures us that by the time we had computed $\mathcal{R}^n(S)$, if not before, we would have turned up $\Box$, despite our ignorance of P. In this sense the Resolution Theorem makes the proof set demon's role unnecessary.

In Section 5 we introduce a little more formal apparatus by a second batch of definitions, and pay off our debts by defining the general notion of resolution and proving the basic Lemma.

5. *Substitution, Unification and Resolution*

The following definitions are concerned with the operation of instantiation, i.e. substitution of terms for variables in well-formed expressions and in sets of well-formed expressions, and with the various auxiliary notions needed to define resolution in general.

5.1 *Substitution components.* A substitution component is any expression of the form T/V, where V is any variable and T is any term different from V. V is called the *variable of* the component T/V, and T is called the *term of* the component T/V.

5.2 *Substitutions.* A substitution is any finite set (possibly empty) of substitution components none of the variables of which are the same. If P is any set of terms, and the terms of the components of the substitution θ are all in P, we say that θ is a substitution over P. We write the substitution whose components are $T_1/V_1, \cdots, T_k/V_k$ as $\{T_1/V_1, \cdots, T_k/V_k\}$, with the understanding that the order of the components is immaterial. We use lower-case Greek letters to denote substitutions. In particular, ϵ is the *empty substitution*.

5.3 *Instantiation.* If E is any finite string of symbols and

$$\theta = \{T_1/V_1, \cdots, T_k/V_k\}$$

is any substitution, then the instantiation of E by θ is the operation of replacing each occurrence of the variable V_i, $1 \leqq i \leqq k$, in E by an occurrence of the term T_i. The resulting string, denoted by $E\theta$, is called the instance of E by θ. I.e., if E is the string $E_0V_{i_1}E_1 \cdots V_{i_n}E_n$, then $E\theta$ is the string $E_0T_{i_1}E_1 \cdots T_{i_n}E_n$. Here, none of the substrings E_j of E contain occurrences of the variables $V_1, \cdots, V_k$, some of the E_j are possibly null, n is possibly 0, and each V_{i_j} is an occurrence of one of the variables $V_1, \cdots, V_k$. Any string $E\theta$ is called an instance of the string E. If C is any set of strings and θ a substitution, then the instance of C by θ is the set of all strings $E\theta$, where E is in C. We denote this set by $C\theta$, and say that it is an instance of C.

5.4 *Standardizations.* If C is any finite set of strings, and $V_1, \cdots, V_k$ are all the distinct variables, in alphabetical order, which occur in strings in C, then the x-standardization of C, denoted by ξ_C, is the substitution $\{x_1/V_1, \cdots, x_k/V_k\}$ and the y-standardization of C, denoted by η_C, is the substitution

$$\{y_1/V_1, \cdots, y_k/V_k\}.$$

5.5 *Composition of substitutions.* If $\theta = \{T_1/V_1, \cdots, T_k/V_k\}$ and λ are any two substitutions, then the set $\theta' \cup \lambda'$, where λ' is the set of all components of λ whose variables are not among $V_1, \cdots, V_k$, and θ' is the set of all components $T_i\lambda/V_i$, $1 \leqq i \leqq k$, such that $T_i\lambda$ is different from V_i, is called the composition of θ and λ, and is denoted by $\theta\lambda$.

It is straightforward to verify that $\epsilon\theta = \theta\epsilon = \theta$ for any substitution θ. Also, composition of substitutions enjoys the associative property $(\theta\lambda)\mu = \theta(\lambda\mu)$, so that we may omit parentheses in writing multiple compositions of substitutions.

The point of the composition operation on substitutions is that, when E is any string, and $\sigma = \theta\lambda$, the string $E\sigma$ is just the string $E\theta\lambda$, i.e. the instance of $E\theta$ by λ.

These properties of the composition of substitutions are established by the following propositions.

5.5.1. $(E\sigma)\lambda = E(\sigma\lambda)$ *for all strings* E *and all substitutions* σ, λ.

PROOF. Let $\sigma = \{T_1/V_1, \cdots, T_k/V_k\}$, $\lambda = \{U_1/W_1, \cdots, U_m/W_m\}$ and $E = E_0V_{i_1}E_1 \cdots V_{i_n}E_n$ as explained in (5.3) above. Then by definition $E\sigma =$

$E_0T_{i_1}E_1 \cdots T_{i_n}E_n$, and $(E\sigma)\lambda = \bar{E}_0\bar{T}_{i_1}\bar{E}_1 \cdots \bar{T}_{i_n}\bar{E}_n$, where each $\bar{T}_{i_j}$ is $T_{i_j}\lambda$, and each $\bar{E}_j$ is $E_j\lambda'$, where λ' is the set of all components of λ whose variables are not among $V_1, \cdots, V_k$ (since none of these variables occur in any E_j). But $\sigma\lambda = \sigma' \cup \lambda'$, where each component of σ' is just $\bar{T}_i/V_i$ whenever $\bar{T}_i$ is different from V_i. Hence $E(\sigma\lambda) = \bar{E}_0\bar{T}_{i_1}\bar{E}_1 \cdots \bar{T}_{i_n}\bar{E}_n$.

5.5.2. *For any substitutions σ, λ: if $E\sigma = E\lambda$ for all strings E, then $\sigma = \lambda$.*

PROOF. Let $V_1, \cdots, V_k$ include all the variables of the components of σ and λ; then $V_j\sigma = V_j\lambda$, for $1 \leqq j \leqq k$. Then all the components of σ and λ are the same.

5.5.3. *For any substitutions σ, λ, μ: $(\sigma\lambda)\mu = \sigma(\lambda\mu)$.*

PROOF. Let E be any string. Then by 5.5.1,

$$\begin{aligned} E((\sigma\lambda)\mu) &= (E(\sigma\lambda))\mu \\ &= ((E\sigma)\lambda)\mu \\ &= (E\sigma)(\lambda\mu) \\ &= E(\sigma(\lambda\mu)). \end{aligned}$$

Hence $(\sigma\lambda)\mu = \sigma(\lambda\mu)$ by (5.5.2).

We shall also have occasion to use the following distributive property.

5.5.4. *For any sets A, B of strings and substitution λ: $(A \cup B)\lambda = A\lambda \cup B\lambda$.*

5.6 *Disagreement sets.* If A is any set of well-formed expressions, we call the set B the disagreement set of A whenever B is the set of all well-formed subexpressions of the well-formed expressions in A, which begin at the first symbol position at which not all well-formed expressions in A have the same symbol. *Example*:

$$A = \{P(x, h(x, y), y), P(x, k(y), y), P(x, a, b)\},$$

$$\text{Disagreement set of } A = \{h(x, y), k(y), a\}.$$

Evidently, if A is nonempty and is not a singleton, then the disagreement set of A is nonempty and is not a singleton.

5.7 *Unification.* If A is any set of well-formed expressions and θ is a substitution, then θ is said to unify A, or to be a unifier of A, if $A\theta$ is a singleton. Any set of well-formed expressions which has a unifier is said to be unifiable.

Evidently, if θ unifies A, but A is not a singleton, then θ unifies the disagreement set of A.

5.8 *Unification Algorithm.* The following process, applicable to any finite nonempty set A of well-formed expressions, is called the Unification Algorithm:

Step 1. Set $\sigma_0 = \epsilon$ and $k = 0$, and go to step 2.

Step 2. If $A\sigma_k$ is not a singleton, go to step 3. Otherwise, set $\sigma_A = \sigma_k$ and terminate.

Step 3. Let V_k be the earliest, and U_k the next earliest, in the lexical ordering of the disagreement set B_k of $A\sigma_k$. If V_k is a variable, and does not occur in U_k, set $\sigma_{k+1} = \sigma_k\{U_k/V_k\}$, add 1 to k, and return to step 2. Otherwise, terminate.

This definition requires justification in the form of a proof that the given process is in fact an algorithm. In fact the process always terminates for any finite nonempty set of well-formed expressions, for otherwise there would be generated an infinite sequence $A, A\sigma_1, A\sigma_2, \cdots$ of finite nonempty sets of well-formed expressions with the property that each successive set contains one less variable than its predecessor (namely, $A\sigma_k$ contains V_k but $A\sigma_{k+1}$ does not). But this is impossible, since A contains only finitely many distinct variables.

5.9 *Most general unifiers.* If A is a finite nonempty set of well-formed expressions for which the Unification Algorithm terminates in step 2, the substitution σ_A then available as output of the Unification Algorithm is called the most general unifier of A, and A is then said to be most generally unifiable.

5.10 *Key triples.* The ordered triple $\langle L, M, N\rangle$ of finite sets of literals is said to be a key triple of the ordered pair $\langle C, D\rangle$ of clauses just in case the following conditions are satisfied.

5.10.1. L and M are nonempty, and $L \subseteq C$, $M \subseteq D$.

5.10.2. N is the set of atomic formulas which are members, or complements of members, of the set $L\xi_C \cup M\eta_D$ (cf. definition (5.4)).

5.10.3. N is most generally unifiable, with most general unifier σ_N.

5.10.4. The sets $L\xi_C\sigma_N$ and $M\eta_D\sigma_N$ are singletons whose members are complements.

Evidently, a pair $\langle C, D\rangle$ of clauses has at most a finite number of key triples, and possibly none at all.

5.11 *Resolvents.* A resolvent of the two clauses C and D is any clause of the form: $(C - L)\xi_C\sigma_N \cup (D - M)\eta_D\sigma_N$ where $\langle L, M, N\rangle$ is a key triple of $\langle C, D\rangle$.

By the remark following definition (5.10) it is clear that two clauses C and D can have at most finitely many resolvents, and possibly none at all.

5.12 *Resolutions.* If S is any set of clauses then the resolution of S, denoted by $\mathcal{R}(S)$, is the set of all clauses which are members of S or resolvents of members of S.

5.13 *N-th resolution.* The nth resolution of S, where S is any set of clauses, is denoted by $\mathcal{R}^n(S)$ and is defined for all $n \geqq 0$ exactly analogously to definition (2.21).

This completes our second group of definitions. The definition of $\mathcal{R}(S)$ as given is adequate for our theoretical argument, but in practice one would not include in it both the resolvents of $\langle C, D\rangle$ and the resolvents of $\langle D, C\rangle$, since these are in fact identical up to a change of variables. When C and D are both ground clauses, the resolvents of $\langle C, D\rangle$ are actually identical with those of $\langle D, C\rangle$, and are precisely the ground resolvents of C and D, as is easily verified.

It now remains to prove the basic Lemma, which will be done after we have first proved the following theorem establishing the basic property of unification, which we need in the proof of the Lemma and elsewhere in our theory:

Unification Theorem. *Let A be any finite nonempty set of well-formed expressions. If A is unifiable, then A is most generally unifiable with most general unifier σ_A; moreover, for any unifier θ of A there is a substitution λ such that $\theta = \sigma_A\lambda$.*

Proof. It will suffice to prove that under the hypotheses of the theorem the Unification Algorithm will terminate, when applied to A, at step 2; and that for each $k \geqq 0$ until the Unification Algorithm so terminates, the equation

5.14. $\theta = \sigma_k \lambda_k$

holds at step 2 for some substitution λ_k. For $k = 0$, (5.14) holds with $\lambda_0 = \theta$, since $\sigma_0 = \epsilon$. Now assume that, for $k \geqq 0$, (5.14) holds at step 2 for some substitution λ_k. Then either $A\sigma_k$ is a singleton, in which case the Unification Algorithm terminates at step 2 with $\sigma_A = \sigma_k$ the most general unifier of A and $\lambda = \lambda_k$ the required substitution; or the Unification Algorithm transfers to step 3. In the latter case, since λ_k unifies $A\sigma_k$, (by (5.14), since θ unifies A) λ_k must also unify the disagreement set B_k of $A\sigma_k$. Hence the V_k and U_k defined in step 3 of the Unification Algorithm satisfy the equation

5.15. $V_k \lambda_k = U_k \lambda_k$.

Since B_k is a disagreement set, the well-formed expressions in B_k cannot all begin with the same symbol; hence they cannot all begin with symbols which are not variables, since B_k is unifiable. Therefore at least one well-formed expression in B_k begins with a variable, and hence is a variable, since it is well-formed. Since variables precede all other well-formed expressions in the lexical order, and since V_k is the earliest well-formed expression in B_k, it follows that V_k is a variable. Now if V_k occurs in U_k, $V_k\lambda_k$ occurs in $U_k\lambda_k$, but since V_k and U_k are distinct well-formed expressions this is impossible because of (5.15). Therefore V_k does not occur in U_k. Hence the Unification Algorithm will not terminate in step 3, but will return to step 2 with $\sigma_{k+1} = \sigma_k\{U_k/V_k\}$. Now let $\lambda_{k+1} = \lambda_k - \{V_k\lambda_k/V_k\}$. Then:

$$\begin{aligned}
\lambda_k &= \{V_k\lambda_k/V_k\} \cup \lambda_{k+1} && \text{by definition of } \lambda_{k+1}, \\
&= \{U_k\lambda_k/V_k\} \cup \lambda_{k+1} && \text{by (5.15)}, \\
&= \{U_k\lambda_{k+1}/V_k\} \cup \lambda_{k+1} && \text{since } V_k \text{ does not occur in } U_k, \\
&= \{U_k/V_k\}\lambda_{k+1} && \text{by definition (5.5).}
\end{aligned}$$

Hence by (5.14) $\theta = \sigma_{k+1}\lambda_{k+1}$. Thus (5.14) holds for all $k \geqq 0$ until the Unification Algorithm terminates in step 2, and the theorem is proved.

We are now in a position to prove the basic Lemma, which we state here again for convenience.

Lemma. *If S is any set of clauses and P is any subset of the Herbrand universe of S, then*: $\mathcal{R}(P(S)) \subseteq P(\mathcal{R}(S))$.

Proof. Assume that $A \in \mathcal{R}(P(S))$. Then either $A \in P(S)$, in which case $A \in P(\mathcal{R}(S))$ since $S \subseteq \mathcal{R}(S)$; or A is a ground resolvent of two ground clauses $C\alpha$, $D\beta$, where $C \in S$, $D \in S$, $\alpha = \{T_1/V_1, \cdots, T_k/V_k\}$, where $V_1, \cdots, V_k$ are all the distinct variables of C in alphabetical order and $T_1, \cdots, T_k$ are in P, and $\beta = \{U_1/W_1, \cdots, U_m/W_m\}$, where $W_1, \cdots, W_m$ are all the distinct variables of D in alphabetical order and $U_1, \cdots, U_m$ are in P. In that case, $A = (C - L)\alpha \cup (D - M)\beta$, where $L \subseteq C$, $M \subseteq D$, L and M are nonempty, and $L\alpha$, $M\beta$ are singletons whose members are complements. Let

$$\theta = \{T_1/x_1, \cdots, T_k/x_k, U_1/y_1, \cdots, U_m/y_m\}.$$

Then it follows that $A = (C - L)\xi_C\theta \cup (D - M)\eta_D\theta$ and that $L\xi_C\theta = L\alpha$ and $M\eta_D\theta = M\beta$. Therefore θ unifies the set N of atomic formulas which are members, or complements of members, of the set $L\xi_C \cup M\eta_D$. Hence by the Unification Theorem N has a most general unifier σ_N, and there is a substitution λ over P such that $\theta = \sigma_N\lambda$. Hence $L\xi_C\sigma_N\lambda = L\alpha$ and $M\eta_D\sigma_N\lambda = M\beta$, and therefore $L\xi_C\sigma_N$ and $M\eta_D\sigma_N$ are singletons whose members are complements. It follows that $\langle L, M, N\rangle$ is a key triple of $\langle C, D\rangle$, and hence that the clause

$$B = (C - L)\xi_C\sigma_N \cup (D - M)\eta_D\sigma_N$$

is a resolvent of C and D; hence $B \in \mathcal{R}(S)$. But since $\theta = \sigma_N\lambda$, it follows by (5.5.4) that $A = B\lambda$ and therefore that $A \in P(\mathcal{R}(S))$. The proof is complete.

The hypotheses of the Lemma do not entail the opposite inclusion $P(\mathcal{R}(S)) \subseteq \mathcal{R}(P(S))$. As a simple counterexample, consider:

$$S = \{\{Q(x, f(y))\}, \{\sim Q(g(y), x)\}\}, \qquad P = \{a\}.$$

A short investigation shows that $P(\mathcal{R}(S))$ contains $\square$ (since $\mathcal{R}(S)$ does) while $\mathcal{R}(P(S))$ does not. Thus S is unsatisfiable, but P is not a proof set for S.

6. *The Resolution Principle: Refutations*

The single inference principle of our system of logic, mentioned in Section 1, is the *resolution principle*, namely: *From any two clauses C and D, one may infer a resolvent of C and D.*

By a *refutation* of the set S of clauses we mean a finite sequence $B_1, \cdots, B_n$ of clauses such that (a) each B_i, $1 \leqq i \leqq n$, is either in S or is a resolvent of two earlier clauses in the sequence, and (b) B_n is $\square$.

It is immediate from the Resolution Theorem that a finite set S of clauses is unsatisfiable if and only if there is a refutation of S. Thus the Resolution Theorem is the completeness theorem for our system of logic.

Two examples of refutations will illustrate the workings of the system.

Example 1. The set containing just the two clauses C_1 and C_2, where

$$C_1 = \{Q(x, g(x), y, h(x, y), z, k(x, y, z))\}$$

$$C_2 = \{\sim Q(u, v, e(v), w, f(v, w), x)\}$$

has the refutation $C_1, C_2, \square$. Note that $\langle C_1, C_2\rangle$ has the key triple $\langle C_1, C_2, N\rangle$, where N is the set

$$\{Q(x_1, g(x_1), x_2, h(x_1, x_2), x_3, k(x_1, x_2, x_3)), \quad Q(y_1, y_2, e(y_2), y_3, f(y_2, y_3), y_4)\}.$$

The reader can verify in a few minutes of computation with the Unification Algorithm that σ_N is the substitution with the components:

$$\begin{array}{ll} y_1/x_1, & h(y_1, e(g(y_1)))/y_3 \\ g(y_1)/y_2, & f(g(y_1), h(y_1, e(g(y_1))))/x_3 \\ e(g(y_1))/x_2, & k(y_1, e(g(y_1)), f(g(y_1), h(y_1, e(g(y_1)))))/y_4 \end{array}$$

and that then $C_1\xi_{C_1}\sigma_N$ and $C_2\eta_{C_2}\sigma_N$ are singletons whose members are complements.

This example illustrates the way in which a proof set is automatically computed as a by-product of the resolution operation. The terms of the above substitution components become those of a proof set for $\{C_1, C_2\}$ when the variable y_1 is replaced throughout by any term of the Herbrand universe of $\{C_1, C_2\}$, e.g. by the individual constant "a." It is interesting to note that the earliest level of this Herbrand universe H to include such a proof set is H_5, which has of the order of 10^{64} members. Consequently $H_5(\{C_1, C_2\})$ has of the order of 10^{256} members. A level-saturation procedure would not find this example feasible.

Example 2. A more interesting example is one which was discussed in [5]. It arises from the following algebraic problem.

Prove that in any associative system which has left and right solutions x and y for all equations $x \cdot a = b$ and $a \cdot y = b$, there is a right identity element.

To formalize this problem in our logic, we deny the alleged conclusion, and try to refute the set containing the clauses (where $Q(x, y, z)$ is to mean $x \cdot y = z$):

C_1 :	$\{\sim Q(x, y, u), \sim Q(y, z, v), \sim Q(x, v, w), Q(u, z, w)\}$	Associativity
C_2 :	$\{\sim Q(x, y, u), \sim Q(y, z, v), \sim Q(u, z, w), Q(x, v, w)\}$	
C_3 :	$\{Q(g(x, y), x, y)\}$	Existence of left and right
C_4 :	$\{Q(x, h(x, y), y)\}$	solutions
C_5 :	$\{Q(x, y, f(x, y))\}$	Closure under ·
C_6 :	$\{\sim Q(k(x), x, k(x))\}$	No right identity.

By adding the following resolvents, we get a refutation:

C_7 : $\{\sim Q(y_1, x_6, y_1), Q(y_2, x_6, y_2)\}$

C_8 : $\{\sim Q(y_1, y_2, y_1)\}$

C_9 : $\square$

Commentary. C_7 is the resolvent of the pair $\langle C_1, C_3\rangle$ for the key triple

$$\langle\{\sim Q(x, y, u), \sim Q(x, v, w)\}, \{Q(g(x, y), x, y)\}, N\rangle$$

where N is the set $\{Q(x_4, x_5, x_1), Q(x_4, x_2, x_3), Q(g(y_1, y_2), y_1, y_2)\}$. The σ_N computed for this N by the Unification Algorithm is

$$\{y_2/x_1, y_1/x_2, y_2/x_3, g(y_1, y_2)/x_4, y_1/x_5\},$$

as is easily verified. C_8 is the only resolvent of $\langle C_6, C_7\rangle$, and $\square$ is the only resolvent of $\langle C_4, C_8\rangle$.

This example illustrates the way in which the single steps in a refutation made with the resolution principle go beyond, in their complexity, the capacity of the human mind to apprehend their correctness in one single intellectual act. By taking larger bites, so to speak, the resolution principle in this case permits a very compact, not to say elegant, piece of reasoning. C_2 and C_5 are not used as premisses in the refutation, although this has nothing to do with the resolution principle. Hence a nonredundant refutation for this example is the sequence: $C_1, C_3, C_4, C_6, C_7, C_8, \square$.

7. *Refutation Procedures, Search Principles*

The foregoing discussion was intended only to establish the theoretical framework, in the form of a special system of logic, for the design of theorem-proving programs, i.e. in the present case, refutation procedures. No attempt has been made thus far to discuss the question of developing efficient refutation procedures, and in this final section of the paper we briefly discuss this question.

The raw implementation of the Resolution Theorem would produce a very inefficient refutation procedure, namely, the procedure would consist of computing, given the finite set S of clauses as input, the sequence of sets S, $\mathcal{R}(S)$, $\mathcal{R}^2(S)$, $\cdots$, until one is encountered, say, $\mathcal{R}^n(S)$, which either contains $\square$ or else does not contain $\square$ but is equal to its successor $\mathcal{R}^{n+1}(S)$. In the former case, a refutation of S is obtained by tracing back the genesis of $\square$; in the latter case the conclusion is that S is satisfiable. By Church's Theorem [1] we know that for some inputs S this procedure, and in general all correct refutation procedures, will not terminate in either of these two ways but will continue computing indefinitely.

In some cases we can foresee the nonterminating behavior. Consider the example of the set S whose members are:

$$C_1: \ \{Q(a)\}, \qquad C_2: \ \{\sim Q(x), Q(f(x))\}.$$

(The reader will recognize this as the formulation, in our logic, of a fragment of Peano's postulates for the natural numbers, with "$Q(x)$" for "x is a natural number," "a" for "0," and "$f(x)$" for "the successor of x".) It is easy to see that for this S the procedure described above would generate successively the resolvents $\{Q(f(a))\}$, $\{Q(f(f(a)))\}$, $\{Q(f(f(f(a))))\}$, $\cdots$, etc., *ad infinitum.*

This example suggests our attempting to formulate a principle which would allow us effectively to recognize the particular indefinite continuation phenomenon which it exhibits, so that we might incorporate the principle into a refutation procedure and cause it to terminate for this S and for other similar examples. Such a principle, which we call the *purity principle*, is available, based on the notion of a literal being pure in a set S of clauses. We define this notion as follows.

7.1 *Pure literals.* If S is any finite set of clauses, C a clause in S, and L a literal in C with the property that there is no key triple $\langle\{L\}, M, N\rangle$ for any pair $\langle C, D\rangle$ of clauses, where D is any clause in $S - \{C\}$, then L is said to be pure in S.

The purity principle is then based on the following theorem.

PURITY THEOREM. *If S is any finite set of clauses, and $L \in C \in S$ is a literal which is pure in S, then S is satisfiable if and only if $S - \{C\}$ is satisfiable.*

PROOF. If S is satisfiable then so is $S - \{C\}$ since it is a subset of S. If $S - \{C\}$ is satisfiable, then there is a model A of $S - \{C\}$, every literal in which occurs in some clause of $H(S)$, where H is the Herbrand universe of S. Let N be the set of all ground literals $L\theta$, where θ is a substitution over H, and let K consist of all complements of members of N. Then the set $P = N \cup (A - K)$ is a model; moreover it is a model of S, since every clause in $H(\{C\})$ contains a member of P (namely a member of N), and every clause in $H(S - \{C\})$ contains a member of P, namely a member of $A - K$; for no clause in $H(S - \{C\})$ contains a member of K, since otherwise, if $D\beta$ were such a clause, with $D \in S - \{C\}$, then there would be an $M \subseteq D$ such that $M\beta$ would be a singleton containing a member of K. Then there would be some substitution α over H such that $\{L\}\alpha$, $M\beta$ contained complementary singletons. Hence by the same argument as in the proof of the Lemma, there would be a key triple $\langle\{L\}, M, N\rangle$ of the pair $\langle C, D\rangle$, contradicting the purity of L in S. The theorem is proved.

The *purity principle* is then simply the following: *One may delete, from a finite set S of clauses, any clause containing a literal which is pure in S.*

When S is the little Peano example given earlier, i.e., is the set containing just the two clauses

$$C_1: \ \{Q(a)\}, \qquad C_2: \ \{\sim Q(x), \underline{Q(f(x))}\}.$$

we see that the underlined literal in C_2 is pure in S. Hence we may delete C_2, obtaining the set $S - \{C_2\}$ whose only clause is

$$C_1: \ \{\underline{Q(a)}\}.$$

But of course the underlined literal is, trivially, pure in $S - \{C_2\}$; hence we may delete C_1, obtaining the set $S - \{C_1\} - \{C_2\}$, which is empty, hence satisfiable. Hence by the Purity Theorem, S is satisfiable.

Thus a refutation procedure incorporating the purity principle as well as the resolution principle "converges" for more finite sets of clauses than a procedure based on the resolution principle alone. Such principles as the purity principle we call *search principles*, to distinguish them from inference principles.

There is another search principle which, though not increasing the range of convergence, does help to increase the rate of convergence, of refutation procedures. We call this principle the *subsumption principle* and base it on the following definition.

7.2 *Subsumption.* If C and D are two distinct nonempty clauses, we say that C subsumes D just in case there is a substitution σ such that $C\sigma \subseteq D$.

The following theorem establishes the basic property of subsumption.

Subsumption Theorem. *If S is any finite set of clauses, and D is any clause in S which is subsumed by some clause in $S - \{D\}$, then S is satisfiable if and only if $S - \{D\}$ is satisfiable.*

Proof. We need only show that if M is a model of $S - \{D\}$, then M is a model of S. Let M be a model of $S - \{D\}$, and suppose that $C \in S - \{D\}$ subsumes D. Then there is a substitution σ such that $C\sigma \subseteq D$. Since $D \in S$, the terms of the components of σ must be formed from function symbols in the functional vocabulary of S, together possibly with variables. Hence every ground instance of $C\sigma$ over H is a ground instance of C over H, and hence contains a member of M. But every ground instance $D\lambda$ of D includes the ground instance $C\sigma\lambda$ of C, and hence contains a member of M. So M is a model of S and the theorem is proved.

The *subsumption principle* is then simply the following: *One may delete, from a finite set S of clauses, any clause D which is subsumed by a clause in* $S - \{D\}$.

In order to make the subsumption principle available for incorporation into a refutation procedure, we must give an algorithm for deciding whether one clause C subsumes another clause D. Such an algorithm is the following Subsumption Algorithm:

Step 1. Let $V_1, \cdots, V_m$ be all the distinct variables, in alphabetical order, of D. Let $J_1, \cdots, J_m$ be distinct individual constants, none of which occur in C or D. Let $\theta = \{J_1/V_1, \cdots, J_m/V_m\}$. Compute $D\theta$ and go to step 2.
Step 2. Set $A_0 = \{C\}$, $k = 0$, and go to step 3.
Step 3. If A_k does not contain $\Box$, let A_{k+1} be the set of all clauses of the form $(K\sigma_N - M\sigma_N)$, where $K \in A_k$, $M \subseteq K$, $N = M \cup \{P\}$, for some $P \in D\theta$, and N is most generally unifiable with most general unifier σ_N; and go to step 4. Otherwise, terminate.
Step 4. If A_{k+1} is nonempty, add 1 to k and return to step 3. Otherwise, terminate.

That this is an algorithm is clear from the fact that each clause in A_{k+1} is smaller, by at least one literal, than the clause in A_k from which it was obtained. Hence, since the only clause in A_0 has but finitely many literals, the sequence $A_0, A_1, \cdots$, must eventually contain a set which contains $\Box$ or is empty.

That the Subsumption Algorithm is correct is shown by the following argument that it terminates in step 3 if and only if C subsumes D.

If C subsumes D, then $C\sigma \subseteq D$ for some σ. Hence $C\sigma\theta \subseteq D\theta$. Hence $C\mu \subseteq D\theta$, for some μ. Now assume, for $k \geqq 0$, that $K \in A_k$ and that, for some μ, $K\mu \subseteq D\theta$. If K is not $\Box$, let P be a literal in $K\mu \cap D\theta$. Then there is an $M \subseteq K$ such that $N = (M \cup \{P\})$ is unified by μ. Therefore by the Unification Theorem N has a most general unifier σ_N, and the clause $G = (K\sigma_N - M\sigma_N)$ is in A_{k+1}. But by the Unification Theorem $\mu = \sigma_N\lambda$, for some λ, hence $K\sigma_N\lambda \subseteq D\theta$. Therefore $G\lambda \subseteq D\theta$. Since $C \in A_0$, this shows that each A_k, $k \geqq 0$, either contains $\Box$ or is nonempty. Hence the Subsumption Algorithm does not terminate in step 4. Therefore it terminates in step 3.

If the Subsumption Algorithm terminates in step 3, for C and D as input, then there is a finite sequence $C_0, C_1, \cdots, C_{n+1}$ of clauses such that $C_0 = C$, $C_{n+1} = \Box$, and, for $0 \leqq j \leqq n$, $C_{j+1} = C_j\sigma_j - M_j\sigma_j$, where $M_j \subseteq C_j$, and σ_j is the most general unifier of $M_j \cup \{P\}$, where $P \in D\theta$. It follows that (since $M_j\sigma_j$ contains no variables, $0 \leqq j \leqq n$) we have

$$C_{n+1} = \Box = C\sigma_0\sigma_1 \cdots \sigma_n - M_0\sigma_0 - M_1\sigma_1 - \cdots - M_n\sigma_n,$$

i.e. that $C\sigma_0\sigma_1 \cdots \sigma_n \subseteq (M_0\sigma_0 \cup M_1\sigma_1 \cup \cdots \cup M_n\sigma_n) \subseteq D\theta$. Hence, for some λ, $C\lambda \subseteq D\theta$. Let σ be the substitution obtained from λ by the replacement, in each component of λ of J_i by V_i, for $1 \leqq i \leqq m$. Then $C\sigma \subseteq D$.

A particularly useful application of the subsumption principle is the following: Suppose a resolvent R of C and D subsumes one of C, D; then in adding R by the resolution principle we may simultaneously delete, by the subsumption principle, that one of C, D which R subsumes. This combined operation amounts to replacing C or D by R; accordingly we name the principle involved the *replacement principle*.

The following example, used by Gilmore [4], Davis-Putnam [2] and Friedman [3], illustrates the utility of these search principles in speeding up convergence. Consider the set S whose members are:

$$C_1: \quad \underset{(6)}{\underline{\{P(x_1, x_2)\}}}$$

$$C_2: \quad \{\underset{(1)}{\underline{\sim P(y_2, f(y_1, y_2))}}, \ \underset{(2)}{\underline{\sim P(f(y_1, y_2), f(y_1, y_2))}}, Q(y_1, y_2)\}$$

$$C_3: \quad \{\underset{(3)}{\underline{\sim P(y_2, f(y_1, y_2))}}, \ \underset{(4)}{\underline{\sim P(f(y_1, y_2), f(y_1, y_2))}}, \ \underset{(5)}{\underline{\sim Q(y_1, f(y_1, y_2))}}, \sim Q(f(y_1, y_2), f(y_1, y_2))\}$$

and we obtain the set S' whose only members are:

$$C_4: \quad \{Q(y_1, y_2)\}$$

$$C_5: \quad \{\sim Q(f(y_1, y_2), f(y_1, y_2))\}$$

in six stages which may be followed through by deleting the underlined literals' and the underlined clause, in the order indicated. This gives the set of clauses at each stage. Deletions (1) through (5) are by virtue of the replacement principle; deletion (6), of the entire clause C_1, is by virtue of the purity principle. The set S' in turn is found immediately to be unsatisfiable, since C_4 and C_5 have $\square$ as their only resolvent.

Gilmore's 704 program failed to converge after 21 minutes' running time, when given this example. The more efficient procedure of Davis and Putnam converges, for this example, in 30 minutes of hand computation.

The application, to a finite set S of clauses, of any of the three search principles we have described, produces a set S' which either has fewer clauses than S or has the same number of clauses as S but with one or more shorter clauses. An obvious method of exploiting these principles in a refutation procedure is therefore never to add new clauses, by the resolution principle, except to a set to which the three principles are no longer applicable. We might call such sets *irreducible*. The way in which such a procedure would terminate, for satisfiable S within its range of convergence, would then be with a set which is either empty (as in the Peano example) or nonempty, irreducible, and with the property that each resolvent of any pair of its clauses is subsumed by some one of its clauses.

There are further search principles of this same general sort, which are less simple than those discussed in this section. A sequel to the present paper is planned in which the theoretical framework developed here will be used as the basis for a more extensive treatment of search principles and of the design of refutation procedures. This section has been merely a sketch of the general nature of the problem, and a brief view of some of the approaches to it.

Acknowledgments. I should like to express my indebtedness to my colleagues Dr. George A. Robinson and Dr. Lawrence T. Wos, of the Argonne National Laboratory, and to Professor William Davidon of Haverford College, for their invaluable insights and criticisms concerning the basic concepts of this paper. My thanks are also due to the ACM referees, and to Dr. T. Hailperin of the Sandia Corporation, whose comments on and criticisms of a prior version of the paper greatly facilitated the writing of the present complete revision.

RECEIVED SEPTEMBER, 1963; REVISED AUGUST, 1964

REFERENCES

1. CHURCH, A. A note on the Entscheidungsproblem. *J. Symb. Logic 1* (1936), 40–41. Correction, ibid., 101–102.
2. DAVIS, M.. AND PUTNAM, H. A computing procedure for quantification theory. *J. ACM 7* (Mar. 1960), 201–215.
3. FRIEDMAN, J. A semi-decision procedure for the functional calculus. *J. ACM 10* (Jan. 1963), 1–24.
4. GILMORE, P. C. A proof method for quantification theory. *IBM J. Res. Develop. 4* (1960), 28–35.
5. ROBINSON, J. A. Theorem-proving on the computer. *J. ACM 10* (Apr. 1963), 163–174.

Automatic Deduction with Hyper-Resolution

J. A. Robinson

*Rice University**

§1. In [5] a specially simplified and standardized system of logic is formulated, involving a single principle of inference called the *resolution* principle. The purpose of [5] was to lay down the theoretical framework within which a new and more efficient family of automatic deduction procedures might be developed. Accordingly, [5] is intended to serve as a reference and prerequisite for a series of later papers, of

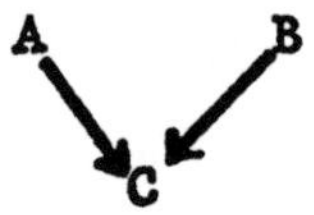

FIG. 1

which this is one; in particular, the reader is asked to refer to sections 2 and 5 of [5] for formal definitions of terminology and notations which are not defined in the present paper.

Within the system of [5] a deduction problem is always posed as the problem to deduce a contradiction from a finite set S of *clauses*. The individual steps of the deduction are all of the form shown in the diagram in Fig. 1: A and B are clauses which are each either in S or already deduced from S, and C is a *resolvent* of A and B. The contradiction takes the form of the *empty clause*, written as: $\square$. Thus a deduction of a clause C from a set S of clauses has the form of a *tree* with a clause at each of its nodes; at each of the initial nodes is a clause in S; at the terminal node is the clause C; and the clause at each non-

* This research was partly supported by National Science Foundation Grant GP 2466.

initial node is a resolvent of the clauses at the two immediately preceding nodes. Such a deduction is illustrated in the diagram of Fig. 2, in which the set S is assumed to include the subset $\{S_1, \ldots, S_6\}$. It should be noted that the same clause may occur at more than one node in a deduction; this is a point which will be of some importance in the sequel.

It is shown in [5] that, for any finite set S of clauses, S is *unsatisfiable* if and only if there is a deduction of $\square$ from S, i.e., that the system is sound and complete. An algorithm is provided for computing, given

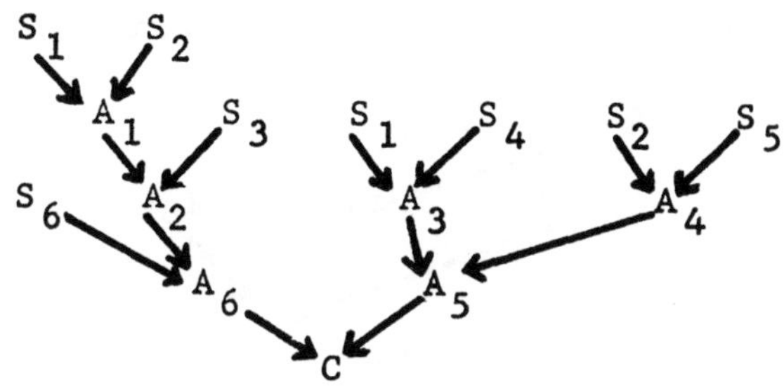

FIG. 2

two clauses A and B as input, all the clauses which are resolvents of A and B. There are never more than finitely many of these and in some cases there are none at all.

In the present paper we wish to exhibit the utility of the theory of [5] by developing within it a derived inference principle, which we call *hyper-resolution*, with the help of which it appears that a very strong family of automatic deduction procedures can be constructed.

§2. The key result of [5] can be stated in terms of the diagram of Fig. 3, in which the solid arrows, as before, represent resolution and the broken arrows now represent *instantiation*. The result is this: *if the clauses A' and B' are instances, respectively, of the clauses A and B, and if C' is a resolvent of A' and B', then there is a resolvent C of A and B such that C' is an instance of C.*

It follows easily by induction from this result that, in general, if there is a deduction D' of a clause C' from a set S' of instances of clauses in the set S, then there is a deduction D of a clause C from S which is *isomorphic* to D', the clause at each node in D' being an instance of the clause at the corresponding node in D. (Two trees D and D' are said to be isomorphic if there is a one-to-one correspondence $X \leftrightarrow X'$ between their nodes which preserves the arrow connections, i.e., an arrow leads from node X to node Y in D if and only if an arrow leads from the corresponding node X' to the corresponding node Y' in D'.)

A word of caution is in order concerning a subtle point. As we pointed out earlier in connection with Fig. 2, the same clause can occur at several distinct nodes in a tree deduction. Isomorphic deductions may or may not preserve this property; *the one-to-one correspondence is between the nodes and not between the clauses.*

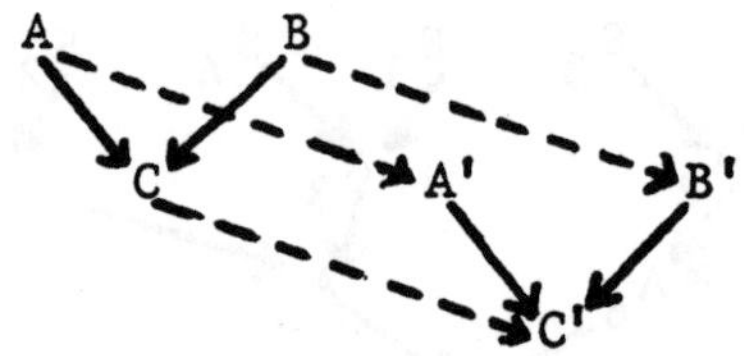

FIG. 3

The utility of this more general principle lies in its allowing us to reason about deductions involving *ground clauses* only, and then to transfer many of our results immediately to deductions involving clauses of any kind. Since reasoning about ground clauses is far simpler than reasoning about clauses in general, this is a distinct advantage. It is this kind of application which will be made in the remainder of the present paper.

§3. If the clause C' is an instance of the clause C, there are certain properties P such that C' must have P if C has P, and conversely. Such properties are called *instantiation invariants* of clauses. Examples of instantiation invariants of clauses are: *being empty, containing no negated literals, containing no unnegated literals, containing at least one negated literal, containing at least one unnegated literal.* More generally, if the deduction D' is an instance of the deduction D (i.e., if D' and D are isomorphic as trees, and the clause at each node of D' is an instance of the clause at the corresponding node of D) there are certain properties P such that D' must have P if D has P, and conversely. Such properties are called *instantiation invariants* of deductions. Examples of instantiation invariants of deductions are: *being a deduction of* $\square$, *being a deduction in which one of the two predecessors of each resolvent contains no negative literals.*

Now suppose that we could establish, for some instantiation invariant P, the theorem: a finite set S' of *ground* clauses is unsatisfiable if and only if there is a deduction D' of $\square$ from S', where D' has the property P. It would immediately follow that *any* finite set S of clauses is unsatisfiable if and only if there is a deduction D from S, where D has the property P. This is because, by Herbrand's Theorem, (see [5], end of §2), a finite set S of clauses is unsatisfiable if and only if some finite set S' of ground instances of clauses in S is unsatisfiable. We can in fact establish such a theorem when, for example, P is the property mentioned above: *being a deduction in which one of the two predecessors of each resolvent contains no negative literals.* Let us call this propery P_1; and to save writing, let us call such resolvents *P_1-resolvents,* and such deductions *P_1-deductions.* Since the 'only if' part of the theorem is trivial, we need only prove the following:

> *Theorem 1. If S is a finite unsatisfiable set of ground clauses, then there is a P_1-deduction of $\square$ from S.*

From this theorem there will immediately follow, by the above remarks:

> *Theorem 2. If S is a finite unsatisfiable set of clauses, then there is a P_1-deduction of $\square$ from S.*

Now in order to prove Theorem 1, it will suffice to prove the following:

> *Lemma. If S is a finite unsatisfiable set of ground clauses which does not contain $\square$, then there is a P_1-resolvent of S which is not in S.*

Here, a P_1-resolvent of S is simply a P_1-resolvent of some pair of clauses in S.

Theorem 1 follows from the Lemma because S, being finite, can involve only finitely many literals, and any resolvent of S will be composed of some subset of those literals. Repeated application of the Lemma then produces a sequence S_0, S_1, . . ., (where S_0 is S, and, for $j \geqq 0$, S_{j+1} is just $S_j \cup \{R_j\}$, R_j being the P_1-resolvent of S_j provided by the Lemma) of unsatisfiable sets; and since only finitely many distinct R_j are possible, this sequence must terminate in a set which contains □. Tracing the genesis of □ back to members of S then provides the required P_1-deduction. It then remains to prove the Lemma. To facilitate our doing so, let us call clauses which do not contain any negated literals *positive*, and clauses which contain at least one negated literal *negative*.

Proof of the Lemma. Let A be the set of positive clauses in S. Let $L_1, \ldots, L_n$ be the distinct atoms which occur, or whose negations occur, in clauses in S. By a *model* we shall mean a sequence $M_1, \ldots, M_n$ in which each M_i is either L_i or $\bar{L}_i$ (the negation of L_i). Note that A is non-empty (for otherwise the model $\bar{L}_1, \ldots, \bar{L}_n$ would satisfy S) and satisfiable (e.g., by the model $L_1, \ldots, L_n$). Now let M be a model which satisfies A and which contains the minimum number of unnegated literals that any such models contain. Note that this number is greater than zero. Then M falsifies one or more clauses in $S-A$. Let K be such a clause, and let K contain the minimum number of negated literals that any such clause contains. Note that this number is greater than zero. Now let $\bar{L}$ be one of the negated literals in K. There must exist at least one clause in A which contains L and in which all other atoms are negated in M. (For otherwise the model $\bar{M}$, obtained from M by replacing L by $\bar{L}$ would be a model of A containing fewer unnegated literals tham M). Let J be such a clause. Then

$$R = (J-\{\bar{L}\}) \cup (K-\{\bar{L}\})$$

is a P_1-resolvent of J and K, hence of S. If R is □ then R is not in S by the hypothesis of the Lemma. If R is not □ and is positive then R is not in A because R is false in M; hence R is not in S. If R is not □ and is negative, then R is not in $S-A$ because R is false in M and contains fewer negated literals than K; hence R is not in S. This completes the proof of the Lemma.

In the next section we introduce the concept of hyper-resolution.

§4. Consider P_1-deductions of the special form illustrated in the diagram of Fig. 4, in which the clauses $A_0, \ldots, A_m$ and C are positive, while $B_1, \ldots, B_m$ and B are negative. The special form consists of the fact that there is only one initial node at which a negative clause occurs. We call a P_1-deduction which has this special form a *P_2-deduction.* For brevity, let us call a node in a deduction *positive* if the clause at that node is a positive clause, and *negative* if the clause at that node is negative.

Now it is not difficult to see that every P_1-deduction D of $\square$ can be uniquely decomposed into subdeductions which are P_2-deductions whose negative initial nodes are also initial nodes of D. This is done simply by entering the tree D at each negative initial node B and

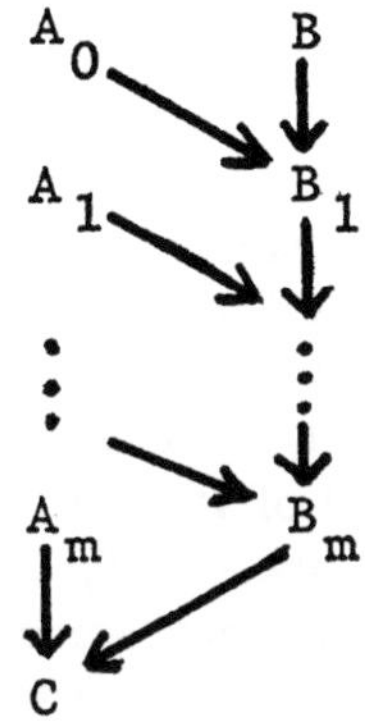

FIG. 4

tracing the chain of arrows downwards from B until the first positive node is reached. Indeed, this shows that there is one such subdeduction for each negative initial node of D.

Now let us say that C is a *hyper-resolvent of* $A_0, \ldots, A_m$, *and* B if C and $A_0, \ldots, A_m$ are positive clauses, B is a negative clause, and there

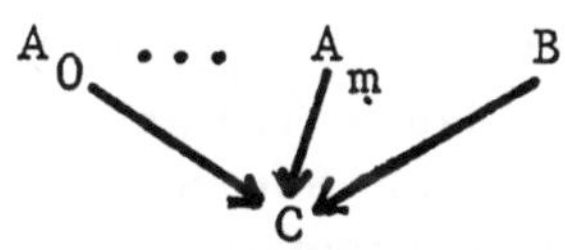

FIG. 5

is a P_2-deduction of C from $A_0, \ldots, A_m$ and B. We have in mind a direct relationship, as indicated in the diagram in Fig. 5, out of which tree deductions can be built. Specifically, we shall say a tree D is a *hyper-deduction* of a clause C from a set S of clauses, just in case C is the clause at the terminal node of D, the clause at each initial node of D is in S, and the clause at each noninitial node of D is a hyper-resolvent of the clauses at its immediate predecessor nodes.

Theorem 2, and the fact that every P_1-deduction can be decomposed into P_2-deductions in the way described above, are sufficient to establish:

Theorem 3. If S is a finite unsatisfiable set of clauses, there is a hyper-deduction of $\square$ *from S.*

Provided that we can give a suitable algorithm for computing the *hyper-resolution* $\tilde{\mathscr{R}}(S)$ of a finite set S of clauses (i.e., the set S together with all the nontrivial hyper-resolvents of members of S) we see that Theorem 3 provides the basis for a family of automatic deduction procedures. Such an algorithm is easy to give: let S_0 be S; then, for $j \geqq 0$, compute the sets B_j and A_j as, respectively, the set of all negative P_1-resolvents of S_j, and the set of all positive P_1-resolvents of S_j which are not mere instances of members of S_j; and let S_{j+1} be $S_j \cup B_j$.

This computation must terminate, since the maximum number of negative literals in any clause in B_j decreases by one as j increases by one, hence, for some j, B_j is empty, and $S_{j+1} = S_j$. For this j, the set $A = A_0 \cup \ldots \cup A_j$ is the set of all nontrivial hyper-resolvents of S (i.e., those that are not mere instances of members of S), and we then have: $\tilde{\mathscr{R}}(S) = S \cup A$. We can then summarize the result of the paper as follows: if the finite set S of clauses is unsatisfiable, then, for some $n \geqq 0$, $\tilde{\mathscr{R}}^n(S)$ contains $\square$.

§5. The motivation behind this result is that of cutting out as much as possible of the unnecessary computing in automatic deduction procedures. It will be noted that, in hyper-resolution computations, we never retain *any* negative clauses. The only negative clauses in $\tilde{\mathscr{R}}^n(S)$, for all n, are those which are in S itself. The role of negative clauses is reduced to that of purely temporary scratchpad entities which exist "inside" the $\tilde{\mathscr{R}}$ computations but are not part of their output.

We conclude with two examples of the complete output of hyper-resolution computations, which give some feel for what reductions are achieved compared with earlier automatic deduction algorithms. Both examples were carried out by hand in a total of fifteen minutes or so.

Example 1. S is the set:

$\{P(x,y)\}$, $\{\bar{P}(y,f(x,y)),\ \bar{P}(f(x,y),f(x,y)),\ Q(x,y)\}$,
$\{\bar{P}(y,f(x,y)),\ \bar{P}(f(x,y),f(x,y)),\ \bar{Q}(x,f(x,y)),\ \bar{Q}(f(x,y),f(x,y))\}$.

$\tilde{\mathscr{R}}(S)$ is obtained by adding $\{Q(x,y)\}$, and $\tilde{\mathscr{R}}^2(S)$ by adding $\square$. This example has been discussed by several authors. See [1, 2, 3, 4, 5].

Example 2. S is the set:

$\{P(a,b,c)\}$, $\{P(x,x,e)\}$, $\{P(x,e,x)\}$, $\{P(e,x,x)\}$, $\{\bar{P}(b,a,c)\}$,
$\{\bar{P}(x,y,u), \bar{P}(y,z,v), \bar{P}(x,v,w), P(u,z,w)\}$,
$\{\bar{P}(x,y,u), \bar{P}(y,z,v), \bar{P}(u,z,w), P(x,v,w)\}$.

We get $\tilde{\mathscr{R}}(S)$ by adding $\{P(a,c,b)\}$, $\{P(c,b,a)\}$; $\tilde{\mathscr{R}}^2(S)$ by adding $\{P(c,a,b)\}$, $\{P(b,c,a)\}$; $\tilde{\mathscr{R}}^3(S)$ by adding $\{P(b,a,c)\}$; and finally we get $\tilde{\mathscr{R}}^4(S)$ by adding $\square$.

This example is the formulation in our logic of the theorem that if a semigroup G is such that the square of each element in G is the identity element of G, then G is commutative. In [6] the authors report that with this example their CDC3600 program took 30 minutes to compute the first and second resolutions of S before the run was abandoned. With some ingenious reductive strategies added to the program they were able to get a deduction of $\square$ very rapidly, but even so 119 clauses (with one strategy) and 71 clauses (with a different strategy) were retained in doing so, in contrast to the 5 clauses retained here when no reductive strategies are added to the basic computation. This example, however, is most effective as an illustration of the difference in efficiency between the basic resolution method and the derived hyper-resolution method.

References

1. Davis, M. and Putnam, H. A computing procedure for quantification theory. *Journal of the Association for Computing Machinery* 7 (1960), 201–215.
2. Friedman, J. A semi-decision procedure for the functional calculus. *Journal of the Association for Computing Machinery* **10** (1963), 1–24.
3. Gilmore, P. C. A proof method for quantification theory. *IBM Journal of Research and Development* **4** (1960), 28–35.
4. Robinson, J. A. Theorem-proving on the computer. *Journal of the Association for Computing Machinery* **10** (1963), 163–174.
5. Robinson, J. A. A machine-oriented logic based on the resolution principle. *Journal of the Association for Computing Machinery* **12** (1965) 23–41. in the January 1965 issue).
6. Wos, L., Carson, D., and Robinson, G. The unit-preference strategy in theorem-proving. 1964 *Fall Joint Computer Conference*, AFIPS.

An Algorithm for a Machine Search of a Natural Logical Deduction in a Propositional Calculus

N. A. Shanin, G. V. Davydov, S. Yu. Maslov, G. E. Mints,
V. P. Orevkov, A. O. Slisenko

Abstract

The problem of computer modelling of complex forms of human intellectual activity is of central importance among various problems of the mathematical cybernetics. Attention is paid in particular to the problem of computer modelling of theorem proof search in mathematical theories. This brochure is a short communication of results of concrete investigations in this direction done at the Leningrade division of the V.A. Steklov Mathematical institute.

Main principles of the algorithm developed by authors for finding natural logical deduction of tested formulas from given initial formulas by means of the classical propositional calculus are presented in the brochure. When the tested formula is derivable from given initial formulas by the mentioned logical means, the algorithm constructs a natural logical deduction (i.e. deduction having the form in general use in mathematics). The brochure contains also brief data concerning computer program implementing this algorithm as well as examples of logical deductions constructed by computer.

The brochure is intended for mathematicians (scientific workers, undergraduate and graduate students) as well as everybody interested in mathematical logic and mathematical cybernetics.

§ 1. This communication presents in brief some data concerning the algorithm for natural deduction search (ALPEV-LOMI-2) and corresponding computer program developed at the Leningrad division of V.A. Steklov Mathematical Institute (LOMI) of the Soviet Academy of Sciences*.**

---

* Initial ideas of the algorithm were suggested by N.A. Shanin in his lecture at the IV Soviet mathematical congress in 1961. A number of improvements in the principal scheme of the algorithm was successively suggested in a series of lectures presented by N.A. Shanin during 1962 to the Leningrad mathematical logic seminar. Further improvement of the principal scheme of the algorithm, detailed development of all its parts and of mechanics of their joint functioning was done by the collective consisting of all authors of this communication.

The principles of the construction of the final version of this algorithm were first time presented to the Leningrad Mathematical Society 12.3., 19.3. and 9.4.1963 and to the V Soviet General Algebra Colloquium in May 1963.

A program implementing developed algorithm was written for URAL-4 computer. It was finally debugged in March 1964. At the same time main experiments with the program were carried out. Program was written by N.A. Beljaeva, Ç.V. Davydov, S.Ju. Maslov, G.E. Minc, V.P. Orevkov, A.O. Slisenko, A.V. Sochilina (Rukolaine), I. Gavel, M. Benesova (Vachsmanova) and M. Krčalova (Hajkova) also took part.

Lectures on theoretical foundations of the program including demonstration of results of some experiments were presented to the mathematical logic seminar of the Moscow University (April 1964) and I Soviet computer proof-search symposium at Trakai (July 1964). Some results of computer experiments were also demonstrated at the meeting of the Leningrad Mathematical Society (November 1964) and at the meeting of the joint seminar of the Leningrad division of V.A. Steklov Mathematical Institute.

** ALPEV is Russian abbreviation for "algorithm for natural deduction search" - Translator's note.

If a list of propositional formulas $Q_1, \dots, Q_n$ is given together with some additional formula R, then the algorithm to be described finds out whether R is derivable from $Q_1, \dots, Q_n$ by logical means of the classical propositional calculus (i.e. whether R is derivable in the calculus which is obtained by adding $Q_1, \dots, Q_n$ as new initial formulas to the classical propositional calculus). If the answer is positive, the algorithm constructs (and this is its main task) a natural deduction of R from $Q_1, \dots, Q_n$. Computer program mentioned above implements this algorithm on the URAL-4 computer under some restrictions on some quantitative characteristics of formula list $Q_1, \dots, Q_n, R$ (see § 3 below).

The problem of constructing computer programs finding out whether R is derivable form $Q_1, \dots, Q_n$ by means of the classical propositional calculus (without producing natural deduction) is much simpler than one considered here. The programs of the former kind have been constructed by many authors.

Main aim of the authors of this communication was the development of the algorithm constructing "as natural as possible" deductions in the logical framework of the classical propositional calculus. The authors of this communication can only partially make precise the requirement of being "as natural as possible" (see below). Satisfaction of this "intuitive" requirement by various specific logical deductions has been established on the base of revealing the claims concerning "quality" of derivations which could have been made by working mathematicians comparing to the "practice" of human derivations. Such claims were sources of successive improvements of the algorithm. * (note see next side)

Speaking of construction of a "natural" logical deduction we have in mind "sufficiently compact" deduction** having a form familiar in mathematics and proceeding according to logical inference rules in general use in mathematics. Formalized version of the way of presenting proofs generally used in classical mathematics was suggested by G. Gentzen [Math. Zeitschr., 39, N 2, 176-210 (1934)] in the form of a specific calculus which he called classical natural deduction calculus. In this calculus such generally used method as assumption introduction is formalized. All logical inference rules of this calculus are familiar for mathematicians.

Classical natural deduction calculus is equivalent to more traditional versions of the classical first order predicate calculus.*** The part of the classical natural deduction calculus where individual variables are absent (both from the language and from the logical deduction apparatus) will be called below propositional natural deduction calculus.

---

* Logical means of the classical propositional calculus form only a part of the logical means used ordinarily in mathematics as a base for constructing axiomatically presented theories. So the problem the authors of this communication tried to solve should be thought of as a preparatory one for much more complicated problem of developing an algorithm searching "as natural as possible" logical derivations in the logical framework of the classical (first order) predicate calculus.

** Compactness of the final result of the algorithm is meant.

*** This means that the sets of formulas derivable in the calculuses mentioned here are equal.

Quantifiers (i.e. logical connectives "for any" and "there exists") are absent from this calculus and it is equivalent to more traditional versions of the classical propositional calculus.

The calculus obtained by adding formulas $Q_1,\dots,Q_n$ to the propositional natural deduction calculus as additional initial formulas will be called propositional natural deduction calculus with specific axioms $Q_1,\dots,Q_n$ (precise description of this calculus is given below). Introduction of such extensions of the natural deduction calculus reflects (to a certain extent) the method of introducing specific axiomatic theories on the base of given logical means.

Speaking below of constructing natural deduction of a formula R from formulas $Q_1,\dots,Q_n$ we always have in mind "sufficiently compact" deduction of R in the propositional natural deduction calculus with specific axioms $Q_1,\dots,Q_n$.

§ 2 Propositional natural deduction calculus with specific axioms $Q_1,\dots,Q_n$ is deseribed as follows.

Primitives of the language of the propositional natural deduction calculus are:

1) propositional variables, i.e. symbolic expressions $t_1,t_2,t_3,\dots$;
2) logical connectives, i.e. signs $\neg$("not"), & ("and"), v ("or"), $\supset$("if...then"), $\equiv$("if and only if");
3) brackets;
4) sequent sign $\rightarrow$.

Propositional Formulas (or simply formulas) are defined by following construction rules:

a) any propositional variable is a formula;

b) if U is a formula then the word $\neg U$ is also a formula;

c) if U, V are formulas then the words $(U \& V)$, $(U \vee V)$, $(U \supset V)$, $(U \equiv V)$ are also formulas.

Statement "Under assumptions $U_1, \ldots, U_K$ holds V" is symbolized by the expression

$$U_1, \ldots, U_K \rightarrow V$$

which is called sequent (the list $U_1, \ldots, U_K$ may be empty here).

Statement "Assumptions $U_1, \ldots, U_n$ are contradictory" is symbolized by the expression

$$U_1, \ldots, U_K \rightarrow$$

which is also called sequent. Objects generated by the calculus we describe (derivable in the calculus) are sequents.

The kinds of sequents employed below in the description of the calculus will be characterized by symbolic expressions called sequent schemata. A sequent scheme becomes specific sequent when some admissible values are substituted in the scheme instead of all scheme variables present in this scheme. As scheme variables we shall use:

1) X,Y,Z [all formulas are admissible values];

2) $\Gamma, \Sigma, \Pi, \Gamma', \Sigma', \Pi', \Gamma'', \Sigma'', \Pi'', \ldots$ [admissible values are all lists of formulas including empty list, that is empty word];

3) Δ [admissible values are all formulas and the empty word].

Initial sequents.

Initial sequents are of the three kinds.

0. Initial sequents corresponding to formulas $Q_1,\ldots,Q_n$: these are sequents

$$\rightarrow Q_1,\ldots,\rightarrow Q_n.$$

1. Trivial initial sequents given by the sequent scheme

$$X \rightarrow X.$$

Note. Introduction of a sequent U→U for some formula U into a derivation reflects the act of introducing assumption U. Sequent U→U is read: "Under assumption U holds U." So the assumption U is introduced in the "working part" of a derivation (to the right of the arrow) as well as in the "memory" of the derivation (to the left of the arrow).

2. Excluded middle kind of initial sequents: These are sequents given by sequent schemata

$$\rightarrow Y \vee \neg Y, \ \rightarrow \neg Y \vee Y.$$

Logical inference rules.

Following rules are logical inference rules.

3. Assumption discharge rule

$$\frac{\Gamma, X, \Sigma \rightarrow Y}{\Gamma, \Sigma \rightarrow (X \supset Y)}$$

4. Premiss addition rule

$$\frac{\Gamma \rightarrow Y}{\Gamma \rightarrow (X \supset Y)}$$

5. Arbitrary conclusion introduction rules

$$\frac{\Gamma \rightarrow \neg X}{\Gamma \rightarrow (X \supset Y)} \qquad \frac{\Gamma \rightarrow X}{\Gamma \rightarrow (\neg X \supset Y)}$$

6. Detachment rule

$$\frac{\begin{matrix}\Gamma \rightarrow X \\ \Sigma \rightarrow (X \supset Y)\end{matrix}}{\Gamma, \Sigma \rightarrow Y}$$

7. Two-premiss cut rule

$$\frac{\begin{matrix}\Gamma \rightarrow X \\ \Sigma, X, \Pi \rightarrow Y\end{matrix}}{\Sigma, \Gamma, \Pi \rightarrow Y}$$

8. Three-premiss cut rule

$$\frac{\begin{array}{c}\Gamma \to X \\ \Sigma \to Y \\ \Pi', X, \Pi'', Y, \Pi''' \to Z\end{array}}{\Pi', \Gamma, \Pi'', \Sigma, \Pi''' \to Z}$$

9. Implication premiss refutation rules

$$\frac{\begin{array}{c}\Gamma \to \neg Y \\ \Sigma \to (X \supset Y)\end{array}}{\Gamma, \Sigma \to \neg X} \qquad \frac{\begin{array}{c}\Gamma \to Y \\ \Sigma \to (X \supset \neg Y)\end{array}}{\Gamma, \Sigma \to \neg X} \qquad \frac{\begin{array}{c}\Gamma \to \neg Y \\ \Sigma \to (\neg X \supset Y)\end{array}}{\Gamma, \Sigma \to X} \qquad \frac{\begin{array}{c}\Gamma \to Y \\ \Sigma \to (\neg X \supset \neg Y)\end{array}}{\Gamma, \Sigma \to X}$$

10. Conjunction-introduction rule

$$\frac{\begin{array}{c}\Gamma \to X \\ \Sigma \to Y\end{array}}{\Gamma, \Sigma \to (X \& Y)}$$

11. Conjunction-elimination rules

$$\frac{\Gamma \to (X \& Y)}{\Gamma \to X} \qquad \frac{\Gamma \to (X \& Y)}{\Gamma \to Y}$$

12. Conjunction negation rules

$$\frac{\Gamma \to \neg(X \& Y)}{\Gamma \to (\neg X \lor \neg Y)} \qquad \frac{\Gamma \to \neg(\neg X \& Y)}{\Gamma \to (X \lor \neg Y)} \qquad \frac{\Gamma \to \neg(X \& \neg Y)}{\Gamma \to (\neg X \lor Y)} \qquad \frac{\Gamma \to \neg(\neg X \& \neg Y)}{\Gamma \to (X \lor Y)}$$

13. Disjunction-introduction rules

$$\frac{\Gamma \to X}{\Gamma \to (X \lor Y)} \qquad \frac{\Gamma \to Y}{\Gamma \to (X \lor Y)}$$

14. The rule of arguing by cases

$$\frac{\begin{array}{c}\Gamma \to (X \lor Y) \\ \Sigma', X, \Sigma'' \to Z \\ \Pi', Y, \Pi'' \to Z\end{array}}{\Gamma, \Sigma', \Sigma'', \Pi', \Pi'' \to Z}$$

15. Disjunctive member detachment rules

$$\frac{\begin{array}{c}\Gamma \to \neg X \\ \Sigma \to X \lor Y\end{array}}{\Gamma, \Sigma \to Y} \qquad \frac{\begin{array}{c}\Gamma \to X \\ \Sigma \to (\neg X \lor Y)\end{array}}{\Gamma, \Sigma \to Y} \qquad \frac{\begin{array}{c}\Gamma \to \neg Y \\ \Sigma \to (X \lor Y)\end{array}}{\Gamma, \Sigma \to X} \qquad \frac{\begin{array}{c}\Gamma \to Y \\ \Sigma \to (X \lor \neg Y)\end{array}}{\Gamma, \Sigma \to X}$$

16. Disjunction negation rules

$$\frac{\Gamma \to \neg(X \vee Y)}{\Gamma \to (\neg X \& \neg Y)} \qquad \frac{\Gamma \to \neg(\neg X \vee Y)}{\Gamma \to (X \& \neg Y)} \qquad \frac{\Gamma \to \neg(X \vee \neg Y)}{\Gamma \to (\neg X \& Y)} \qquad \frac{\Gamma \to \neg(\neg X \vee \neg Y)}{\Gamma \to (X \& Y)}$$

17. Equivalence-introduction rule

$$\frac{\begin{array}{l}\Gamma \to (X \supset Y)\\ \Sigma \to (Y \supset X)\end{array}}{\Gamma, \Sigma \to (X \equiv Y)}$$

18. Equivalence-elimination rules

$$\frac{\Gamma \to (X \equiv Y)}{\Gamma \to (X \supset Y)} \qquad \frac{\Gamma \to (X \equiv Y)}{\Gamma \to (Y \supset X)}$$

19. Equivalence negation rule

$$\frac{\Gamma \to \neg(X \equiv Y)}{\Gamma \to (\neg(X \supset Y) \vee \neg(Y \supset X))}$$

20. Rule of refuting by contradiction

$$\frac{\Gamma, X, \Sigma \to}{\Gamma, \Sigma \to \neg X}$$

21. Rule of inference ad absurdum

$$\frac{\Gamma, \neg X, \Sigma \to}{\Gamma, \Sigma \to X}$$

22. Rule of stating the contradiction

$$\frac{\begin{array}{l}\Gamma \to X\\ \Sigma \to \neg X\end{array}}{\Gamma, \Sigma \to}$$

23. Contraction rule

$$\frac{\Gamma, X, \Sigma, X, \Pi \to \Delta}{\Gamma, X, \Sigma, \Pi \to \Delta}$$

24. Permutation rule

$$\frac{\Gamma, X, \Sigma, Y, \Pi \to \Delta}{\Gamma, Y, \Sigma, X, \Pi \to \Delta}$$

We denote the calculus described above by $И_2[Q_1,\dots,Q_n]$. Formula U is derivable in the calculus $И_2[Q_1,\dots,Q_n]$ if the sequent $\rightarrow U$ is derivable there. The calculus $И_2[Q_1,\dots,Q_n]$ has some redundances: it is equivalent to calculus $И_1[Q_1,\dots,Q_n]$ containing only part of initial sequents and inference rules of $И_2[Q_1,\dots,Q_n]$, namely initial sequents of the kinds (o) and (1) and inference rules (3), (4), (6), (10), (11), (13), (14), (17), (18), (20), (21), (22), (23).

Every sequent having the form of the law of excluded middle is derivable in $И_1[Q_1,\dots,Q_n]$ and an application of any one of the rules (5), (7), (8), (9), (12), (15), (16), (19) can be replaced by a chain of the inferences according to rules of $И_1[Q_1,\dots,Q_n]$. Inclusion of the latter rules allows however to avoid excessive fragmentation of logical derivation constructed by algorithm. Rule (24) was introduced for purely technical reasons: it makes programming easier eliminating the necessity to observe definite order of members in assumption lists.

§ 3 Computer program implementing ALPEV-LOMI-2 in the URAL-4 computer* contains over 15 000 instructions in the machine code. The program uses 3 magnetic drums in addition to operative memory. Input consists of the code of a "proof search strategy" (see below; considerable number of "strategies" is provided for in the algorithm), initial formulas and the formula to be derived. All input formulas are first of all rewritten into bracket-free (Polish) notation. The concluding step of the programm execution consists in rewriting the output into usual notation.

---

* The programmers have been listed in § 1.

The sum of lengths (in letters) of bracket-free translations of all input formulas should not exceed 63. Total number of different propositional variables in these formulas is not greater than 7. The number of initial formulas is not greater than 4. Input formulas should not contain successive negation signs. All these restrictions are of purely technical nature and are not connected with the principal scheme of the algorithm.

If $Q_1,\ldots,Q_n$ are initial formulas, $n \leq 4$, R is the formula to be tested and the conditions above are satisfied, then for any given "strategy" the program constructs a deduction of R in the calculus $И_2[Q_1,\ldots,Q_n]$ if R is derivable in this calculus, and gives the answer "THE FORMULA IS NOT DERIVABLE" otherwise.

The steps according to contraction rule (23) and permutation rule (24) are not listed explicitly when logical deductions are printed by computer. Every reference to some logical inference rule is to be understood as follows: the sequent printed at the given step is obtained from previously derived sequents listed in this reference by application of the rule mentioned in the reference and possibly single or multiple applications of rules (23), (24). If the logical inference rule referred to has several modifications, one of them suitable in this case is understood. Computer produces a reference to the rule without specifying modification of the rule applied at the given step.

An abbreviation is used in recording of sequents in deductions: to the left of $\rightarrow$ sign (i.e. in the assumption list) special notations for formulas are used instead of formulas themselves. Any notation consists of a letter Д (from Допущение = assumption) and the number of the given assumption in the sequence of the introduced assumptions.

An experiment was conducted with the program. During the experiments some inputs differed only by the choice of a "strategy". In some of these cases the deductions turned out to be identical, in other cases they were different. Total number of deductions obtained during the experiment is more than 50.

At the end of this communication 5 uncomplicated examples of logical derivations constructed by computer are presented (see Supplement).

Some results obtained during experiment with the program show that the deduction printed may contain some outright "unnatural features". It may turn out for example that some of the assumptions and initial sequents introduced into the deduction are used only after a considerable number of steps. This is explained to a considerable degree by the difficulty of purposeful transformation of a complicated graph into a column (the deduction is constructed by the algorithm in the form of a graph). Examples presented in the Supplement illustrate well the character of "unnatural features" occurring in some deduction constructed by computer.

§ 4 Let us pass to brief description of the principal foundation of the algorithm. Starting point of the algorithm development was the calculus $Э_В 1$ (the calculus of "heuristical [ЭВРИСТИЧЕСКОГО] proof-search")* presented below. Description of this calculus uses the same symbolism as one used in the descirption of $И_2[Q_1,...,Q_n]$.

---

* Reasons suggesting this name of the calculus will be mentioned below.

Initial sequents of $\partial\mathcal{B}1$ are given by the following sequent schemata:

$$\alpha_1)\quad \Gamma, X, \quad \Sigma \to X;$$

$$\alpha_2)\quad \Gamma, X, \Sigma, \neg X, \Pi \to \Delta \qquad\qquad \alpha_3)\quad \Gamma, \neg X, \Sigma, X, \Pi \to \Delta.$$

Logical inference rules of $\partial\mathcal{B}1$ are the following rules:

$$\beta_1)\quad \frac{\Gamma, X \to}{\Gamma \to \neg X} \qquad\qquad \beta_2)\quad \frac{\Gamma, X, \Sigma \to \Delta}{\Gamma, \neg\neg X, \Sigma \to \Delta}$$

$$\gamma_1)\quad \frac{\Gamma \to X \qquad \Gamma \to Y}{\Gamma \to (X \& Y)}$$

$$\gamma_2)\quad \frac{\Gamma, X, Y, \Sigma \to \Delta}{\Gamma, (X \& Y), \Sigma \to \Delta} \qquad\qquad \gamma_3)\quad \frac{\Gamma, \neg X, \Sigma \to \Delta \qquad \Gamma, \neg Y, \Sigma \to \Delta}{\Gamma, \neg(X \& Y), \Sigma \to \Delta}$$

$$\delta_1)\quad \frac{\Gamma, \neg X \to Y}{\Gamma \to (X \vee Y)}$$

$$\delta_2)\quad \frac{\Gamma, X, \Sigma \to \Delta \qquad \Gamma, Y, \Sigma \to \Delta}{\Gamma, (X \vee Y), \Sigma \to \Delta} \qquad\qquad \delta_3)\quad \frac{\Gamma, \neg X, \neg Y, \Sigma \to \Delta}{\Gamma \neg (X \vee Y), \Sigma \to \Delta}$$

$$\varepsilon_1)\quad \frac{\Gamma, X \to Y}{\Gamma \to (X \supset Y)}$$

$$\varepsilon_2)\quad \frac{\Gamma, \neg X, \Sigma \to \Delta \qquad \Gamma, Y, \Sigma \to \Delta}{\Gamma, (X \supset Y), \Sigma \to \Delta} \qquad\qquad \varepsilon_3)\quad \frac{\Gamma, X, \neg Y, \Sigma \to \Delta}{\Gamma, \neg(X \supset Y), \Sigma \to \Delta}$$

$$\zeta_1)\quad \frac{\Gamma \to X \supset Y \qquad \Gamma \to Y \supset X}{\Gamma \to X \equiv Y}$$

$$\zeta_2)\quad \frac{\Gamma, (X \supset Y), (Y \supset X), \Sigma \to \Delta}{\Gamma \to (X \equiv Y), \Sigma \to \Delta}$$

$$\zeta_3)\quad \frac{\Gamma, \neg(X \supset Y), \Sigma \to \Delta \qquad \Gamma, \neg(Y \supset X), \Sigma \to \Delta}{\Gamma, \neg(X \equiv Y), \Sigma \to \Delta}$$

Below we shall use following terminology concerning sequents, sequent schemata and logical inference rules.

The part of sequent or of a sequent scheme situated to the left (to the right) of the $\rightarrow$ sign is called antecedent (or respectively succedent) of the sequent or sequent scheme. Consider some logical inference rule of the calculus $\mathcal{JB}$ 1. Sequent schemata situated over the line are called premisses of the rules and the sequent scheme below the line is called the conclusion of the rule. One and only one member of the conclusion of a rule is changed when one passes from conclusion of a rule up to premis(ses). This member is called main formula of the rule considered. *

A rule is antecedent (succedent) one if the only occurrence of the main formula which is transformed (during the passage mentioned above) is situated in the antecedent (respectively, succedent) of the conclusion of the rule. Counterinference according to a given rule applied to some sequent $\mathcal{S}$ is a construction of a finite set of sequents (one-element set if the rule has one premiss and two-element set in the case of two-premiss rule) from which $\mathcal{S}$ is obtained by inference according to this rule (if the set contains two elements, then the sequents from which the set consists are placed above the same horizontal line).

Realization of given rule is a figure obtained by substitution of all scheme variables of the rule by some admissible values of these variables. The definitions and terms introduced above for the rules are extended in a natural way to realizations of the rules. **

---

* Note that the term "main formula" refers here not to a propositional formula but to a formula scheme.

** We shall use similar definitions in appropriate situations also relative to rules of other calculi to be introduced below.

A sequent $\mathcal{S}$ is derivable in $\exists B1$ if and only if it is derivable in $И_2[\Lambda]$, where Λ is the empty list of formulas. Moreover, all logical inference rules of $\exists B1$ are invertible* and inference according to any one of this rules can be replaced by several inferences according to rules of $И_2[\Lambda]$. These properties of $\exists B1$ allow one (very simply in principle) to recognize whether a given formula R is derivable from initial formulas $Q_1,\ldots,Q_n$ by means of the classical propositional calculus, and when the answer is positive these properties allow one to obtain some deduction of R in $И_2[Q_1,\ldots,Q_n]$.

Indeed, R is derivable in $И_2[Q_1,\ldots,Q_n]$ if and only if the sequent $Q_1,\ldots,Q_n \rightarrow R$ is derivable in $И_2[\Lambda]$, and moreover, any derivation of the sequent $Q_1,\ldots,Q_n \rightarrow R$ can be extended to a derivation of R in $И_2[Q_1,\ldots,Q_n]$ by several inferences according to the rule (7). We shall construct succesive proof-search trees in $\exists B1$ for the sequent $Q_1,\ldots,Q_n \rightarrow R$ starting from this sequent itself. It is initial proof-search tree for itself. Any next construction step consists of counterinference according to one of the rules of calculus $\exists B1$ applied to a sequent situated at the end of one of branches in the tree already constructed, and in listing the result of this counterinference immediately above the end of the branch under consideration. **

---

* A logical inference rule is invertible if for any realization of this rule the sequent situated under the line is derivable in the calculus under consideration if and only if all sequents situated over the line are derivable.

** Each passage of this kind corresponds to the following "heuristical" train of thought: we shall be able to construct the derivation in $И_2[\Lambda]$ of the sequent considered at the given step, if we shall be able to construct derivations in $И_2[\Lambda]$ of simpler (in some respect) sequents situated above the line in the employed realization of a rule of $\exists B1$.

The proof-search tree construction process is finished when some initial sequent of ЭВ1 is situated at the end of each branch or it turns out that at the end of at least one branch a sequent is situated which is neither initial sequent of ЭВ1 nor argument for applying counterinference according to some rule of ЭВ1 (the latter is the case if and only if every member of the sequent is either propositional variable or of the form $\neg W$ where W is a propositional variable). Sequent $Q_1,\ldots,n \to R$ is derivable in $И_2[\Lambda]$ if and only if this process of successive proof-search tree construction is finished in the first of the ways described above. If this process is finished in the first way, then the tree obtained will be deduction tree in ЭВ1 and by means of insertions of some sequents if can be transformed in a tree-form deduction of the sequent $Q_1,\ldots,Q_n \to R$ in $И_2[\Lambda]$. As pointed out above this latter deduction can be extended to a derivation of R in the calculus $И_2[Q_1,\ldots,Q_n]$.

The deduction obtained in this primitive way will however have as a rule a number of defects as to naturality and compactness. Such a deduction can contain in particular a lot of redundant branches, the sequent of this deduction can contain redundant formulas, a lot of redundant assumptions can be introduced and some segments of the deduction can look very unnatural. In this connection the principal scheme forming the foundation of the algorithm ALPEV-LOMI-2 has been obtained by means of considerable qualitative changes and improvements of the process just described.

§ 5 The aim of the introduced improvements is both to achieve the most possible compactness of the derivations output by the algorithm and to remove such segments of deduction which can seem unnatural to a mathematician. Let us mention some of these improvements.

a) To remove some of the possible "unnaturalnesses" of some segments of the output derivation we use instead of the rules (γ_3), (δ_2), (ε_2), (ζ_3) in some cases during the tree extension procedure for empty-succedent conclusion the rules *

γ_3') $\dfrac{\Gamma,\Sigma\to X \qquad \Gamma,\Sigma\to Y}{\Gamma,\neg(X\&Y),\Sigma\to}$ $\qquad$ δ_2') $\dfrac{\Gamma,\Sigma\to\neg X \qquad \Gamma,\Sigma\to\neg Y}{\Gamma,(X\vee Y),\Sigma\to}$

ε_2') $\dfrac{\Gamma,\Sigma\to X \qquad \Gamma,\Sigma\to\neg Y}{\Gamma,(X\supset Y),\Sigma\to}$ $\qquad$ ζ_3') $\dfrac{\Gamma,\Sigma\to X\supset Y \qquad \Gamma,\Sigma\to Y\supset X}{\Gamma,\neg(X\equiv Y),\Sigma\to}$

To restrict applications of the rule (β_2) during the construction of proof-search trees the algorithm uses together with the rules of $\mathfrak{F}B1$ calculus and the rules just introduced also some modifications of these rules. Together with the rule (γ_2) one uses for example rule

$$\frac{\Gamma,\neg X,\Sigma\to\Delta \qquad \Gamma,Y,\Sigma\to\Delta}{\Gamma,\neg(X\&\neg Y),\Sigma\to\Delta}$$

as well as two other similar modifications of the rule (γ_3). Together with the rule (δ_1) one uses rules

$$\frac{\Gamma,X\to Y}{\Gamma\to(\neg X\vee Y)} \qquad\qquad \frac{\Gamma,Y\to X}{\Gamma\to X\vee\neg Y}$$

---

* Footnote is on the following page

By $\ni_B 2$ we denote below the calculus obtained by adding to $\ni_B 1$ rules (γ_3'), (δ_2'), (ε_2'), (ζ_3') as well as the rules obtained by means of the just mentioned modifications of the rules (γ_3), (δ_1), (δ_3), (ε_2), (ε_3), (δ_2') and (ε_2'). All rules of the calculus $\ni_B 2$ are invertible and so suitable for extending proof-search trees. A sequent is derivable in $\ni_B 2$ if and only if it is derivable in $\ni_B 1$.

---

* Formation of the final output of the algorithm is made by the "Assembling" block [see section e)]. The principles of its work are such that 1) introduction of similar modifications of one-premiss rules of the calculus $\ni_B 1$ when the succedent is empty, turns out to be superfluous (and even inexpedient); 2) if a proof-search tree is extended by a counterinference according to one of the rules (γ_3), (δ_2), (ε_2), (ζ_3) applied to a sequent with empty succedent and at least one of the sequents obtained as a result turns out to be initial sequent of $\ni_B 1$ of the type (α_2) or (α_3), then the change of the employed rule by a corresponding rule from the list (γ_3'), (δ_2'), (ε_2'), (ζ_3') is not expedient and is not made in our algorithm.

If in the construction of a proof-search tree we fail to use one of the rules (γ_3'), (δ_2'), (ε_2'), (ζ_3') instead of corresponding rule of $\ni_B 1$, then in the final output derivation one can obtain a segment consisting of inferences from some "contradictions" to other "contradictions". Such inferences are not ordinarily encountered in mathematical proofs.

Let us present an example of a proof-search tree in the calculus Эв 2. Here as well as in all the following examples the letters A, B, C, E stand for some pairwise distinct propositional variables and the letter Ω stands for some list of formulas.

Example 1

6. $C,\Omega,A,E\to A$ 7. $B,C,\Omega,A,E\to$ 13. $B,E,\Omega,A,E\to$

5. $\neg(A\&\neg B),C,\Omega,A,E\to$ 10. $\neg A,E,\Omega,A\to\neg E$ 12. $B,E,\Omega,A\to\neg E$

4. $\neg(A\&\neg B),C,\Omega,A\to\neg E$ 9. $\neg A,E,\Omega\to A\supset\neg E$ 11. $B,E,\Omega\to A\supset\neg E$

3. $\neg(A\&\neg B),C,\Omega\to(A\supset\neg E)$ 8. $\neg(A\&\neg B),E,\Omega\to A\supset\neg E$

2. $\neg(A\&\neg B),(C\vee E),\Omega\to(A\supset\neg E)$

1. $(\neg(A\&\neg B)\&(C\vee E));\Omega\to(A\supset\neg E)$

This example will be used below for illustration. *

---

* Numbers situated to the left of sequent occurrences in a tree will be used to denote these occurrences. The l-th antecedent member of k-th sequent will be denoted by a pair [k,l]. (Sequent members are enumerated from left to right, and the formula list denoted by Ω will be considered as a whole for the purposes of enumeration.) The succedent of the k-th sequent will be denoted by a pair (k,C). Such definitions will be used also in similar cases below.

Describing further improvements of the "primitive" proof-search algorithm presented at the end of § 4 we shall systematically have to deal with proof-search trees and later also with proof-search graphs. These trees and graphs are "growing above" from the unique "root" which is the sequent whose deduction is looked for. Let us fix terminology to be used in the treatment of such trees and graphs.* At this moment we shall restrict attention fo proof-search trees. When proof-search graphs will be introduced, the reader will have no difficulty in extending to them notions introduced below.

Consider some proof-search tree. Occurrences of sequents in this tree from which this tree structurally consists are called nodes of the tree. An end node is one having no overstructure above. The lowermost node is called initial one. An end node is called 1) closed if it is initial sequent of the calculus $\ni_{B}$ 2; 2) blind if it is not closed and at the same time cannot serve as initial datum for a counterinference according to any rule of the calculus $\ni_{B}$ 2; open, if it is neither blind nor closed.

In the tree from the example 1 the nodes 6, 7, 10, 13 are end nodes and 6 and 10 are closed end nodes.**

---

* This terminology is partially borrowed form the graph theory and partially suggested by specific features of proof-search trees and graphs as well as by the central role played in the later developments by processes of tree (graph) extension.

** The information about nodes 7 and 13 is insufficient to determine their type.

A tree is called closed (open) if all its end nodes are closed (or respectively if it has no blind end nodes but has at least one open end node).

Let B be some node of the tree considered which is not initial node, and let Φ be some member of the node B. Some node is situated immediately below B. This lower node will be called the predecessor of B in the tree considered. Let us find a member Ψ of the node B from which Φ "originated" as a result of a counterinference according to some rule of the calculus $\exists B$ **2**. Ψ will be called a predecessor of Φ. We shall distinguish following cases: 1) Φ is equal (literally) to its predecessor (i.e. the predecessor of Φ was not broken up during the mentioned counterinference): in this case Φ will be called an old member of the node B; 2) Φ was obtained as a result of breaking up its predecessor* during a counterinference: in this case Φ will be called a new member of the node B.

In the tree of example 1 the nodes 2, 3, 4, 5, 6, 7, 8, 9, 10, 11, 12, 13 have as their predecessors nodes 1, 2, 3, 4, 5, 5, 2, 8, 9, 8, 11, 12 respectively. Examples of old members of tree nodes: (2,C), (4,1), (13,2); they have as predecessors (1,C), (3,1), (12,2) respectively. Examples of new nodes of the tree are (2,1), (2,2), (3,2), (4,4), (4,C), (6,C), (7,1), (8,2), (13,5); they have as predecessors (1,1), (1,1), (2,2), (3,C), (3,C), (5,1), (5,1), (12,C) respectively.

---

* The term "breaking up of a formula" is applied also in cases when the transformation of the formula consists in removing one or two outermost occurrences of $\neg$ [see inference rules (β_1), (β_2)].

Let B be some non-end node of the tree under considera-tion and Φ be a member of B. A node or a pair of nodes is situated immediately above B. This node or any member of the pair of nodes will be called successor of the node B. Let B' be one of successors of B and Φ' be some member of B'. We shall say that Φ' is a successor of Φ in B' if Φ is the predecessor of Φ'. Main member of B is the member which is broken up when one passes from B to its successor (or successors).*

In the tree from the example 1 the nodes 1, 2, 3, 4, 5, 8, 9, 11, 12 have as their successors the nodes 2, 3 and 8, 4, 5, 6 and 7, 9 and 11, 10, 12, 13 respectively and main members of these nodes are (1,1), (2,2), (3,C), (4,C), (5,1), (8,1), (9,C), (11,C), (12,C) respectively. Successors of node members (1,1), (1,C), (2,1), (2,2), (2,C), (3,1), (3,2), (3,C), (4,C), (5,1), (5,2) are (2,1) and (2,2), (2,C), (3,1) and (8,1), (3,2) and (8,2), (3,C) and (8,C), (4,1), (4,2), (4,4) and (4,C), (5,5), (6,C) and (7,1), (6,1) and (7,2).

Below we consider trees where inferences are made according to rules of calculi having the same initial sequents as Эв 2 but different from Эв 2 as whole. However the rules of these calculi and the trees considered are such that for any passage in the tree, corresponding rule is unique and moreover for any member of a node one can find a unique member of a node situated immediately below from which the node considered was obtained during counterinference accor-ding to the rule considered. So all notions just introduced are extended automatically to such trees.**

---

* New members of successors of B are those members who have main member of B as a predecessor.

** These trees are different from ones considered up to now: some member of a node considered may turn out to have no descendants in descendants of this node or it may have descendant only in one of the descendants of the node.

b) Block "Pruning" is introduced to simplify proof search trees (under suitable conditions). Starting points for "pruning" are: 1) closed end nodes and 2) end nodes having repetitions of some formula.* [A formula X is said to have repetitions in a sequent $\mathcal{S}$ if $\mathcal{S}$ is of the one of the following forms: $\Gamma, X, \Sigma, X, \Pi \to \Delta$; $\Gamma, X, \Sigma \to \neg X$; $\Gamma, \neg X, \Sigma \to X$. The occurrences of X and $\neg X$ in $\mathcal{S}$ justifying the statement that X has repetitions in $\mathcal{S}$, are called repetitions of X in $\mathcal{S}$.] In the process of "pruning" all formulas in $\Gamma, \Sigma, \Pi, \Delta$ are deleted from initial sequents [i.e. ones of the form (α_1), (α_2) and (α_3)]; all repetitions of a formula except one are deleted from sequent having repetitions of this formula.** Then the formula deletion process is extended down. At each stage the next sequent is taken and those of its members which turn out to be redundant as a result of previous simplifications are deleted. The whole parts of the tree may turn out to be redundant as a result of such process; they are then deleted (as a whole). Let us describe induction step of the "pruning" process in more detail.

---

* In proof search graphs arising during the work of the block "Joining" [see section d) below] situations of the third kind calling the block "Pruning" can arise. This will be discussed below.

** If the node considered was obtained during the tree construction in the process of counterinference according to some one-premiss rule, then usually we delete, "old" repetition (i.e. the occurrence of the formula having repetitions or its negation which is the old member of the node); exceptions are only some cases involving the work of the block "Joining" and the calculus $\ni\!\mathcal{B}$ 2+ [see sections d) and e) below]. If the sequent considered is result of a counterinference according to a two-premiss rule, then "new" repetition is deleted in all cases.

Suppose the "pruning" process has been already executed with respect to the part of the tree situation over node B. Speaking below of descendants of the node B and of members of these descendants we shall mean the tree resulting from the execution of the "pruning" over the node B in the tree input to the "Pruning" block. Main member of a node B is deleted only in the following cases: (i) B has only one descendant, and this descendant has no new members; (ii) B has two descendants and at least one of these descendants has no new member (in this case redundant descendant of B is by the definition the descendant having a new member; if no descendant has a new member then any of them can be chosen as a redundant one). A member of B which is not a main member is deleted if it either has no descendants or has a descendant only in the redundant descendant of B. After deletion of all members of B which have to be deleted according to the rules indicated above, we delete from the tree also the redundant descendant of B as well as the whole part situated above it.

The tree obtained as a result of this process is subjected to some further transformation. If at least one member is deleted from the initial sequent of the tree input in the "Pruning" block, then everything situated above initial node is deleted and the resulting single sequent is considered to be the result of the "Pruning" block. The aim of such a radical operation is to retain the possibility of improving the "quality" of final output derivation applying the same "proof search strategy" [see section c) below] to the sequent containing only part of members of the tested sequent (the proof search process may take another more efficient turn).* If the "pruning" process described above stops before some node, then the transformation of the tree obtained is done as follows. An obstacle node [which is by definition the image of the node of input tree before which "pruning" have stopped] is being found as well as the main successor of the obstacle node [which is by definition the successor which is literally different from its preimage in the input and is not initial sequent of the calculus ЭВ 2).** If the obstacle node has no main successor, the execution of "Pruning" block is finished. If main successor exists, then the "Pruning" block deletes the whole part of the tree situated over the main successor of the obstacle node.*** If main successor has empty succedent, one more step is made in some cases: main successor is transformed by transferring its new member into succedent (with adding outer negation sign or deleting one)****.

---

* Such a radical deletion is not always expedient (more over it prolongs the proof search). A version of the algorithm without radical deletion is also possible. In this version "succedent filling" is done for some nodes having empty succedent (see below) instead of deleting part of the tree.

** Obstacle node is unique. Its main successor, if one exists, is also unique. This follows from the following features of the algorithm: 1) the passage from a proof search tree to its immediate successor is executed with the help of extension by means of a counterinference of a rule of ЭВ 2 applied to unique node; 2) "Pruning" block is run whenever possible.

*** Everything said in the footnote * is valid also for deletions of this kind.

**** Such a correction of the main successor of obstacle node has the same purpose as the introduction of rules (γ_3'), (δ_2'), (ε_2'), (ζ_3'). The principles of execution of "Assembling" block are such that this correction is useless (and is not done) if the preimage of the main successor in the input tree has empty succedent.

After executing "Pruning" block one obtains a tree with a bottom sequent which is derivable in ЭВ 2 if and only if the bottom sequent of the input tree is derivable. Some passages in the output tree may turn out to be made in disagreement with the rules of ЭВ 2. However the bottom sequent of the output tree is derivable in $И_2[\Lambda]$ (or in ЭВ 2, which is equivalent) if and only if all sequents situated at end nodes of the tree are derivable. Moreover any passage in the output tree as well as the passage from the bottom sequent of this new tree to the bottom sequent of the input tree can be replaced by several inferences in $И_2[\Lambda]$. Having this in mind one can say that the output tree is a proof search tree both for its own bottom sequent and for the bottom sequent of input tree.

The execution of "Pruning" block is illustrated by the following three examples. (Input proof search trees in ЭВ 2 are presented in the left column, the corresponding output trees resulting from execution of "Pruning" block are in the right column. Formulas deleted by "pruning" are underlined by dotted lines and the parts of a tree which turned out to be redundant are framed.)

Example 2

Input tree (left column):

A,B,E→A
———————
A,B→E⊃A   [A,C→E⊃A] (framed)
———————————————
A,(BvC) → E⊃A
———————————————
(A&(BvC))→(E⊃A)

Output tree (right column):

A → E ⊃ A
———————————————
A & (BvC)→ E ⊃A

Example 3

```
¬A→A        [¬C→A]
──────────────────
¬(A&C)→A        [B→A]
─────────────────────
 ((A&C)⊃B)→A                          → A
─────────────────                     ──────────────────
→(((A&C)⊃B)⊃A)                        →(((A&C)⊃B)⊃A)
```

Example 4

```
¬A,¬C→A
───────
¬(AvC)→A           B→A                →AvC          B→A
──────────────────────                ──────────────────
 ((AvC)⊃B)→A                           ((AvC)⊃B)→A
```

In the last example the intermediate tree (i.e. one subjected to operation of "succedent filling" for the main descendant of the obstacle node) is of the form

```
¬(AvC)→           B→A
─────────────────────
     ((AvC)⊃B)→A
```

c) Assume that some proof search tree has been constructed and some of its open end nodes is distinguished. This node we shall call the distinguished node. The tree can be extended over distinguished node in several ways because various members of this node can be chosen for counterinference according to a rule of the calculus ∃B 2. This coice *)
is executed by a special "Strategy" block using the proof-search strategy input into the algorithm before it is run as well as the information about distinguished node collected by the algorithm.

---

* The "quality" of the final output deduction can depend (sometimes in essential way) from the choice made.

Important ingredient of the information about distinguished node collected by the algorithm are the data concerning literal equality of subformulas of sequent's members. Let the "weights" of connectives $\neg, \&, \vee, \supset, \equiv$ be 0,1,1,1,1 1/2 respectively. Let the depth of given occurrence of the formula $\mathcal{V}$ in the formula $\mathcal{U}$ be the sum of "weights" of logical connectives which have to be "torn" to "uncover" considered occurrence of $\mathcal{V}$ in $\mathcal{U}$. The block "Data collection" finds out for every member of distinguished node whether some formula has two different occurrences in this member with sum of weights not exceeding 3. If the answer is positive for the member considered we say there is an identification in this member and the mentioned formula is called identification kernel. In this case the occurrences of the identification kernel justifying identification are marked in a special way. In addition to this the "Data collection" block finds out for every pair of members of distinguished node whether there is a formula having occurences both in the first and in the second member satisfying condition: the sum of depths of these occurrences is no more than 2. If the answer is positive for the pair considered we say that the members of this pair are related and the mentioned formula is called the relationship kernel. In this case the occurrences of the relationship kernel into both members justifying the relationship are marked in a special way. If several versions of marking or various choices of kernel are possible for some member (in the case of identification) or for pair of members (in the case of relationship), then each version is recorded separately.

Proof search strategies are based on a number of characteristics of sequent members and pairs of sequent members. This include in particular: the type of the rule of $\exists B$ 2 for which the member considered can be main formula (one- and two-premiss rules are distinguished); the place of the member considered (in antecedent or in succedent); the type of the outermost logical connective (negation of a binary logical connective is treated as an independent connective; outermost connectivesare taken into account together with the place of the member considered and favourable and unfavourable ones are distinguished; implication-antecedent and negation of implication-antecedent are considered unfavourable *); literal length of the member considered; whether this member is a fragment of the succedent of lowermost node of the tree **);

---

* If a counterinference breaks unfavourable logical connective then, generally speaking, the chain of inferences corresponding to this counterinference in the final output deduction will have a form not very familiar to mathematicians.

** By fragments of given member of a node B we mean here descendants of this member in descendants of B, descendants of these descendants in corresponding nodes etc. Taking suitable account of succedent fragments in a proof search strategy can help to avoid the situation where only antecedent members of the lowest node and their fragments are processed for considerable time (and during various stages of the proof search) without orientation into making explicit the connections between the structure of initial formulas $Q_1,\ldots,Q_n$ and one of the formula R to be derived from them.

the presence of identification in the member considered and the type of this identification (the type is given by pair consisting of depths of the marked identification kernel occurrences); literal length of the identification kernel; the presence and type of relationship between two members of sequent (the type of relationship is given like one of identification); literal length of relationship kernel; maximum and average of literal lengths of formulas forming pair of related formulas.

Specification of a concrete proof search strategy consists in establishing of some ordering among certain combinations of characteristics of sequent members and pairs of sequent members. Any admissible ordering defines preference rule for members and pairs of sequent members of the distinguished node during choice of the member or of the member pair taken as input for the next stage of the work of algorithm **. Considerable number of strategies is provided for in the algorithm. Strategy is fixed before algorithm is run and cannot be changed during execution.

When the proof-search strategy is specified, the "Strategy" block processes given proof-search tree and distinguished open end node in the following way *. It analyses the information concerning members and member pairs of the distinguished node and chooses the

---

* Distinguished node input into the "Strategy" block is always chosen among end nodes which either were constructed during immediately preceeding stage of algorithm execution, or became end nodes during the same stage as a result of deleting part of the tree (or the graph). So the process of constructing final tree (graph) is organized so that no new branch is formed until the processed branch is finished. (But completely finished branch can be deleted during subsequent "prunings".)

** We shall not present the method of forming the proof search strategy in this short communication.

most prefered version (or one of such versions, if there are several) according to preference rule specified in the proof-search strategy. In some cases a member of the distinguished node without identifications is chosen, in other cases a member containing identification is chosen and in some cases two related members are chosen.

In the first case the algorithm extends proof search tree over distinguished node by a counterinference according to a rule of $\ni B$ 2 corresponding to the chosen member. In the second and third cases the execution of the algorithm involves generally the "Merging" block.

d) "Merging" block uses identification in sequent members and relationships among sequent members to reduce the number of branches in constructed proof-search trees and to simplify these trees in other ways.

If the "Strategy" block will choose in the distinguished end node some member containing identification or some related pair, then the algorithm makes successively a number of counterinferences according to rules of $\ni B$ 2 so that to "strip" both marked occurrences of the identification (or relationship) kernel. This process can be further entangled by the work of "Pruning" block. If "Pruning" block will delete no members of the distinguished end node of the initial tree, then the set of the new end nodes *) (let us denote this set by K) is analysed to find out possibilities of its simplification.

---

* This set contains at most three nodes.

We have in mind simplifications which can be realized by means of a very restricted stock of logical inferences which is called below the "narrow" stock of logical means. This "narrow" stock is fixed by the following obvious statements concerning derivability in the calculus $\mathfrak{U}_2[\Lambda]$:

(I) a sequent of the form $\Gamma \to \neg X$ [of the form $\Gamma \to X$] is derivable if and only if the sequent $\Gamma, X \to$ [respectively $\Gamma, \neg X \to$] is derivable;

(II) if both sequent $\Gamma', X, \Gamma'' \to$ and $\Sigma', \neg X, \Sigma'' \to$ are derivable, then also $\Gamma', \Gamma'', \Sigma', \Sigma'' \to$ is derivable;

(III) if $\Gamma', \Gamma'' \to$ is derivable then $\Gamma', X, \Gamma'' \to$ is derivable;

(IV) sequent of the form $\Gamma, X, \Sigma, X, \Pi \to$ [of the form $\Gamma, X, \Sigma, Y, \Pi \to$] is derivable if and only if $\Gamma, X, \Sigma, \Pi \to$ [$\Gamma, Y, \Sigma, X, \Pi \to$ respectively] is derivable.

It is easy to formulate the method of analysing the set K allowing one to answer following questions.

1) Are some sequents from K literally equal?
2) Does K contain a "subsuming" sequent, i.e. one implying at least one of the remaining sequents by the "narrow" stock of logical means?
3) Does K contain a "subsuming" sequent pair, which imply the third sequent from K by the "narrow" stock of logical means?
4) Does K contain a pair of sequent which imply by the "narrow" stock of logical means some new sequent which does not belong to K but implies at least one of sequents in K by the "narrow" stock of logical means?

The answer to at least one of these questions turns out to be affirmative for the vast majority of relationship and identification types. During the analysis made to answer these questions the statement (I) allows one to reduce arbitrary sequents to ones with empty succedent;

the statement (II) allows [sometimes in conjunction with (I) and (IV)] in suitable cases to give affirmative answer to the third question or to form the new sequent mentioned in the forth question; the statement (III) allows [sometimes in conjunction with (I) and (IV)] to give affirmative answer to the second question or to establish the derivability by "narrow" logical means of some sequent belonging to K from the newly formed sequent mentioned in the fourth question.

If the answer to the first question is affirmative, then one node is picked up among duplicated nodes in K, and its remaining copies are "joined" to it. If the answer to the second question is affirmative, then every node in K derivable from the "subsuming" node* by "narrow" stock, is "joined" to subsuming node. If the answer to the third question is affirmative, then the node from K which does not belong to the "subsuming" pair is "joined" to this pair. If the answer to the fourth question is affirmative, then a new node is superimposed over K. The newly formed sequent is placed into this node and all nodes belonging to K and derivable from this node by "narrow" stock of logical means are "joined" to this node. "Joining" of a node to another node will be represented by arrow from that another node to the "joined" node. "Joining" to a pair of nodes will be represented by pair of arrows: from every sequent of the pair separate arrow is drawn to "joined" sequent.

---

* Expressions "Node B is derivable from a node B'" and "Node B is derivable from nodes B'and B" " are used to stand for corresponding statements concerning sequents situated in these nodes.

After "joining" process is accomplished, the resulting graph is not a tree. Some nodes in this graph can have two predecessors* ("pruning" process proceeds from such nodes down to both predecessors). The nodes from K which are "joined" to some other node are not counted as end nodes of the graph obtained.

Concerning passages (down) in graphs resulting from "Joining" block or from joint work of "Pruning" and "Merging" blocks one can say the same, what was said above about passages in graphs resulting from the sole "Pruning" block. The result of adding to $\ni\!\mathcal{B}\ 2$ all inference rules actually used in passages employed in proof search graphs formed by two blocks just mentioned, will be called the calculus $\ni\!\mathcal{B}\ 3$. Not all rules of this computation are invertible. Examples are

$$\frac{\Gamma, X, \Sigma \to Z}{\Gamma, (X\&Y), \Sigma \to Z}, \quad \frac{\Gamma \to X \qquad \Sigma \to Y}{\Gamma, \Sigma \to (X\&Y)}, \quad \frac{\Gamma, \Sigma \to \Delta}{\Gamma, X, \Sigma \to \Delta}, \quad \frac{\Gamma \to}{\Gamma \to X},$$

$$\frac{\Gamma, \Sigma \to Y}{\Gamma, \neg Y, \Sigma \to X}, \quad \frac{\Gamma, \Sigma \to \neg Y}{\Gamma, Y, \Sigma \to X}, \quad \frac{\Gamma, Y, \Sigma \to}{\Gamma, X, \Sigma \to \neg Y}, \quad \frac{\Gamma, \neg Y, \Sigma \to}{\Gamma, X, \Sigma \to Y}.$$

Last six rules play special part in "joining" process. One of these rules turns out to be used in all cases, when "joining" to a single node is employed but the joined node is not literally equal to it. Every such passage is the starting point of the third kind, for the "pruning" process (Formula X is redundant, one can delete it). **

---

* The notions "predecessor of a node", "successor of a node" etc. are extended to graphs with understanding that if a node B' is joined to a node B" or to a pair including B", then B' is situated immediately below B", and B" is situated immediately above B'.

** See two asterisks ** on page 22.

Examples 5 - 9 below illustrate joint operation of "Pruning" and "Merging" blocks. In these examples marked occurrences of identification (relationship) kernels are underlined. The left column contains graphs obtained after "Stripping" marked occurrences of identification (relationship) kernels, introducing (if necessary) new nodes and "joining". Formulas and sequents which have been present before the "pruning" preceeding the "joining" are framed. Dotted lines are placed over formulas and sequents deleted during "pruning" which follows the "joining". Sequents implying (by "narrow" stock of logical means) the sequent added as a new node during "joining" are marked by (*). Results of "pruning" after "joining" of respective graphs are found in the right column.

Example 5

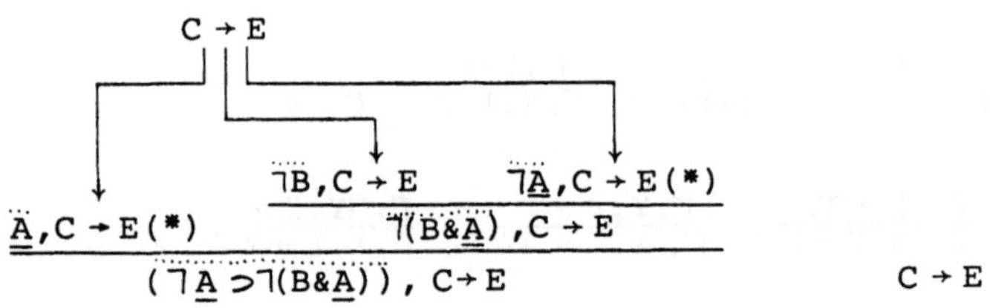

Example 6

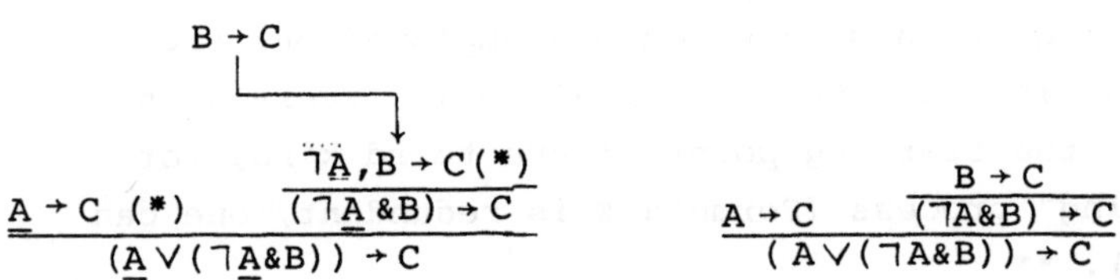

Example 7

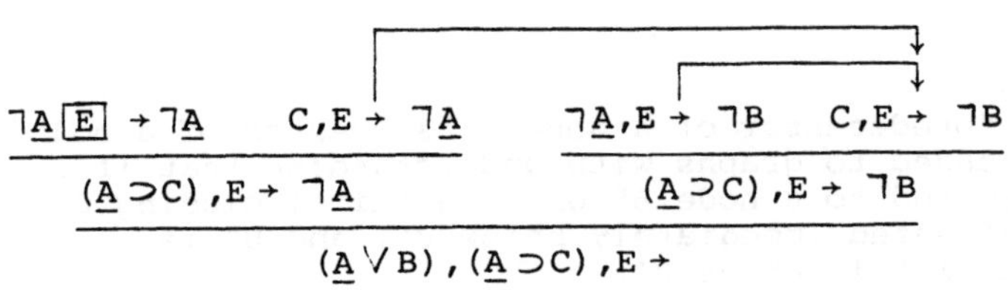

Example 8

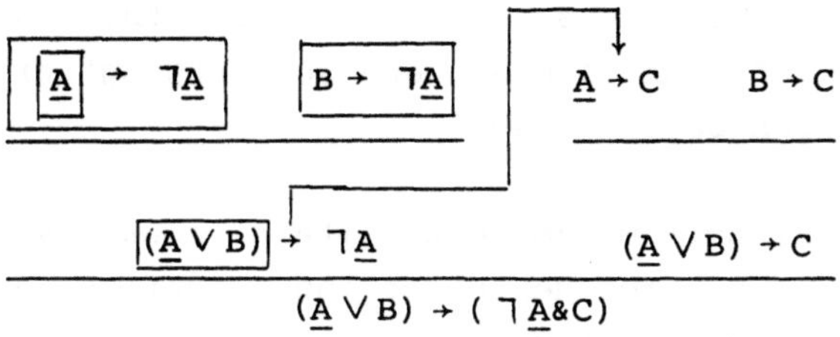

"Pruning" during formation of proof search graphs can change some "joining" , made in previous stages.

Example 9

Consider proof search graph.

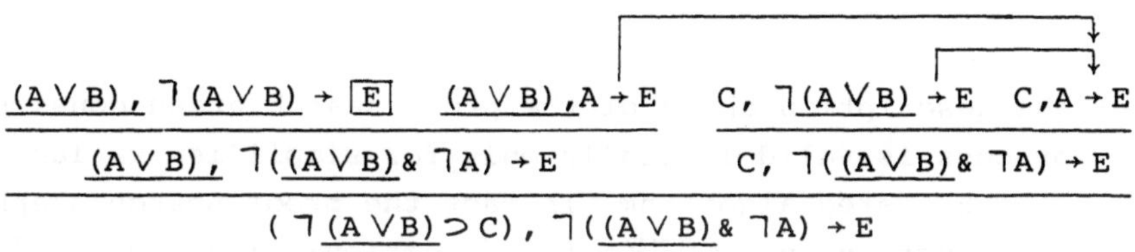

Let us separate "subsuming" pair of nodes and the node "joined" to them:

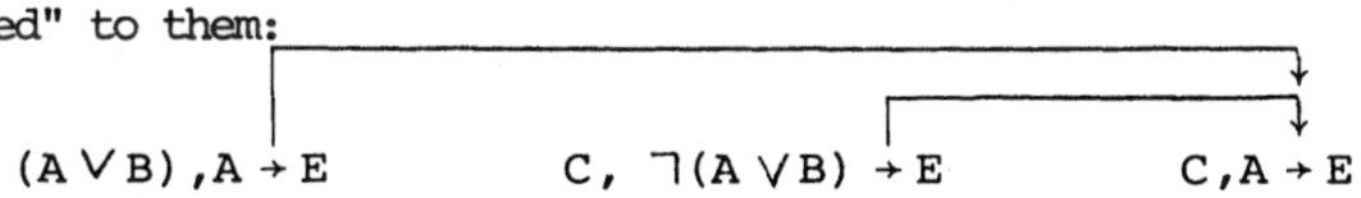

As a result of a counterinference according to suitable rule of $\mathcal{B}$ 2 to the leftmost sequent and the subsequent "pruning" we have:

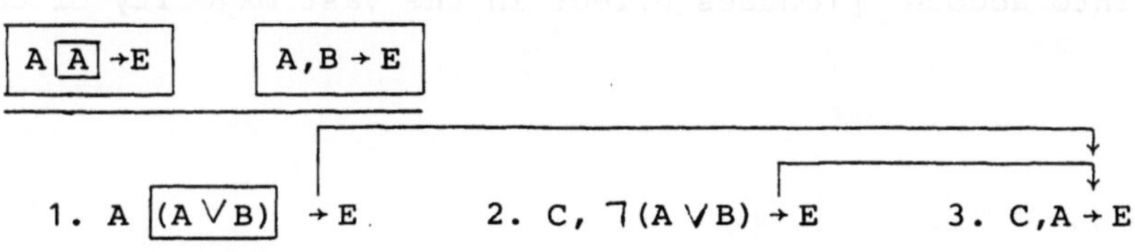

Now one can see, that the sequent 3 "joined" to sequents 1 and 2 is derivable from the sole node 1 with the aid of "narrow" stock of logical means and so "merging" of node 3 to node 2 can be cancelled.

Our proof search algorithm "notices" literal equality in sequent members only for depths of input subformulas, whose sum does not exceed bounds mentioned in section c. Extension of "joining" method to higher bounds gives in some cases some effect, improving quality of proof, and in some cases does not give expected effect.* In АЛПЕВ-ЛОМИ-2 the "Joining" block is developed for bounds mentioned in section c because the effect is perceptible only in some cases and because obtaining this partial effect would require making "Joining" block much more complicated and increase considerably the running time of the algorithm as a whole.

Д) = e) The description above of the proof search graph construction process is valid literally only for simplified variant of considered algorithm. In fact the proof search graph construction in АЛПЕВ-ЛОМИ-2 employs instead of ЭВ 2 another calculus ЭВ 2^+ obtained from ЭВ 2 by means of following changes.

---

* Under our bounds taking identification and relationship into account produces effect in the vast majority of cases.

1) The notion of sequent is changed a little: succedent members marked with a special sign ⊗ are allowed.

2) Initial sequents of ∃B 2+ are sequents of the same type as ones of ∃B 2, but now "sequents" described in preceeding sentence are allowed.

3) Antecedent rules of ∃B 2 are changed as follows: main formula of the rule considered is added as antecedent member to any premiss of the rule and this "additional" member of the premiss is marked by ⊗ (main formulas in conclusions of rules of ∃B 2+ are not marked by ⊗)*. For example rules (δ_2) and (δ'_2) are replaced respectively by the rules

$$\frac{\Gamma, X, \Sigma, \otimes(X \vee Y) \to \Delta \qquad \Gamma, Y, \Sigma, \otimes(X \vee Y) \to \Delta}{\Gamma, (X \vee Y), \Sigma \to \Delta}$$

$$\frac{\Gamma, \Sigma, \otimes(X \vee Y) \to \neg X \qquad \Gamma, \Sigma, \otimes(X \vee Y) \to \neg Y}{\Gamma, (X \vee Y), \Sigma \to}$$

Succedent rules are not changed.

Sequent members marked by ⊗ are not split during proof-search graphs construction.** They however are taken into consideration when one is looking for initial sequents of ∃B 2+ or for repetitions of sequent members or for pairs

---

* "Additional" member of proof-search tree (graph) node is considered to be successor of the main member of the predecessor of the node considered. "Additional" member is considered to be old member of the node.

** Actually the member marked with ⊗ can be deleted as soon as the process of splitting non-marked members proceeds so far as this marked member cannot turn out to be a sub-formula of some marked member.

of related members*. As a result of passage to calculus ЭВ 2^+ both the class of cases when the end node of graph turns out to be closed and the class of cases when blocks "Pruning" or "Joining" are used, is extended. But application of "Pruning" and "Joining" to sequents containing ⊗ has some specific features. These features have not been presented in the description of "Pruning" and "Joining" blocks above.

Passages in the proof-search graphs obtained on the base of ЭВ 2^+ are made according to rules of some calculus ЭВ 3^+ which has the same relation to ЭВ 3, as ЭВ 2^+ has to ЭВ 2.

e) After "pruned" and "joined" deduction of sequent $Q_1,\ldots,Q_n \to R$ in the calculus ЭВ 3^+ is obtained, this deduction is transformed into deduction of sequent $\to R$ in calculus $И_2[Q_1,\ldots,Q_n]$. If this would be done by means of insertions as mentioned at the end of § 4 the final derivation would contain redundant assumption introductions (i.e. sequents of the form $X \to X$) as well as a number of other "unnatural" features. The block "Assembling" introduced to avoid these defects transforms ЭВ 3^+ - deduction of the sequent $Q_1,\ldots,Q_n \to R$ into deduction of the sequent $\to R$ in the calculus $И_2[Q_1,\ldots Q_n]$ in much more economic and purposeful way. Presentation of ideas of the "Assembling" block will be restricted to the case when the ЭВ 3^+-deduction of sequent $Q_1,\ldots,Q_n \to R$ contains no "joinings", i.e. it is a tree.

---

* Of interest for us are only those pairs of related members, where either both members are not marked by ⊗, or only one of them is marked and this member (or the result of dropping outer negation sign) is the relationship kernel.

We shall use for illustration the tree shown in table 1 below. This tree was formed in the process of solving the problem: derive formula $(\neg(E\supset\neg A)\vee(B\&C))$ from initial formulas $(A\vee B)$ and $((\neg A\&C)\vee(E\&(A\vee C)))$. It was obtained using some bad proof-search strategy (which is so bad that it is not included in the stock used by the algorithm). The choice of strategy was made to illustrate in the sole simple example various situations arising when the block "Assembling" is executed. The tree shown in table 1 is deduction both in the calculus $\exists B\,3^+$ and in $\exists B\,3$ (all sequent members marked ⊗ were deleted as a result of "pruning").*

Let us introduce several notons concerning "pruned" proof search trees in $\exists B\,3^+$ (i.e. trees formed on the base of the calculus $\exists B\,2^+$ and "Pruning" block).** Let B be some node of a "pruned" proof-search tree in $\exists B\,3^+$ different from lowest node and let Φ be some new member of the node B. Let B_0 be the predecessor of B. We say that Φ is a bound new member of the first kind if 1) Φ is in the antecedent of B and 2) the predecessor of Φ is in the antecedent of B_0 and 3) B is the only successor of B_0.

---

* In this particular problem we obtain the same tree using the described algorithm (based on the calculus $\exists B\,2^+$) and using simplified version of the algorithm (based on the calculus $\exists B\,2$).

** All definitions introduced below have the same formulations for "pruned" proof search trees in the calculus $\exists B\,3$ obtained by the simplified version of the algorithm using calculus $\exists B\,2$ to extend trees.

We say that Φ is a bound new member of the second kind if 1) Φ is in the antecedent of B and 2) B_0 has second successor and 3) the second successor of B_0 is an initial sequent of $\exists B\, 3^+$ having the type (δ_2) or (δ_3). If Φ is a bound new member of the second kind then a quasipredecessor is associated to Φ together with a predecessor. By a quasipredecessor we understand the predecessor (in B_0) of the old member of the second sucessor of B_0 (any initial sequent of $\exists B\, 3^+$ in a "pruned" deduction contains exactly two members, so the quasipredecessor is unique). We say that Φ is a free new member if it is in the antecedent and is not a bound new member of the first or second kind.*

For all non-end nodes of the tree shown in table 1 all bound new members of the first or second kind (with quasipredecessors) and all free new members are listed.

With any antedecedent member of any node a certain list of formulas called projection of this member is associated. Projection of any antecedent member of the lowest node of the tree is empty by definition. Consider some node of the tree and its antecedent member Φ. Assume projections to be already defined for any antecedent member of the predecessor of the node considered. If Φ is a free new member

* A free new member can arise as a result of a counterinference according to two-premiss antecedent rule or according to succedent one-premiss rule.

then the projection of Φ is defined to be Φ itself. If Φ is old member or bound new member of the first kind then the projection of Φ is defined to be the projection of its predecessor. If Φ is bound new member of the second kind then the projection of Φ is defined to be the union of the projections of the predecessor of Φ and of the quasi-predecessor of Φ.

In the table 1 projections are shown for all antecedent members of non-end nodes.

Projection of a node in the tree is defined to be the sequent whose antecedent is the union of projections of all antecedent members of this node and whose succedent is one of the node considered. To every antecedent member of any node containing at least one new antecedent member,* a new sequent is assigned, which is called a counterprojection of the node considered associated with given antecedent member. It is by definition the sequent whose antecedent is the projection of the given antecedent member and the succedent is the new antecedent member itself.

---

* All antecedent members of lowest sequent are considered to be new ones.

Table 1

```
                                                                              B→B            C→C
                                                                              ____________________
                                                                 A,¬A→        13. B,C→(B&C)
                                                                 ______       _____________________
                                                          E,¬E→        12. B,(AvC),¬A→(B&C)
                                                          ______________________________________
          B→B           C→C              E→E                           11. B,E,(AvC),(E⊃¬A)→(B&C)
¬A→¬A     ___________________            ________    ¬A→¬A             __________________________
_____     6. B,C→B&C                     9. E&(AvC)→E                  10. B,(E&(AvC)),(E⊃¬A)→(B&C)
4. (¬A&C)→¬A    5. B,(¬A&C)→B&C          8. (E&(AvC)),(E⊃¬A)→¬A
_______________________________          ____________________________________________________
3. (AvB),(¬A&C)→B&C                              7. (AvB),(E&(AvC)),(E⊃¬A)→(B&C)
_________________________________________________________________________________
            2. (AvB),((¬A&C)v(E&(AvC))),(E⊃¬A)→(B&C)
            _____________________________________________
            1. (AvB),((¬A&C)v(E&(AvC)))→(¬(E⊃¬A)v(B&C))
```

Bound new members of the first kind
(in non-end nodes): (6,2),(11,2),(11,3).

Bound new members of the second kind:
(12,3),(13,2); their quasipredecessors are
(11,2),(12,3) respectively.

Free new members (in non-end nodes):
(2,3),(3,2),(5,1),(7,2),(10,1).

Projections of antecedent members of non-end nodes:

for (1,1),(1,2),(2,1),(2,2),(3,1),(7,1) - empty list;
for (2,3),(7,3),(8,2),(10,3),(11,4) - formula (E⊃¬A);
for (3,2),(4,1),(5,2),(6,2) - formula (¬A&C);
for (5,1),(6,1),(10,1),(11,1),(12,1),(13,1) - formula B;
for (7,2),(8,1),(9,1),(10,2),(11,2),(11,3),(12,2)-formula (E&(AvC))
for (12,3) - the list (E⊃¬A),(E&(AvC));
for (13,2) - the list (E&(AvC)),(E⊃¬A),(E&(AvC))

Table 2

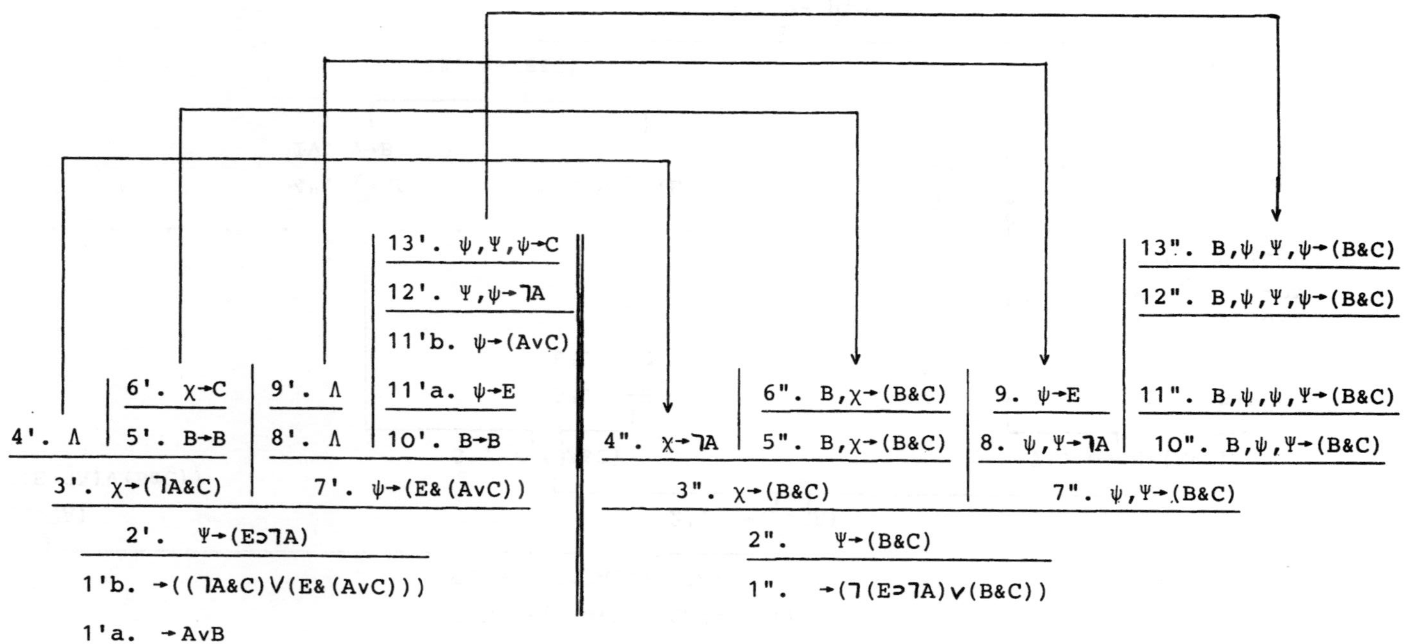

Following notation is used: Ψ stands for (E⊃ᒣA), χ for (ᒣA&C) and ψ for (E&(AvC)).

Table 3

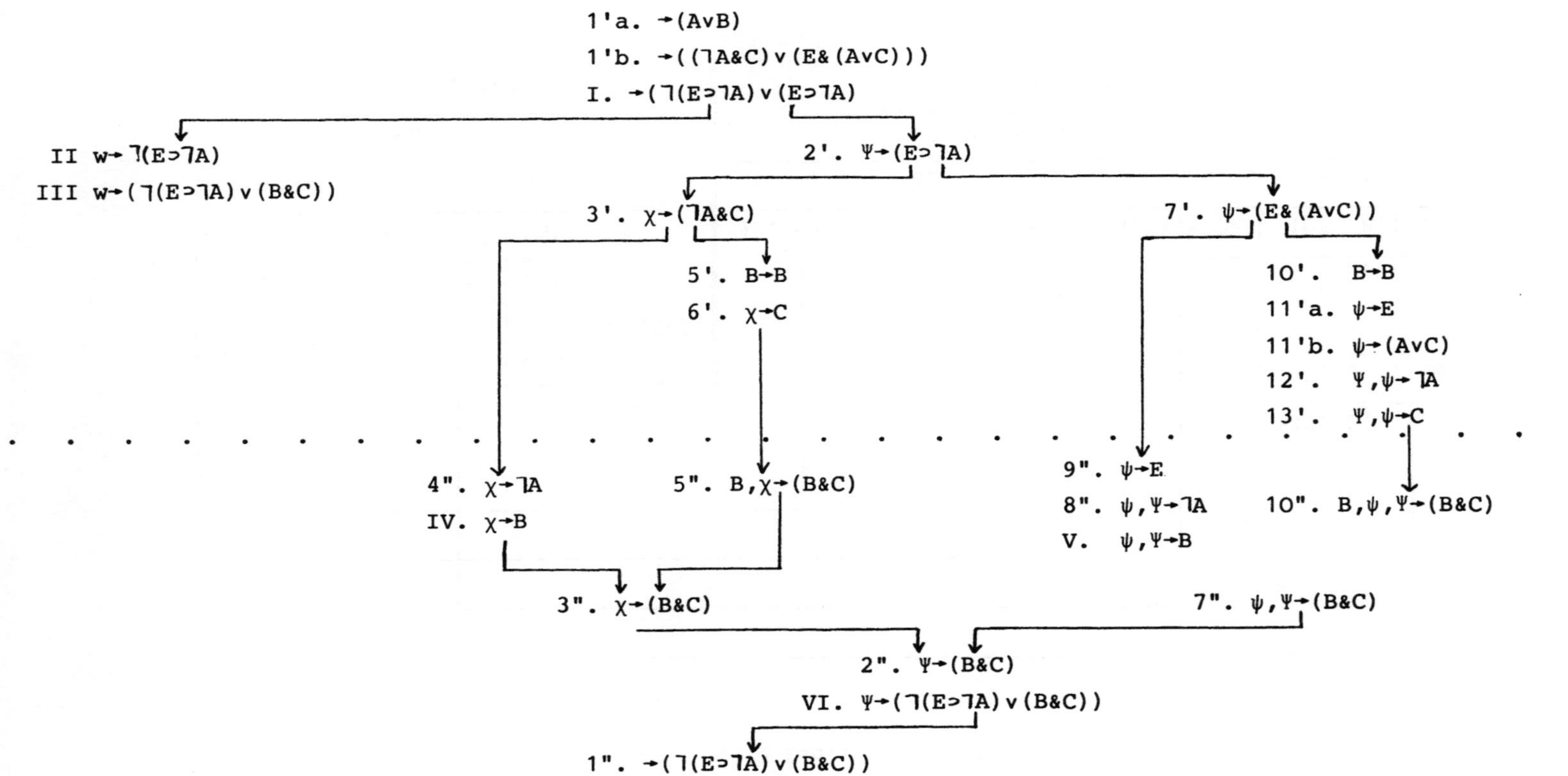

1'a. →(AvB)
1'b. →((¬A&C)v(E&(AvC)))
I. →(¬(E⊃¬A)v(E⊃¬A)
II w→¬(E⊃¬A)
III w→(¬(E⊃¬A)v(B&C))
2'. Ψ→(E⊃¬A)
3'. χ→(¬A&C)
7'. ψ→(E&(AvC))
5'. B→B
6'. χ→C
10'. B→B
11'a. ψ→E
11'b. ψ→(AvC)
12'. Ψ,ψ→¬A
13'. Ψ,ψ→C
4". χ→¬A
IV. χ→B
5". B,χ→(B&C)
9". ψ→E
8". ψ,Ψ→¬A
V. ψ,Ψ→B
10". B,ψ,Ψ→(B&C)
3". χ→(B&C)
7". ψ,Ψ→(B&C)
2". Ψ→(B&C)
VI. Ψ→(¬(E⊃¬A)v(B&C))
1". →(¬(E⊃¬A)v(B&C))

We use following notation: Ψ stands for $(E\supset\neg A)$, χ for $(\neg A\&C)$, ψ for $(E\&(A\vee C))$,   for $\neg(E\supset\neg A)$.

1'a and 1'b are initial sequents corresponding to initial formulas; I is the excluded middle kind of initial sequent; II, 2', 3', 5', 7', 10' are trivial initial sequents ("assumption introductions").

III is obtained from II by the rule (13); 4" from 3' by the rule (11); IV from 4" and 1'a by the rule (15); 6' from 3' by the rule (11); 5" from 5' and 6' by the rule (10); 3" from IV and 5" by the rules (7) and (23); 9" from 7' by the rule (11); 8" from 9" and 2 by the rule (6); V from 8" and 1'a by the rule (15); 11" from 7' by the rule (11); 11'b from 7' by the rule (11); 12' from 11'a and 2' by the rules (6) and (24); 13' from 12' and 11'b by the rules (15), (23) and (24); 10" from 10' and 13' by the rule (10); 7" from V and 10" by the rules (7) and (23); 2" from 1'b, 3" and 7" by the rule (14); VI from 2" by the rule (13); 1" from I, III and VI by the rule (14).

Projection of a closed proof search tree is by definition the figure obtained by means of deleting all end nodes and replacing all remaining nodes by their projections. Counterprojection of a closed tree is by definition the figure obtained by means of deleting all end nodes and replacing any remaining nodes by the list of all its counterprojections * (if any) or by a special sign (say Λ) if there are none.

Let us write down the counterprojection of the tree considered and to the right of it the projection of the same tree. Mark those occurences of sequents and Λ in both figures which have no superstructure over them. The number of such marked occurrences is the same for both figures. Draw arrow from every marked occurrence in the left figure to corresponding (i.e. having the same number in the enumeration of marked occurrences from left to right) marked occurrence in the right figure. The result is a graph (see table 2). Let us "rectify" this graph: delete all occurrences of Λ, remove repetitions of members in antecedents (by deleting them) and remove duplications ** of sequents situated in the same branch in succession. The graph obtained in this way will be called a natural deduction carcass corresponding to given closed tree. It turns out to be possible to obtain the deduction of the sequent R in the calculus $\mathcal{U}_2[Q_1,\ldots,Q_n]$ by means

---

* A node different from the lowest one cannot have more than two counterprojections.

** Duplicates are sequents differing only by the order of antecedent members.

of inserting some additional nodes (both between existing nodes and on the additional branches if we consider the natural deduction carcass from the left lower corner upward and, after passing to the right figure along the arrow, down to the right lower corner (i.e. if we introduce corresponding partial ordering of nodes in the graph).

Table 3 presents derivation of sequent

$$\rightarrow(\neg(E\supset\neg A)\vee(B\&C))$$

in the calculus

$$\mathcal{U}_2[(A\vee B),((\neg A\&C)\vee(E\&(A\vee C)))]$$

obtained in this way. "Inserted" nodes are numbered by Roman numerals.

Formation of sequents inserted into natural deduction carcass and the insertion itself (in particular the junction of end nodes of projection and of counterprojection) as well as writing down analyses showing how (by which rule) and from which preceeding nodes the given node is obtained, is made according to special rules on the base of information collected in the tree input into the "Assembling" block. The process of "inserting" additional nodes and analyses formation is beginning from the lowermost line of the "counterprojection-projection" figure. It proceeds upwards one line at a time and for each line the insertions into the left and right parts of figure as well as additional branches are formed simultaneously.

The final graph may contain deduction branching points: some nodes can be followed by two branches (all such nodes "originate" in counterprojections of the initial tree nodes).

Ordinarily one of the branches in the deduction branching begins with introduction of some assumption (i.e. the introduction of the sequent X→X). Two or three branches are merged at some nodes ("originating" in counter-projections of the initial tree nodes). Every such node is obtained from some preceeding nodes by two-premiss or three-premiss logical inference rule of the calculus $И_2[Q_1,...,Q_n]$.

Concluding stage of "Assembling" is the transformation of the graph into a column preserving the ordering of nodes comparable in the graph (in the sense of partial ordering introduced for nodes).

If the closed graph input into the " Assembling" block is not a tree, then the principal features of "Assembling" are preserved but some details became more complicated. This is because a member of a node to which another node is joined, can have more than one predecessor.

All blocks mentioned above before the "Assembling" block are working in ALPEV-LOMI-2 in complicated interaction with each other and with a number of other blocks which have not been mentioned. It is impossible in a short communication to describe main blocks of the algorithm in sufficient detail, to say nothing of the process of their joint functioning.

Supplement

Examples of logical deductions constructed by the "URAL-4" computer on the base of ALPEV-LOMI-2.

Texts above differ from actual printouts only in the form of some symbols (and of course in translation of printed supplementary texts into English). Letters A,B,C,E,H,K,P are used as propositional varaibles.

Example 1.

Problem. Derive from the initial formula

1. (((B⊃C)&(E⊃H))&(K⊃P))

the formula

(((A⊃B)&(C⊃E))⊃((H⊃K)⊃(A⊃P)))

Solution. According to formulation of the problem we have initial sequent

1. →(((B⊃C)&(E⊃H))&(K⊃P))

Assume

(A 1) ((A⊃B)&(C⊃E))

We have trivial initial sequent.

2. A1 →((A⊃B)&(C⊃E))

From 1 we obtain by conjunction-elimination rule (11)

3. →((B⊃C)&(E⊃H))

From 1 we obtain by conjunction-elimination rule (11)

4. →(K⊃P)

Assume

(A 2) (H⊃K)

We have trivial initial sequent

5. $\mathcal{A}2 \rightarrow (H \supset K)$

From 2 we obtain by conjunction-elimination rule (11)

6. $\mathcal{A}1 \rightarrow (A \supset B)$

From 2 we obtain by conjunction-elimination rule (11)

7. $\mathcal{A}1 \rightarrow (C \supset E)$

From 3 we obtain by conjunction-elimination rule (11)

8. $\rightarrow (B \supset C)$

From 3 we obtain by conjunction-elimination rule (11)

9. $\rightarrow (E \supset H)$

Assume

$\mathcal{A}3$ A

We have trivial initial sequent

10. $\mathcal{A}3 \rightarrow A$

From 10 and 6 we obtain by detachment rule (6)

11. $\mathcal{A}3, \mathcal{A}1 \rightarrow B$

From 11 and 8 we obtain by detachment rule (6)

12. $\mathcal{A}3, \mathcal{A}1 \rightarrow C$

From 12 and 7 we obtain by detachment rule (6)

13. $\mathcal{A}3, \mathcal{A}1 \rightarrow E$

From 13 and 9 we obtain by detachment rule (6)

14. $\mathcal{A}3, \mathcal{A}1 \rightarrow H$

From 14 and 5 we obtain by detachment rule (6)

15. $\mathcal{A}3, \mathcal{A}1, \mathcal{A}2 \rightarrow K$

From 15 and 4 we obtain by detachment rule (6)

16. $\mathcal{A}3, \mathcal{A}1, \mathcal{A}2 \rightarrow P$

From 16 we obtain by assumption discharge rule (3)

17. $\mathcal{A}1, \mathcal{A}2 \rightarrow (A \supset P)$

From 17 we obtain by assumption discharge rule (3)

18. $\mathcal{A}1 \rightarrow ((H \supset K) \supset (A \supset P))$

From 18 we obtain by assumption discharge rule (3)

19. $\rightarrow (((A \supset B) \& (C \supset E)) \supset ((H \supset K) \supset (A \supset P)))$

which had to be proved.

Example 2.

Problem. Derive from the initial formulas

1. $(H \vee \neg(A \& K))$
2. $(H \supset (\neg A \vee B))$

the formula

$(\neg(K \supset A) \vee (K \supset B))$

Solution. According to formulation of the problem we have initial sequents

1. $\rightarrow (H \vee \neg(A \& K))$
2. $\rightarrow (H \supset (\neg A \vee B))$

By the law of excluded middle we have

3. $\rightarrow (\neg(K \supset A) \vee (K \supset A))$

Assume

($\mathcal{A}$ 1) $\neg(K \supset A)$

(the first member of disjunction in conclusion of sequent 3)

We have trivial initial sequent

4. $\mathcal{A}\,1 \rightarrow \neg(K \supset A)$

From 4 we obtain by disjunction-introduction rule (13)

5. $\mathcal{A}\,1 \rightarrow (\neg(K \supset A) \vee (K \supset B))$

Assume

($\mathcal{A}$ 2) $(K \supset A)$

(the second member of disjunction in conclusion of sequent 3)

We have trivial initial sequent

6. $\mathcal{A}\,2 \rightarrow (K \supset A)$

Assume

($\mathcal{A}$ 3) K

We have trivial initial sequent

7. $\mathcal{A}\,3 \rightarrow K$

From 6 and 7 we obtain by detachment rule (6)

8. $\mathcal{A}\,3, \mathcal{A}\,2 \rightarrow A$

Assume

($\mathcal{A}$ 4) H

We have trivial initial sequent

9. $\mathcal{A}4 \rightarrow H$

From 9 and 2 we obtain by detachment rule (6)

10. $\mathcal{A}4 \rightarrow (\neg A \vee B)$

From 8 und 10 we obtain by disjunctive member detachment rule (15)

11. $\mathcal{A}4, \mathcal{A}3, \mathcal{A}2 \rightarrow B$

From 8 and 7 we obtain by conjunction-introduction rule (10)

12. $\mathcal{A}2, \mathcal{A}3 \rightarrow (A \& K)$

From 12 and 1 we obtain by disjunctive member detachment rule (15)

13. $\mathcal{A}2, \mathcal{A}3 \rightarrow H$

From 13 and 11 we obtain by two-premiss cut rule (7)

14. $\mathcal{A}2, \mathcal{A}3 \rightarrow B$

From 14 we obtain by assumption discharge rule (3)

15. $\mathcal{A}2 \rightarrow (K \supset B)$

From 15 we obtain by disjunction-introduction rule (13)

16. $\mathcal{A}2 \rightarrow (\neg(K \supset A) \vee (K \supset B)$

From 3, 5 and 16 we obtain by the rule of arguing by cases (14)

17. $\rightarrow (\neg(K \supset A) \vee (K \supset B))$

which had to be proved.

Example 3.

Problem. Derive from the initial formulas

1. $(C \supset (\neg A \vee B))$
2. $\neg(\neg(C \supset A) \vee (C \supset B))$

the formula

$\neg(C \vee \neg(A \& E))$

Solution. According to formulation of the problem we have initial sequents

1. $\rightarrow (C \supset (\neg A \vee B))$

2. $\rightarrow \neg(\neg(\beta A) \vee (C \supset B))$

From 2 we obtain by disjunction-negation rule (16)

3. $\rightarrow ((C \supset A) \,\&\, \neg(C \supset B))$

From 3 we obtain by conjunction-elimination rule (11)

4. $\rightarrow (C \supset A)$

From 3 we obtain by conjunction-elimination rule (11)

5. $\rightarrow \neg(C \supset B)$

Assume

($\mathcal{A}$1) $\neg C$

We have trivial initial sequent

6. $\mathcal{A}1 \rightarrow \neg C$

From 6 we obtain by arbitrary conclusion-introduction rule (5)

7. $\mathcal{A}1 \rightarrow (C \supset B)$

conclusions of sequents 7 and 5 are mutually inconsistent so by the rule of stating the contradiction (22) we obtain

8. $\mathcal{A}1 \rightarrow$

From 8 we obtain by rule of inference ad absurdum (21)

9. $\rightarrow C$

Assume

($\mathcal{A}$2) B

We have trivial initial sequent

10. $\mathcal{A}2 \rightarrow B$

From 10 we obtain by premiss-addition rule (4)

11. $\mathcal{A}2 \rightarrow (C \supset B)$

Conclusions of sequents 11 and 5 are mutually inconsistent, so by the rule of stating the contradiction (22) we obtain

12. $\mathcal{A}2 \rightarrow$

Assumptions in 12 lead to contradiction, so by the rule of refuting by contradiction (20) we obtain

13. $\rightarrow \neg B$

From 9 and 1 we obtain by detachment rule (6)

14. $\rightarrow (\neg A \vee B)$

From 13 and 14 we obtain by disjunctive member detachment rule (15)

15. $\rightarrow \neg A$

From 9 and 4 we obtain by detachment rule (6)

16. $\rightarrow A$

From 16 we obtain by arbitrary conclusion-introduction rule (5)

17. $\rightarrow (\neg A \supset \neg(C \vee \neg(A \& E)))$

From 15 and 17 we obtain by detachment rule (6)

18. $\rightarrow \neg(C \vee \neg(A \& E))$

what had to be proved.

Example 4.

Problem. Derive from the initial formulas

1. $((\neg A \supset (B \vee \neg C)) \supset (\neg E \supset H))$
2. $((\neg E \& \neg H) \vee (C \supset (A \vee B)))$

the formula

$$(((\neg A \supset \neg(\neg B \& C)) \vee C) \supset ((\neg C \supset H) \vee E))$$

Solution. According to formulation of the problem we have initial sequents

1. $\rightarrow ((\neg A \supset (B \vee \neg C)) \supset (\neg E \supset H))$
2. $\rightarrow ((\neg E \& \neg H) \vee (C \supset (A \vee B)))$

By the law of excluded middle we have

3. $\rightarrow E \vee \neg E$

Assume

($\mathcal{A}$1) E

(the first member of disjunction in conclusion of sequent 3)

We have trivial initial sequent

4. $\mathcal{A}1 \rightarrow E$

From 4 we obtain by disjunction-introduction rule (13)

5. $\mathcal{A}1 \rightarrow ((\neg C \supset H) \vee E)$

Assume

($\mathcal{A}$2) $\neg E$

(the second member of disjunction in conclusion of sequent 3)

We have trivial initial sequent

6. $\mathcal{A}2 \rightarrow \neg E$

By the law of excluded middle we have

7. $\rightarrow ((\neg A \supset (B \vee \neg C)) \vee \neg(\neg A \supset (B \vee \neg C)))$

Assume

($\mathcal{A}$3) $(\neg A \supset (B \vee \neg C))$

We have trivial initial sequent

8. $\mathcal{A}3 \rightarrow \neg A \supset (B \vee \neg C))$

From 8 and 1 we obtain by detachment rule (6)

9. $\mathcal{A}3 \rightarrow (\neg E \supset H)$

Assume

($\mathcal{A}$4) $(\neg E \supset H)$

We have trivial initial sequent

10. $\mathcal{A}4 \rightarrow (\neg E \supset H)$

From 6 and 10 we obtain by detachment rule (6)

11. $\mathcal{A}2, \mathcal{A}4 \rightarrow H$

Assume

($\mathcal{A}$5) $\neg(\neg A \supset (B \vee \neg C))$

(the second member of disjunction in conclusion of sequent 7)

We have trivial initial sequent

12. $\mathcal{A}5 \rightarrow \neg(\neg A \supset (B \vee \neg C))$

Assume

($\mathcal{A}$6) $(B \vee \neg C)$

We have trivial initial sequent

13. $\mathcal{A}6 \rightarrow (B \vee \neg C)$

From 13 we obtain by premiss-addition rule (4)

14. $\mathcal{A}6 \rightarrow (\neg A \supset (B \vee \neg C))$

Conclusions of sequents 14 and 12 are mutually inconsistent, so by the rule of stating the contradiction (22) we have

15. $\mathcal{A}6, \mathcal{A}5$

Assumptions in 15 lead to contradiction, so by the rule of refuting by contradiction (20) we obtain

16. $\mathcal{A}5 \rightarrow \neg(B \vee \neg C)$

From 16 we obtain by disjunction negation rule (16)

17. $\mathcal{A}5 \rightarrow (\neg B \& C)$

From 17 we obtain by conjunction-elimination rule (11)

18. $\mathcal{A}5 \rightarrow C$

From 18 we obtain by arbitrary conclusion-introduction rule (5)

19. $\mathcal{A}5 \rightarrow (\neg C \supset H)$

From 11 we obtain by premiss-addition rule (4)

20. $\mathcal{A}2, \mathcal{A}4 \rightarrow (\neg C \supset H)$

From 9 and 20 we obtain by two-premiss cut rule (7)

21. $\mathcal{A}2, \mathcal{A}3 \rightarrow (\neg C \supset H)$

From 7, 21 and 19 we obtain by the rule of arguing by cases (14)

22. $\mathcal{A}2 \rightarrow (\neg C \supset H)$

From 22 we obtain by disjunction-introduction rule (13)

23. $\mathcal{A}2 \rightarrow ((\neg C \supset H) \vee E)$

From 3, 5 and 23 we obtain by the rule of arguing by cases (14)

24. $\rightarrow ((\neg C \supset H) \vee E)$

From 24 we obtain by premiss-addition rule (4)

25. $\rightarrow (((\neg A \supset \neg(\neg B \& C)) \vee C) \supset ((\neg C \supset H) \vee E))$

what had to be proved.

Example 5.

Problem. Derive the formula

$(((A \equiv B) \equiv A) \equiv B)$

Solution.

Assume

($\mathcal{A}1$) $((A \equiv B) \equiv A)$

We have trivial initial sequent

1. $\mathcal{A}1 \rightarrow (A \equiv B) \equiv A$

From 1 we obtain by equivalence-elimination rule (18)

2. $\mathcal{A}1 \rightarrow (A \equiv B) \supset A$

From 1 we obtain by equivalence-elimination rule (18)

3. $\mathcal{A}1 \rightarrow A \supset (A \equiv B)$

By the law of excluded middle we have

4. $\rightarrow A \vee \neg A$

Assume

($\mathcal{A}$ 2) A

(the first member of disjunction in conclusion of sequent 4)

We have trivial initial sequent

5. $\mathcal{A}2 \rightarrow A$

From 5 and 3 we obtain by detachment rule (6)

6. $\mathcal{A}1, \mathcal{A}2 \rightarrow (A \equiv B)$

Assume

($\mathcal{A}$ 3) $A \equiv B$

We have trivial initial sequent

7. $\mathcal{A}3 \rightarrow (A \equiv B)$

From 7 and 2 we obtain by detachment rule (6)

8. $\mathcal{A}3, \mathcal{A}1 \rightarrow A$

Assume

($\mathcal{A}$ 4) $\neg A$

(the second member of disjunction in conclusion of sequent 4)

We have trivial initial sequent

9. $\mathcal{A}4 \rightarrow \neg A$

From 9 and 2 we obtain by the implication premiss refutation rule (9)

10. $\mathcal{A}4, \mathcal{A}1 \rightarrow \neg(A \equiv B)$

From 10 we obtain by equivalence negation rule (19)

11. $\mathcal{A}4, \mathcal{A}1 \rightarrow (\neg(A \supset B) \vee \neg(B \supset A))$

Assume

($\mathcal{A}$ 5) $\neg(B \supset A)$

We have trivial initial sequent

12. $\mathcal{A}5 \rightarrow \neg(B \supset A)$

Assume

($\mathcal{A}$ 6) $\neg B$

We have trivial initial sequent

13. $\mathcal{A}6 \rightarrow \neg B$

From 13 we obtain by the arbitrary conclusion-introduction rule (5)

14. $\mathcal{A}6 \rightarrow (B \supset A)$

Conclusions of sequents 14 and 12 are mutually inconsistent, so by the rule of stating the contradiction (22) we have

15. $\mathcal{A}6, \mathcal{A}5$

From 7 we obtain by equivalence-elimination rule (18)

16. $\mathcal{A}3 \rightarrow (A \quad B)$

Assume

($\mathcal{A}7$) B

We have trivial initial sequent

17. $\mathcal{A}7 \rightarrow B$

Assume

($\mathcal{A}8$) $(A \equiv B)$

We have trivial initial sequent

18. $\mathcal{A}8 \rightarrow (A \equiv B)$

From 18 we obtain by equivalence elimination rule (18)

19. $\mathcal{A}8 \rightarrow (B \supset A)$

Assume

($\mathcal{A}9$) A

We have trivial initial sequent

20. $\mathcal{A}9 \rightarrow A$

From 20 we obtain by premiss addition rule (4)

21. $\mathcal{A}9 \rightarrow (B \supset A)$

From 17 we obtain by premiss addition rule (4)

22. $\mathcal{A}7 \rightarrow (A \supset B)$

From 21 and 22 we obtain by equivalence-introduction rule (17)

23. $\mathcal{A}7, \mathcal{A}9 \rightarrow (A \equiv B)$

From 23 we obtain by assumption discharge rule (3)

24. $\mathcal{A}7 \rightarrow (A \supset (A \equiv B))$

From 17 and 19 we obtain by detachment rule (6)

25. $\mathcal{A}8, \mathcal{A}7 \rightarrow A$

From 25 we obtain by assumption discharge rule (3)

26. $\mathcal{A}7 \rightarrow ((A \equiv B) \supset A)$

From 26 and 24 we obtain by equivalence-introduction rule (17)

27. $\mathcal{A}7 \rightarrow ((A \equiv B) \equiv A)$

From 27 we obtain by assumption-discharge rule (3)

28. $\rightarrow B \supset ((A \equiv B) \equiv A))$

From 8 and 16 we obtain by detachment rule (6)

29. $\mathcal{A}3, \mathcal{A}1 \rightarrow B$

From 15 we obtain by the rule of inference ad absurdum (21)

30. $\mathcal{A}5 \rightarrow B$

From 9 we obtain by arbitrary conclusion-introduction rule (5)

31. $\mathcal{A}4 \rightarrow (A \supset B)$

From 31 and 11 we obtain by disjunctive member detachment rule (15)

32. $\mathcal{A}4, \mathcal{A}1 \rightarrow \neg(B \supset A)$

From 30 and 32 we obtain by two-premiss cut rule (7)

33. $\mathcal{A}1, \mathcal{A}4 \rightarrow B$

From 6 and 29 we obtain by two-premiss cut rule (7)

34. $\mathcal{A}1, \mathcal{A}2 \rightarrow B$

From 4, 34 and 33 we obtain by the rule of arguing by cases (14)

35. $\mathcal{A}1 \rightarrow B$

From 35 we obtain by assumption discharge rule (3)

36. $\rightarrow (((A \equiv B) \equiv A) \supset B)$

From 36 and 28 we obtain by equivalence-introduction rule (17)

37. $\rightarrow (((A \equiv B) \equiv A) \equiv B)$

what had to be proved.

Efficiency and Completeness of the Set of Support Strategy in Theorem Proving

L.T. Wos, G. A. Robinson, D. F. Carson

Abstract. One of the major problems in mechanical theorem proving is the generation of a plethora of redundant and irrelevant information. To use computers effectively for obtaining proofs, it is necessary to find strategies which will materially impede the generation of irrelevant inferences. One strategy which achieves this end is the *set of support strategy*. With any such strategy two questions of primary interest are that of its efficiency and that of its logical completeness. Evidence of the efficiency of this strategy is presented, and a theorem giving sufficient conditions for its logical completeness is proved.

1. Introduction

In a previous paper on theorem proving [1] a strategy based on the concept of set of support was briefly discussed. This concept, its logical significance, and the theorem establishing the logical completeness of the corresponding strategy are the focus of attention in the first part of the present paper.

The remainder of the paper is concerned with the efficiency and sensitivity of the strategy. Evidence is supplied of its effect on time and memory requirements and of its effect on the importance of the choice of parameter values. Evidence is also given of the value of the strategy in obtaining proofs of theorems whose mathematical depth is measurably greater than that of those previously provable by mechanical means.

The evidence was obtained with the theorem-proving program PG1 implemented for the Control Data 3600. The language employed by PG1 is first-order predicate calculus, and the chief rule of inference employed therein is resolution [1, 2, 3].

2. Definitions

The following definitions are given in preparation for defining *set of support.*

Definition 1. For a set S of clauses, S^0 is the set of clauses B such that B is in S or there is a clause C in S such that B is a factor [1][1] of C. For $i > 0$, S^i is the set of clauses B such that B is in S^{i-1} or there exist clauses C and D in S^{i-1} such that B is a resolvent [1, 3] of C and D or a factor of a resolvent of C and D.

Definition 2. A deduction D of the clause A from the set S of clauses is a finite sequence A_i of clauses such that: (1) $A = A_i$ for some i, and (2) every A_i is either in S, or there exists $j < i$ with A_i a factor of A_j, or there exist $j < i$ and $k < i$ with A_i a resolvent of A_j and A_k.

Definition 3. D is a *deduction of contradiction* from the set S of clauses if D is a deduction from S of two unit clauses A and B such that A and B give unit conflict.

Work performed under the auspices of the US Atomic Energy Commission.

[1] In connection with the search strategies employed herein it is most useful to consider separately as in [1] the operators of resolution and factoring. These two operators are unified into a single rule of inference in [3]; however, the term *resolution* as used therein does not refer to an operator.

Article from: Journal of the Association for Computing Machinery, Volume 12, Number 4, October 1965.

A and B give *unit conflict* if A and B are unit clauses (have but one literal), are opposite in sign, and if there exist instantiations of A and B which make them otherwise identical. Such a deduction D is called a *proof* of the theorem corresponding to S, or just a *proof*. Two unit clauses which give unit conflict are called *contradictory units*.

Definition 4. For $T \subseteq S$, $T_S{}^0$ is the set of clauses B such that B is in T or there is a clause $C \in T$ such that B is a factor of C. For $i > 0$, $T_S{}^i$ is the set of clauses B such that B is in T_S^{i-1} or there exist C in T_S^{i-1} and D in $S^0 \cup T_S^{i-1}$ such that B is a resolvent of C and D or a factor of a resolvent of C and D.

Definition 5. The *T-level* of a clause B is the smallest j for which $B \in T_S{}^j$. If no such j exists, the T-level of B is undefined.

Definition 6. The clause B is said to have *T-support* if and only if for some $i \geq 0$, $B \in T_S{}^i$. T is said to be a *set of support* for B and B is said to be *supported* by T.

To illustrate the concept of set of support, let T contain a single clause B. Where A is a given element of S, the following are some of the clauses having T as set of support: B, any factor B' of B, any resolvent C of A and B', any resolvent of A and C' where C' is a factor of C.

Definition 7. The set $T \subseteq S$ is a *set of support for the deduction D* if every A_i of D has T-support, is in S itself, or is a factor of some element in S.

3. Set of Support Strategy

The *set of support strategy* for theorem proving consists of selecting a subset T of the given set S of clauses and an integer k and generating the elements of $S^0 \cup T_S{}^k$. The integer k is called the *level bound*. Equivalently, the adjunction of a clause is permitted if and only if it is a factor of an element of S or it has T-support and its T-level does not exceed the level bound.

To extract the full value of this strategy requires judicious choice of the set T. It is easy to construct unsatisfiable sets S of clauses which allow for an unwise choice of T as set of support—unwise in the sense that unsatisfiability cannot be established through use of the set of support strategy with that choice of T. A poor choice of T could in fact prevent one from discovering a proof even when seeking one for an already established theorem. A rather natural choice of T would seem to be the join of two sets of clauses: the subset of clauses within S which corresponds to the denial of the purported theorem, and that subset within S which corresponds to the special hypotheses of the theorem. The special hypothesis of the theorem refers to that which is given in the theorem but which is outside the basic set of axioms for the theory in question. (It is rather natural from the mathematical viewpoint to choose T in this fashion since the mathematician often seeks proof by contradiction, assuming the theorem false while focusing attention on the additional hypotheses.) The logical significance of this choice for T lies in the fact that the basic axioms of a theory are assumed to form a satisfiable set. The set of support strategy avoids seeking a proof, a demonstration of unsatisfiability, within that subset which is assumed to be itself satisfiable. The corresponding gain from the viewpoint of mechanical theorem proving lies in the fact that many clauses which correspond to trivial lemmas of the theory will not be generated, and many of the possible alternate proofs of the theorem under study will not be begun. An additional aid in avoiding trivial lemmas is provided by that feature of PG1 which allows control of literal length. One of the input parameters establishes a bound on the number of symbols which may occur in any literal of any clause generated by the program.

4. Completeness

The question now arises connecting the given choice of T as set of support and the possible loss of logical completeness. Put in broader terms, where S is a finite unsatisfiable set of clauses, what condition is sufficient to force the choice of $T \subseteq S$ as set of support to have the desired property: unsatisfiability will be established after generating a finite number of clauses when employing the set of support strategy? An answer to this question is contained in the theorem on completeness, below. In particular the choice of T discussed above has this property. This is equivalent to saying that, for a valid theorem, there exists a proof D obtainable by generating a finite number of clauses which has as set of support the join of the special hypotheses and the denial of the conclusion of the theorem.

LEMMA 1. *If S is a finite unsatisfiable set of clauses containing no variables,*[2] *and if $T \subseteq S$ is such that $S-T$ is satisfiable, then there exists a proof D with T as set of support.*[3]

PROOF. Let P be the set of atoms of S. If P contains a single element, say Q, then among the elements of S there exist the units Q and $-Q$. The two-line deduction Q, $-Q$ will suffice. Assume that Lemma 1 holds[4] when P contains i elements, $1 \leq i \leq n$. To complete the induction, consider P such that P contains exactly $n+1$ elements. Let Q be any element of P. Let S_1 be that set obtained from S by deleting those clauses containing the literal Q and by deleting $-Q$ from the remaining clauses. Let T_1 be obtained from T similarly. Let S_2 be that set obtained from S by deleting those clauses containing $-Q$ and by deleting Q from the rest. T_2 is obtained from T similarly. S_1 and S_2 are unsatisfiable since S is. $S_1 - T_1$ or $S_2 - T_2$ (or both) are satisfiable since $S - T$ is. Assume without loss of generality that $S_1 - T_1$ is satisfiable. Where the induction hypothesis applies there exists a deduction of contradiction from S_1 with T_1 as set of support. (The only case in which it does not apply is that in which $-Q$ is a unit clause in S.) Let R and $-R$ be contradictory units contained in that deduction. The corresponding deduction from S has T-support. At best it contains R and $-R$, and the proof of Lemma 1 is complete. At worst, the clauses $-Q \vee R$ and $-Q \vee -R$ are part of that deduction, and by adding their resolvent one has a deduction D_1 of $-Q$ as a unit, and the deduction has T-support. In the case in which the induction hypothesis does not apply, the T-supported deduction D_1 of $-Q$ consists of a single line $-Q$. Each of the remaining cases can be extended to a deduction D_1 with T-support of the unit $-Q$. Similarly, if $S_2 - T_2$ is satisfiable at worst there exists a deduction D_2 of Q with T-support. The juxtaposition of D_1 and D_2 is a deduction D of contradiction. One case remains—that in which $S_2 - T_2$ is unsatisfiable. Let U be the set of those clauses of $S_2 - T_2$ which are obtained from S by deletion of the literal Q. Since the

[2] In the hypothesis of Lemma 1 each clause in S is tacitly assumed to have at least one literal. Thus the so-called empty clause is excluded from S.

[3] The authors are indebted to J. A. Robinson for his invaluable assistance in obtaining proofs of Lemma 1 and the corresponding completeness theorem.

[4] See footnote 2.

elements of $(S_2 - T_2) - U$ are elements of $S - T$, $(S_2 - T_2) - U$ is satisfiable. If the induction hypothesis does not hold in $S_2 - T_2$, then the unit clause Q is a member of S. In that case form the deduction D by adding the line Q to D_1. Such a deduction D is a deduction of contradiction with T as set of support. When the induction hypothesis holds in $S_2 - T_2$, there exists a deduction D_3 of contradiction with U-support. Let U_1 be the set of those clauses of U which appear in D_3. Let V be that subset of S from which U_1 is obtained by deletion of the literal Q. Equivalently the elements of U_1 are obtainable by successive resolution of $-Q$ with the elements of V. Since $-Q$ has T-support the elements of U_1 do also. Therefore, the deduction formed by juxtaposing D_1, V, and D_3 is a deduction D of contradiction with T-support.

Remark. If S and T satisfy the hypothesis of Lemma 1, then there exists a j such that S^j contains contradictory units which have T-support. This follows from the fact that a proof, by definition, is finite in length and that resolution is the only rule of inference which is employed in Lemma 1.

The following two lemmas are given without proof.

LEMMA 2. *If C' is a resolvent of A' and B' where A' and B' are obtained from A and B respectively by instantiation, then there exists a C such that C' is obtainable from C by instantiation and such that C is a resolvent of F and E, where $F = A$ or F is a factor of A and $E = B$ or E is a factor of B. Conversely, if C is a resolvent of A and B, and if C' is obtainable from C by instantiation, then there exist A' and B' such that C' is a resolvent of A' and B' and such that A' and B' are obtainable from A and B respectively by instantiation.*

LEMMA 3. *If A' is obtainable from A by instantiation and has strictly fewer literals than A, then there exists a factor B of A such that A' is obtainable from B by instantiation and such that A' and B have the same number of literals.*

With Lemmas 1, 2, and 3, the proof of the completeness theorem is available.

THEOREM. *If S is a finite unsatisfiable set of clauses, and if $T \subseteq S$ is such that $S - T$ is satisfiable, then there exists a T-supported proof D and, moreover, a j such that the clauses of D are members of S^j.*

PROOF. By a theorem of Herbrand there exists a finite set H of constants such that, when S is instantiated over H, a set S' of clauses is obtained which is finite, free of variables and still unsatisfiable. T', the instantiation of T over H, is such that $S' - T'$ is satisfiable since $S' - T'$ is contained in the set obtained by instantiating $S - T$ over H. By Lemma 1 there exists a proof D' from S' with T'-support. The object is to construct a corresponding proof D from S with T-support. The following procedure will suffice. When A' appears in D' and is an instance of some $A \in S$: if A' and A have the same number of literals, replace A' by A; if A' and A do not, replace A' by the sequence A, B, where A and B are determined as in Lemma 3. When C' appears in D' and is obtained by resolving A' and B' and where A and B were previously obtained from A' and B': replace C' by the sequence F, E, C as determined by Lemma 2; if the number of literals of C' is strictly less than the number of C, replace C by the two-element sequence determined by Lemma 3. This procedure will replace the proof D' by a proof D from S. D will have T-support, and the clauses of D will be members of S^{n-2}, where n is the number of terms in D.

Since every S^i, and hence every $T_S{}^i$, is finite, the completeness theorem guarantees the existence of a proof procedure based on the set of support strategy under proper

TABLE 1. DATA ON FOUR EXAMPLES RUN WITH PG1
(When the value j appears in the column headed "Level-Proof," use of j as level bound is sufficient to obtain a proof.)

Example	*Set of Support*	*Level*		*Literal Length*		*Clauses*		*Time (sec)*
		Bound	*Proof*	*Bound*	*Proof*	*Generated*	*Retained*	
1	$K_2 \cup K_3$	4	4	6	6	1130	198	5.18
1	$K_2 \cup K_3$	4	4	None	6	1471	514	37.5
1	S	4	4	5	5	1908	235	6.26
1	S	4	4	8	5	14664	1443	411
1	S	4	—	None	—	6299	1999	919
1	K_3	4	—	6	—	188635	1894	1183
1	K_3	6	6	6	6	4100	712	30.9
2a	K_3	5	5	10	4	135	72	.494
2a	K_3	7	5	10	4	302	142	1.78
2a	S	3	3	10	6	303	145	1.91
2a	S	5	5	10	10	9803	1042	191
2b	K_3	6	6	10	5	465	214	3.17
2b	K_3	9	9	10	7	1772	706	29.9
2b	S	3	3	10	6	337	156	2.11
2b	S	6	—	10	—	44529	1999	655
3	K_3	11	11	7	6	1299	371	7.36
3	S	5	5	7	6	40094	1329	188
4	K_3	5	5	7	7	6462	1035	49.0
4	S	5	—	7	—	66256	1999	444

— indicates no proof was obtained.

choice of T as set of support. For a given T-supported proof $D = A_1, \cdots, A_n$, not only are all the A_i in S^{n-2}, but in fact all the A_i are in T_S^{n-2}. An easily proven but important corollary is:

COROLLARY 1. *Resolution coupled with factoring is complete.*

PROOF. Let $T = S$ in the theorem.

COROLLARY 2. *Let S be the join of K_1, K_2 and K_3, where K_1 corresponds to a characterization of the basic axioms in some area of mathematics, K_2 to the special hypotheses (if any) of some valid theorem, and K_3 to the denial of the conclusion of that theorem. A proof of the theorem can be found with the set of support strategy when the set of support T is either $K_2 \cup K_3$ or K_3 alone. (Finiteness of S and the consistency of $K_1 \cup K_2$ are assumed.)*

5. Examples

If a theorem-proving program is actually to be used by the mathematician to find proofs, prior knowledge as to the best choice for such parameters as level bound and literal bound will be unavailable. The examples which have been run with PG1,[5] a program implemented for the Control Data 3600, indicate that use of set of support strategy as suggested by Corollary 2 materially reduces the sensitivity to the choice of these bounds. Example 1 (below) illustrates the potential sensitivity of the choice of literal bound when the set of support strategy is abandoned. Example 2b gives the corresponding illustration for the choice of level bound.

[5] In addition to the set of support strategy, PG1 employs certain search strategies, notably the unit preference strategy [1].

Although use of the suggested set of support may raise the level bound required to obtain a proof of a given theorem, the efficiency of the strategy more than compensates for this. In a number of examples there is a marked improvement both in the total number of clauses generated and in the number of clauses retained in memory at the time proof is obtained. (Deletions due to subsidiary strategies account for the difference between the number of clauses generated and retained.) The large number of clauses generated in the absence of the set of support strategy indicates the potential combinatoric explosion, which is sharply retarded in its presence. There appear to be theorems for which a proof with PG1 is unattainable without the use of this strategy. In particular, Example 4 appears to be such a theorem (see also Table 1).

Example 1. In a group, if $x^2 = e$ for every x, the group is commutative.

Example 2. In a group, the axioms of right identity and right inverse are dependent on the remaining. Example 2a refers to the right identity problem and 2b to the right inverse problem.

Example 3. In a ring, $x \cdot 0 = 0$ for every x.

Example 4. In a ring, $-x \cdot -y = x \cdot y$ for every x and y.

6. Summary

The set of support strategy with a set of support dictated by the completeness theorem is complete in the logical sense. Its use as dictated by Corollary 2 has both logical and mathematical significance. Logically speaking, the strategy takes advantage of the known satisfiability of the basic axioms of the mathematical theory under study. The mathematical significance comes from immediately focussing attention on the so-called special hypotheses of the theorem and the denial of the conclusion of the theorem. Use of the set of support strategy enables the program PG1 to avoid generating many of the lemmas not germane to the theorem being attempted; consequently, proofs of a number of theorems of some mathematical depth have been easily obtained.

RECEIVED MARCH, 1965; REVISED APRIL 1965

REFERENCES

1. WOS, L., CARSON, D., AND ROBINSON, G. The unit preference strategy in theorem proving. *AFIPS Conference Proceedings 26*, Spartan Books, Washington, D. C., 1964, pp. 615–621.
2. ROBINSON, G. A., WOS, L. T., AND CARSON, D. F. Some theorem-proving strategies and their implementation. AMD Tech. Memo. No. 72, Argonne Nat. Laboratory, 1964.
3. ROBINSON, J. A. A machine-oriented logic based on the resolution principle. *J. ACM 12* (Jan., 1965), 23–41.

1966

Theorem-Proving for Computers: Some Results on Resolution and Renaming

B. Meltzer

It is shown that J. A. Robinson's P_1—deduction is a special case of a large class of types of deduction by resolution, an optimum choice from which should be possible for any particular theorem to be proved. Some further results, based on the operation of renaming literals by means of their negations, are obtained and suggest an alternative approach to automatic deduction.

A considerable step forward in the development of theorem-proving by machine was taken by Robinson (1965) with the introduction of the resolution method. In this method the conjunction of the axioms and the negation of the theorem to be proved are in the usual way (cf. Davis, 1963) converted into a conjunction of so-called clauses, each clause being a disjunction of atoms (i.e. atomic predicates) which may or may not be negated. The arguments of these predicates are variables or constants or functions of these. For the theorem to be valid the conjunction of this set of clauses must be shown to be unsatisfiable.

Previous methods (cf. Davis, 1963) used Herbrand's theorem directly by explicitly instantiating these clauses over a finite subset of the Herbrand universe of constants, and then attempting to show that the conjunction of the resulting set of clauses, the so-called ground clauses, led to a truth-functional contradiction. (A *ground clause* is any clause in which each variable has been replaced by a constant belonging to the Herbrand universe.)

In Robinson's method, the test for unsatisfiability is carried out directly on the clauses, and not on their instantiated or ground versions. This is effected by the iteration of a single operation termed *resolution*. To describe this operation, let us—as is customary—term any atom or any negated atom a *literal*, and term two literals which are negations of each other *complements*. Resolution operates on two clauses, when one of them contains at least one literal whose complement is either contained in the other or can be generated in the other by some substitution for its arguments. When this is the case, a new clause known as the *resolvent* is formed from the two parent clauses, which consists of all the literals in the parents except the matched complementary pairs. (The arguments in some of the literals may, of course, have been changed if a substitution had been required for the matching.) Robinson's basic result is that if resolutions are carried out on the original set of clauses and the ones generated by these operations, then the original set is unsatisfiable if and only if an empty clause can be generated. So the theorem is proved when an empty clause has been generated.

The superiority of this method to others described in the literature is clear. To use it for a systematic proof procedure on a computer, however, does still in general demand a great deal of data processing, for since one does not know to start with which chain of resolutions is going to lead to an empty clause, one would appear to have to try out systematically all possible resolutions. Put another way, one does not know—to start with—which subtree of the full tree of deductions by resolution to select.

Some computer programs have therefore been written which use resolution but incorporate heuristic devices which restrict the extent of the full deduction tree traversed.

However, in as yet unpublished work, Robinson has pointed to a non-heuristic way of achieving this end, which appears to be very powerful. This is by means of what he terms P_1-deduction, which is described below.

The purpose of the present article is, first, to extend this result by showing that P_1-deduction is only a special case of a large family of types of deduction termed P_p-deductions, from which one should be able to select —for any given problem—the most suitable one for restricting the extent of the deduction tree traversed. The demonstration of this result is based on a simple operation, termed *renaming*, which consists merely in replacing the use of a given atom, A say, by the use of another one A' which is its negation, so that $A = \bar{A}'$ and $\bar{A} = A'$.

Simple, and apparently trivial as the operation of renaming is, the second part of the present article shows how by its aid some interesting and potentially powerful results in proof theory can be obtained. It will be seen that there are even indications that it may be used as a basis for an automatic deduction procedure, alternative to resolution itself.

Robinson's P_1-deduction theorem

Let us term a clause which has no negated atoms a *positive clause* (and one which has only negated atoms a *negative clause*). Any resolution in which one of the parents is a positive clause is termed a P_1-resolution, and the resulting clause a P_1-resolvent. (It is obvious that the other parent cannot be a positive clause too.) A chain of P_1-resolutions is termed a P_1-deduction.

Robinson has proved the following:

Theorem 1

If S is a finite unsatisfiable set of clauses then there is a P_1-deduction of the empty clause from S.

Metamathematics Unit, University of Edinburgh, 5, *Buccleuch Place, Edinburgh* 8.

It is immediately clear how very greatly, in general, this theorem will allow us to restrict the number of resolutions to be tried out systematically in attempting to prove a theorem.

Definition of renaming

Let $A_1, A_2, \ldots, A_k$ be any set of atoms appearing in a set S of clauses. The replacing in S of A_1 by $\bar{A}_1'$, $\bar{A}_1$ by A_1', A_2 by $\bar{A}_2'$, $\bar{A}_2$ by $A_2', \ldots, \bar{A}_k$ by A_k', is termed a *renaming* of S.

Definition of P_p-resolution, P_p-resolvent and P_p-deduction

Let all the atoms appearing in a set of clauses be partitioned by a partition p into two sets p_1 and p_2. A clause in which every atom belonging to p_1 is negative and every one belonging to p_2 is positive we shall term a *p*-clause. Clearly if one of the parent clauses of a resolution is a *p*-clause the other is not. Any resolution in which one of the parents is a *p*-clause is called a P_p-resolution and the resolvent a P_p-resolvent. Any deduction consisting of a chain of P_p-resolutions is called a P_p-deduction.

Theorem 2

Let S be a finite unsatisfiable set of clauses and p any partition of all the atoms occurring in S. Then there is a P_p-deduction of the empty clause from S.

Proof:

Consider the full tree T of all deductions by resolution from S. Consider the isomorphic tree T' in which atoms have been re-named by changing all literals A_i and $\bar{A}_i$ derived from the set p_1 into $\bar{A}_i'$ and A_i', respectively, the set S being transformed into the set S', say. Since S is unsatisfiable, so is S', and therefore by Theorem 1, there is a subtree in T' ending in the empty clause, in which one parent of every resolution consists of positive literals only. Consider the image of this subtree in the tree T. There, every atom deriving from the set p_1 must clearly be negative (while all atoms deriving from the set p_2—since these were not changed—will be positive). This image subtree in T therefore provides a P_p-deduction of the empty clause from S.

It may be noted that P_1-deduction is the special case of P_p-deduction, where the partition of the set of atoms is into itself and the empty set; and that the number of P_p-deduction types is 2^n, where n is the number of atoms appearing in the clauses.

Some further theorems on unsatisfiable sets of clauses

Theorem 3

Any finite unsatisfiable set of clauses S not containing the empty clause must contain at least one positive clause and at least one negative clause. (This theorem is contained implicitly in Robinson's proof of Theorem 1.)

Proof:

Following Robinson we shall term a set of literals M a model of a set of ground clauses, Sg, if M contains no complementary pairs and shares at least one literal with each clause of Sg. Then, clearly, Sg will be unsatisfiable if and only if it has no model.

If S is unsatisfiable, then by Herbrand's theorem there is a set of ground clauses Sg derivable from it, which is unsatisfiable. If no clause of S is positive no clause of Sg is positive, i.e. every clause of Sg contains at least one negative literal. Therefore the set of all atoms appearing in Sg, negated, would be a model and Sg would be satisfiable.

Similarly, if no clause of S is negative, it would follow that the set of all atoms in Sg, un-negated, would constitute a model of Sg.

Hence the theorem follows.

Note that this theorem provides a sufficient condition, which can be checked by mere inspection, for the satisfiability of a set of clauses, such as the set of axioms of a theory. That is to say, if such a set does not contain a positive clause or does not contain a negative clause the set must be satisfiable.

Theorem 4

Let S be any finite unsatisfiable set of clauses not containing the empty set, and let S' be the set that results on any renaming. Then S' contains at least one positive clause and one negative clause.

Proof:

Obviously re-naming does not affect the satisfiability of the set of clauses. Hence the theorem follows from Theorem 3.

This theorem provides an even more powerful sufficient condition for the satisfiability of a set of clauses: namely, if any renaming fails to produce either a positive clause or a negative one the set is satisfiable.

The possibility of a proof procedure based on renaming

It would be very convenient if the converse of Theorem 4 were true. This would mean that if a set S maintained the property of containing at least one positive and one negative clause for every possible renaming then it would be unsatisfiable. We would then have a universally effective decision procedure; if on all possible renamings—and there would for finite sets of finite clauses be only a finite number of them—the set retained a positive and a negative clause, our theorem would be proved, and if it did not, then by Theorem 4 it would be disproved. This would imply that a recursive decision procedure was possible for the predicate calculus—which we know by Church's theorem not to be the case.

However, that a useful proof procedure may yet possibly be based on renaming is suggested by the following result:

Theorem 5

Let Sg be a finite set of ground clauses not containing the empty clause. If for every renaming of its literals Sg retains at least one positive clause then it is unsatisfiable. This also holds if we substitute "negative" for "positive".

Proof:

Suppose Sg were satisfiable. Then Sg has a model M. Some of the atoms in M will in general be negated and some not. Let us now apply a renaming which converts all the un-negated atoms in M into negated ones, thus transforming the set Sg into the renamed set S_g', say. Since S_g' has a model consisting entirely of negative literals, every clause of S_g' must have at least one negative literal.

Hence we have the result that if S_g is satisfiable there is a renaming under which no clause is positive. From this it follows that if in every renaming some clause is positive, then Sg is unsatisfiable.

The second part of the theorem can be proved by applying a renaming which converts all the negated atoms of M into un-negated ones.

It is instructive to note why when this theorem holds for a set Sg of ground clauses it does not necessarily hold for a set S of clauses from which Sg has been derived by instantiation. Consider the following example, in which S consists of a single clause and Sg of two:

$$S = \{P(x, a) \vee Q(y, f(y))\}$$

$$Sg = \{P(a, a) \vee Q(a, f(a))\}$$

$$\& \{P(f(a), a) \vee Q(a, f(a))\}$$

We see that while some of the renamings of Sg correspond to renamings of S,
e.g. $P(a, a)$ renamed $\bar{P}'(a, a)$ and $P(f(a), a)$ renamed $\bar{P}'(f(a), a)$, others do not,
e.g. only $P(a, a)$ renamed $\bar{P}'(a, a)$.

In fact (if for convenience we treat identity as a renaming too) S has $2^2 = 4$ renamings, while Sg has $2^3 = 8$, and only 4 of the latter arise from the former.

Thus, in general, all possible renamings of a set S of clauses, do not, by instantiation, generate all possible renamings of a set of ground clauses Sg derived from it, and for this reason Theorem 5 cannot be taken over to sets of non-ground clauses.

Since, however, Theorem 4 obviously applies to ground clauses too, sets of ground clauses can be tested definitively for satisfiability by renaming.

The possibility of a useful proof procedure arises if one could design an algorithm, which—working directly on a set S of clauses—would determine the effects of renaming for all possible substitution instances, rather in the way Robinson's unification algorithm (Robinson, 1965) effects all possible matchings of instances of clauses without actually explicitly generating the instances.

References

DAVIS, M. (1963). "Eliminating the irrelevant from mechanical proofs," *Proceedings of Symposia in Applied Mathematics*, American Mathematical Society, p. 15.

ROBINSON, J. A. (1965). "A machine-oriented logic based on the resolution principle," *Journal of the Association for Computing Machinery*, Vol. 12, p. 23.

Note added in proof: Robinson's results on P_1-deduction will be found in a forthcoming article in the *International Journal of Computer Mathematics.*

Bibliography on Computational Logic

To the best knowledge of the editors, this bibliography is complete up to and including 1970. Exceptions were made where a work quoted as being published after 1970 definitely appeared before, perhaps elsewhere in a different form.

Included is work with direct relevance to computational logic, in particular automated theorem proving, and its applications. Excluded are purely logical papers (with no direct relevance to computers) and work on program semantics and verification which has been extensively referenced elsewhere.

Aandreaa, S.: A Deterministic Proof Procedure. Technical Report, Harvard, May 1964

Aandreaa, S., Andrews, P., Dreben, B.: False Lemmas in Herbrand. Bull. of The American Maths. Soc., Vol. 68, pp. 699-706, 1963

Abrahams, P.W.: Machine Verification of Mathem. Proof. Ph.D. Thesis in Maths, MIT, May 1963

Abrahams, P.W.: Machine Verification of Mathematical Proofs. Maths Algorithms, Vol. 1 (1966), Vol. 2 (1967), Vol. 3 (1968)

Abrahams, P.W., Rode, W.: A Proposal for a Proof-Checker for Certain Axiomatic Systems. SRI-MEMO-41, Stanford Research Institute, Stanford, USA, 1964

Allen, J., Luckham, D.: An Interactive Theorem-Proving Program. Machine Intelligence 5 (Meltzer and Michie, eds.), American Elsevier Publishing Co., New York, pp. 321-326, 1970

Amarel, S.: An Approach to Problem-Solving by Computer. Final Report AFCRL-62-367, Pt. 2, Air Force Cambridge Res. Lab., Cambridge, Mass., May 1962

Amarel, S.: An Approach to Heuristic Problem Solving and Theorem-Proving in Propositional Calculus. In: Hart, J.F., Takasu, S. (eds.), Systems and Computer Science, University of Toronto, PR, 1967

Amarel, S.: On Representations of Problems of Reasoning about Actions. Machine Intelligence 3 (Meltzer and Michie, eds.), American Elsevier Publishing Co., New York, pp. 131-171, 1968

Anderson, B.: An Investigation of some Clause Indexing Schemes. MIP-R-80, Edingburgh University, Technical Report, 1970

Anderson, R.: Completeness of the Locking Restriction for Paramodulation. Dept. Computer Science, Univ. of Texas, Austin, Technical Report 1970

Anderson, R.: Completeness Results for E-Resolution. AFIPS 1970 Spring Joint Comput. Conf., pp. 652-656, 1970

Anderson, R.: Some Theoretical Aspects of Automatic Theorem Proving. Ph.D. Thesis, Univ. of Texas, Austin, 1970

Anderson, R., Bledsoe, W.W.: A Linear Format for Resolution with Merging and a New Technique for Establishing Completeness. J. ACM 17, pp. 525-534, July 1970

Andrews, P.B.: A Transfinite Type Theory with Type Variables. North-Holland Publ., Amsterdam, 1965

Andrews, P.B.: Resolution with Merging. J. ACM 15, pp. 367-381, July 1968

Andrews, P.B.: On Simplifying the Matrix of a Wff. J. Symbolic Logic, Vol. 33, No. 2, pp. 180-192, 1968

Andrews, P.B.: Resolution in Type Theory. J. Symbolic Logic, Vol. 36, No. 3, pp. 414-432, 1971

Andrews, P.B.: General Models and Extensionality. J. Symbolic Logic, Vol. 37, No. 2, pp. 395-397, 1972

Anufriev, F.V., Fedjurko, V.V., Leticevskii, A.A., Asel'derov, Z.M., Diduh, J.J.: On a Certain Algorithm for Search of Proofs of Theorems in the Theory of Groups. Kibernetica (Kiev), No. 1, pp. 23-29, 1966

Backer, P., Sayre, D.: The Reduced Model for Satisfiability for two Decidable Classes of Formulae in the Predicate Calculus. IBM Research Report, RC 1083, Yorktown Heights, 1963

Ballantyne, M., Bennett, J.H.: Semi-Automated Mathematics. J. ACM 16,1, January 1969

Bennett, J.H., Easton, W.B., Guard, J.R., Mott, T.H.: Introduction to Semi-Automated Mathematics. AFCRL 63-180, Air Force Cambridge Res. Lab., Cambridge, Mass., April 1963

Bennett, J.H., Easton, W.B., Guard, J.R., Mott, T.H.: Toward Semi-Automated Mathematics: The Language and Logic of SAM III. Sci. Rep. No. 2, AFCRL 64-562, Air Force Cambridge Res. Lab., Cambridge, Mass., May 1964

Bennett, J.H., Easton, W.B., Guard, J.R., Loveman, J., Mott, T.H.: Semi-Automated Mathematics: SAM IV. Sci. Rep. No. 3, AFCRL 64-827, Air Force Cambridge Res. Lab., Cambridge, Mass., October 1964

Bennett, J.H., Easton, W.B., Guard, J.R., Settle, L.G.: CRT-Aided Semi-Automated Mathematics. Final Report. AFCRL 67-017, Air Force Cambridge Res. Lab., Cambridge, Mass., January 1967

Beth, E.W.: La Crise de la Raison et la Logique. Paris et Louvain, 1957

Beth, E.W.: On Machines which prove Theorems. Simon Stevin Wis-en Naturkundig Tijdschrift, Vol. 32, pp. 49-60, 1958

Beth, E.W.: Semantical Entailment and Formal Decidability. Med. D. Koen. Akad. von Wetensch, 18, No. 13, 1955

Beth, E.W.: Observations concerning Computing, Deduction and Heuristics. Comp. Progs. and Formal Systems. North-Holland (Braffort, Hirschberg, eds.), 1963

Bing, K.: Natural Deduction with Few Restrictions on Variables. Inf. Sciences 1, 4, pp. 381-402, October 1969

Binkley, R.W., Clark, R.L.: A Cancellation Algorithm for Elementary Logic. Theoria 33, pp. 79-97, 1967
Corrected in: Theoria, p. 85, 1968

Black, F.: A Deductive Question-Answering System. Semantic Information Processing (M. Minsky, ed.), MIT Press, Cambridge, Mass., pp. 354-402, 1964

Bledsoe, W.W.: Splitting and Reduction Heuristics in Automatic Theorem Proving. Artif. Intelligence 2, No. 1, pp. 57-78, 1971

Bliss, K., Chien, R., Stohl, F.: R2 a Natural Language Question-Answering System. Proc. AFIPS, pp. 303-308, 1971

Bowden, B.V. (ed.): Faster than Thought. London 1953

Boyer, R.S.: Locking: A Restriction of Resolution. Ph.D. Thesis, Univ. of Texas, Austin, Texas, 1971

Boyer, R.S., Moore, J.S.: The Sharing of Structure in Resolution Program. Metamathematics Unit, Univ. of Edinburgh, Edinburgh, Scotland, 1971

Brice, C., Derksen, J.: A Heuristically Guided Equality Rule in a Resolution Thorem Prover. Tech. Note 45, Stanford Res. Inst., Artificial Intelligence Group, Menlo Park, CA, 1971

Brown, T.C.: Resolution with Covering Strategies and Equality Theory. Calif. Institute of Technology, California, 1968

Bruce, B.C.: A Model for Temporal References and its Application in a Question Answering Program. Artif. Intelligence 3, No. 1, pp. 1-26, 1972

Bruijn, N.G. de: The Mathematical Language AUTOMATH, Usage and Extensions. Symp. on Automat. Demonstration 1968, Springer 1970

Bruijn, N.G. de: Formulas with Indications for Establishing Definitional Equivalence. Notitie 15, Techn. Hochschule Eindhoven, 1970

Bruijn, N.G. de: On the Use of Bound Variables in AUTOMATH. Notitie 9, Techn. Hochschule Eindhoven, 1970

Bundy, A.: Counterexamples and Conjectures. There is no Best Proof Procedure. Metamathematics Unit, Univ. of Edinburgh, Scotland, Technical Report, 1971

Bursky, P., Slagle, J.: Experiments with a Multipurpose Theorem-Proving Heuristic Program. J. ACM, Vol. 15, 1, pp. 85-99, January 1968

Burstall, R.M.: A Scheme for Indexing and Retrieving Clauses for a Resolution Theorem-Prover. MIP-R-45, Univ. of Edinburgh, Edinburgh, Scotland, 1968

Burstall, R.M.: Formalising Semantics of First Order Logic in First Order Logic, and Application to Planning for Robots. MIP-R-73, Dept. of Machine Intelligence, Univ. of Edinburgh, Edinburgh, Scotland, March 1970

Burstall, R.M.: Formal Description of Program Structure and Semantics in First Order Logic. In: Meltzer and Michie (eds.), Machine Intelligence 5, Edinburgh U. Press, pp. 78-98, 1970

Burstall, R.M., London, R.J.: Programs and their Proofs: An Algebraic Approch. In: Machine Intelligence 4 (Meltzer and Michie (eds.), American Elsevier, New York, pp. 17-43, 1969

Cantarella, R.G.: Efficient Maximal Semantic Resolution Proofs Based Upon Binary Semantic Trees. TR-69-3, Electrical Engineering Dept., Syracuse Univ., Syracuse, N.Y., Ph.D. Thesis, June 1969

Carson, D., Wos, L., Robinson, G.: The Unit Preference Strategy in TP. Proc. AFIPS, Vol. 26, 1964
Spartan Books, Washington, D.C., pp. 615-621, 19

Carson, D., Wos, L., Robinson, G.: Some TP Strategies and Their Implementation. Argonne Nat. Lab., Technical Memo No. 72, 1964

Carson, D., Wos, L., Robinson, G.: Efficiency and Completeness of the Set-of-Support Strategy in Theorem Proving. J. ACM 14, pp. 536-541, 1965

Carson, D., Robinson, G., Wos, L.: Automatic Generation of Proofs in the Language of Mathematics. Proc. IFIP Congress, Spartan Books, Washington, D.C., pp. 325-326, 1965

Carson, D., Robinson, G., Shalla, L., Wos, L.: The Concept of Demodulation in TP. J. ACM 14, pp. 698-709, 1967

Chang, C.L.: The Unit Proof and the Input Proof in Theorem Proving. J. ACM 17, pp. 698-707, October 1970

Chang, C.L.: Renamable Paramodulation for Automatic Theorem Proving with Equality. Artif. Intelligence 1, pp. 247-256, Winter 1970

Chang, C.L., Lee, R.C.T.: Some Properties of Fuzzy Logic. Div. of Comp. Res. and Tech., National Institutes of Health, Bethesda, MD., 1970

Chang, C.L., Lee, R.C.T., Slagle, J.R.: Completeness Theorems for Semantic Resolution in Consequence-Finding. Proc. of IJCAI, Tbilis, USSR, 1969

Chang, C.L., Lee, R.C.T., Slagle, J.R.: A New Algorithm for Generating Prime Implicants. IEEE Trans. on Computers, Vol. C-19, 4, pp. 304-310, 1970

Chang, C.L., Lee, R.C.T.: Notes on Theorem-Proving. Div. of Computer Res. and Tech., National Institutes of Health, Bethesda, MD., 1970

Chang, C.L., Lee, R.C.T., Dixon, J.: Specialization of Programs by Theorem-Proving. Div. of Comput. Res. and Technology, National Institutes of Health, Bethesda, MD., 1971

Chang, C.L., Slagle, J.R.: Completeness of Linear Refutation for Theories with Equality. J. ACM 18, 1, pp. 126-136, January 1971

Chinlund, T.J., Davis, M., Hineman, P.G., McIlroy, M.D.: Theorem Proving by Matching. Bell Laboratory, Technical Report, 1964

Cinman, L.L.: The Role of the Principle of Induction in a Formal Arithmetic System. Mat. Sb. (N. S.), 77 (119), pp. 71-104, 1968

Cohen, J., Rubin, A.: An Interactive System for Proving Theorems in the Predicate Calculus. Second Symp. on Symbolic and Algebraic Manipulation, 1970

Collins, G.E.: Computational Reductions in Tarski's Decision Method for Elementary Algebra. IBM Corp., Yorktown Heights, 1962

Cook, S.A.: Algebraic Techniques and The Mech. of Number Theory. RM-4319-PR, Santa Monica, California, Rand Corp., 1965

Cook, S.A.: The Complexity of TP Procedures, Proc. 3rd Ann. ACM Symp. Theory Comp., pp. 151-158, 1971

Cooper, W.S.: Fact Retrieval and Deductive Question Answering Information Retrieval Systems. J. ACM 11, 2, pp. 117-137, 1964

Cooper, D.C.: Theorem Proving in Computers. Advances in Programming and Non-numerical Computation (L. Fox, ed.), pp. 155-182, 1966

Cooper, D.C.: Mathematical Proofs about Computer Programs. Machine Intelligence, Vol. 1 (N.L. Collins and D. Michie, eds.), American Elsevier, New York, pp. 17-30, 1967

Cooper, D.C.: Programs for Mechanical Program Verification. Machine Intelligence, Vol. 6 (B. Meltzer and D. Michie, eds.), American Elsevier, New York, pp. 43-59, 1971

Copi, K., Beard, M.: Programming an Idealized General Purpose Computer to Decide Questions of Truth and Falsehood. Rep. 2144-402-T, Willow Run Lab., Univ. of Michigan, 1959

Craig, W.: Linear Reasoning: A New Form of the Herbrand-Gentzen Theorem. J. Symbolic Logic 22, pp. 250-268, 1957

Craig, W.: Three Uses of the Herbrand-Gentzen Theorem Relating Model Theorem to Proof Theorem. J. Symbolic Logic 22, pp. 269-285, 1957

Darlington, J.L.: A Comit Program for Davis-Putnam Algorithm. Research Laboratory, Electron. Mech. Translation Group, M.I.T., 1962

Darlington, J.L.: Translating Ordinary Language into Symbolic Logic. MAC-M-149, M.I.T., Cambridge, Mass., March 1964

Darlington, J.L.: Machine methods for proving logical arguments expressed in English. Mech. Transl. 8, 41-67, 1965

Darlington, J.L.: Automatic theorem-proving with equality substitutions and mathematical induction. Machine Intelligence 3 [Dale and Michie, eds.], Oliver and Boyd, Edinburgh, 113-127, 1968

Darlington, J.L.: Some theorem-proving strategies based on the resolution principle. Machine Intelligence. Vol. 2 (F. Dale and D. Michie, eds), American Elsevier, New York, 57-71, 1968

Darlington, J.L.: Theorem proving and information retrieval. Machine Intelligence. Vol. 4 (B. Meltzer and D. Michie, eds.), American Elsevier, New York, 173-181, 1969

Darlington, J.L.: Theorem Provers as Question Answerers. In: Walker and Norton (eds.), Proc. Int. Joint Conference on Artificial Intelligence, Washington, D.C., 317, May 1969

Davis, M.: A Computer Program for Presburger's Algorithm. Summer Inst. for Symbolic Logic, Cornell U.V., 215-233, 1957

Davis, M., Logemann, G. and Loveland, D.: A machine program for theorem proving.CACM 5 394-397, July 1962

Davis, M.: Eliminating the irrelevant from mechanical proofs. Proc. Symp. Appl. Math. 15, 15-30, 1963

Davis, M.: Invited commentary on new directions in mechanical theorem-proving. Proc. IFIP Congress, Vol. 1, North-Holland Publ., Amsterdam, 67-68, 1968

Davis M., Putnam,H.: A Computational Proof Procedure. AFOSR TR 59-124, Rensselaer Polytechn. Institution, Troy, N.Y., 1959

Davis M., Putnam, H.: A computing procedure for quantification theory. J. ACM 7, No. 3, 201-215, 1960

Davydov, G.V.: On the Correction of Unprovable Formulas. SIM (Translation: Seminars in mathematics V. A. Steklov Mathem. Institute, Leningrad, Consultants Bureau, New York-London) 4, 5-8, 1969

Davydov, G.V.: A Proof Method for the Classical Predicate Calculus. In: Zapiski Nauchnyh Seminarov Lomi, 4, 8-17 (translated), 1967

Davydov, G.V.: Method of Establishing Deducibility in Classical Predicate Calculus. SIM (Translation: Seminars in mathematics V. A. Steklov Mathem. Institute, Leningrad, Consultants Bureau, New York-London), 5, 1-4, 1969

Davydov, G.V.: Some Remarks on Proof Search in the Predicate Calculus. SIM (Translation: Seminars in mathematics V. A. Steklov Mathem. Institute, Leningrad, Consultants Bureau, New York-London), 4, 1-6, 1970

Davydov, G.V., Maslov, S. Yu., Mints, G.E., Orevkov, V.P., Slisenko, G.O.: A Computer Algorithm for the Determination of Deducibility on the Basis of the Inverse Method. SIM (Translation: Seminars in mathematics V. A. Steklov Mathem. Institute, Leningrad, Consultants Bureau, New York-London), 16, 1-6, 1971

Dawson, D.: A Note on the Feasability of the Davis/Putnam Procedure to Elementary Theory. M.I.T., Memo 40, Cambridge, Mass., 1969

Dixon, J.: An Improved Method for Solving Deductive Problems on a Computer by Compiled Axioms. Ph.D. Thesis, University of California, Dept. of Comp. Science, California, 1970

Dreben, B.: Systematic Treatment of the Decision Problem. Talks at Summer Institute of Symbolic Logic, Cornell University, p. 363, 1957

Dreben, B., Wang, H.: A Refutation Procedure and its Model-Theoretic Justification. Harvard University, Cambridge, Massachusetts, 1964

Dunham, B., Fridshal, R., Sward, G.L.: A non-heuristic program for proving elementary logical theorems. Proc. IFIP Congr., 282-285, 1959

Dunham, B., Fridshal, R., North, J.H.: Exploratory Mathematics by Machine. In: Information and Decision Processes, (ed) R.E. Machol, P. Grey, Macmillan New York, 1962

Dunham, B., North, J.H.: Theorem testing by computer. Proc. Sympos. Math. Theory Automata, Polytechnic Press, Brooklyn, N.Y., 173-177, 1963

Elliot, R.W.: A Model for a Fact Retrieval System. Ph.D. Thesis,TNN-42, Computation Center, University of Texas, Austin, Texas, May 1965

Ernst, G.W.: GPS and Decision Making: An Overview. In: Banerji (ed) Theoretical Approaches to Non-Numerical Problem Solving, Berlin-München-Heidelberg, 1970

Ernst, G.W.: The Utility of Independent Subgoals in Theorem Proving. Systems Research Centre, Case Western Reserve University, Cleveland, 1970

Evans, T.G.: A Heuristic Program to Solve Geometry Analogy Problems. In: Semantic Inform. Proc. (ed) Minsky, M.I.T. Press, 1968

Fikes, R.E., Nilsson, N.J.: STRIPS: a new approach to the application of theorem proving to problem solving. Proc. 2nd Internat. Joint Conf. Artificial Intelligence, London, 608-620, 1971

Fishman, D.H.: The Application of Theorem-Proving Techniques to Question-Answering Systems. Scholarly Paper 8, Computer Science Center, University of Maryland, College Park, MD., May 1970

Friedman, J.: A semi-decision procedure for the functional calculus. J. ACM 10, No. 1, 1-24, 1963

Friedman, J.: A computer program for a solvable case of the decision problem. J. ACM 10, No. 3, 348-356, 1963

Friedman, J.: A New Decision Procedure in Logic and Its Computer Realization. Ph.D. Thesis, Harvard University, Cambridge, Mass., 1954

Friedman, J.: Computer realization of a decision procedure in logic. Proc. IFIP Congr. 65, 327-328, 1965

Garvey, T.D.: User's Guide to QA 3,5 Question-Answering System. Techn. Note 15, Artificial Intelligence Group, Stanford Research Inst., Menlo Park, Calif., Dec. 1969

Gelernter, H.: A Note on Syntactic Symmetry and the Manipulation of Formal Systems. Information and Control, Vol. 2, March 1959

Gelernter, H., Rochester, N.: Intelligent Behaviour in Problem Solving Machines. IBM Journal 2, 336-345, 1958

Gelernter, H.: Realization of a geometry-theorem proving machine. Proc. Intern. Conf. on Inform. Processing, UNESCO House, 273-282, 1959. Reproduced in Computers and Thought [Feigenbaum and Feldman, eds.] McGraw-Hill, New York, 134-152, 1963

Gelernter, H.: Machine-generated problem-solving graphs. Proc. Sympos. Math. Theory Automa, Polytechnic Press, Brooklyn, N.Y., 179-203

Gelernter, H., Hansen, J.R., and Loveland, D.W.: Empirical explorations of the geometry-theorem proving machine. Proc. West. Joint Comp. Conf., May 1960, 1143-147. Reproduced in Computers and Thought [Feigenbaum and Feldman, eds.], McGraw-Hill, New York, 153-167, 1963

Gilmore, P.C.: A procedure for the production of proofs from axioms, for theories derivable within the first order predicate calculus. Proc. IFIP Congr., 265-273, 1959

Gilmore, P.C.: A proof method for quantification theory: its justification and realization. IBM J. Res. Develop 4, 28-35, January 1960

Gilmore, P.C.: An examination of the geometry theorem machine. Artif. Intelligence 1 (Fall 1970), 171-187, 1970

Golota, J.J.: Nets of Marks and Deducibility in Intuitionistic PC. Zupishi Nauchnyh Seminarov, Lomi, 16, 28-43, 1969

Golota, J.J.: Some Techniques for Simplifying the Construction of Nets of Marks. SIM (Translation: Seminars in mathematics V. A. Steklov Mathem. Institute, Leningrad, Consultants Bureau, New York-London), 16, 20-25, 1971

Gould, W.E.: A matching procedure for omega logic. Air Force Cambridge Res. Lab., Rep. AFCRL-66-781, Princeton, New Jersey, 1966

Green, C.C., Raphael, B.: Research on Intelligent Question Answering Systems. Scientific Rep. 1, Contract AF 19(628)-5919. SRI Project 4641, SRI, Menlo Park, Calif., May 1967

Green, C.C.: Theorem-proving by resolution as a basis for question-answering systems. Machine Intelligence 4 [Meltzer and Michie, ed.], Edinburgh University Press, Edinburgh, 183-205, 1969

Green, C.C.: The application of Theorem Proving to Question Answering Systems. Ph.D. Thesis, Stanford University at Stanford, Cal., 1969

Green, C.C.: Application of theorem proving to problem solving. Proc. 1st Internat. Joint Conf. Artificial Intelligence, 219-239, 1969

Green, C.C., Raphael, B.: The use of theorem-proving techniques in question-answering systems. Proc. of the 23rd National Conf. of ACM, Brandon Systems Press, Princeton, 169-181, 1968

Green, B. Jr., Wolf, A.K., Chomsky, C. and Laughary, K.: Baseball: an automatic question answerer. "Computers and Thought" (E.A. Feigenbaum and J. Feldman, eds.), McGraw-Hill, New York, 207-216, 1963

Guard, J.R.: Automated Logic for Semi-Automated Mathematics. Sci. Rep. No. 1, AFCRL 64-411, Air Force Cambridge Res. Lab., Cambridge, Mass., March 1964

Guard, J.R.: The Arbitrarily-Large Entities in Man-Machine Mathematics. In: Mesarovic, D. Mihajlo and R.B. Banerji (eds.), Formal Systems and Nonnumerical Problem Solving by Computers, 1970

Guard,J.R.,Oglesby,F.C.,Bennett,J.H.,Settle,L.G.:Semi-automated mathematics. J. ACM 18, 49-62, January 1969

Gurevic, Yu. S.: Effective Recognition of Realizability of Formulae of the Restricted Predicate Calculus. Algebra i Logika Sem., 5, No. 2, 25-55, 1966

Hall, D.: Semantic Heuristics in a First-Order Problem Solver. Ph.D. Thesis, Carnegie Inst. of Techn., Pittsburgh, 1966

Hart, T.P.: A Proposal for a Geometry TP Program. M.I.T., Memo 56, Cambridge, Mass., 1963

Hart, T.P.: A Useful Algebraic Property of Robinson's Unification Algorithm. A. I. Memo 91, Artificial Intelligence Project, M.I.T., Cambridge, Mass., 1965

Hayes, P.: A Machine-Oriented Formulation of the Extended Functional Calculus. Stanford Artif. Intelligence Project, Memo 86, Stanford University, Stanford, Cal., 1969

Hayes, P.: Robotologie. "Machine Intelligence" Vol. 5 (B. Meltzer and D. Michie, eds.), American Elsevier, New York, 533-554, 1970

Hayes, P.: A Logic of Actions. DCL Memo No. 35, University of Edinburgh, Edinburgh, 1970

Hayes, P., Kowalski, R.: Automatic Theorem-Proving. DCL Memo No. 40, University of Edinburgh, Edinburgh, 1971

Hearn, A.C.: The Problem of Substitution. IFIP 68, SRI-Memo-70, 1968

Hearn, A.C.: REDUCE: A User-Oriented Interactive System for Algebraic Simplification. In: Interactive Systems for Experimental Applied Mathematics, Academic Press New York, 79-90, 1967

Henschen, L.J.: Some new results on resolution in automated theorem-Proving. Rep. No. 261, Dept. Computer Science, University of Ill., May 1968

Henschen, L.J.: Resolution, merging, set of support and tautologies. Rep. No. 817, Dept. of Computer Science, University of Ill., Dec. 16, 1969

Henschen, L.J.: A Resolution Style Proof Procedure for Higher Order Logic. Ph.D. Thesis, University of Ill. at Urbana-Champaign, Ill., 1971

Hewitt, C.: PLANNER: A language for proving theorems in robots. Proc. First Intern. Joint Conf. on Artif. Intelligence, Washington, D.C., 295-301, 1969

Horn, A.: On Sentences which are True of Direct Unions of Algebras. J. Symbolic Logic 16, 14-21, March 1951

Hunt, E.H., Marin, J., Stone, P.J.: Experiments in Induction. Academic Press, New York, 1966

Hunt, F.M., Quinland, J.R.: A Formal Deductive Problem Solving System. JACM 15, No. 4, 625-646, 1968

Jutting, L.S.: Definition of the language AUTOMATH. Report, Techn. Hochschule Eindhoven, 1970

Jutting, L.S.: Example of a text written in AUTOMATH. Report, Techn. Hochschule Eindhoven, 1970

Kahr, A.S., Moore, E.F., Wang, H.: Entscheidungs-problem reduced to the AEA case. Proc. Nat. Acad. Sci. 48, pp. 365-377, 1962

Kallick, B.: Automatic TP and Game Playing, a Dispassionate View. TIT Res. Inst., Chicago, Ill., January 1965

Kallick, B.: Theorem-Proving by Computer. TIT Res. Inst., Chicago, Ill., January 1965

Kallick, B.: Theorem-Proving by Computer. TIT Res. Inst., Chicago, Ill., April 1966

Kallick, B.: A Decision Procedure Based on the Resolution Method. IFIP Congress 1968, 1, pp. 269-275, 1968

Kallick, B.: Proof Procedures and Decision Procedures Based on the Resolution Method. Ph.D. Thesis, Northwestern Univ., August 1968

Kanger, S.: A simplified Proof Method for Elementary Logic. In: Computer Programming and Formal Systems. (P. Braffort, D. Hirschberg, eds.), North-Holland Publ., Amsterdam, pp. 87-94, 1963

King, J., Floyd, R.W.: An Interpretation-oriented Theorem Prover over Integers. In: Second Annual ACM Symposium on Theory of Computing, Northampton, Mass., pp. 169-179, May 1970

Knuth, D.E., Bendix, P.B.: Simple Word Problems in Universal Algebras. In:Computational Problems in Abstract Algebra (Leech, ed.), Pergamon Press, New York, pp. 263-267, 1970

Kowalski, R.: An Exposition of Paramodulation with Refinements. Univ. of Edinburgh, Edinburgh, Scotland, DCL Memo No. 19, October 1968

Kowalski, R.: The Case for Using Equality Axioms in Automatic Demonstration. Symp. on Automatic Demonstration, Lecture Notes in Math. 125, Springer Berlin, pp. 112-127, 1970

Kowalski, R.: Studies in the Completeness and Efficiency of Theorem-Proving by Resolution. Ph.D. Thesis, University of Edinburgh, 1970

Kowalski, R.: Search Strategies for Theorem-Proving. Machine Intelligence 5 (Meltzer and Michie, eds.), Edinburgh University Press, Edinburgh, pp. 181-201, 1970

Kowalski, R., Hayes,P.J.:Semantic Trees in Automatic Theorem-Proving. Machine Intelligence 4 (Meltzer and Michie, eds.), Edinburgh University Press, Edinburgh, pp. 87-101, 1969

Kowalski, R., Kuehner, D.: Linear Resolution with Selection Function. Metamathematics Unit, Edinburgh Univ., Scotland, Research Report 34, 1970

Kuehner, D.: Bi-Directional Search with Horn Clauses. DCL Memo No. 20, University of Edinburgh, Scotland, 1969

Kuehner, D.: A Note on the Relation Between Resolution and Maslov's Inverse Method. DCL Memo No. 36, University of Edinburgh, 1971

Kuhns, J.L.: Answering Questions by Computer: A Logical Study. Memo RM-5428-PR, The Rand Corp., Santa Monica, California, 1967

Kuvoda, S.: An Investigation of the Logical Structure of Maths. XIII, A Method of Programming Proofs in Maths for Electr. Computers. Magoya Mathem. Journal, Vol. 16, pp. 195-203, 1960

Lederberg, J., Feigenbaum, E.A.: Mechanization of Inductive Inference in Organic Chemistry. A.I. Memo No. 54, Stanford Univ., Palo Alto, California, August 1967

Lee, R.C.T.: A Completeness Theorem and a Computer Program for Finding Theorems Derivable from Given Axioms. Ph.D. Thesis, Univ. of California, Berkeley, California, 1967

Lee, R.C.T., Chang, C.L.: Program Analysis and Theorem Proving. Div. of Comput. Res. and Technol., Nat. Institute of Health, Bethesda, Maryland, 1971

Lee, R.C.T., Waldinger, R.J.: PROW: A Step Toward Automatic Program Writing. In: (Walker and Norton, eds.), Proc. Int. Joint Conf. on A.I., Washington, D.C., pp. 241-252, May 1969

Lehmer, D.H.: Some High-Speed Logic. Proc. Symp. in Appl. Maths, Amer. Maths Soc., Vol. 15, 1963

Levine, R.E., Maron, M.E.: A Computer System for Inference Execution and Data Retrieval. Comm. ACM 10, 11, pp. 715-721, 1967

Lifshits, V.A.: Specialization of the Form of Deduction in the Predicate Calculus with Equality and Functional Symbols. I, Trudy Matem. Inst. Steklov: Translation: Proc. Steklov Inst. Math., pp. 5-25, 1968

Lindsay, R.K.: Inferential Memory as the Basis of Machines which Understand Natural Language. Computers and Thought (E. Feigenbaum and J. Feldman, eds.), MyGraw-Hill, New York, pp. 217-236, 1963

Loveland, D.W.: Mechanical Theorem Proving by Model Elimination. J. ACM 15, pp. 236-251, April 1968

Loveland, D.W.: A Simplified Format for the Model Elimination Procedure. J. ACM 16, pp. 349-363, July 1969

Loveland, D.W.: Theorem-Provers Combining Model Elimination and Resolution. Machine Intelligence 4 (Meltzer and Michie, eds.), Edinburgh University Press, Edinburgh, pp. 73-86, 1969

Loveland, D.W.: A Linear Format for Resolution. Symp. on Automatic Demonstration. Lecture Notes in Math 125, Springer Berlin, pp. 147-162, 1970

Loveland, D.W.: Some Linear Herbrand Proof Procedures: An Analysis. Dept. of Comput. Sci., Carnegie-Mellon Univ., Pittsburgh, Pennsylvania, 1970

Luckham, D.: The Resolution Principle in Theorem-Proving. Machine Intelligence 1 (N.L. Collins and D. Michie, eds.), American Elsevier, New York, pp. 47-61, 1967

Luckham, D.: The Ancestry Filter Method in Automatic Demonstration. Stanford Artificial Intelligence Project Memo, Stanford, CA, 1968

Luckham, D.: Some Tree-Paring Strategies for Theorem-Proving. Machine Intelligence 3 (Dale and Michie, eds.), Oliver and Boyd, Edinburgh, pp. 95-112, 1968

Luckham, D.: Refinement Theorems in Resolution Theory. Symp. on Automatic Demonstration, Lecture Notes in Math. 125, Springer Berlin, pp. 163-190, 1970

Luckham, D., Nilsson, N.J.: Extracting Information from Resolution Proof Trees. Artif. Intelligence 2, No. 1, pp. 27-54, 1971

Manna, Z., Waldinger, H.J.: Towards Automatic Program Synthesis. Stanford Artificial Intelligence Project Memo AIM-127, Stanford Univ., Menlo Park, CA, July 1970

Maslov, S.Yu: An Inverse Method for Establishing Deducibility in Classical Predicate Calculus. In: Dokl. Akad. Nauk, SSR 159, pp. 17-20, 1964

Maslov, S. Yu, Mints, G.E., Orevkov, V.P.: Unsolvability in the Constructive Predicate Calculus of Certain Classes of Formulas Containing only Monadic Predicate Variables. Soviet Math. Dokl. 163, No. 2, pp. 295-297, 1965

Maslov, S. Yu: Application of the Inverse Method of Establishing Deducibility to the Theory of Decidable Fragments of Classical Predicate Calculus. In: Dokl. Akad. Nauk SSSR, 171, pp. 1282-1285, 1966 (translated)

Maslov, S. Yu: An Inverse Method for Establishing Deducibility of Nonprenex Formulas of the Predicate Calculus. In: Dokl. Akad. Nauk SSSR, 172, pp. 22-25, 1967 (translated)

Maslov, S. Yu: The Inverse Method of Establishing Deducibility for Logical Calculus. Trudy Matemat. Inst. An SSSR, 98, pp. 26-87, 1968

Maslov, S. Yu: A Connection between Tactics of the Resolution and Inverse Methods. Zapiski Nauchnyh Seminarov Loni 16, pp. 137-146,1969

Maslov, S. Yu: Invertible Sequential Variant of Constructive Predicate Calculus. Seminars in Mathematics, V.A. Steklov Mathem. Institute, Leningrad, Consultants Bureau, New York-London, Vol. 4, pp. 36-42, 1969

Maslov, S. Yu: Proof-Search Strategies for Methods of the Resolution Type. Machine Intelligence, Vol. 6 (Meltzer and Michie, eds.), American Elsevier, New York, pp. 77-90, 1971

Maslov, S. Yu: Deduction-Search Tactics Based on Unification of the Order of Members in a Favourable Set. Seminars in mathematics V.A. Steklov Math. Institute, Leningrad, Consultants Bureau, New York - London, Vol. 16, pp. 64-68, 1971

Maslov, S. Yu: Relationship between Tactics of the Inverse Method and the Resolution Method. Seminars in Mathematics, V.A. Steklov Mathem. Institure, Leningrad, Consultants Bureau, New York-London, Vol. 16, pp. 69-73, 1971

Matulis, V.A.: Variants of the Classical Predicate Calculus with Unique Deduction Tree. Soviet Math. Dokl. 148, No. 4, pp. 768-770, 1963

McCarthy, J.: Computer Programs for Checking Mathematical Proofs. AMS Symp. Recursive Function Theory, New York, 1961

McCarthy, J.: Situations, Actions and Causual Laws. Stanford AI Project, Memo No. 2, 1963

McCarthy, J.: A Tough Nut for Proof Procedures. Stanford Artificial Intelligence Project, Memo No. 16, Stanford Univ., Stanford, CA, 1964

Meltzer, B.: Theorem-Proving for Computers. Memo 24, Metamathematics Unit, Univ. of Edinburgh, Edinburgh, 1965

Meltzer, B.: Theorem-Proving for Computers: Some Results on Resolution and Renaming. Comput. Jour. 8, pp. 341-343, January 1966

Meltzer, B.: Logic and the Formalization of Mathematics. Science Progress, Vol. 55, Oxford, pp. 583-595, 1967

Meltzer, B.: Mathematics, Logic and Undecidability. Memo No. 9, Metamathematics Unit, Univ. of Edinburgh, Edinburgh, 1967

Meltzer, B.: Some Recent Developments in Complete Strategies for Theorem Proving by Computers. Zeitschr. fuer Math. Logik und Grundl. der Math., Ed. 14, pp. 377-382, 1968

Meltzer, B.: A new Look at Mathematics and its Mechanization. Machine Intelligence, Vol. 3 (Michie ed.), American Elsevier, New York, pp. 63-70, 1968

Meltzer, B.: Some Notes on Resolution Strategies. Machine Intelligence, Vol. 3 (Michie ed.), American Elsevier, New York, pp. 71-76, 1968

Meltzer, B.: The Use of Symbolic Logic in Proving Mathematical Theorems by Means of a Digital Computer. Foundations of Maths. 39-44, Springer New York, 1969

Meltzer, B.: Generation of Hypothesis and Theories. Nature, London, Vol. 225, p. 972, 1970

Meltzer, B.: The Semantics of Induction and the Possibility of Complete Systems of Inductive Inference. Artif. Intelligence 1, No. 3, pp. 189-192, 1970

Meltzer, B.: Power Amplification for Theorem-Provers. Machine Intelligence, Vol. 5 (Meltzer and Michie, eds.), American Elsevier, New York, pp. 165-179, 1970

Meltzer, B.: Prolegomena to a Theory of Efficiency of Proof Procedures. Proc. of Nato Adv. Study Inst. In: Art. Int. and Heuristic. Progr., American Elsevier (Findler and Meltzer, eds.), 1970

Meltzer, B.: The Programming of Deduction and Induction. Proc. of Nato Symp. on Human Thinking, Comp. Techn., 1971

Meltzer, B.: Proof, Abstraction and Semantics in Mathematics and Artificial Intelligence. DCL Memo No. 85, Edinburgh Univ., Edinburgh, 1973

Meltzer, B., Poggi, P.: An Improved Complete Strategy for TP by Resolution Report. Metamathematics Unit, University of Edinburgh, Edinburgh, 1966

Michie, D.: Notes on G-Deduction. Memo MIP-R-93, Univ. of Edinburgh, Edinburgh, 1970

Millstein, R.: The Logic Theorist in LISP. Intern. J. Computer Math. 2, pp. 111-122, April 1968

Minsky, M.: Notes on the Geometry Problem. AI-Project, Dartmouth College, Hanover, 1956

Minsky, M.: Working Paper on a Proof-Checker for Set-Theory.

Mints, G.E.: Choice of Terms in Quantifier Rules of Constructive Predicate Calculus. Seminars in mathematics V.A. Steklov Mathem. Institute, Leningrad, Consultants Bureau, New York-London, Vol. 4, pp. 43-46, 1969

Mints, G.E.: Variation in the Deduction Search Tactics in Sequential Calculi. Seminars in mathematics V.A. Steklov Mathem. Institute, Leningrad, Consultants Bureau, New York-London, Vol. 4, pp. 52-59, 1969

Morris, J.B.: E-Resolution: Extensions of Resolution to include the Equality Relation. Proc. International Joint Conf. on Artif. Intelligence, Washington, D.C., pp. 287-294, 1969

Morris, J.B.: Working Paper on a Proof-Checker for Set Theory. Univ. of Texas, Dept. of Mathematics, Report, May 1969

Nederpelt, R.P.: AUTOMATH, a Language for Checking Mathematics with a Computer. Report, Techn. Hochschule Eindhoven, 1970

Newell, A., Shaw, J.C., Simon, H.A.: The Logic Theory Machine. IRE Trans. Information Theory, IT-2, No.3, pp. 61-79, 1956
Also: Computers and Thought, McGraw-Hill, 1963

Newell, A., Simon, H.A.: GPS, a Program that Simulates Human Thought. Lernende Automaten, Munich, 1961.
Reproduced in: Computers and Thought (Feigenbuam and Feldman, eds.), McGraw-Hill, New York, pp. 279-296, 1963

Newell, A., Shaw, J.C., Simon, H.A.: Empirical Explorations with the Logic Theory Machine. Proc. West. Joint Comp. Conf., pp. 218-239, 1937.
Reproduced in: Computers and Thought (Feigenbaum and Feldman, eds.), McGraw-Hill, New York, pp. 109-133, 1963

Newell, A., Ernst, G.: The Search for Generality. Proc. IFIP Congress 1965, Vol. 1, Spartan Books, Washington, D.C., pp. 17-24, 1965

Niethammer, W., Veenker, G.: Maschinen und Mathematische Beweise. Math. Phys. Semesterberichte, Bd. 16, Heft 2, 1969

Nilsson, N.: A Mobile Automaton: An Application of Artificial Intelligence Techniques. In: Walker and Norton (eds.), AI Proc. International Joint Conf. on AI., Washington, D.C., May 1969

Nilsson, N.J.: Predicate Calculus Theorem Proving. Stanford Research Inst., Menlo Park, CA, 1969

Norton, L.M.: ADEPT - A Heuristic Program for Proving Theorems of Group Theory. Ph.D. Thesis, MIT, Boston, Report MAC-TR-33, Project MAC, MIT, September 1966

Norton, L.M.: Experiments with a Heuristic Theorem-Proving Program for Predicate Calculus with Equality. Heuristics Lab., Div. of Comp. Res. and Tech., National Institute of Health, Bethesda, MD, 1971

Orekov, V.P.: On Nonlengthening Applications of Equality Rules. Seminars in mathematics V.A. Steklov Mathem. Institute, Leningrad, Consultants Bureau, New York-London, 16, pp. 77-79, 1971

Owen, R.H.: Some Experiments with a Computer Realization of a Theorem-Proving Method. Dept. of Machine Intelligence and Perception, Univ. of Edinburgh, Res. Memo, No. MIP-R-43, November 1968

Phyushevichene, A. Yo: Elimination of Cut-Type Rules in Axiomatic Theories with Equality. Seminars in mathematics V.A. Steklov Mathem. Institute, Leningrad, Consultants Bureau, New York-London, 16, pp. 90-94, 1971

Pietrzykowski, T.: A Language for Computer Assisted Theorem-Proving. Dept. of Appl. Analysis and Comp. Sci, Research Report C5RR 2009, Univ. of Waterloo, Canada, 19

Pitrat, J.: Realization of a Program which chooses the Theorems it Proves. Proc. IFIP Congress 1965, pp. 324-325, 1965

Pitrat, J.: Realisation de Programmes de Demonstration de Theorems Utilisant des Methodes Heuristiques. Ph.D. Thesis, Univ. of Paris, 1966

Plotkin, G.D.: Lattice Theoretic Properties of Subsumption. Memo, MIP-R-77, Dept. of Machine Intelligence and Perception, Univ. of Edinburgh, June, 1970

Plotkin, G.D.: A Note on Inductive Generalization. Machine Intelligence, Vol. 5, (Meltzer and Michie, eds.), American Elsevier, New York, 1970

Plyushkevichus, R.A.: On a Variant of the Constructive Predicate Calculus without Structural Deduction Rules. Soviet Math. Dokl. 161, No. 2, pp. 292-295, 1965

Plyushkevichus, R.A.: Sequent-Variant of the Calculus of Constructive Logic for Normal Formulas. Trudy Matem. Inst. Steklov: Translation: Proc. Steklov Inst. Math., 98, pp. 155-202, 1968

Plyusikevichus, R.A.: Kanger's Variant of Predicate Calculus with Symbols for Functions That are not Everywhere Defined. Seminars in mathematics V.A. Steklov Mathem. Institute, Leningrad, Consultants Bureau, New York-London, 8, pp. 103, 109, 1970

Pohl, I.: Bi-Directional and Heuristic Search in Path-Problems. Ph.D. Thesis, Stanford University.
Also: Slac-Report, No. 104, 1969

Pohl, I: First Results on the Effect of Error in Heuristic Search. Machine Intelligence 5, 1970

Popplestone, R.J.: Beth Tree Methods in Theorem Proving. Machine Intelligence, Vol. 1, (Collins and Michie, eds.), American Elsevier, New York, pp. 31-46, 1967

Popplestone, R.J.: Experiments with Automatic Induction. In: Meltzer and Michie (eds.), Machine Intelligence, Vol. 5, American Elsevier, New York, pp. 203-206, 1970

Prawitz, D.: Mekanisk bevisföring i predikatkalkyling. Stockholm, Techn. Report, 1957

Prawitz, D.: An Improved Proof Procedure. Theoria 26, pp. 102-139, 1960

Prawitz, D., Prawitz, H., Voghera, N.: A Mechanical Proof Procedure and Its Realization in an Electronic Computer. J. ACM 7, No. 1-2, pp. 102-128, 1960

Prawitz, D.: Natural Deduction: A Proof Theoretical Study. Almquist and Wiksell, Stockholm, 1965

Prawitz, D.: Completeness and Hauptsatz for Second Order Logic. Theoria 33, pp. 246-254, 1967

Prawitz, D.: Hauptsatz for Higher Order Logic. J. Symbolic Logic 33, pp. 452-457, 1968

Prawitz, D.: Advances and Problems in Mechanical Proof Procedures. In: Machine Intelligence 4 (Meltzer and Michie, eds.), Edinburgh Univ. Press, Edinburgh, pp. 73-89, 1969

Prawitz, D.: A Proof Procedure with Matrix Reduction. Lecture Notes in Maths 125, Springer Berlin, 1970

Psenicnikova, S.V.: On an Algorithm for the Automatic Proof of Certain Theorems in Analysis. Izv. Akad. Nauk Azerbaidzan. SSR. Ser. Fiz-Teh. Mat. Nauk, No. 4, pp. 65-71, 1964

Putnam, H.: Degree of Confirmation and Inductive Logic. The Phil. of R. Casnap. Open Court Publ. Comp., Illinois, 1963

Quinland, J.R., Hunt, E.B.: A Formal Deductive Problem-Solving System. J. ACM 15, 4, pp. 625-646, 1968

Quinland, J.R., Hunt, E.B.: An Experience Gathering Problem-Solving System. Techn. Report 68-1-03, Comp. Sci Group, Univ. of Washington, Seattle, May 1968

Raphael, B.: A Computer Program which Understands. Proc. AFIPS Full Joint Comput. Conf. 1964. Spartan Press, Baltimore, Maryland, pp. 577-589, 1964

Raphael, B.: SIR: A Computer Program for Semantic Information Retrieval. Ph.D. Thesis, Mathematics Dept., MIT, Cambridge, Mass., 1964

Raphael, B.: Programming a Robot. Proc. Fourth IFIP Congress, North-Holland Publ. Co., Amsterdam, pp. 135-139, 1968

Raphael, B.: Some Results about Proof by Resolution. SIGART Newsletter, No. 14, pp. 22-25, February 1969

Raphael, B.: The Frame Problem in Problem-Solving Systems. Tech. Note 33, A.I. Group, Stanford Reserach Inst., Menlo Park, CA, June 1970

Reiter, R.: The Predicate Elimination Strategy in Theorem-Proving. In: Second Annual ACM Symposium on Theory of Computing. Northampton, Mass., pp. 180-183, May 1970

Reiter, R.: Two Results on Ordering for Resolution with Merging and Linear Format. T.R., Dept. of Computer Science, Univ. of British Columbia, Vancouver, B.C., Canada, July 1970

Reynolds, J.C.: A Generalized Resolution Principle Based upon Context-Free Grammars. Proc. IFIP Congress 1968, pp. 1405-1411, 1968

Reynolds, J.C.: Transformational Systems and the Algebraic Structure of Atomic Formulas. Machine Intelligence, Vol. 5 (Meltzer and Michie, eds.), American Elsevier, New York, pp. 135-151, 1970

Robinson, A.: Proving Theorems,(as Done by Man, Logician, or Machine). Summaries of Talks Presented at the Summer Institute for Symbolic Logic. Communications Res. Div., Inst. for Defense Analysis, Princeton, New Jersey, 1957. 2nd Edition 1960

Robinson, A.: On the Mechanization of the Theory of Equations. Bull. Res. Council Israel, 9F, pp. 47-70, 1960

Robinson, A.: A Basis for the Mechanization of the Theory of Equations. Computer Programming and Formal Systems (P. Braffort and D. Hirshberg, eds.), North-Holland Publ., Amsterdam, pp. 95-99, 1967

Robinson, G.: Dependancy of Equality Axioms in Elementary Group Theory. Comp. Group, Techn. Memo No. 53, Stanford Lin. Acc. Center, 1967

Robinson, G., Wos, L.T., Carson, D.F.: Some Theorem-Proving Strategies and Their Implementations. AMD Tech. Memo No. 72, Argonne Nat. Laboratory, 1964

Robinson, G., Shalla, L., Wos, L.T.: Two Inference Rules for FOPC with Equality. AMD Techn. Memo No. 142, Argonne Nat. Laboratory, 1967

Robinson, G., Wos, L.T.: Paramodulation and Theorem-Proving in First Order Theories with Equality. Machine Intelligence, Vol. 4 (Meltzer and Michie, eds.), Edinburgh University Press, Edinburgh, pp. 135-150, 1969

Robinson, G., Wos, L.T.: The Maximal Model Theorem. JSL 34, pp. 159 - 160, 1969

Robinson, G., Wos, L.T.: Completeness of Paramodulation. JSL 34, p. 160, 1969
Also: Spring Meeting Ass. For Symbolic Logic 1968, 1969

Robinson, G., Wos, L.T.: Paramodulation and Set-of-Support. Symp. on Autom. Dem., Lecture Notes in Maths 145, Springer, 1970

Robinson, G., Wos, L.T.: Axiom Systems in Automatic Theorem Proving. Symp. on Automatic Demonstration. Lecture Notes in Maths 125, Springer Berlin, pp. 215-236, 1970

Robinson, J.A.: A General Theorem-Proving Program for the IBM 704. Argonne Nat. Laboratory, Memo No. 6447, November 1961

Robinson, J.A.: A Machine-Oriented First-Order Logic. J. Symb. Logic 28, p. 302 (abstract), 1963

Robinson, J.A.: Theorem-Proving on the Computer. J. ACM 10, 2, pp. 163-174, 1963

Robinson, J.A.: On Automatic Deduction. Rice University, Studies 50, pp. 69-89, 1964

Robinson, J.A.: A Machine-Oriented Logic Based on the Resolution Principle. J. ACM 12, pp. 23-41, January 1965

Robinson, J.A.: Automatic Deduction with Hyper-Resolution. Intern. Jour. of Computer Math. 1, pp. 227-234, 1965

Robinson, J.A.: A Review of Automatic Theorem-Proving. Annual Symposium in Applied Math. XIX, American Math. Society, Providence, pp. 1-18, 1967

Robinson, J.A.: The Generalized Resolution Principle. Machine Intelligence 3 (Dale and Michie, eds.), Oliver and Boyd, Edinburgh, pp. 77-93, 1968

Robinson, J.A.: Heuristic and Complete Processes in the Mech. of Theorem Proving. In: Hart and Tahusu (eds.), Systems and Computer Science, University of Toronto Press, pp. 116-124, 1967

Robinson, J.A.: The Present State of Mechanical Theorem Proving. Proc. of the Fourth Systems Symp., 1968

Robinson, J.A.: New Directions in Mechanical Theorem Proving. Proc. IFIP Congress 1968, North-Holland Publ. Co., Amsterdam, pp. 63-68, 1968

Robinson, J.A.: Mechanizing Higher Order Logic. Machine Intelligence 4 (Meltzer and Michie, eds.), American Elsevier, New York, pp. 151-170, 1969

Robinson, J.A.: A Note on Mechanizing Higher Order Logic. Machine Intelligence 5 (Meltzer and Michie, eds.), American Elsevier, New York, pp. 123-133, 1970

Robinson, J.A.: Computational Logic: The Unification Computation. Machine Intelligence 6 (Meltzer and Michie, eds.), American Elsevier, New York, 1970

Robinson, J.A.: Building Deduction Machines. Art. Int. and Heuristic Programming. In: Findler and Meltzer (eds.), American Elsevier, New York, 1971

Robinson, J.A.: An Overview of Mechanical Theorem Proving. In: Banerij and Mesarovic (eds.), Theoretical Approaches to Non-Numerical Problem Solving, New York, pp. 2-20, 1970

Rogova, M.G.: On Sequential Modifications of Applied Predicate Calculi. Seminars in mathematics V.A. Steklov Methem. Institute, Leningrad, Consultants Bureau, New York-London, 4, pp. 77-81, 1969

Sandewall, E.J.: A Property-List Representation for Certain Formulas in Predicate Calculus. Report Nr. 18, Uppsala University, Computer Sciences Dept., Uppsala, Sweden, Jan. 1969

Shanin, N.A., Davydov, G.V., Maslov, S. Yu., Mints, G.E., Orevkov, V.P., Slisenko, A.O.: An algorithm for a machine search of a natural logical deduction in a propositional calculus. Izdat. "Nauka", Moscow, 1965

Sharonov, P., Zamov, N.K.: On a strategy which can be used to establish decidability by the resolution principle. National Lending Lib., Russ. Transl. Program, 5857, Boston Spa, Yorkshire, 1969

Sibert, E.E.: A machine-oriented logic incorporating the equality relation. Machine Intelligence 4 [Meltzer and Michie, eds.], Edinburgh University Press, Edinburgh, 103-133, 1969

Siltere, M. Ya.: Mechanical Deduction of Arithmetical Identities. Automatics and Telemechanics", No. 6, 110-114, 1969

Simmons, R.: Answering English questions by computer, a survey. Comm. ACM 8, No. 1, 53-70, 1965

Slagle, J.R.: A Proposed Preference Strategy Using Sufficiency Resolution for Answering Question. UCRL-14361, Lawrence Radiation Lab., Livermore, Cal., 1965

Slagle, J.R.: A multipurpose, theorem proving, heuristic program that learns. Proc. IFIP Congr. 2, 323-328, 1965

Slagle, J.R.: Experiments with a deductive question-answering program. Comm. ACM 8, 792-798, 1965

Slagle, J.R.: Automatic theorem proving with renamable and semantic resolution. J. ACM 14, 687-697, October 1967

Slagle, J.R.: Heuristic search programs. "Theoretical Approaches to Non-numerical Problem Solving" (R. Banerji and M. Mesavoric, eds.), Springer-Verlag, New York, 246-273, 1970

Slagle, J.R.: Interpolation theorem for resolution in lower predicate calculus. J. ACM 17, No. 3, 535-542, 1970

Slagle, J.R., Bursky, P.: Experiments with a multipurpose, theorem-proving, heuristic program. J. ACM 15, No. 1, 85-99, 1968

Slagle, J.R., Chang, C.L., Lee, R.C.T.: Completeness theorems for semantic resolution in consequence finding. Proc. 1st Internat. Joint Conf. Artificial Intelligence, 281-285, Washington, D.C., Walker and Norton (eds.), 1969

Tauts, A.: Solution of Logical Equations in the First Order Predicate Calculus by the Iteration Method. Proc. of the Inst. Phys. Astr. of the Acad. Sci. Est., No. 24, 17-24, 1964

Travis, L.G.: Experiments with AI theorem utilizing program. AFIPS Conf. Proc., Vol. 25, 1964, SJCL, Spartan Books, Baltimore, Mo., 339-358, 1964

Tseitin, G.S.: On the Complexity of Derivation in Propositional Calculus. Seminars in mathematics V.A. Steklov Math. Institute, Leningrad, Consultants Bureau, New York-London, Vol. 8, 115-125, 1970

VanderBrug, G.J., Fishman, D.H., Minker, J.: Outline, bibliography and KWIC index on mechanical theorem proving and its applications. Tech. Report TR-159, Computer Science Center, University of Maryland, 51 pp., June 1971

Veenker, G.: Untersuchungen über das Beweisen mathematischer Sätze durch das Zusammenwirken von Mensch und Rechenmaschine. Arbeitsbericht, Tübingen, 1969

Veenker, G.: Beweisverfahren für die Prädikatenlogik. Computing 2, 263-283, 1967

Veenker, G.: A proof procedure with spec. reliance on the equality relation. Proc. Intern. Symp., Bonn, 1970

Veenker, G.: Beweisverfahren für den Prädikatenkalkül. Ph.D. Thesis, Universität Tübingen, 1967

Veenker, G.: Ein Entscheidungsverfahren für den Aussagenkalkül und seine Realisierung in einem Rechenautomaten. Grundlagen der Kybern. Geisteswiss. 4, 127-136, 1969

Veenker, G.: Maschinelles Beweisen. Angewandte Informatik 6, 1971

Vorob'ev, N.N.: A New Decision Algorithm in the Constructive Propositional Calculus. Trydy Matem. Inst. Steklov; Transl.: Proc. Steklov Inst. Math., Vol. 2, 52, 193-225, 1958

Waldinger, R.: Constructing Programs Automatically Using Theorem Proving. Ph.D. Thesis, Carnegie-Mellon University, Pittsburgh, Penn., 1969

Waldinger, R., Lee, R.C.T.: PROW: a step toward automatic program writing. Proc. 1st Internat. Joint Conf. Artificial Intelligence, 241-252, 1969

Wang, H.: The Axiomatization of Arithmetic. J. Symbolic Logic 22, 1957

Wang, H.: Circuit synthesis by solving sequential Boolean equations. Zeit. Math. Logik und Grundlagen der Mathematik 5, 291-322, 1959

Wang, H.: Toward mechanical mathematics. IBM J. Res. Develop. 4, 2-22, 1960

Wang, H.: Proving theorems by pattern recognition I. C. ACM 3, 220-234, April 1960

Wang, H.: Proving theorems by pattern recognition II. Bell System Techn. J. 40, 1-41, January 1961

Wang, H.: Dominoes and the AEA case of the decision problem. Symp. Math. Theory Machines, Brooklyn Polytechnic Inst. 23-56, 1962

Wang, H.: The mechanization of mathematical arguments. Proc. Symp. Appl. Math. 15, 31-40, Americ. Math. Soc., Providence, 1963

Wang, H.: Games, logic and computers. Scientific American, 98-107, 1965

Wang, H.: Formalization and automatic theorem-proving. Proc. IFIP Congr., 51-58, 1965

Wang, H.: Mechanical mathematics and inferential analysis. "Computer Programming and Formal Systems" (P. Braffort and D. Hirschberg, eds.), North-Holland Publ., Amsterdam, 1-20, 1967

Wang, H.: Remarks on machine, sets, and the decision problem. "Formal Systems and Recursive Functions" (J.N. Crossley and M. Dummett, eds.), North-Holland Publ., Amsterdam, 304-320, 1967

Wos, L.T., Robinson, G.A., Carson, D.F.: Some theory proving strategies and their implementation. Argonne Nat. Lab., Techn. Memo, No. 72, Argonne, Ill., 1964

Wos, L.T., Carson, D.F., Robinson,G.A.: The unit preference strategy in theorem proving. AFIPS Conf. Proc. 26, Spartan Books, Washington, D.C., 615-621, 1964

Wos, L.T., Robinson, G.A., Carson, D.F.: Efficiency and completeness of the set of support strategy in theorem proving. J. ACM 12, No.4, 536-541, 1965

Wos, L.T., Robinson, G.A., Carson, D.F.: Automatic generation of proofs in the language of mathematics. Proc. IFIP Congr. 1965 2, 325-326, 1965

Wos, L.T., Robinson, G.A., Carson, D.F., Shalla, L.: The concept of demodulation in theorem proving. J. ACM 14, 698-709, October 1967

Wos, L.T., Robinson, G.A., Carson, D.F.: Automatic generation of proofs in the language of mathematics. Proc. IFIP Congress, Spartan Books, Washington, D.C., 325-326, 1965

Wos, L.T., Robinson, G.A.: The maximal model theorem. Abstract, Spring 1968 Meeting of Association for Symbolic Logic, J. of Symbolic Logic, 1968

Wos, L.T., Robinson, G.A.: Paramodulation and set of support. Symp. on Automatic Demonstration. Lecture Notes in Math. 125, Springer-Verlag Berlin, 276-310, 1970

Wos, L.T., Robinson, G.A.: Maximal Models and Refutations Completeness: Semidec. Proc. in ATP. In: Boone (ed.): Word Problems, North-Holland Publ. Co., 1973

Yates, R., Raphael, B., Hart, T.: Resolution graphs. Artificial Intelligence 1, 257-289, 1970

Zamov, N.K., Sharonov, V.J.: On a class of strategies which can be used to establish decidability by the resolution principle. Issled, Po Konstruktivnoye Matematikye i Matematicheskoie Logikye 3,16, 54-64, Nat. Lend. Libr. Boston Spa., Yorkshire, 1969

Zamov, N.K., Sharonov, V.J.: A Class of Strategies for the Determination of Provability by the Resolution Method. Seminars in mathematics V.A. Steklov Mathem. Institute, Leningrad, Consultants Bureau, New York-London, Vol. 16, 26-31, 1971

Symbolic Computation

Managing Editors:

J.L. Encarnação
(TH Darmstadt, Fed. Rep. of Germany)

P. Hayes
(University of Rochester, USA)

Devoted to topics in non-numeric computation, this new series currently embraces artificial intelligence and computer graphics, but will be open to fields ranging from simulation and modeling to information retrieval and text processing. The treatment of these topcis will emphasize general computational concepts rather than concentrate on specific applications, but will also provide the necessary tools for solving practical problems.

As each speciality grows in importance – as is already the case with artificial intelligence and computer graphics – a subseries will be created to afford it adequate coverage.

The textbooks and monographs in **Symbolic Computation** will prove a reliable and up-to-date source of information for researchers, students and practitioners in all the fields served.

MIX
Papier aus verantwortungsvollen Quellen
Paper from responsible sources
FSC® C105338

If you have any concerns about our products,
you can contact us on
ProductSafety@springernature.com

In case Publisher is established outside the EU,
the EU authorized representative is:
Springer Nature Customer Service Center GmbH
Europaplatz 3, 69115 Heidelberg, Germany

Printed by Libri Plureos GmbH
in Hamburg, Germany